SEEPAGE IN SOILS

SEEPAGE IN SOILS

Principles and Applications

LAKSHMI N. REDDI
Professor of Civil Engineering
Kansas State University

WILEY
JOHN WILEY & SONS, INC.

This book is printed on acid-free paper. ♾

Published by John Wiley & Sons, Inc., Hoboken, New Jersey
Published simultaneously in Canada

Library of Congress Cataloging-in-Publication Data

Reddi, Lakshmi N.
Seepage in soils : principles and applications / Lakshmi N. Reddi.
p. cm.
Includes bibliographical references (p.).
ISBN 0-471-35616-6 (cloth)
1. Seepage. 2. Groundwater flow. I. Title.

TC176 .R434 2003
624.1'5136—dc21 2002191010

Printed in the United States of America

10 9 8 7 6 5 4 3 2 1

To my father,
Sri Reddi Venkata Suryanarayana,
who was first in his community to bring about
the use of groundwater for irrigation purposes,
and whose intellect and intuition, although not
encouraged to flourish in his place and time,
have been a constant source of inspiration to me.

CONTENTS

PREFACE

Seepage in soils is an essential element of study in civil engineering. Its importance is recognized in almost all major areas of civil engineering: (1) geotechnical and structural, in the context of providing drainage in earth dams, slopes, retaining structures, and so on, to enhance their stability; (2) transportation, in providing subsurface pavement drainage; (3) geoenvironmental, in the design of subsurface seepage barriers and waste collection systems; and (4) water resources, where a fundamental study of seepage is an essential prerequisite to local/regional water distribution and management studies.

With such a prominent role in engineering, it is no surprise that seepage is a well-established course of instruction in civil engineering curricula. A course on groundwater hydrology or hydrogeology is usually the first place where most engineering students get their understanding of seepage. However, new courses are increasingly being developed to teach seepage, with the specific goal of presenting its relevance in the stability of the infrastructure systems mentioned above. Where this is done, seepage is currently taught in two different ways: either in a quantitatively rigorous approach involving mathematical, analytical, or numerical modeling of seepage, or in a qualitative or descriptive approach, where a variety of field problems are presented and solved with minimal mathematical rigor using concepts no more complicated than flow nets. In this book I combine these two modes of presentation so that readers may gain insight into the fundamental principles and mathematical solutions of seepage while appreciating their relevance in a number of field applications.

In presenting principles and solutions, I chose to minimize the use of computer software and numerical models. I did this with the intention that the high priority should be to give students a sound knowledge of fundamental principles and available closed-form solutions. Equipped with this knowledge, students should easily understand and confidently use canned software programs later, and more important, will be in a position to interpret with confidence the results yielded by these programs. The early phases of development of the study of seepage in soils attracted the attention of several eminent applied mathematicians. As a result, we have a wealth of closed-form analytical solutions available to solve problems even with complicated flow domains. It is painfully obvious at times that these solutions are not being exploited in industry. Often, numerical models that require extensive computing times are used to solve problems for which simple analytical solutions exist in the literature. Numerical solutions obtained using a discretization of flow domain, although required in a number of complicated cases, are no match for analytical closed-form solutions (where available) in providing an insight into the nature of the problem.

The book is intended to serve as a textbook for advanced undergraduate students in civil engineering and for graduate students pursuing degrees in any of the specialty areas of civil engineering. As described in Chapter 1, the contents of the book could be viewed in terms of three modules: principles and mathematical solutions (Chapters 2 to 6), filters and drainage layers (Chapters 7 and 8), and applications (Chapters 9 to 11). The presentation does not assume that students have advanced backgrounds in mathematics, soil mechanics, or any other specialized subject. However, familiarity with complex variables is essential in understanding Chapter 5. In an elementary coverage of the subject, Chapter 5 could be skipped without loss of continuity. Ideally, Chapter 6, on seepage forces, is best presented when students have a concurrent course in soil mechanics. The examples and problems are designed to convey sensitivity analyses and design principles and to clarify difficult concepts; it is not the intent to show how to plug numbers into every equation or formula discussed.

My job of preparing this book was made easy by several pioneering researchers and teachers. To name a few, Casagrande's excellent treatise on seepage through dams, published in 1937, Harr's comprehensive work on mathematical solutions, published in 1962, the excellent presentation of Mansur and Kaufman on dewatering, published in 1962, and Moulton's report on highway subdrainage, published in 1980, have all been extremely helpful. These are time-tested resources, and their relevance in instruction will not fade away with the emergence of new books. It is these authors' footprints that I followed in presenting the various specialized topics covered in this book. I suggest that every instructor of seepage make these resources required reading for students.

I was also fortunate to get very competent help from a number of people at Kansas State University. Patrick Hackenberg, graphic designer at the Kansas State Research and Extension, drafted all the illustrations in a very timely fashion, with some initial help from Fred Anderson. Tammineedi Ramakrishna shouldered the difficult job of editing the equations and putting together all the pieces of the manuscript, which at times felt never-ending. Wei Jin contributed to the example and exercise problems and assisted in solving the problems. Sai Prasad Kakuturu helped in putting together Chapter 6. I express my profound gratitude to all these people, without whose help the project could not have been completed.

I would sincerely appreciate receiving comments from readers—students and instructors alike. Feedback, including any errors, criticisms, and suggestions for revision, will be gratefully received.

LAKSHMI N. REDDI

CHAPTER 1

INTRODUCTION

Life on Earth is a result of, and is very much dependent on, the availability of water. Undeniably, water is the most precious resource on Earth. Civilizations prospered because of its availability and perished because of its lack. Combine this resource with soils—another resource that is often taken for granted, although we derive from it our physical sheath, a myriad of medical substances to maintain that sheath, and those glamorous silicon chips that we are all proud of "creating"—we then have the key players in the material ahead of us.

The subject of this book, seepage of water in soils, is one of several branches of study dealing with the flow of water in soils. Before we define our scope of study, it is important that we understand how our subject is related to other branches of study, the major disciplines dealing with water being:

- *Hydrology*: a study of the quantity and quality of the various bodies of water on Earth, with special emphasis on occurrence, movement, circulation, storage, and management on an aquifer or drainage basin scale
- *Surface-water hydrology:* a study limited to surface-water sources and their interactions
- *Groundwater hydrology or hydrogeology:* a study that deals with the occurrence, distribution, storage, and management of waters available in the subsurface portion of Earth
- *Contaminant hydrology or hydrogeology:* a study that focuses on the transfer and transport of pollutants in surface- and groundwater sources

All of these branches are concerned with the study of water as a resource. The distribution and management of water, with or without an emphasis on its quality, are the central themes in these disciplines. These branches of science and engineering are now mature fields of study following decades of cross-disciplinary interactions among geology, mathematics, meteorology, oceanography, physics, and several other basic science disciplines.

Although an extremely valuable resource on this planet, water presents some challenges in the construction and maintenance of civil infrastructure systems involving soil. Its abundance on Earth makes its occurrence common at construction sites for civil infrastructure. Water is encountered in one way or another in pavements, tunnels, foundations, embankments and slopes, and earth- and water-retaining structures (dams). The movement of water in soils, as it relates to the construction, maintenance, and stability of these infrastructure systems, creates another dimension of study—the subject of this book. With regard to the principles governing flow of water in soils, our subject shares a common ground with the other branches of study listed above. However, with regard to the application of these principles, we take an approach that is more "engineering" than "science" and focus on the design of infrastructure elements in the presence of water. We thus face the following questions in our treatment of seepage principles and applications:

1. How does the presence and movement of water influence the stability of the various civil infrastructure systems?
2. How do we control the movement of water (and forces associated with this movement) in designing these systems?
3. How do we engineer the flow so that the energies of water are safely dissipated without affecting adversely the stability of such structures as earth dams?

These questions, which define the subject matter of this book, presuppose that when left uncontrolled, water may affect the stability of structures adversely. Indeed, this is the case when, for example, a construction activity intervenes with the natural flow of water, such as in the case of a pavement constructed in a cut saturated hillslope. Thus we have as our central theme in this study: control of water flow in soils as it pertains to safe construction and performance of aboveground or subsurface infrastructure systems.

1.1 NEED FOR DRAINAGE AND SEEPAGE CONTROL

Recognition of the need for drainage and seepage control is not new. As we will see in Section 1.3, historical evidence shows that the need for drainage around subsurface structures was well recognized as early as in 3000 B.C. Extensive documentation is available in the literature on successes and failures with respect to handling seepage. One need go no further than a few clicks on the Internet or visit a few issues of an engineering magazine or newspaper to come across examples of structural failures that can be related, directly or indirectly, to inadequate drainage. Table 1.1 summarizes the key objectives of providing drainage and seepage in various areas of activity.

TABLE 1.1 Need for Drainage and Seepage Control

Area of Activity	Objective for Seepage Consideration
Foundations	Dewatering: water level must be brought down sufficiently to facilitate construction; permanent cutoff of seepage may be necessary at times to keep the foundation dry.
Embankments and slopes	Water must be drained away from the slope to keep the slope dry and protect from slides; high pore-water pressures reduce the shear strength of soils.
Pavements	Seepage must be controlled or cut off to keep the pavement bases or subbases dry; drainage layers must be provided underneath the pavement slabs to drain away water infiltrating through the slabs.
Dams	High energies of upstream water must be controlled by providing adequate filter and drainage layers within earth dams to protect against internal erosion and piping; in addition, seepage through foundation materials underneath the dams may need to be minimized or cut off. Seepage forces and uplift pressures due to flow of water are important considerations.
Waste containment systems	Seepage should be minimized through the barriers isolating the waste; drainage layers should be provided within the waste containment to drain away leachate.
Earth-retaining structures	Hydrostatic and hydrodynamic pressures due to water must be relieved by providing adequate drainage to maintain structure stability.

Perhaps the strongest accounts in the recent past are in the areas of earth dams, pavement stability, and slope and retaining wall stability.

Extensive statistical data assembled by the International Commission on Large Dams (ICOLD) on failures of earth dams reveal that 30 to 50% of accidents continue to involve poor or inadequate drainage provisions within the dams. The U.S. Bureau of Reclamation identified piping (progressive loss of soil material as a result of internal erosion) as a predominant cause of dam failure. During an April 1999 workshop, the Association of State Dam Safety Officials identified research relating to seepage and piping as a high-priority research need. The safety of embankment dams depends to a large extent on proper control of water and minimization of its potential to erode soil particles. The high energies of water erode soil particles, causing "pipes," which may breach the dam and release suddenly the high energies of the reservoir water.

Uncontrolled presence and migration of water beneath pavement slabs is a major issue believed to be costing billions of dollars annually in the United States. When water is accumulated in the base or subbase layers beneath a pavement slab, it may be pressurized due to traffic loads. Cracks in the pavement will become outlets for pumping of the pressurized water. Repeated pumping of this water will cause some of the base materials to erode, leading to slab instabilities and failures.

The presence of water in soils alters force balance and may in some cases disturb the static equilibrium of soils. In addition to the hydrostatic stress conditions, flow of water induces seepage stresses in the direction of flow. The hydrostatic and hydrodynamic forces may together be significant enough to make slopes and retaining structures unstable. Very often, landslides and retaining structure failures occur shortly after a rainfall event. When poorly drained, the accumulated water will surely find a path to release its energy. If the engineer did not identify this path in the design stages, nature will identify it for him or her according to its own time schedule.

1.2 SCOPE AND ORGANIZATION OF THE BOOK

As stated earlier, our treatment of seepage shares a common ground with other disciplines of hydrology with regard to the fundamental principles and mathematical solutions. The principles involved in the flow of soils were developed in the field of *groundwater hydrology,* which in turn was a blend of geology and hydrology. An understanding of these principles is essential in all applications of seepage. We therefore start with these principles in Chapter 2, introducing along the way a very basic law, Darcy's law, which governs the flow of water in soils. This law allows us to postulate a governing equation for flow in multiple dimensions. The solution of this equation gives us the graphical tool of flow nets (Chapter 3), useful in solving a range of seepage problems. Simple analytical solutions for seepage are possible when a couple of assumptions (Dupuit–Forchheimer assumptions) are made with regard to gradients in an unconfined flow domain. This is the topic of Chapter 4. The use of conformal mapping techniques allows us to expand the range of mathematical solutions available, as discussed in Chapter 5. Chapters 2 to 5 thus give us all the essential principles and mathematical solutions useful in analyzing seepage in a specific area of activity. Forces associated with the flow of water are treated in Chapter 6. The seepage forces, uplift pressures on structures, and effective stresses in soils provide a basis on which to assess the stability of hydraulic structures, sheet piles, slopes, and retaining structures.

In our efforts to control the energies of water, we will find it necessary to use engineered seepage layers in several types of infrastructure systems. These are the filters and drainage layers placed around groundwater wells, underneath pavement structures, and within zoned dams. They are designed to dissipate pore pressures of water safely and to prevent erosion and transport of soil particles. Design of these layers is an important task, as structural failure of dams and pavements is often linked directly to inadequate performance of these layers. Chapters 7 and 8 deal with filters and drainage layers.

The specific applications of seepage principles in the areas of dewatering and seepage control, pavement drainage systems, and waste containment systems are dealt with in Chapters 9, 10, and 11, respectively. The purpose of these chapters is to apply the seepage principles and mathematical solutions (covered in Chapters 2 to 6) to design seepage control and dewatering systems, drainage layers, and seepage barriers.

To summarize, the contents of this book can be viewed in terms of three modules:

1. *Principles and mathematical solutions* (Chapters 2 to 6), which give us tools to use in designing seepage control systems or in assessing the stability of infrastructure systems

2. *Design of filters and drainage layers* (Chapters 7 and 8), which deals with criteria for designing engineered layers (both soil and synthetic fabrics) for safe dissipation of excess water pressures and prevention of soil particle erosion and transport
3. *Applications to specific infrastructure systems* (Chapters 9 to 11), which deal with application of the first two modules to design seepage control and dewatering systems, pavement drainage layers, and seepage barriers

1.3 PIONEERS IN SEEPAGE

Civilizations, ancient and current, are intimately tied to hydrologic engineering. It should be no surprise that history is rich in its recordings of the major milestones in the evolution of earth dam construction and in the engineering activities involving control of surface and ground waters. Table 1.2 presents a chronology of recorded hydrologic

TABLE 1.2 Chronology of Recorded Hydrologic Engineering Prior to 600 B.C.

Date (B.C.)[a]	Event
3200	Reign of King Scorpion; first recorded evidence of water resources work.
3000	Damming of the Nile, thus diverting its course, by King Menes.
3000	Nilometers used to record fluctuations of the Nile.
2850	Failure of Sadd el-Kafara dam.
2750	Origin of Indus Valley water supply and drainage systems.
2200	Various waterworks of the "Great Yu" in China.
2200	Water from spring conveyed to the Palace of Cnossos (Crete); dams at Mahkai and Lakorian in Persia.
1950	Connection of the Nile River and Red Sea by a navigational canal during the reign of Seostris I.
1900	Sinnor constructed at Gezer (Palestine).
1850	Lake Moeris and other works of Pharaoh Amenemhet III.
1800	Nilometers at second Cataract in Semna.
1750	Water codes of King Hammurabi.
1700	Joseph's Well near Cairo, nearly 325 ft in depth.
1500	Two springs joined by a sinnor in the city of Tell Ta'annek in Palestine;
?	Marduk dam on Tigris near Samarra (destroyed in A.D. 1256).
1300	Irrigation and drainage systems of Nippur; Quatinah dam on the Orontes River in Syria constructed under the reign of Sethi I or Ramses II.
1050	Water meters used at the Gadames oasis in North Africa.
750	Marib and other dams on River Wadi Dhana in Yemen.
714	Destruction of qanat systems of Ulhu (Armenia) by King Saragon II; qanat system spread gradually to Persia, Egypt, and India.
690	Construction of Sennacherib's channel.
600	Dams in the Murghab River in Persia (destroyed in A.D. 1258).

Source: Adapted from Biswas (1970).

[a]In the absence of accurate information, many of these dates are approximate.

engineering prior to 600 B.C. The importance of water diversion using dams seemed to be well recognized as early as in 3000 B.C., when King Menes dammed the Nile and diverted the course of the river to a newly dug channel between two hills. The failure of Sadd el-Kafara dam, the Dam of the Pagans, built sometime between 2950 and 2750 B.C. is perhaps the oldest documented account of dam performance. The dam apparently failed during its first flood season as a result of complete filling of the reservoir and overtopping. Unlike in modern times, the study of water and its use was undertaken by scientists, philosophers, and esotericists alike. Even Hippocrates (450 B.C.), the father of modern medicine, formulated ideas on the constitution of water and conducted experiments on how water evaporates. Kautilya, regarded in the East for his work *Arthasastra*, a treatise on economics, politics, and administration, appeared to be one of the first to design a rain gauge.

Most historians believe that the quantitative era of groundwater hydrology begun in the year 1856, when the French engineer Henry Philibert Gaspard Darcy (1803–1858) published his report on the water supply of the city of Dijon, France. Darcy is undoubtedly the father of the science of fluid flow in soils, for his experiments on the flow of water through a sand column resulted in the most basic cause-and-effect relationship, named after him as Darcy's law (Chapter 2). Perhaps due to a tradition of conducting scientific experiments at hospitals, it is documented that Darcy conducted his experiments at a hospital (Freeze, 1994). Apparently, the water hammer effects from the continual turning on and off of the numerous taps at the hospital created noise in the manometer levels of the experiment. Although Darcy is known for his experiments in water supply and flow through soils, it is often forgotten that his work in Dijon included supervision and construction of bridges and a railway tunnel. Perhaps of importance to us in this materialistic era is the generosity displayed by

Darcy
(*Source:* Philip, 1995.)

Darcy in the conduct of his engineering duties. In his study on Darcy, "Desperately Seeking Darcy in Dijon," Philip (1995) states: "As a designer of the scheme and director of the works, Darcy was entitled to claim fees of the order of 55,000 francs, which was a considerable sum. At the time, an artisan earned 800 francs a year. The fees waived by Darcy represented some 70 artisan-years and nearly 100 years of [the labor of] an ordinary workman. In today's world these 55,000 francs would correspond roughly to $1.5 million." Darcy was also said to have rejected completely the then-prevalent hypothesis that rainwater was unable to penetrate more than a few feet into the soil. He conceptualized the aquifer as a large pipe connecting two reservoirs at different levels.

The pioneering work conducted by Darcy was extended by another French engineer, Arsène Jules Emile Juvenal Dupuit (1804–1866), and an engineer from Vienna, Philip Forchheimer (1852–1933). As we note in Chapter 4, much of what we now know about flow toward wells in porous media could be traced to Dupuit's work. Unlike the empirical treatment of Darcy, Dupuit sought theoretical expressions for flow rate, including the expression for flow rate toward wells. The assumptions he made on the nature of gradients in a free flow, named after him the Dupuit assumptions (Chapter 4), greatly simplified the mathematical treatment of flow toward wells. He was also one of the first to view soils at a pore scale and to conceptualize flow paths in sands as tiny microchannels. Forchheimer is known for his contributions in the area of mathematics applied to groundwater problems. Using Darcy's law and the Dupuit assumptions, he derived the Laplace equation as the equation governing flow in soils. For the first time, he used complex variables to analyze groundwater flow toward wells and paved the way for advanced mathematical solutions for seepage using conformal mapping techniques. Because of his excellent mathematical background, he

Dupuit
(*Source:* Rouse and Ince, 1957.)

Forchheimer
(*Source:* Rouse and Ince, 1957.)

was able to provide quantitative solutions for several complicated flow problems using mirror images, potential theory, and other concepts (Chapters 4 and 9). It was the mathematical relationship developed by Forchheimer between equipotential lines and streamlines that gave us the useful graphical tool, flow nets (Chapter 3). Speaking of Forchheimer's contributions to this field, Terzaghi (1942) says: "In my opinion, his contribution accomplished more in the line of clarifying our ideas concerning the movement of ground water than those of all the other contemporaneous hydrologists of Europe combined."

In the twentieth century, Pelageia Yakovlevna Polubarinova-Kochina (1899–1999), a native of Astrakhan, Russia, extended, perhaps more than any one, the work of Philip Forchheimer. She used her excellent mathematical background in complex analysis and free boundary layer problems to solve difficult issues in fluid mechanics and groundwater movement. The textbook covering her major contributions, *Theory of Motion of Ground Water* (Polubarinova-Kochina, 1952), is a guidepost to researchers in the West. Much of what we study in Chapter 5 is due to her work in the area of groundwater movement, and to Western scholars such as Milton Harr, who expanded on her studies and familiarized the West with her work and that of her Russian contemporaries. Kochina also contributed to the literature on the history of mathematics. She is a prolific writer, contributing such technical articles as "Some Properties of a Fractional-Linear Transformation" when she was 100 years old.

The twentieth century has also seen a marriage between the mechanics of soils and groundwater dynamics. Karl Terzaghi, the "father" of soil mechanics, proposed the effective stress principle (Chapter 6), which gives us a rational basis to account for the effect of water on mechanical and engineering properties of soils. His studies led to a formal approach of incorporating seepage principles in the design of such

Polubarinova-Kochina
(*Source:* www.groups.dcs.stand.ac.uk/~history/Mathematicians/Kochina.html.)

systems as slopes, retaining structures, and earth dams. Assessing the importance of seepage control through the foundations of dams, Karl Terzaghi (1929) states: "To pass judgement on the quality of a dam foundation is one of the most difficult and responsible tasks. It requires both careful consideration of the geological conditions and the capacity for evaluating the hydraulic importance of the geological facts which can only be obtained by a thorough training in the hydraulics of seepage." Terzaghi (1939) also emphasized the importance of taking water presence into account in any engineering endeavor when he stated: "In engineering practice, difficulties with soils are almost exclusively due not to the soils themselves but to the water contained in their voids. On a planet without any water there would be no need for soil mechanics." Terzaghi's original studies were extended by other pioneers in the field, most notably, Arthur Casagrande. Among his several contributions to the field of soil mechanics and foundations engineering are the development of procedures for identifying, classifying, and testing soils, and analyses on the control of seepage through foundations and dam abutments. He provided a practical outlet for the early theoretical works of Dupuit and Forchheimer in the area of seepage control through earth dams.

The twentieth century also saw several passionate workers who promoted the importance of seepage considerations in the design of civil infrastructure systems. Most notable among them is Harry Cedergren, who lamented that the United States is wasting billions of dollars annually because of lack of or improper drainage provisions in pavements. He was alone for a long time in his crusade to introduce well-drained subgrades in pavement construction, but his eloquent presentation of drainage principles has finally given him some followers. His characterization of U.S. pavements as the *world's longest bathtubs* (Cedergren, 1994) is perhaps a needed exaggeration, and he was certainly successful in convincing the pavement designers of his time to follow

Terzaghi
(*Source:* Courtesy of Ralph B. Peck.)

the Romans of 2000 years ago and design well-drained pavements. Cedergren has made several contributions in the area of using flow nets to solve practical problems effectively in zoned dams and pavements.

These are only some of the pioneers in seepage research; many more were and are dedicating their lifetimes to advancement of our knowledge of this subject. History is full of examples of uncelebrated heroes. Perhaps several brilliant rays of intuitive thought passed (and are currently passing) without receiving attention from acknowledged specialists. And perhaps these intuitive thoughts may not realize intellectual expression for a long time, or never. But to the extent possible, it is important for an earnest student in any given field to be thankful for the vast repository of knowledge developed through the ages.

CHAPTER 2

PRINCIPLES OF FLOW

Like any matter in nature, water is associated with energy. Under the grand scheme of nature's design, there is a constant tendency for energy to be balanced and conserved. In fulfillment of this design, water flows from locations of higher energy to those of lower energy. Thus, the key to understanding the flow of water in soils is to know the energy associated with the water at various locations. We first define these energies by way of understanding the driving forces causing the flow of water, and then proceed to express the law governing flow using a cause-and-effect relationship. This will lead us to a study of one of the most important properties of soils, hydraulic conductivity.

2.1 POTENTIAL AND KINETIC ENERGIES OF WATER MASS: BERNOULLI'S EQUATION

The energy possessed by a water mass at a given location is of three types:

1. *Potential/elevation energy* E_z, which is due to the elevation of a water mass above a specified datum
2. *Strain/pressure energy* E_p, which is due to the pressure in a water mass
3. *Kinetic energy* E_v, which is due to movement of a water mass

These three energies can be expressed as

$$E_z = Mgz \tag{2.1}$$

$$E_p = \frac{pM}{\rho} \tag{2.2}$$

$$E_v = \frac{Mv^2}{2} \tag{2.3}$$

where M is the mass of water body, g the acceleration due to gravity, z the elevation of water mass, p the pressure in water, ρ the density of water, and v the velocity of water. The three energies can be expressed per unit mass of the fluid to avoid using a finite mass of water body. Or in a more common form, they can be expressed per unit weight by dividing each term in Eqs. (2.1) to (2.3) by Mg. When written thus, they are known as *heads*, which have the units of length. The *total head* h associated with water mass may therefore be expressed as

$$h = z + \frac{p}{\gamma_w} + \frac{v^2}{2g} \tag{2.4}$$

where the unit weight of water γ_w is used in the place of ρg in the second term. Equation (2.4), commonly referred to in fluid mechanics as the *Bernoulli equation,* gives us a convenient way of expressing the total energy of moving water mass and comparing the energies at different locations. Note that we take this expression as valid not only for a free body of water flowing in a pipe but also for water flowing in the interstices of a soil mass. When applied to a soil mass, we usually consider the velocity heads to be negligible compared to the elevation and pressure heads. Using examples given later in the chapter, the reader may verify this to be generally valid.

The driving force responsible for the flow of water from point A to Point B in a soil mass is the difference in total hydraulic head at these two points (Fig. 2.1). Water

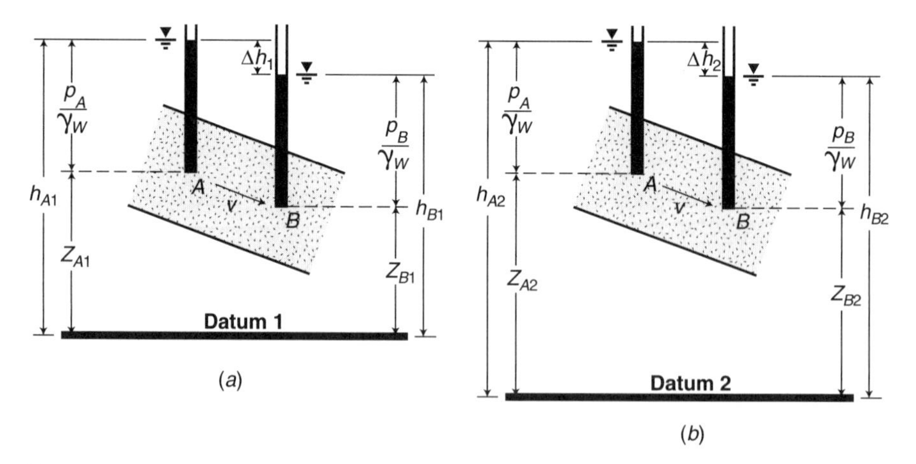

Fig. 2.1. Elevation, pressure, and total heads with reference to two different datum elevations.

flows from the point with the higher head (A) toward locations with the lower head (B). Note that the total h at a given point in soil depends on the choice of the datum elevation. The elevation heads and therefore the total heads at any specific point (say, A) in Fig. 2.1a are different from the corresponding heads in Fig. 2.1b; the pressure heads remain the same, however. If we are to compare the total heads at two different points, A and B, we must use a common datum. When this is done, the difference in the heads (Δh) will remain the same regardless of the choice of the datum. Thus, Δh_1 in Fig. 2.1a is equal to Δh_2 in Fig. 2.1b.

We should also note here that there are other forms of energy associated with water. Differences in the chemical composition of water create osmotic gradients. Also, differences in temperature and application of electric gradients between two masses of water body create thermal and electrical energies. We will, however, restrict our attention to the elevation and pressure energies, which dominate many of the problems in seepage studies.

2.2 HYDRAULIC GRADIENT

Hydraulic gradient i is a nondimensional quantity used to express the head loss between two points. It is expressed as the ratio of the head difference Δh (between the two points under consideration) to the length of flow between the points,

$$i = \frac{\Delta h}{L} \tag{2.5}$$

Two important points that should be borne in mind regarding this equation are that the head difference refers to the difference in total heads (elevation and pressure heads combined), and L refers to the length of travel, which may not necessarily be the horizontal length between the two points. It is often convenient to draw head profiles along the length of flow to be able to identify Δh accurately. Figure 2.2 shows the example head profiles for horizontal and vertical soil columns subjected to different pressure heads at their ends. The reader will note that gradient computation is one of the key elements in seepage analyses, and rigorous practice using diverse flow scenarios will bring much benefit later.

Example 2.1 Evaluate hydraulic gradients i across the soil specimens in Fig. 2.2.

SOLUTION: (a) From Fig. 2.2a, head difference $\Delta h = 6$ ft $- 3$ ft $= 3$ ft; length of flow $L = 8$ ft $- 2$ ft $= 6$ ft. Therefore, the gradient $i = \Delta h/L = 3$ ft/6 ft $= 0.5$.

(b) From Fig. 2.2b, head difference $\Delta h = 10$ ft $- 0$ ft $= 10$ ft; length of flow $L = 6$ ft $- 0$ ft $= 6$ ft. Therefore, the gradient $i = \Delta h/L = 10$ ft/6 ft ≈ 1.67.

(c) From Fig. 2.2c, head difference $\Delta h = 16$ ft $- 12$ ft $= 4$ ft; length of flow $L = 8$ ft $- 2$ ft $= 6$ ft. Therefore, the gradient $i = \Delta h/L = 4$ ft/6 ft ≈ 0.67.

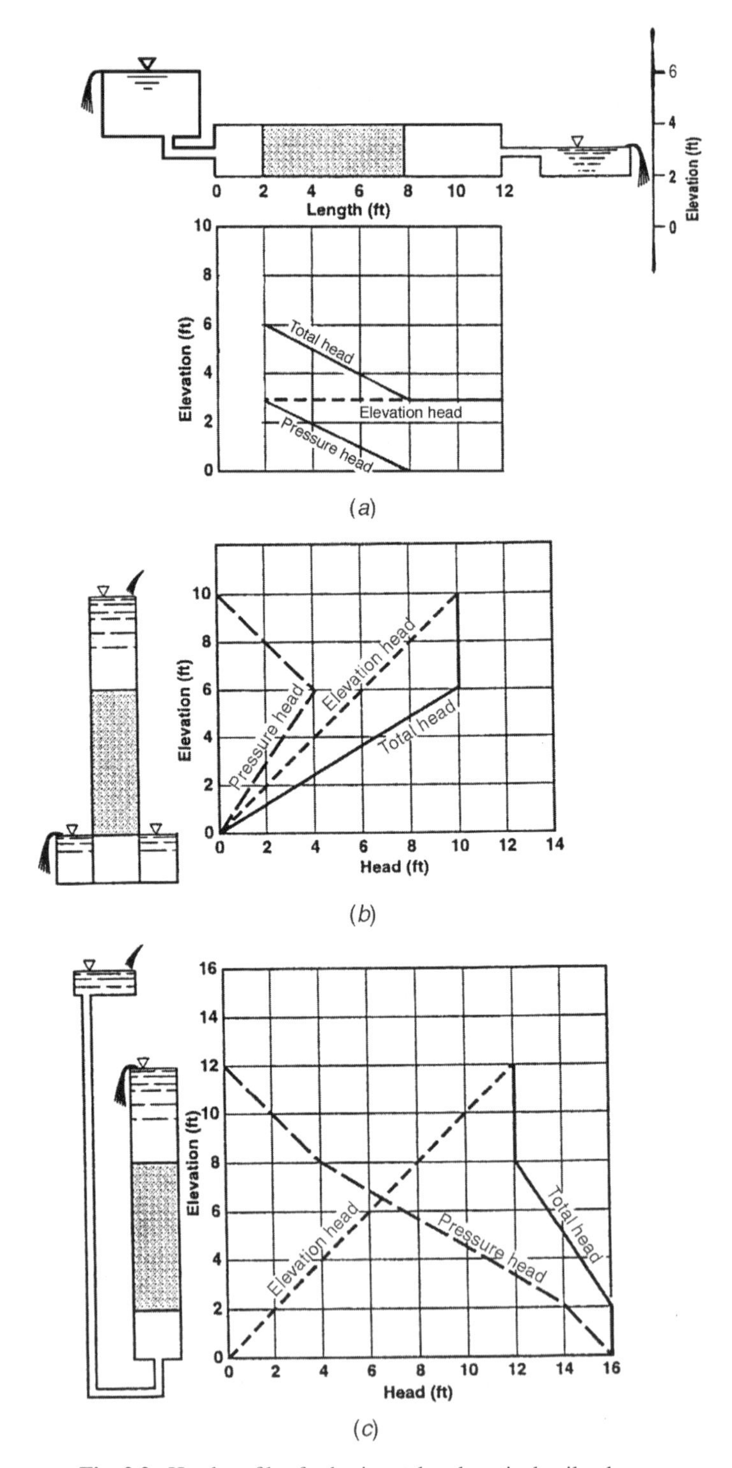

Fig. 2.2. Head profiles for horizontal and vertical soil columns.

2.3 DARCY'S LAW

Much of what we now know about seepage of water in soils is an outgrowth of the experimental studies conducted by the French engineer Henry Darcy in the 1850s. Darcy used an experimental setup similar to the one shown in Fig. 2.3 and observed how the rate of flow through a soil specimen changed in relation to the head difference. He found out experimentally that the flow rate q through the soil was proportional to the gradient i. This proportionality relation has become one of the most commonly cited laws, known as *Darcy's law:*

$$q = kiA \tag{2.6}$$

where k is a constant of proportionality, which came to be known as the *coefficient of hydraulic conductivity* when referring to water as the permeating liquid, or the *coefficient of permeability* in a general sense when no permeating liquid is specified;

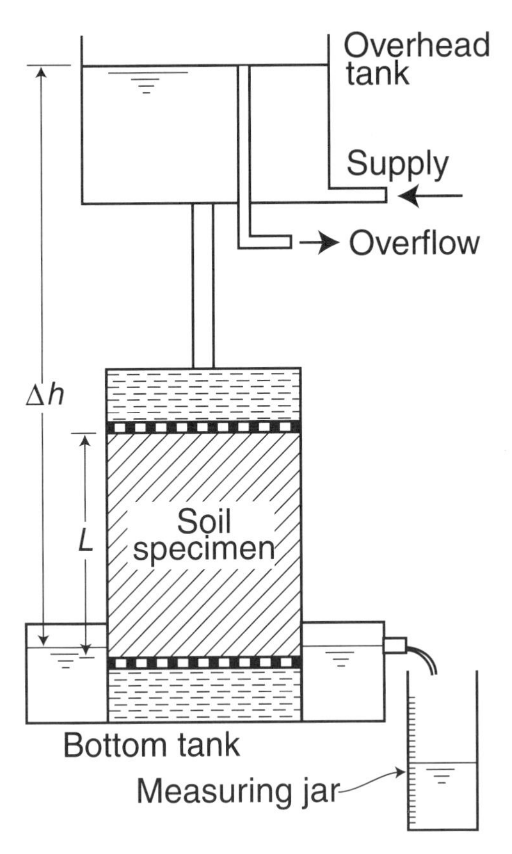

Fig. 2.3. Apparatus similar to the one used by Darcy to determine proportionality between flow rate and hydraulic gradients.

and A is the cross-sectional area of the soil sample (the inside cross-sectional area of the soil container). In the context of seepage of water in soils, the terms *coefficient of hydraulic conductivity* and *coefficient of permeability* are often used synonymously. Equation (2.6) can be rewritten to express the velocity of flow of water v as the ratio of flow rate and cross-sectional area,

$$v = \frac{q}{A} = ki \tag{2.7}$$

Note that we chose to express velocity in terms of the total cross-sectional area of the sample, although flow occurs only in the pore space. For this reason, this velocity is often referred as the *discharge velocity* or *superficial velocity*. The actual velocity may never be determinable with this experimental observation, because of the complicated architecture of soil pores in the sample. However, we can partition the solid volume and pore volume according to the porosity of the soil, as shown in Fig. 2.4, and express an *average effective velocity* or *seepage velocity* v_s as

$$v_s = \frac{q}{A_v} = \frac{q/A}{A_v/A} = \frac{v}{n} \tag{2.8}$$

where A_v is the cross-sectional area filled with void space and n is the porosity of the soil. Thus, because of the reduced cross-sectional area of the pore space, the seepage

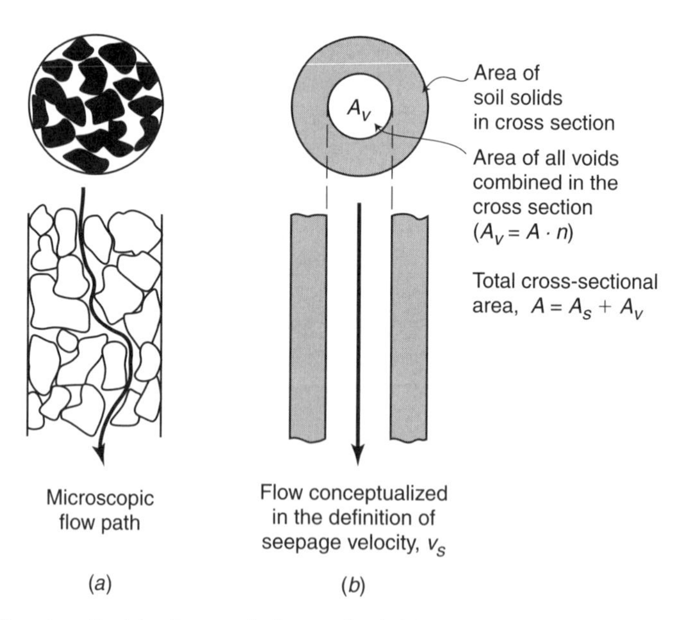

Fig. 2.4. Flow in soils: (*a*) microscopic flow path; (*b*) idealization of flow through a combined area of void space.

velocity is always greater than the discharge velocity. This is of particular relevance in geoenvironmental engineering, where the actual travel velocities of contaminants in pore water are of importance, not the discharge velocities.

The proportionality between the flow rate or velocity and the hydraulic gradient was examined in several investigations subsequent to Darcy's experimental studies. It is generally accepted that Darcy's law is valid for laminar flow regimes characteristic of flow velocities in soils. Although some studies suggest that a minimum threshold gradient is needed to initiate flow in certain fine-grained soils, below which the proportionality is not valid, results from such studies have remained speculative.

Example 2.2 Determine the flow rate q through the soil specimens for the three cases in Example 2.1. Use $k = 1 \times 10^{-5}$ cm/s, $A = 50\ \text{cm}^2$. Also determine discharge velocities and seepage velocities, assuming that porosity $n = 0.3$.

SOLUTION: (a) Flow rate $q = kiA = 1 \times 10^{-5}$ cm/s $\times 0.5 \times 50\ \text{cm}^2 = 2.5 \times 10^{-4}\ \text{cm}^3/\text{s}$, discharge velocity $v = ki = 1 \times 10^{-5}$ cm/s $\times 0.5 = 5 \times 10^{-6}$ cm/s, seepage velocity $v_s = v/n = 5 \times 10^{-6}$ cm/s/0.3 $\approx 1.67 \times 10^{-5}$ cm/s.

(b) Flow rate $q = kiA = 1 \times 10^{-5}$cm/s $\times 1.67 \times 50\ \text{cm}^2 = 8.35 \times 10^{-4}\ \text{cm}^3/\text{s}$, discharge velocity $v = ki = 1 \times 10^{-5}$cm/s $\times 1.67 = 1.67 \times 10^{-5}$ cm/s, seepage velocity $v_s = v/n = 1.67 \times 10^{-5}$ cm/s/0.3 $\approx 5.57 \times 10^{-5}$ cm/s.

(c) Flow rate $q = kiA = 1 \times 10^{-5}$cm/s $\times 0.67 \times 50\ \text{cm}^2 = 3.35 \times 10^{-4}\ \text{cm}^3/\text{s}$, discharge velocity, $v = ki = 1 \times 10^{-5}$cm/s $\times 0.67 = 6.7 \times 10^{-6}$ cm/s, seepage velocity $v_s = v/n = 6.7 \times 10^{-6}$ cm/s/0.3 $\approx 2.23 \times 10^{-5}$ cm/s.

2.4 HYDRAULIC CONDUCTIVITY

The constant of proportionality in Darcy's law, known as the *hydraulic conductivity,* is an important property of soils that characterizes the nature of their seepage. It may be interpreted as the discharge velocity corresponding to a unit gradient, since $k = v/i$ and $k = v$ for a gradient equal to 1. Thus the units of k are the same as those of v (i.e., length per time); k depends on both water and soil properties. With regard to the effects of water properties, k is directly proportional to the density of water and inversely proportional to the viscosity:

$$k \propto \frac{\rho}{\eta} \tag{2.9}$$

where η is the viscosity of water. Since these properties are in turn dependent on the temperature, k could be expressed as a function of temperature. Of the two parameters, density variation due to temperature is generally considered to be negligible. The relation between viscosity and temperature for water is well known (Fig. 2.5); hence k at one temperature can be translated to those at other temperatures using this relationship. It is customary to report the hydraulic conductivities at a temperature of 20°C. The following expression can be used for this purpose:

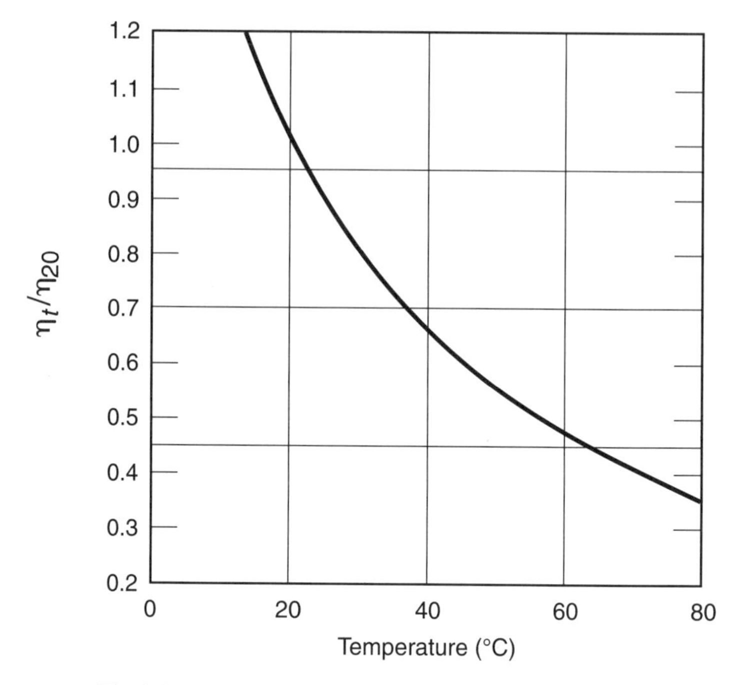

Fig. 2.5. Variation of viscosity of water with temperature.

$$k_{20^\circ\mathrm{C}} = \frac{\eta_t}{\eta_{20^\circ\mathrm{C}}} k_t \tag{2.10}$$

where k_t and η_t are the hydraulic conductivity and water viscosity corresponding to t°C. To eliminate the role of permeant water characteristics, *specific* or *absolute permeability* K (with units of length^2) is often used:

$$K = \frac{k\eta}{\gamma_w} \tag{2.11}$$

where k is the coefficient of hydraulic conductivity, η the viscosity of water, and γ_w the unit weight of water. Although not common, the absolute permeability is sometimes expressed in terms of darcys (1 *Darcy* $= 0.987 \times 10^{-8}\mathrm{cm}^2$). The conversion chart for k and K is shown in Fig. 2.6.

Void ratio and soil structure are the two important soil properties controlling k. These properties, in turn, are dependent on the particle size and gradation for coarse-grained soils such as sands and gravel. Naturally, soils with larger particles are associated with larger pore channels, yielding higher values for k. The wide range of pore architectures possible in soils causes k to vary by several orders of magnitude. As shown in Fig. 2.7, the range of k between gravel and clay (Fig. 2.7*a*) is several orders of magnitude greater than the ranges of two other common engineering properties

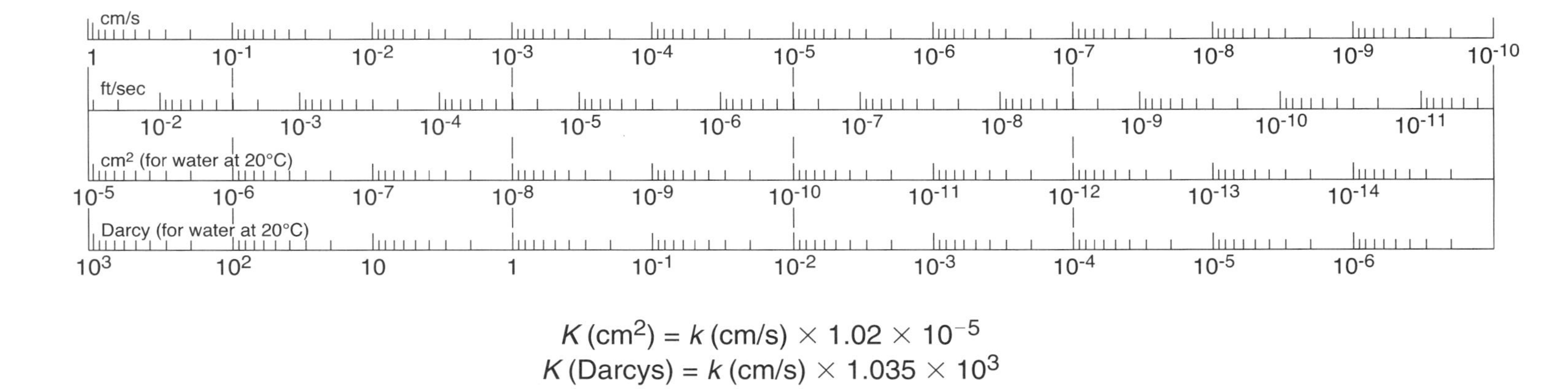

Fig. 2.6. Conversion chart for permeability of soils when water is used as the permeant.

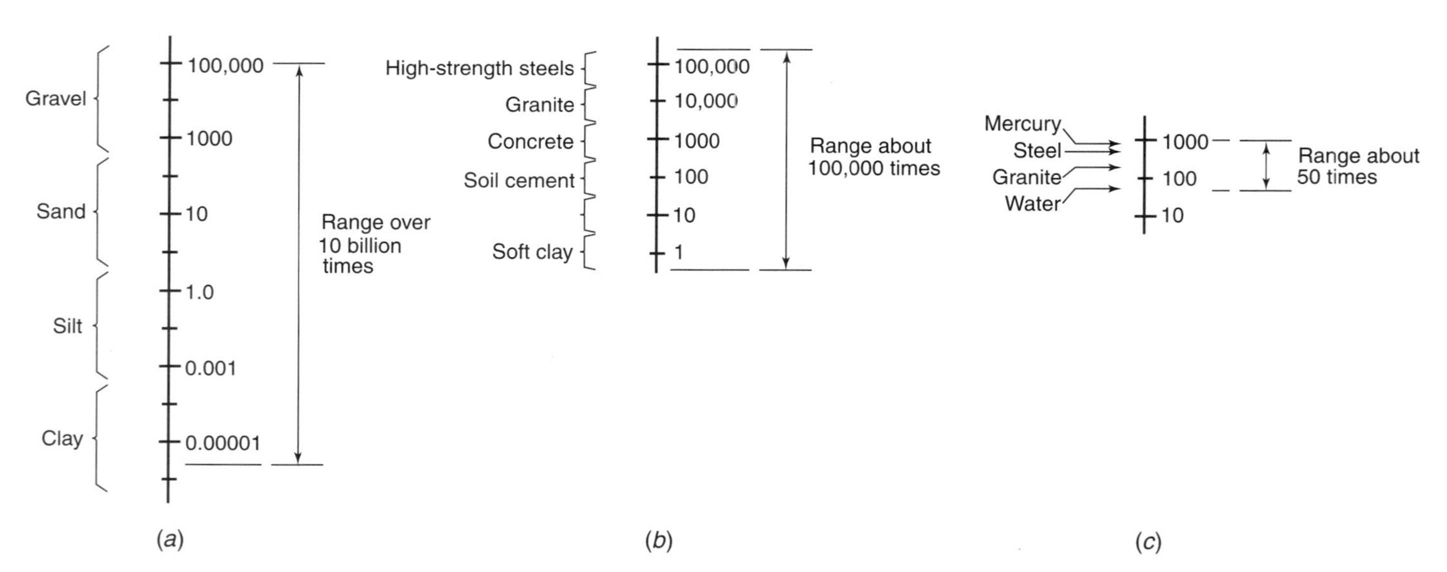

Fig. 2.7. Variability of permeability compared with other engineering properties: (*a*) permeability, ft/day; (*b*) strength, lb/in^2; (*c*) unit weight, lb/ft^3. (Adapted from Cedergren, 1989; copyright © John Wiley & Sons; reprinted with permission.)

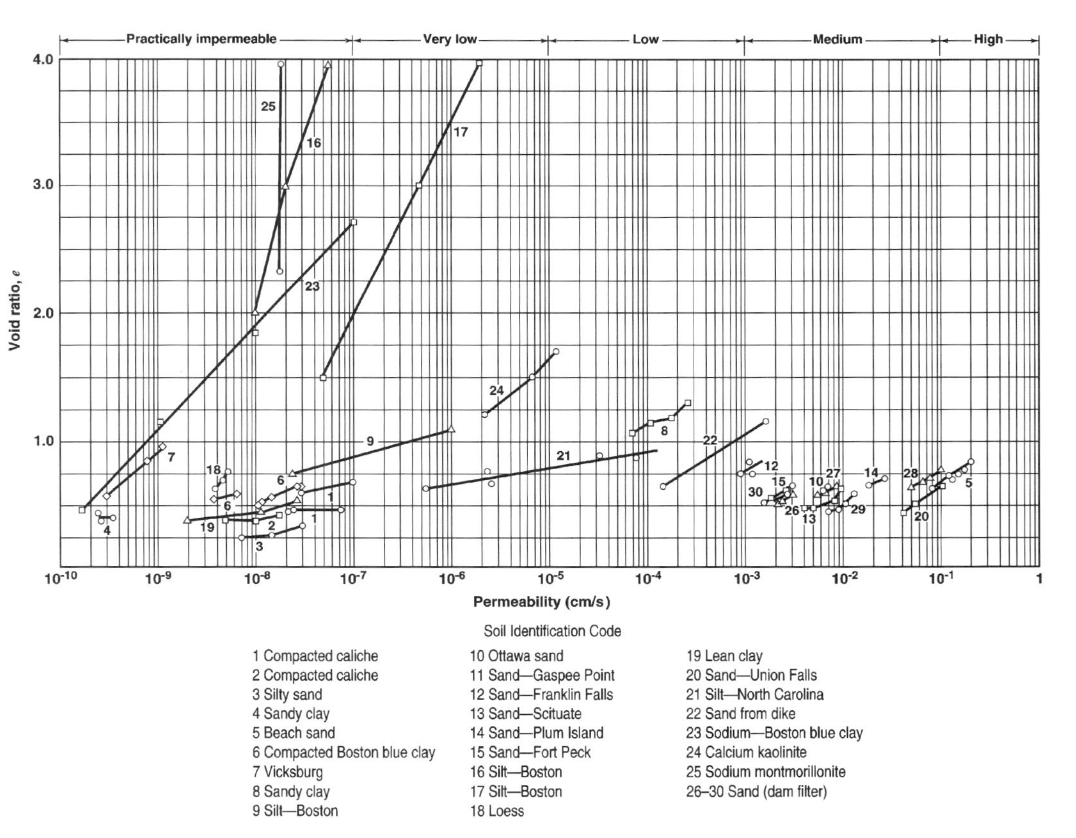

Fig. 2.8. Data showing the variation in permeability with respect to void ratio for a variety of soil types. (Adapted from Lambe and Whitman, 1969; copyright © John Wiley & Sons; reprinted with permission.)

strength and unit weight (Fig. 2.7*b* and *c*). Figure 2.8 demonstrates the wide range of magnitude for k as a function of void ratios for various soil types.

In the case of compacted clays, k is a function of the molding water content and the type and energy of compaction. The complex pore structures associated with compacted clays allow a wide range of values for k even for the same molding water content and dry density. As shown in Fig. 2.9, the permeability may vary by several

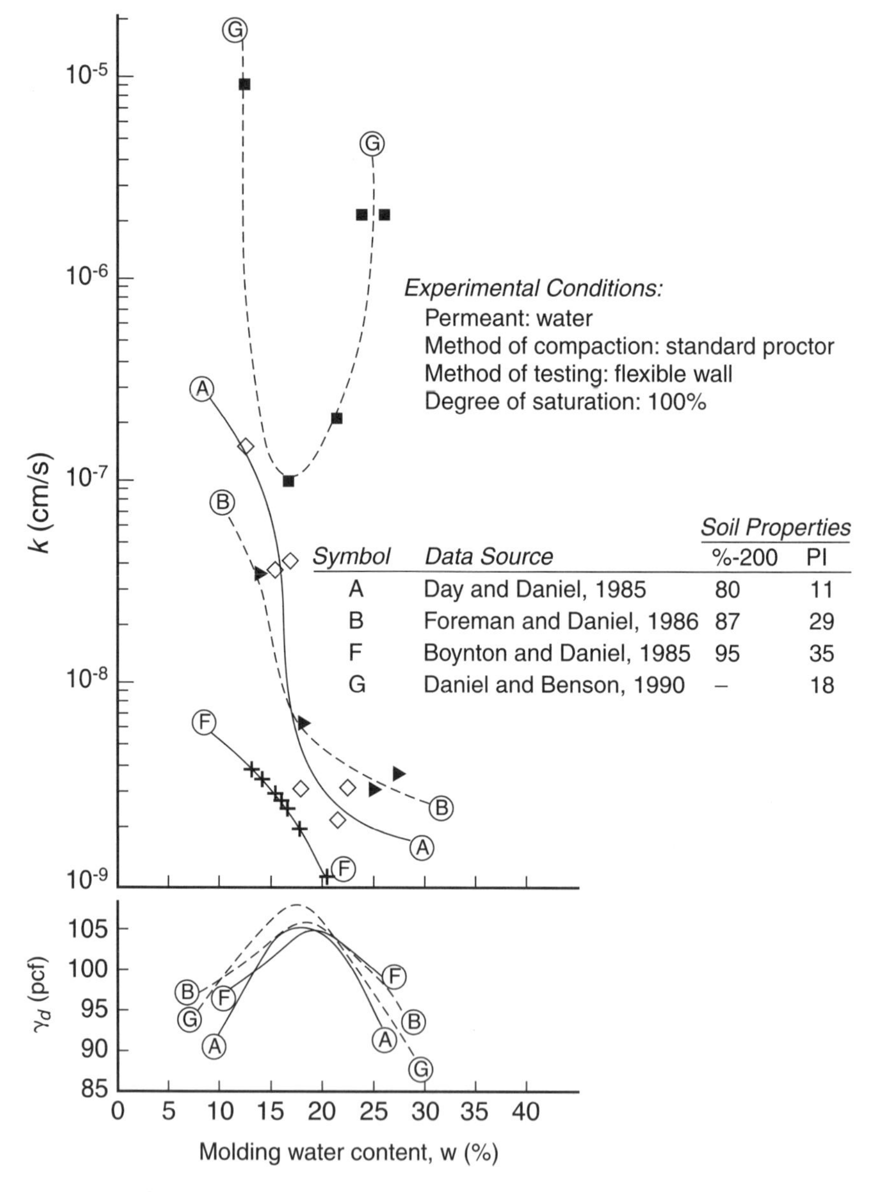

Fig. 2.9. Variation in hydraulic conductivity of compacted clays.

orders of magnitude even when the clays are prepared at nearly the same conditions of molding water content and maximum dry density. This is because of the fact that the microstructure of compacted clays is dependent on a number of other factors, such as the clay mineral type, and type and energy of compaction, in addition to the molding water content and dry density.

The effect of compaction variables on the microstructure and k of compacted clays was first studied by Mitchell et al. (1965) and is shown in Fig. 2.10. As shown in the figure, reduction of two to three orders of magnitude in hydraulic conductivity may result as the structure of clay changes from *dry of optimum* to *wet of optimum.* The lowest k occurs wet of optimum, beyond which a slight rebound is possible. A replot of the same data in Fig. 2.11 brings out the sharp contrast between sands and compacted clays in that the k of the compacted clays does not depend uniquely on the dry density (or the void ratio). For the molding water content of 19%, the permeability changed almost three orders of magnitude, although the dry density

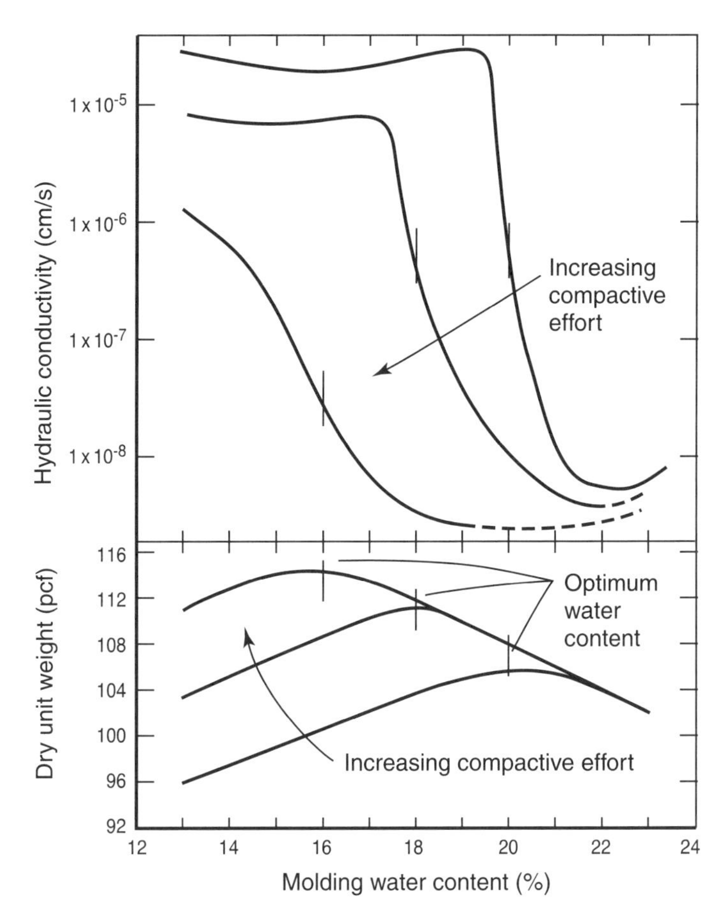

Fig. 2.10. Effect of compaction variables on hydraulic conductivity of compacted clays. (Adapted from Mitchell et al., 1965.)

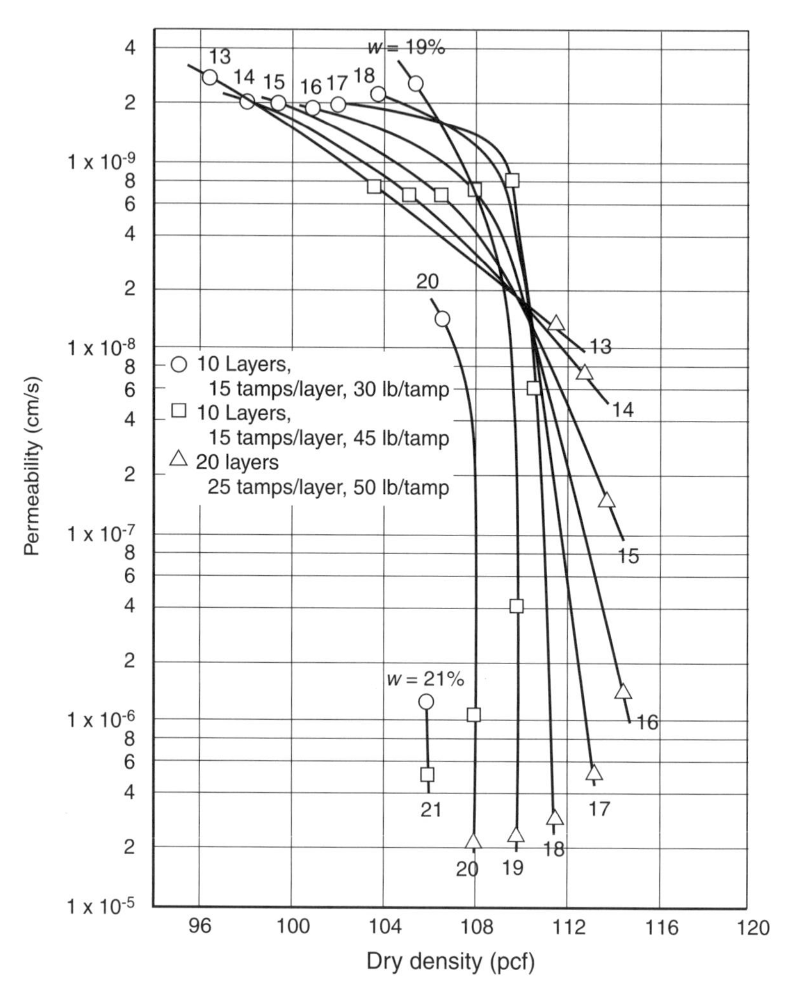

Fig. 2.11. Permeability versus dry density of compacted clays. (Adapted from Mitchell et al., 1965.)

is constant. More recent studies established the effects of such variables as type and energy of compaction and size of clods used in the compaction process. Field compaction of soils can be accomplished using different methods, including impact, static, kneading, and vibratory methods. The compaction method adopted greatly influences the structure and hence the k of compacted clays. For a detailed description of these issues, the reader is referred to Day and Daniel (1985), Daniel (1987), and Benson and Daniel (1990).

Several methods exist for hydraulic conductivity determination of soils. We study these in the next three sections in three groups: empirical, laboratory, and field methods. Not all methods are suitable for a specific project or for a given type of soil; hence the reader should screen the methods, considering the importance of the project and type of soils at hand.

2.5 EMPIRICAL METHODS FOR DETERMINING k

Several studies reported in the literature attempt to relate k directly to other known properties of soils. Hydraulics of flow in open pipes and tubes offered some of the earliest clues on what parameters of soils might influence k. In general, the empirical methods available at present for determining k can be grouped under three categories: those relating k to soil parameters descriptive of the particle sizes; those relating k to soil parameters descriptive of the pore volume, such as void ratio and porosity; and those relating k to the pore size distribution of soils. The methods are discussed below in that order.

Soils can be idealized as bundles of pore tubes of varying cross sections. This idealization will allow us to apply pipe hydraulics and obtain direct formulations for hydraulic conductivity. For a single tube or pipe, the average velocity of flow (averaged over the cross section of the open pipe) can be expressed using *Poiseuille's law:*

$$v_t = \frac{R_t^2}{8\eta} i \tag{2.12}$$

where v_t and R_t are the average velocity and radius of the pore tube, respectively. Equation (2.12) is analogous to (2.7) for soils; hence the fractional term in (2.12) can be viewed as the *hydraulic conductivity of the pore tube.* The hydraulic conductivity of a pore opening is thus directly proportional to the square of the radius of the opening when the opening can be idealized as circular in cross section. In soils with uniform particle sizes, the radius of the pore opening is governed by the particle size. This enabled several investigators to express k of soils as a function of a particle size descriptor. The following expressions belong to this category of empirical functions:

$$k\ (\text{cm/s}) = cD_{10}^2 \qquad \text{Hazen (1930)} \tag{2.13}$$

$$K\ (\text{mm}^2) = (0.05 \text{ to } 1.0)D_5^2 \qquad \text{Kenney et al. (1984)} \tag{2.14}$$

where c is a constant varying from 1.0 to 1.5, D_{10} the effective size of the soil (mm; the diameter of the sieve opening through which 10% of the soil passes), and D_5 the diameter (mm) corresponding to 5% finer soil particles. It should be emphasized that these expressions were given based on experimental observations on sandy soils. The former expression was based on experimental results from loose and clean sands. The latter was based on experimental results on sandy soils with uniformity coefficients C_u ranging from 1.04 to 12. Another empirical correlation belonging to this category was given by the U.S. Department of the Navy (1971) based on a number of experiments on soils with C_u ranging from 2 to 12 and $D_{10}/D_5 < 1.4$. This correlation is shown in Fig. 2.12.

Hydraulic conductivity of coarse-grained granular materials is an important parameter in the design of pavement drainage systems (Chapter 10) since the subbases in pavements are comprised of these materials. For design purposes, data from a large number of samples were used in statistical analyses to identify key parameters controlling k and to determine the correlation between k and these parameters (Moulton,

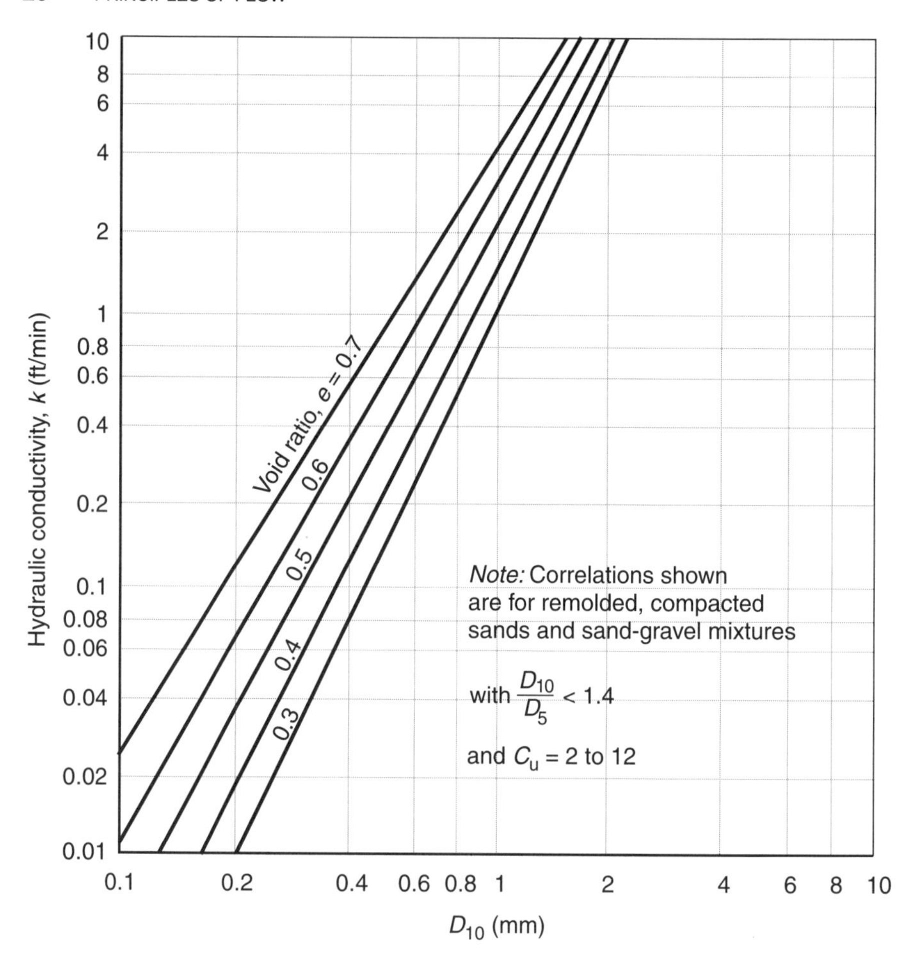

Fig. 2.12. Permeability of granular soils. (Adapted from U.S. Department of the Navy, 1971.)

1980). These correlations suggested that the significant properties controlling k were the effective grain size D_{10}, the porosity n, and the percent passing the No. 200 sieve, P_{200}. The resulting expression for k was

$$k \text{ (ft/day)} = \frac{6.214 \times 10^5 (D_{10})^{1.478} (n)^{6.654}}{(P_{200})^{0.597}} \tag{2.15}$$

Equation (2.15) was given in the form of a nomogram in Fig. 2.13. Note that porosity n was expressed in terms of the dry density γ_d. Figure 2.14 shows yet another correlation useful in pavement drainage system design between k and the gradation curves of granular subbase materials.

Poiseuille's expression in Eq. (2.12) can be extended for the case of nonregular cross sections of pore tubes. This is accomplished by expressing the size of the pore

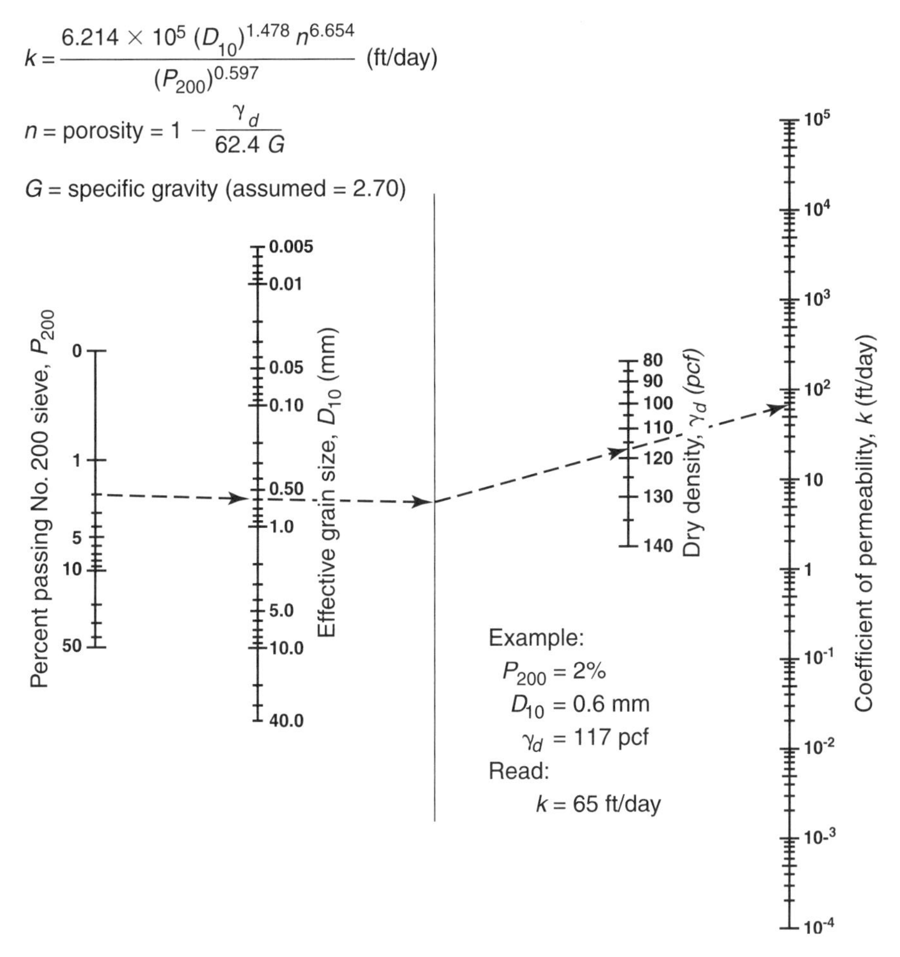

Fig. 2.13. Nomogram for estimating the coefficient of permeability of granular drainage and filter materials. (Adapted from Moulton, 1980.)

tube in terms of hydraulic radius as opposed to radius of a circular cross section. An advantage of doing this is that the hydraulic conductivity of soils can be expressed in terms of void ratio, which is a far more measurable parameter of soils than the sizes of individual pore tubes. This leads us to the well-known *Kozeny–Carman equation,* given by

$$k = C_s \frac{1}{\eta S_0^2} \frac{e^3}{1+e} \tag{2.16}$$

where C_s is a shape constant describing the irregular cross section of the pore tubes, S_0 the wetted surface area per unit volume of solid particles, and e the void ratio of

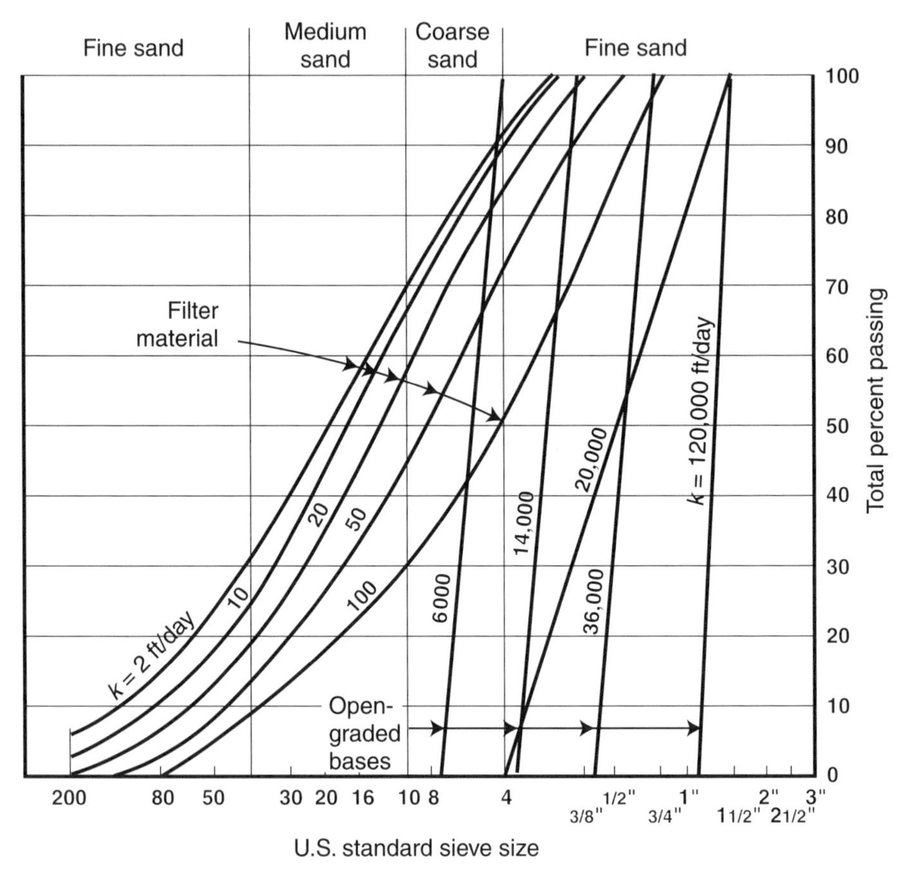

Fig. 2.14. Permeabilities corresponding to various gradations of open-graded bases and filter materials. (Adapted from Moulton, 1980.)

the soil. Equation (2.16) is valuable in that it expresses the proportionality between k and a void ratio function,

$$k \propto \frac{e^3}{1+e} \tag{2.17}$$

A second class of empirical functions, available in the literature, is based on this proportionality. This proportionality was generally found to be valid for a range of granular soils. For the typical ranges of e of granular soils, the following proportionalities were also suggested in various studies:

$$k \propto \frac{e^2}{1+e} \tag{2.18}$$

$$k \propto e^2 \tag{2.19}$$

$$\log k \propto e \tag{2.20}$$

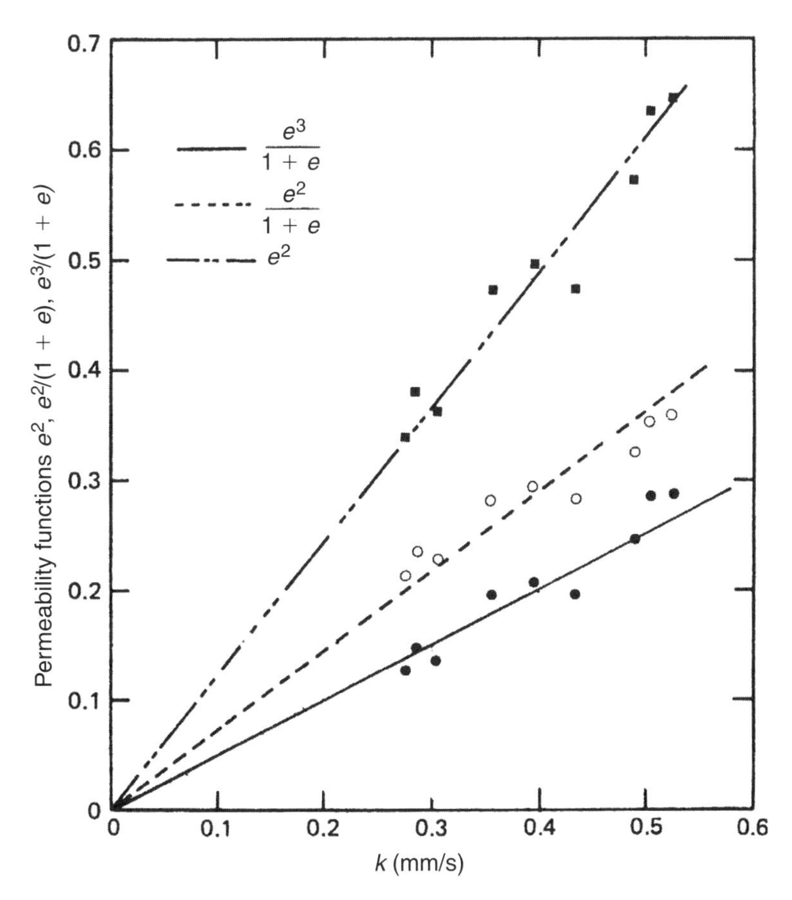

Fig. 2.15. Coefficient of permeability k versus void ratio function. (Adapted from Das, 1983.)

The linear relationship between k and the three functions of void ratio, Eqs. (2.17), (2.18), and (2.19), is demonstrated in Fig. 2.15. The data in Fig. 2.15 are that of a uniform Madison sand with void ratio ranging from 0.58 to 0.80. For this range it is seen that all three functions fit equally good. Figure 2.8 demonstrates the empirical relation expressed by Eq. (2.20). Another empirical expression in use relating k and void ratio is the one given by Casagrande:

$$k = 1.4k_{0.85}e^2 \tag{2.21}$$

where $k_{0.85}$ is the hydraulic conductivity corresponding to a void ratio of 0.85.

A third group of empirical functions for hydraulic conductivity are based on the pore-size distribution of soils. Although not commonly measured in geotechnical engineering, the pore-size distribution of soils provides a useful quantitative measure of the pore structure of soils and is being used increasingly to study soil fabric. A technique known as the *mercury intrusion method* is used to determine the pore-size distribution. Garcia-Bengochea et al. (1979) gave us a correlation useful to obtain hydraulic conductivity when a pore-size parameter PSP is known:

$$k = C_s'(\text{PSP})^b \tag{2.22}$$

where C_s' and b are empirical constants and PSP can be expressed using any of the following three models:

$$\text{PSP} = \begin{cases} E\left(d^2\right) n & \mu\text{m}^2 \quad \text{Capillary model} \quad (2.23) \\ \sum_i^n \sum_j^n (\tilde{d})^2 f(d_i) f(d_j) & \mu\text{m}^2 \quad \text{Marshall model} \quad (2.24) \\ \left[\dfrac{1}{4\sum_i^1 f(d_i)/d_i}\right]^2 n & \mu\text{m}^2 \times 10^{-4} \quad \text{Hydraulic Radius model} \quad (2.25) \end{cases}$$

where d is the pore diameter, $E(d^2)$ the second moment about the origin of the pore-size distribution, n the porosity, $\tilde{d}$ the smaller of the two pore diameters d_i and d_j, and $f(d_i)$ and $f(d_j)$ are the volumetric frequency of occurrence of pore diameters d_i and d_j, respectively. Note that C_s' and b in Eq. (2.22) are dependent on the type of model used to determine PSP. Garcia-Bengochea et al. (1979) used the three models for mixtures of silt and kaolin and obtained the following ranges for the two parameters: Capillary model, $C_s' = 0.194$ to 3.487 and $b = 2.42$ to 5.88; Marshall model, $C_s' = 17.41$ to 4.6×10^5 and $b = 1.62$ to 4.95; Hydraulic Radius model, $C_s' = 0.015$ to 0.166 and $b = 2.43$ to 8.91. The correlation coefficients obtained when the three models were fit to experimental observations ranged from 0.55 to 0.91.

Example 2.3 The coefficient of hydraulic conductivity k for a granular soil with $e = 0.43$ was determined to be 3×10^{-2} cm/s; D_{10} and D_5 for the soil are 0.3 mm and 0.2 mm, respectively; uniformity coefficient $C_u = 4.7$; 2% of the material passes through the No. 200 sieve. Determine k of the soil for $e = 0.70$ using all relevant empirical methods.

SOLUTION: (a) According to the Hazen equation, $k = cD_{10}^2$ cm/s. If $c = 1$, then $k = cD_{10}^2 = 1 \times 0.3^2$ cm/s $= 0.09$ cm/s. If $c = 1.5$, then $k = cD_{10}^2 = 1.5 \times 0.3^2$ cm/s $= 0.135$ cm/s. So the estimate of the coefficient of hydraulic conductivity according to the Hazen equation is in the range 0.09 to 0.135 cm/s.

(b) According to the FHWA (Federal Highway Administration) equation, the permeability k is estimated as

$$k = \frac{6.214 \times 10^5 (D_{10})^{1.478} (n)^{6.654}}{(P_{200})^{0.597}} \text{ ft/day}$$

The porosity n relates to the void ratio of soil e as

$$n = \frac{e}{1+e}$$

The estimate of n for $e = 0.70$ is

$$n = \frac{e}{1+e} = \frac{0.70}{1+0.70} \approx 0.412$$

The estimate of the permeability

$$k = \frac{6.214 \times 10^5 (0.3)^{1.478} (0.412)^{6.654}}{(2)^{0.597}} \text{ ft/day} \approx 189.831 \text{ ft/day} \approx 6.697 \times 10^{-2} \text{ cm/s}$$

(c) Using the proportionality between the permeability and void ratio function, we can estimate the hydraulic conductivity at $e = 0.70$ based on the hydraulic conductivity at $e = 0.43$.

(1) Using the function

$$k \propto \frac{e^3}{1+e}$$

we estimate the permeability as

$$k = \frac{0.7^3/(1+0.7)}{0.43^3/(1+0.43)} \times 3 \times 10^{-2} \text{ cm/s} \approx 0.11 \text{ cm/s}$$

(2) Using the function

$$k \propto \frac{e^2}{1+e}$$

we estimate the permeability as

$$k = \frac{0.7^2/(1+0.7)}{0.43^2/(1+0.43)} \times 3 \times 10^{-2} \text{ cm/s} \approx 6.69 \times 10^{-2} \text{ cm/s}$$

(3) Using the function

$$k \propto e^2$$

we estimate the permeability as

$$k = \frac{0.7^2}{0.43^2} \times 3 \times 10^{-2} \text{ cm/s} \approx 7.95 \times 10^{-2} \text{ cm/s}$$

(4) Using the function

$$\log k \propto e$$

which is equivalent to

$$k \propto \exp(e)$$

we estimate the permeability as

$$k = \frac{\exp(0.7)}{\exp(0.43)} \times 3 \times 10^{-2}\ \text{cm/s} \approx 3.93 \times 10^{-2}\ \text{cm/s}$$

Thus, the empirical estimate of k ranges from 3.93×10^{-2} to 0.135 cm/s.

2.6 LABORATORY METHODS FOR DETERMINING k

Laboratory methods for determining k can be grouped under two categories based on the hydraulic control used to induce flow in soils; and the type of cell used to contain the soil sample. We discuss below the methods falling under these two categories.

2.6.1 Hydraulic Control Methods

To induce flow in soils, a hydraulic gradient is needed across the length of the sample. Three methods are available to create the necessary hydraulic gradient: (1) constant head test, (2) variable head test, and (3) constant flow rate test. In the first two methods, different hydraulic heads are applied at the influent and effluent ends of the sample to create a hydraulic gradient. In the third method, a flow pump is used to discharge water at a constant flow rate through the sample. Thus, the first two methods are gradient-controlled, whereas the third method is flow-rate controlled. All three methods use Darcy's law in one form or another. The dimensions of the soil sample used in the laboratory (length L and cross-sectional area A) are considered known. The test methods give us information on the relationship between q and the hydraulic gradient i. This will leave us with only one unknown, the hydraulic conductivity, which can be determined using Darcy's law. Detailed description of these tests follows.

Constant Head Test In this test, a *constant* hydraulic gradient is applied across the length of the sample. A continuous water supply is used to maintain the water level in the influent tank at a constant elevation while flow takes place through the sample. A typical setup of this test is shown in Fig. 2.16. After a steady flow rate is established, the amount of water exiting the soil column is collected in a graduated cylinder for a known time period Δt. Darcy's law, Eq. (2.6), can then be used to express the total volume Q collected as

$$Q = q\ \Delta t = kiA\ \Delta t \tag{2.26}$$

where the hydraulic gradient i is given by

$$i = \frac{\Delta h}{L} \tag{2.27}$$

The unknown k in Eq. (2.26) can now be related to all known quantities:

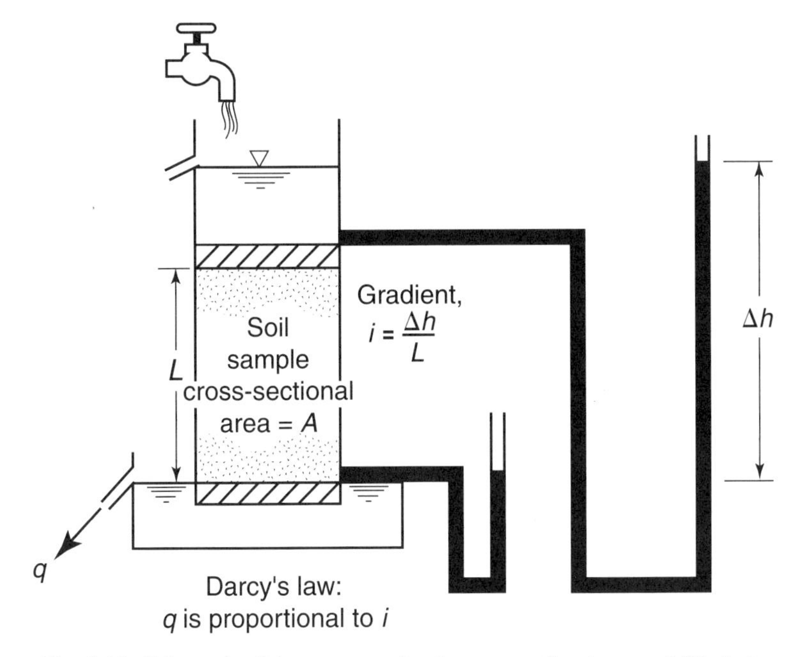

Fig. 2.16. Schematic of the apparatus for the constant head permeability test.

$$k = \frac{QL}{A\,\Delta h\,\Delta t} \tag{2.28}$$

Variable Head Test This test (Fig. 2.17) is designed for the purpose of capturing low flow rates associated with fine-grained soils. The constant head test may not yield measurable flow rates during the period of testing for such soils. A small-diameter standpipe is used as the influent chamber. Although the flow volume through the soil sample is small, the drop in the water levels in the standpipe is measurable because of the small diameter of the pipe. At the beginning of the test, the head difference h_1 is recorded. After a time period Δt is elapsed, the head difference h_2 is again recorded, which will be less than h_1. These head observations can now be used in the Darcy's law to determine k, as described below.

The flow rate q in the soil sample, as given by the Darcy's law, is the same as the rate at which volume of water is lost in the standpipe. Therefore,

$$q = k\frac{h}{L}A = -a\frac{dh}{dt} \tag{2.29}$$

where a is the cross-sectional area of the standpipe. Separation of variables in Eq. (2.29) yields

$$dt = \frac{aL}{Ak}\left(-\frac{dh}{h}\right) \tag{2.30}$$

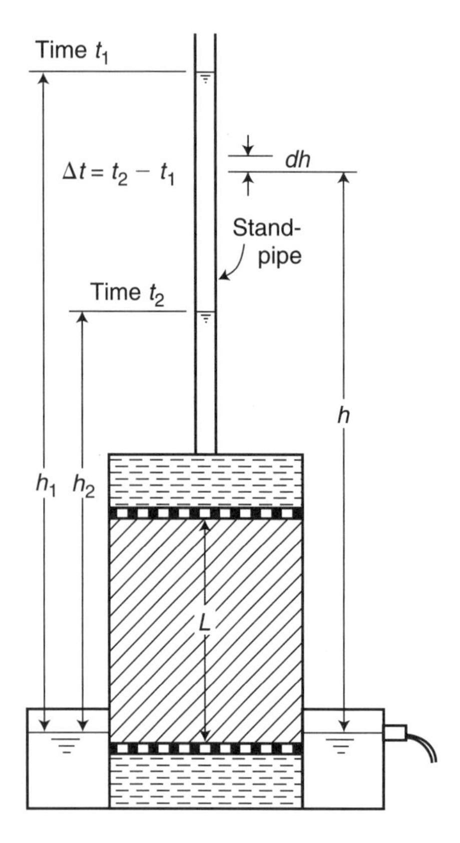

Fig. 2.17. Schematic of the apparatus for the variable head permeability test.

Integration of both sides in Eq. (2.30) within the limits of h_1 and h_2 for the time period Δt gives us

$$\Delta t = \frac{aL}{Ak} \ln \frac{h_1}{h_2} \tag{2.31}$$

Equation (2.31) can be rearranged to express the unknown k in terms of known quantities as

$$k = \frac{aL}{A\ \Delta t} \ln \frac{h_1}{h_2} \tag{2.32}$$

Constant Flow Rate Test The schematic of this test is shown in Fig. 2.18. Water is delivered into the soil sample at a constant predetermined flow rate q. A programmable pump, which is capable of delivering flow rates within the range of expected pressure differences across the sample, may be used. Once the pump is set to deliver a known flow rate, the head difference across the sample length can

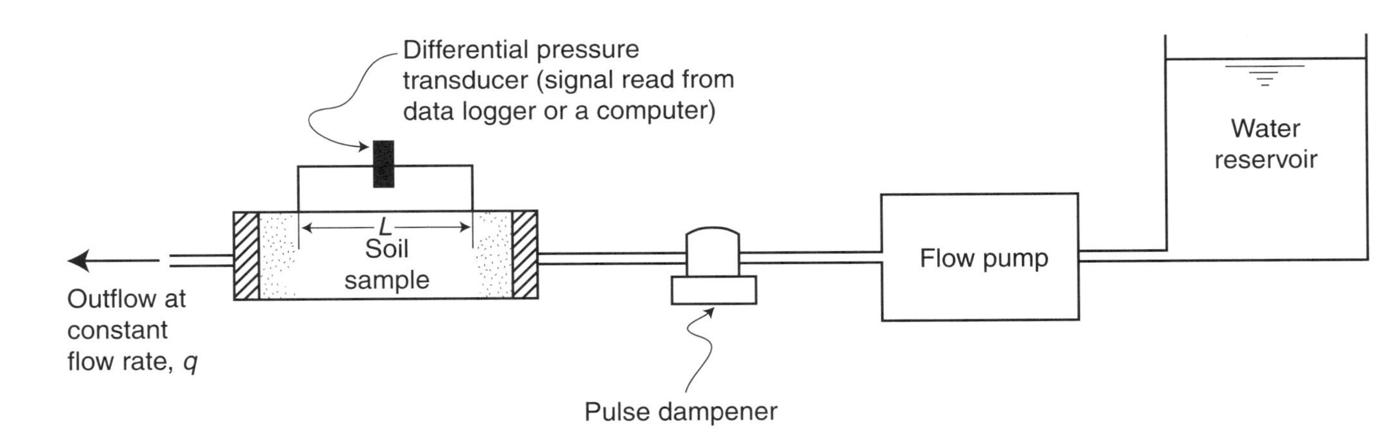

Fig. 2.18. Schematic of the apparatus for the constant flow rate test.

be measured using a pressure gauge or electronic pressure transducer. The pressure difference Δp can be transformed to Δh in length units using

$$\Delta h = \frac{\Delta p}{\gamma_w} \tag{2.33}$$

and Darcy's law can be used to express k in terms of known quantities,

$$k = \frac{qL}{\Delta h\ A} \tag{2.34}$$

Example 2.4 In a laboratory equipment, a pump is used to maintain a steady flow rate, $q = 100$ mL/min, through the soil sample shown in Fig. E2.4. The gradient is read using a pressure gauge attached at the influent end. Determine the permeability k of the soil sample if the gauge shows a pressure of 12 cm of water. The cross-sectional area of the sample is 50 cm^2. The soil is supported in the column with screens on either end of the sample, which have very high permeability relative to that of the soil.

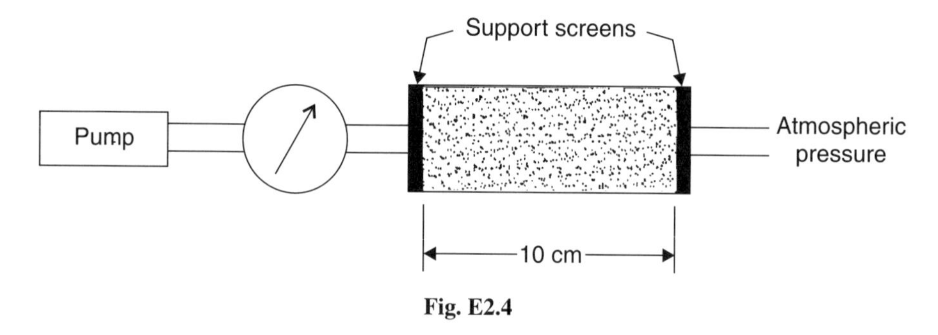

Fig. E2.4

SOLUTION: The permeability k is

$$k = \frac{q}{A}\frac{L}{\Delta h} = \frac{100\ \text{mL/min}}{50\ \text{cm}^2} \times \frac{1\ \text{cm}^3}{\text{mL}} \times \frac{10\ \text{cm}}{12\ \text{cm}} \approx 1.67\ \text{cm/min} = 2.78 \times 10^{-2}\ \text{cm/s}$$

2.6.2 Methods Based on Sample Containment

The types of cells used to contain the soil samples can be grouped under the following categories:

- Rigid-wall permeameters
  - Compaction-mold permeameters
  - Consolidation-cell permeameters
  - Sampling-tube permeameters
- Flexible-wall permeameters

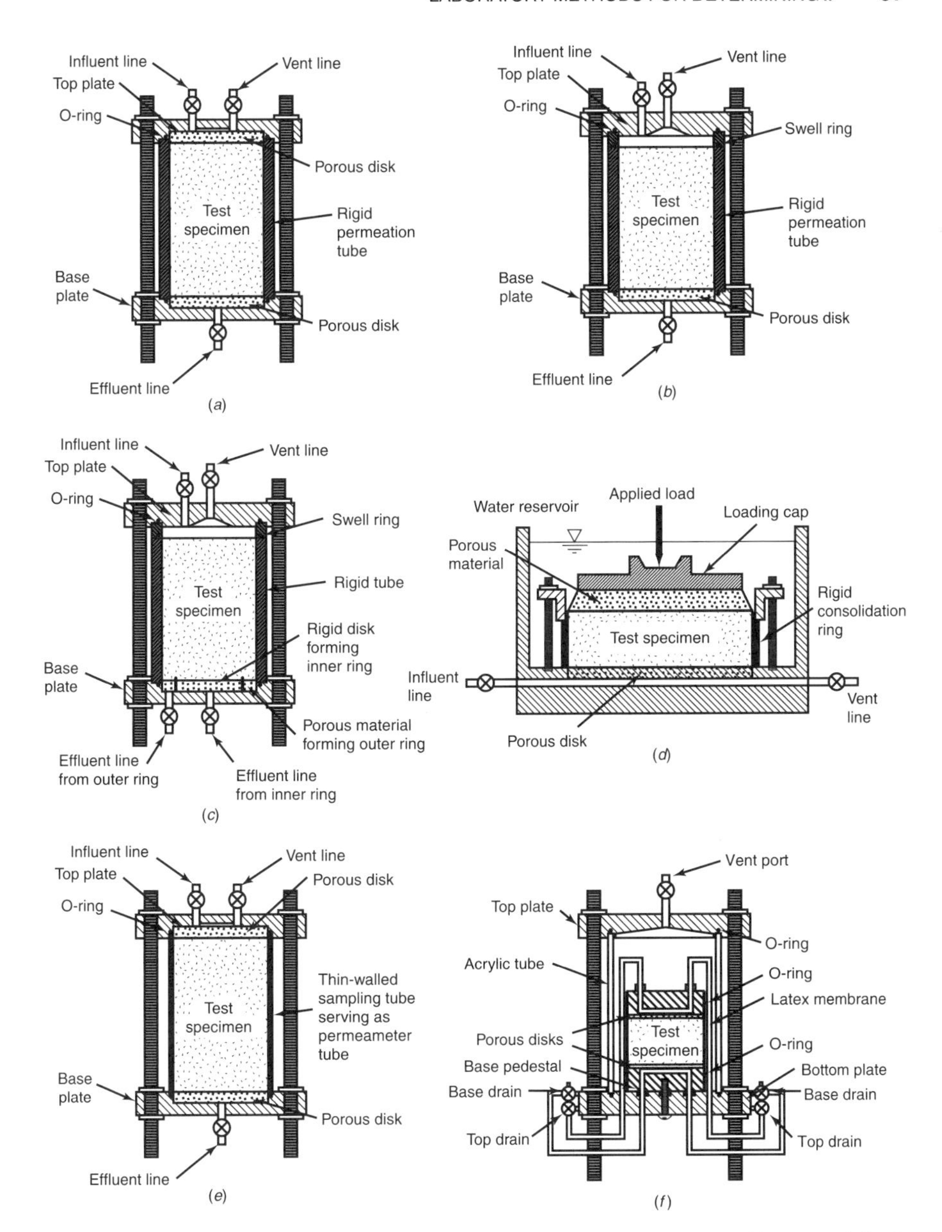

Fig. 2.19. Types of rigid- and flexible-wall cells. (Adapted from Daniel, 1994; copyright © ASTM International; reprinted with permission.)

Rigid-wall permeameters consist of cylindrical molds made of metal, glass, or Plexiglas. Lateral expansion of the soil specimen is restrained; however, swelling in the longitudinal dimension may be allowed. Flow takes place from top to bottom or bottom to top in the longitudinal direction of the cylindrical sample. Upward flow is often preferred to dislodge air bubbles entrapped in the soil sample at the beginning of testing. One of the principal disadvantages of rigid-wall permeameters is the sidewall leakage between the cylindrical sample and the rigid wall. Flexible-wall permeameters, on the other hand, contain soil samples in latex membranes and allow application of confining and vertical stresses on the samples. Because of the sample placement in a latex membrane, sidewall leakage is unlikely. However, permeability determination using flexible-wall permeameters takes considerably longer time periods and is relatively more expensive than use of rigid-wall permeameters.

The various types of rigid- and flexible-wall cells are shown schematically in Fig. 2.19. Compaction mold peremeameters, the most commonly used type of permeameters, are used to determine the k values of a wide variety of materials, ranging from gravel to clay. Whereas materials such as gravel are simply poured into the rigid mold, clays are compacted using Proctor methods in the same permeameters used for water delivery. The sample is confined with porous disks at top and bottom in the case where the soil specimen is not allowed to swell (Fig. 2.19*a*). When soil swelling is allowed to take place during permeability determination, no porous disk is used at the top of the specimen (Fig. 2.19*b*). To minimize the effect of sidewall leakage on determination of k, a double-ring permeameter is sometimes used (Fig. 2.19*c*). The flow rate in this cell is measured only from the inner ring of the cross section. The macropores that may be present near the perimeter of the test specimen are therefore not allowed to contribute to the flow rates measured. The consolidation cell permeameters (Fig. 2.19*d*) allow the samples to be consolidated and the hydraulic conductivity determined indirectly from the rate of consolidation. The sampling-tube permeameters (Fig. 2.19*e*) allow experiments to be conducted with the same tube as that used to obtain undisturbed soil samples. Thus the effect of soil disturbances associated with soil extrusion and cutting is minimized in this method. In flexible-wall permeameters (Fig. 2.19*f*), the sample is confined between porous disks and end caps on the top and bottom and by a latex membrane on the sides. The cell is filled with water or other liquids and pressurized to a desired stress state. The pressure of the liquid will keep the latex membrane pressed against the soil sample, thus minimizing sidewall leakage. For additional discussion on the permeameter cells, the reader is referred to Daniel (1994).

2.7 FIELD METHODS FOR DETERMINING k

The field methods for determining the hydraulic conductivity of soils rely on information available from pumping out of a well or from measurements of groundwater levels taken in boreholes or piezometers. Based on the nature of field data used to obtain hydraulic conductivity, the field methods can be grouped under pump tests, borehole tests, or variable head tests using piezometers or observation wells. We consider these methods in separate subsections.

2.7.1 Pump Tests

In pump tests, water is pumped out of the well at a constant rate, and the drawdown of the water table is monitored using observation wells. In Fig. 2.20 are shown two

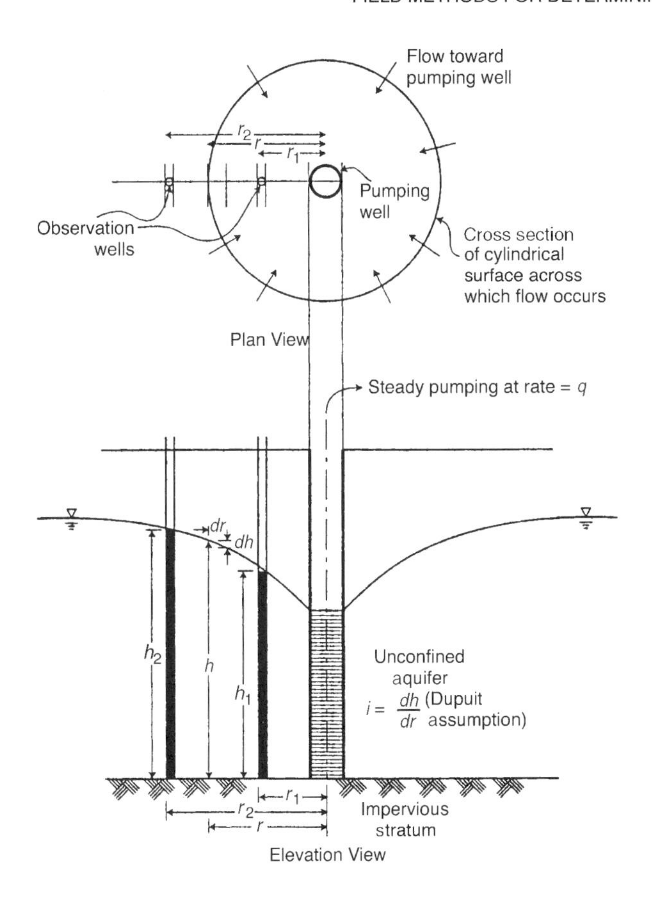

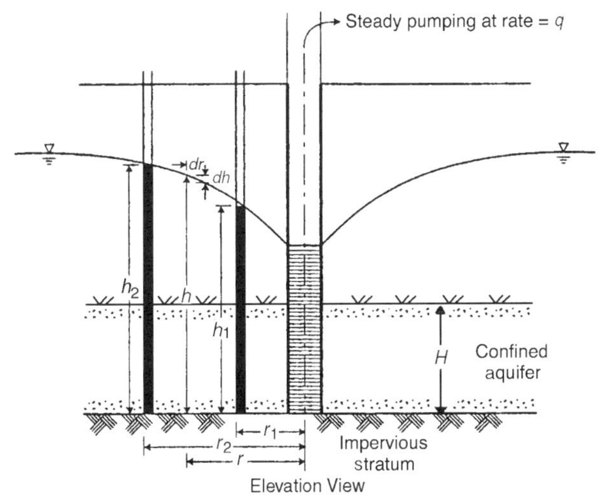

Fig. 2.20. Hydraulic conductivity determination using pump tests.

simple cases of wells fully penetrating unconfined and confined aquifers. Assuming that steady-state conditions are reached, the rate of discharge due to pumping can be expressed using Darcy's law. We consider an annular section bounded by two radii, r_1 and r_2, which correspond to the locations of observation wells. Considering a cylindrical surface in this annular section for the case of an unconfined aquifer,

$$q = k\frac{dh}{dr}(2\pi rh) \tag{2.35}$$

Note that we chose to express hydraulic gradient as dh/dr, which is not the true gradient since dr does not represent the actual path of water flow. This is often referred as *Dupuit's assumption* and is commonly used to simplify mathematical treatment of flow problems. This assumption is discussed further in Chapter 4. Separating the variables in Eq. (2.35) yields

$$\int_{r_1}^{r_2} \frac{dr}{r} = \frac{2\pi k}{q} \int_{h_1}^{h_2} h\,dh \tag{2.36}$$

which allows us to express k in terms of the known quantities as

$$k = \frac{q \ln (r_2/r_1)}{\pi \left(h_2^2 - h_1^2\right)} \tag{2.37}$$

In the case of the well fully penetrating a confined aquifer, the height of the cross-sectional surface through which flow occurs is constant, equal to the thickness of the confined aquifer, H. The reader can verify that this would lead to the following expression for k in the case of a confined aquifer:

$$k = \frac{q \ln (r_2/r_1)}{2\pi H (h_2 - h_1)} \tag{2.38}$$

2.7.2 Borehole Tests

Pump tests require observations from monitoring wells and continuous well pumping, which are expensive tasks. Borehole tests are based on observations that could be made on single wells. Based on the casing provided and the nature of flow from boreholes, several types of tests could be designed, which are discussed briefly below. Detailed descriptions of these methods were given by the U.S. Bureau of Reclamation in their *Earth Manual,* Part 2 (U.S. Department of the Interior, 1990).

Well Permeameter Method This method consists of measuring the rate at which water flows out of an uncased well under a constant gravity head. The types of soil for which the test is applicable range from mixtures of sand, silt, and clay with coefficients of hydraulic conductivity greater than 1×10^{-5} cm/s to relatively clean sands or sandy gravels with k less than 1×10^{-1} cm/s. The test is usually performed

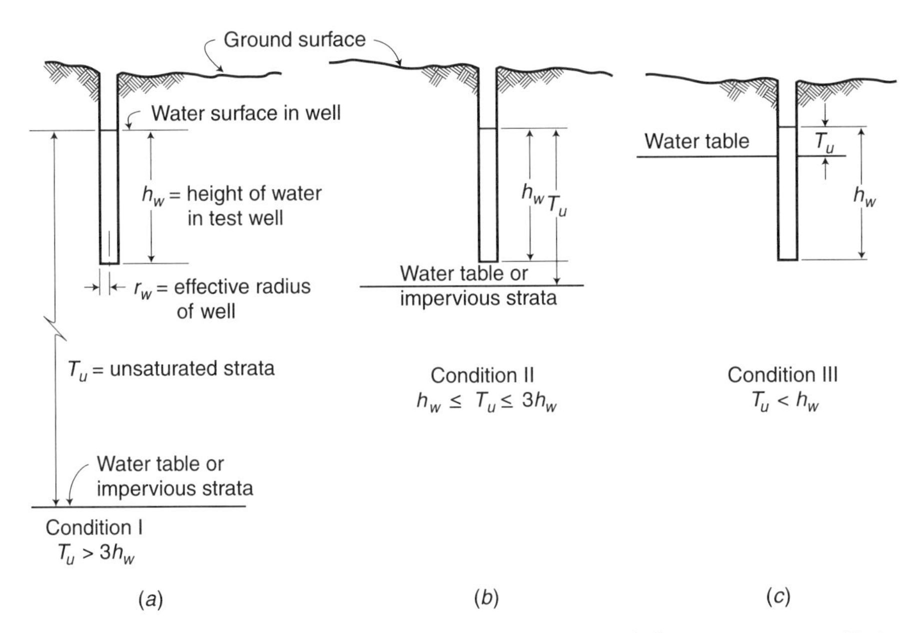

Fig. 2.21. Relationship between depth of water in a test well and distance to a water table in the well permeameter test. (Adapted from U.S. Department of the Interior, 1990.)

in auger holes. For a low-water-table condition (Fig. 2.21*a*), the depth of the well may be of any desired dimension provided that the ratio of water height h_w in the well to well radius r_w is greater than 1. For other elevations of the water table (Fig. 2.21*b* and *c*), h_w/r_w should be greater than $10r_w$. The minimum time duration for conducting the test should be such that a saturated envelope of hemispherical shape is formed around the well, and the maximum duration should be such that a water mound is not built up around the well. The volumes of water corresponding to these time durations are given as

$$V_{\min} = 2.09S\left\{h_w\sqrt{\ln\left[\frac{h_w}{r_w}+\sqrt{\left(\frac{h_w}{r_w}\right)^2+1}\right]-1}\right\}^3 \tag{2.39}$$

and

$$V_{\max} = 2.05V_{\min} \tag{2.40}$$

where S is the specific yield of the soil, which varies from about 0.1 for fine-grained soils to 0.35 for coarse-grained soils, and h_w and r_w are as described in Fig. 2.21. The hydraulic conductivities for the three conditions shown in Fig. 2.21 are given as follows:

Condition I:

$$k_{20} = \frac{q\eta_t}{\eta_{20}\left(2\pi h_w^2\right)}\left\{\ln\left[\frac{h_w}{r_w} + \sqrt{\left(\frac{h_w}{r_w}\right)^2 + 1}\right] - \frac{\sqrt{1 + (h_w/r_w)^2}}{h_w/r_w} + \frac{r_w}{h_w}\right\} \tag{2.41}$$

Condition II:

$$k_{20} = \frac{q\eta_t}{\eta_{20}\left(2\pi h_w^2\right)} \frac{\ln(h_w/r_w)}{\frac{1}{6} + \frac{1}{3}(h_w/T_u)^{-1}} \tag{2.42}$$

Condition III:

$$k_{20} = \frac{q\eta_t}{\eta_{20}\left(2\pi h_w^2\right)} \frac{\ln(h_w/r_w)}{(h_w/T_u)^{-1} + \frac{1}{2}(h_w/T_u)^{-2}} \tag{2.43}$$

where k_{20} is the hydraulic conductivity at 20°C, q the discharge rate of water from the well for steady-state condition determined in the field for a head h_w, η_t and η_{20} the viscosity at the field temperature t (in °C) and at 20°C, respectively, and all other terms are as described in Fig. 2.21.

Shallow-Well Permeameter Method This method was developed for the purpose of evaluating the k value of natural or compacted soil layers, such as canal liners, overlying more pervious soils (Fig. 2.22*a*). No water table is present within the layer for which k is sought. The test was conducted for a sufficient period of time to saturate an envelope of soil around and below the well and to maintain a steady flow rate out of the well. Based on the relative dimensions of H_w, r_w, and h_w (as indicated by α and β in Fig. 2.22*a*), a well coefficient C may be obtained from Fig. 2.22*b*. An example of calculation for coefficient C is shown in Fig. 2.22*b*. The hydraulic conductivity is determined using

$$k_{20} = \frac{\eta_t}{\eta_{20}} C \frac{q}{H_w^2} \tag{2.44}$$

where q is the average flow rate during the last 24 hours of test, and all other terms are as described earlier.

Example 2.5 A shallow well permeameter was installed in a compacted soil liner for determination of k. The dimensions (Fig. 2.22*a*) are $H_w = 60$ cm, $\alpha = 0.35$, $\beta = 0.5$. The average q from the shallow well into the liner was 800 cm^3/hr. (a) Determine k assuming that the test was conducted at 20°C. (b) If the test were to be repeated with $\beta = 0.6$, what would be the expected rate of flow into the liner?

SOLUTION: (a) To determine the permeability of the compacted soil liner using Eq. (2.44), we need to evaluate the coefficient C first. From Fig. 2.22*b*, we find that

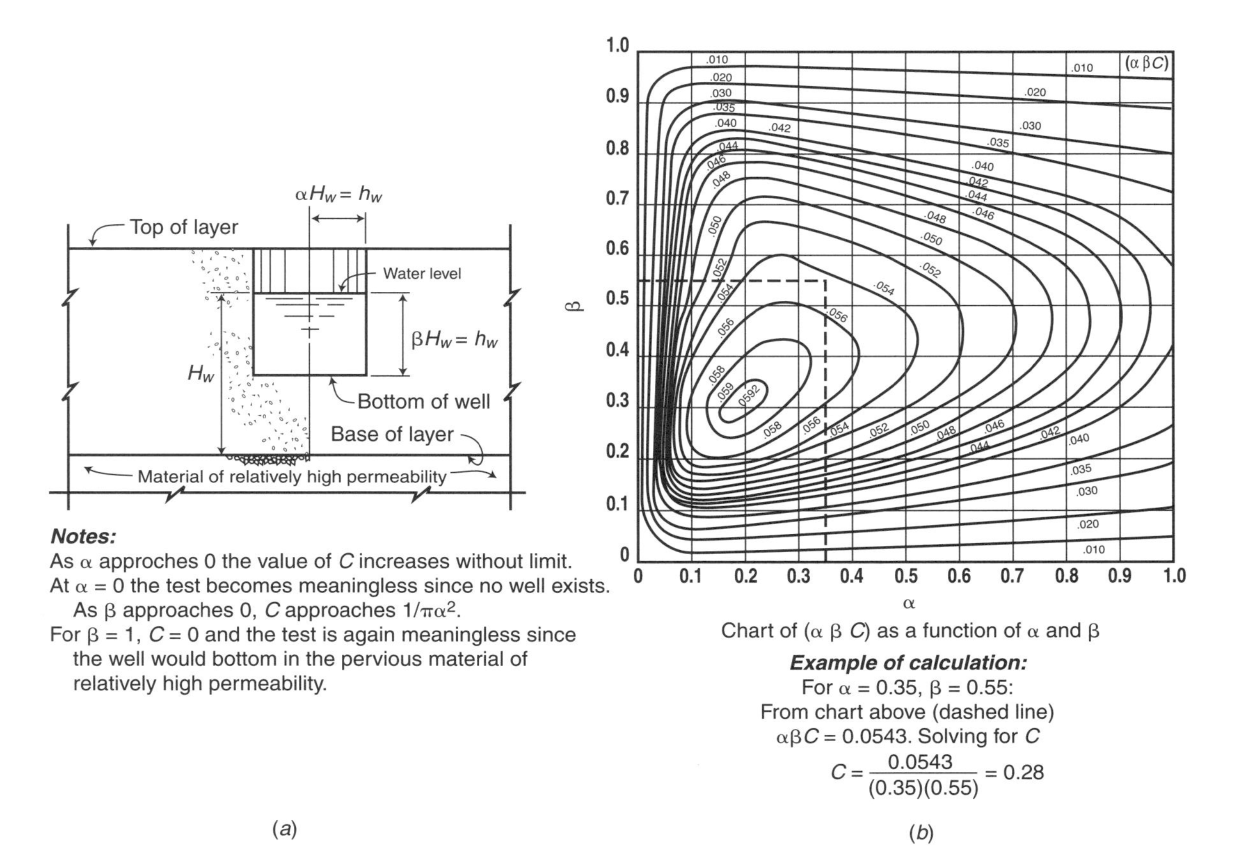

Fig. 2.22. Determination of hydraulic conductivity using the shallow-well permeameter test: (*a*) schematic of field installation; (*b*) determination of factor *C*. (Adapted from U.S. Department of the Interior, 1990.)

$\alpha\beta C = 0.0556$, and the coefficient $C = 0.0556/(0.35 \times 0.5) = 0.318$. Because the test was conducted at 20°C, $\eta_t/\eta_{20} = 1$. Using Eq. (2.44) yields

$$
\begin{aligned}
k &= \frac{\eta_t}{\eta_{20}} C \frac{q}{H_w^2} \\
&= 1 \times 0.318 \times \frac{800\ \text{cm}^3/\text{h}}{(60\ \text{cm})^2} \times \frac{1\ \text{h}}{3600\ \text{s}} \\
&\approx 1.96 \times 10^{-5}\ \text{cm/s}
\end{aligned}
$$

(b) If the test were to be repeated with $\beta = 0.6$, from Fig. 2.22*b* we find that $\alpha\beta C = 0.053$, and the coefficient $C = 0.053/(0.35 \times 0.5) = 0.303$. According to Eq. (2.44), we can see that Cq remains constant when the other parameters remain the same, so the expected rate of flow into the liner is

$$
q = 800\ \text{cm}^3/\text{h} \times \frac{0.318}{0.303} \approx 839.6\ \text{cm}^3/\text{h}
$$

Packer Tests In these tests, the hydraulic conductivity is determined for the strata isolated by packers placed in the borehole either above or below the water table (Fig. 2.23). It is assumed that the soils in the strata isolated by the packers are strong enough to keep the hole open. Water is injected into the isolated interval of the drill

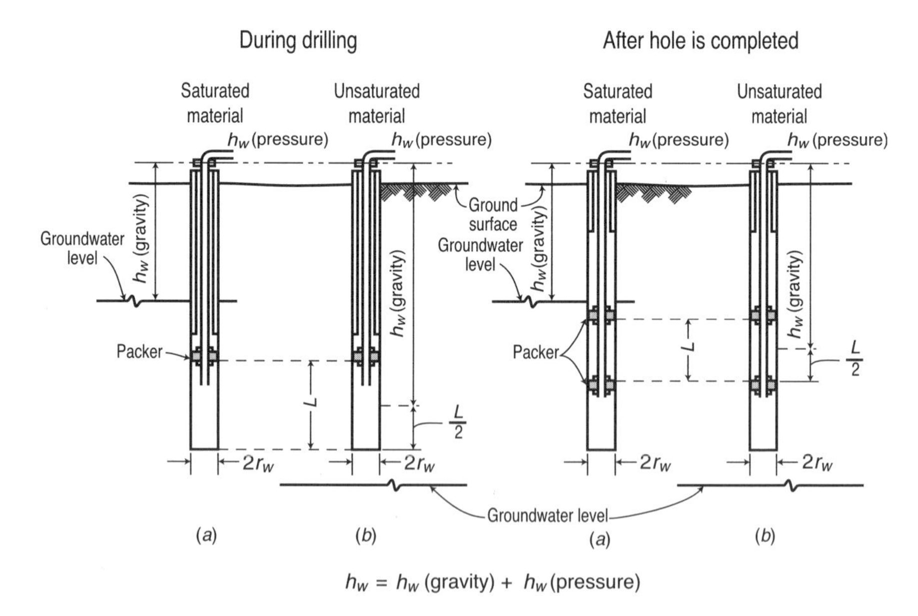

Fig. 2.23. Packer tests for soil permeability. (Adapted from U.S. Department of the Interior, 1990.)

hole, and the volume of water injected is determined for a specific period of time. A hydraulic pump may sometimes be used to create additional pressure head and facilitate injection. Tests could be conducted either during drilling or after the well is completed. The hydraulic conductivity k is obtained using the following expressions:

$$k = \begin{cases} \dfrac{q}{2\pi L h_w} \ln \dfrac{L}{r_w} & L \geq 10r_w \quad (2.45) \\[2ex] \dfrac{q}{2\pi L h_w} \sinh^{-1} \dfrac{L}{2r_w} & 10r_w > L \geq r_w \quad (2.46) \end{cases}$$

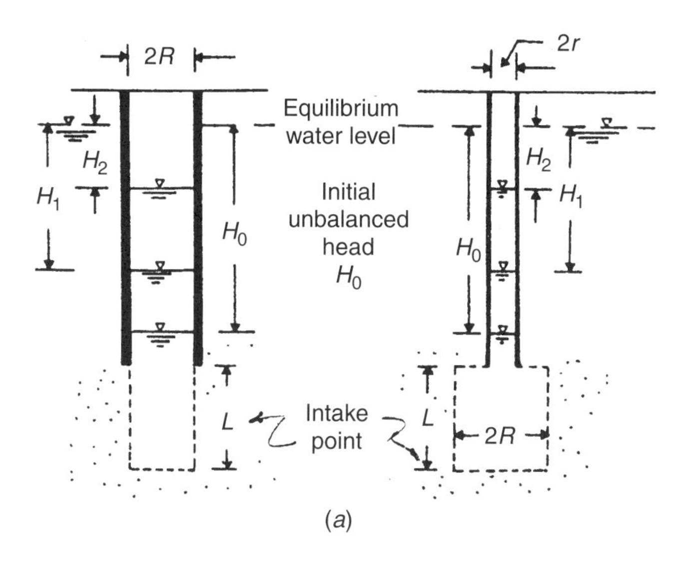

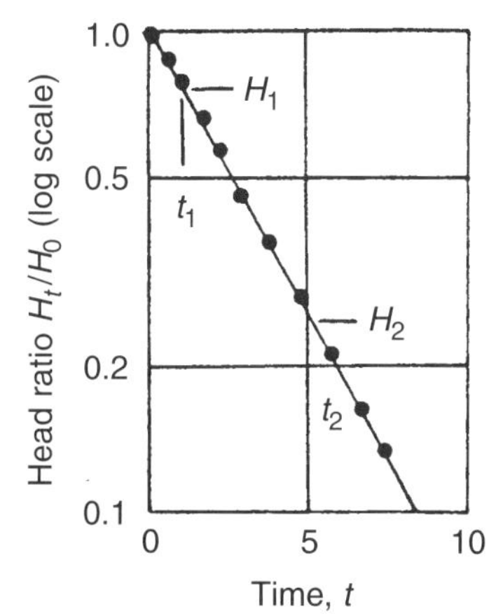

Fig. 2.24. Permeability determination using variable head tests: (*a*) schematic of a well and piezometer; (*b*) plot of observations. (Adapted from U.S. Department of the Navy, Naval Facilities Engineering Command, 1982.)

TABLE 2.1 Computation of Permeability from Variable Head Tests When the Wells or Piezometers Are Located in a Saturated Isotropic Stratum of Infinite Depth

Condition	Diagram	Shape Factor F	Permeability k	Applicability
(A) Uncased hole		$F = 16\pi DS'R$	$k = \frac{R}{16DS'} \times \frac{H_2 - H_1}{t_2 - t_1}$ for $\frac{D}{R} < 50$	Simplest methods for permeability determination. Not applicable in stratified soils. For values of S', see Fig. 2.25*a*.
(B) Cased hole, soil flush with bottom		$F = \frac{11R}{2}$	$k = \frac{2\pi R}{11(t_2 - t_1)} \ln \frac{H_1}{H_2}$ for 6 in. (0.1524 m) $<$ $D <$ 60 in. (1.524 m)	Used for permeability determination at shallow depths below the water table. May yield unreliable results in falling head test with silting of bottom of hole.
(C) Cased hole, uncased or perforated extension of length L		$F = \frac{2\pi L}{\ln(L/R)}$	$k = \frac{R^2}{2L(t_2 - t_1)} \ln \frac{L}{R} \ln \frac{H_1}{H_2}$ for $\frac{L}{R} > 8$	Used for permeability determination at greater depths below water table.
(D) Cased hole, column of soil inside casing to height L		$F = \frac{11\pi R^2}{2\pi R + 11L}$	$k = \frac{2\pi R + 11L}{11(t_2 - t_1)} \ln \frac{H_1}{H_2}$	Principal use is for permeability in vertical direction in anisotropic soils.

Source: Adapted from U.S. Department of the Navy, Naval Facilities Engineering Command (1982).

TABLE 2.2 Computation of Permeability from Variable Head Tests When the Wells or Piezometers Are Located in an Aquifer with an Impervious Upper Layer

Condition	Diagram[a]	Shape Factor F	Permeability k	Applicability
(A) Case hole, opening flush with upper boundary of aquifer of infinite depth	2R, H1, H2, Casing	$F = 4R$	$k = \frac{\pi R}{4(t_2 - t_1)} \ln \frac{H_1}{H_2}$	Used for permeability determination when surface impervious layer is relatively thin. May yield unreliable results in falling head test with silting of bottom of hole.
(B) Cased hole, uncased or perforated extension into aquifer of finite thickness:	2R, H2, H1, Casing, L1, T, L2, L3, R0			
(1) $\frac{L_1}{T} < 0.20$		(1) $F = C_S R$	$k = \frac{\pi R}{C_S(t_2 - t_1)} \ln \frac{H_1}{H_2}$	Used for permeability determination at depths greater than about 5 ft (1.524 m). For values of C_S, see Fig. 2.25*b*.
(2) $0.2 < \frac{L_2}{T} < 0.85$		(2) $F = \frac{2\pi L_2}{\ln(L_2/R)}$	$k = \frac{R^2 \ln(L_2/R)}{2L_2(t_2 - t_1)} \ln \frac{H_1}{H_2}$ for $\frac{L}{R} > 8$	Used for permeability determination at greater depths and for fine-grained soils using porous intake point of piezometer.
(3) $\frac{L_3}{T} < 1.00$		(3) $F = \frac{2\pi L_3}{\ln(R_0/R)}$	$k = \frac{R^2 \ln(R_0/R)}{2L_3(t_2 - t_1)} \ln \frac{H_1}{H_2}$	Assume value of $R_0/R = 200$ for estimates unless observation wells are made to determine actual value of R_0.

Source: Adapted from U.S. Department of the Navy, Naval Facilities Engineering Command (1982).

[a] R_0 is the effective radius to source at constant head.

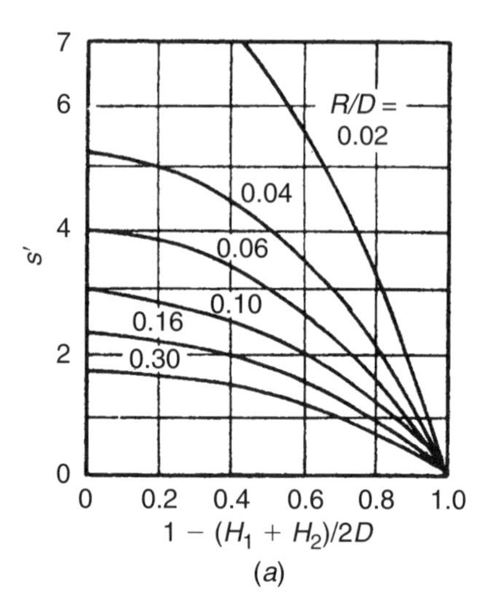

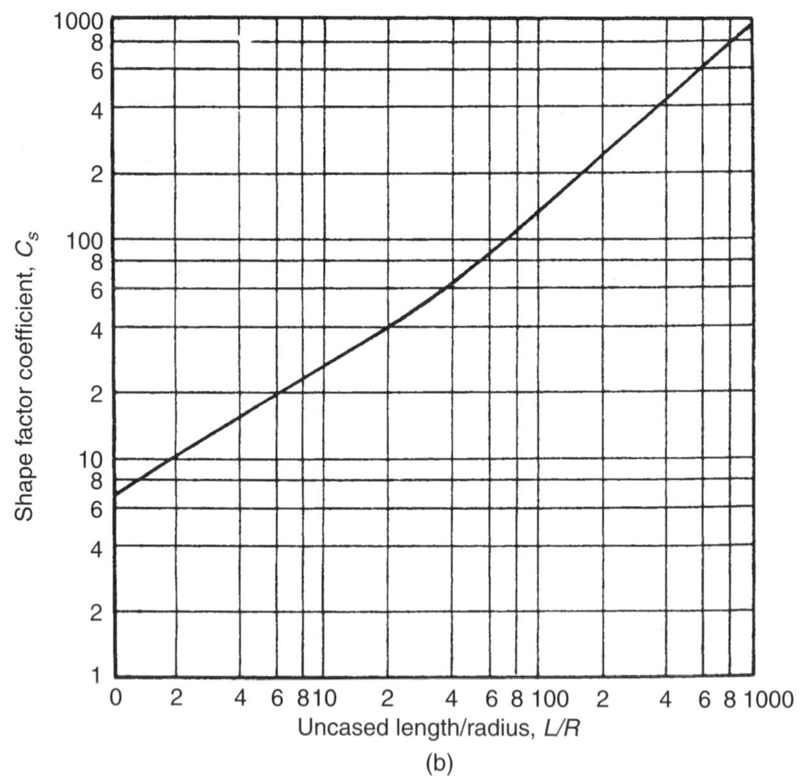

Fig. 2.25. Shape factors (*a*) S' and (*b*) C_s for determining permeability using variable head tests. (Adapted from U.S. Department of the Navy, Naval Facilities Engineering Command, 1982.)

2.7.3 Variable Head Tests

The rate at which water level in a piezometer or an observation well changes is representative of the hydraulic conductivity of soils around the piezometer. The U.S. Department of the Navy, Naval Facilities Engineering Command (1982), developed standard methods to determine k using the information on head variations in piezometers or observation wells. Referring to Fig. 2.24*a*, the permeability is expressed in its general form as

$$k = \frac{A}{F(t_2 - t_1)} \ln \frac{H_1}{H_2} \tag{2.47}$$

where k is the mean permeability, A the standpipe area, F the shape factor of the intake point, and $\ln(H_1/H_2)$ and $(t_2 - t_1)$ are obtained from a plot of observations (Fig. 2.24*b*). Tables 2.1 and 2.2 list the shape factors and the particular forms of permeability equations for various configurations of uncased and cased holes. The shape factor coefficients needed for case A in Table 2.1 and case B-1 in Table 2.2 are shown in Fig. 2.25*a* and *b*, respectively.

2.8 HETEROGENEITY AND ANISOTROPY

Much of our discussion above assumed that hydraulic conductivity was constant throughout the domain under consideration. The equations developed for the pump tests, Eqs. (2.37) and (2.38), were based on the assumption that k was constant in the flow domain. For the cases where the soil types vary in the domain, the k obtained using these equations will reflect only an effective and "averaged" value. Both natural and engineered soils are known to exhibit spatial variability in their k. In the case of natural soils, the variability comes from the fact that soil strata/layers were subjected to different compression forces in the horizontal and vertical directions during their formation. In the case of engineered soils such as embankment and earth dam materials, layered placement and compaction result in different properties in the vertical and horizontal directions. In general, the horizontal permeability is larger than the vertical permeability because of higher compressive stresses in the vertical direction.

The spatial variability in k may be characterized in terms of *location dependency* and *directional dependency*. The terms *homogeneity* and *heterogeneity* are used to refer to location dependency. Similarly, *isotropy* and *anisotropy* are used to refer to directional dependency. When k is independent of location in the flow domain under consideration, the soil is termed *homogeneous* with respect to k; otherwise, it is *heterogeneous*. Similarly, when k is independent of the direction, the soil is *isotropic* with respect to k; otherwise, it is *anisotropic*. Heterogeneity implies that k varies in the Cartesian coordinate system, and anisotropy implies that k varies with the angular coordinate. Figure 2.26 demonstrates the spatial variability of k in terms of these two factors.

Natural and engineered soils can be characterized using a layered system. Each of the layers may have a different k. It is often convenient to represent such a layered

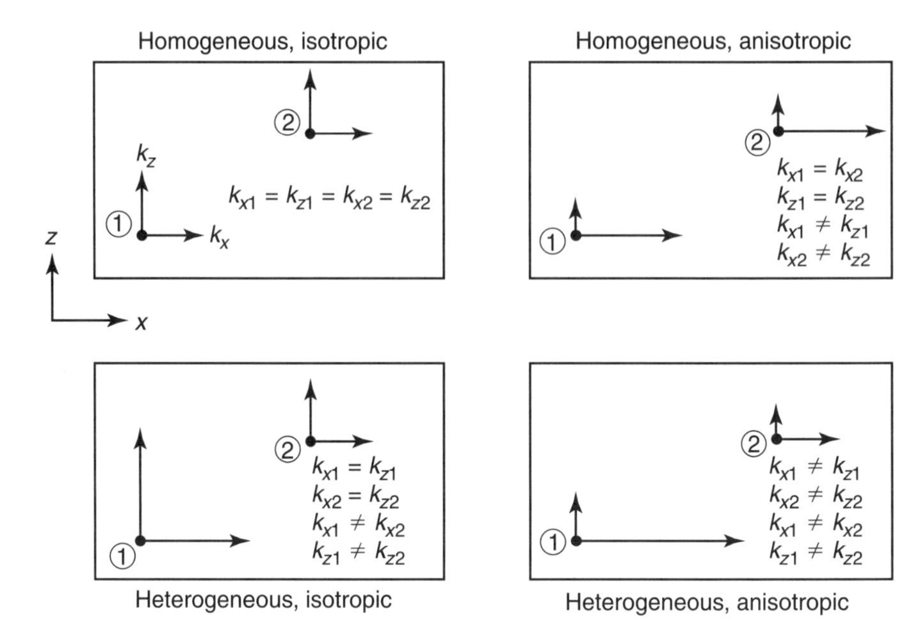

Fig. 2.26. Spatial and directional variation of permeability.

system using a single *equivalent* hydraulic conductivity in the vertical or horizontal dimension. Consider two layers with hydraulic conductivities k_1 and k_2 as shown in Fig. 2.27. Let us consider flow in the x direction (Fig. 2.27*a*). The flow rate q in this direction is the sum of flow rates in the two layers. Thus,

$$q = k_1 i_1 H_1 + k_2 i_2 H_2 \tag{2.48}$$

Note that we considered unit dimension of the layers perpendicular to the cross section shown in Fig. 2.27*a*. Since the difference in pressure heads is same for all the layers, the gradients are equal (i.e., $i_1 = i_2 = i$). Considering an equivalent single layer of hydraulic conductivity $k_{x(\text{eq})}$,

$$k_{x(\text{eq})} i H = k_1 i H_1 + k_2 i H_2 \tag{2.49}$$

which in turn results in

$$k_{x(\text{eq})} = \frac{1}{H}(k_1 H_1 + k_2 H_2) \tag{2.50}$$

Thus, when flow is parallel to the layer orientation, the layered system could be represented by an equivalent single medium whose hydraulic conductivity in the direction of flow is equal to the weighted average of the individual layer hydraulic conductivities. The reader can verify that this can be extrapolated for n layers as

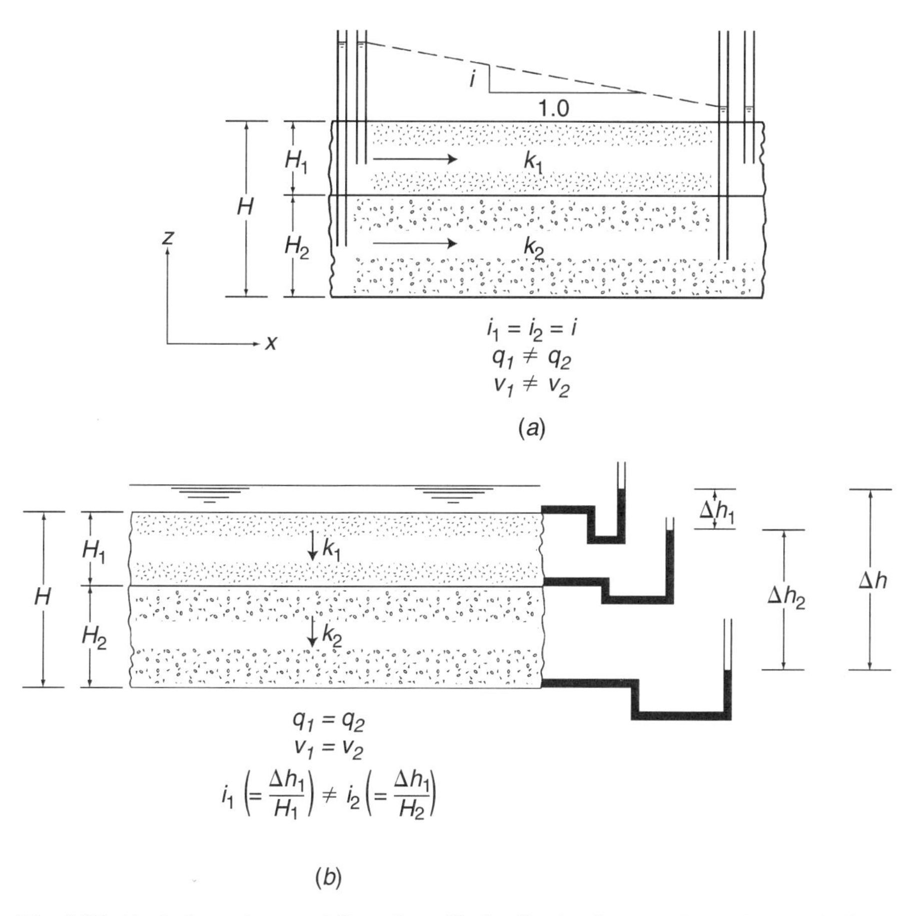

Fig. 2.27. Equivalent of permeability of stratified soils: (*a*) flow parallel to layers; (*b*) flow perpendicular to layers.

$$k_{x(\text{eq})} = \frac{1}{H}\,(k_1H_1 + k_2H_2 + k_3H_3 + \cdots + k_nH_n) \tag{2.51}$$

Let us now consider flow in the z direction (Fig. 2.27*b*). In this case, the flow rate q must be same in all the layers. Considering flow through a unit cross-sectional area, the product of k and i must therefore be same for all the layers. Equating the products for the individual layers and the equivalent medium gives us

$$k_{z(\text{eq})}\frac{\Delta h}{H} = k_1 i_1 = k_2 i_2 \tag{2.52}$$

where Δh represents the total head lost through both of the layers and $\Delta h/H$ is the gradient of the equivalent single medium. The head lost through the layers is the sum of the individual head losses through the two layers; therefore,

$$\Delta h = \Delta h_1 + \Delta h_2 \tag{2.53}$$

or

$$\Delta h = i_1 H_1 + i_2 H_2 \tag{2.54}$$

Substituting i_1 and i_2 from Eq. (2.52) in (2.54),

$$k_{z(\mathrm{eq})} = \frac{H}{(H_1/k_1) + (H_2/k_2)} \tag{2.55}$$

Thus, Eq. (2.55) provides us with an expression for k of a single medium hydraulically equivalent to a layered system when flow is perpendicular to the layer orientation. The reader can verify that Eq. (2.55) can be extrapolated for n layers as

$$k_{z(\mathrm{eq})} = \frac{H}{(H_1/k_1) + (H_2/k_2) + (H_3/k_3) + \cdots + (H_n/k_n)} \tag{2.56}$$

Example 2.6 The experiment in Example 2.4 is repeated replacing the support screens with porous stones 1 cm in thickness (Fig. E2.6). The pressure gauge shows a reading of 20 cm of water for $q = 100$ mL/min. What is the permeability of the porous stones?

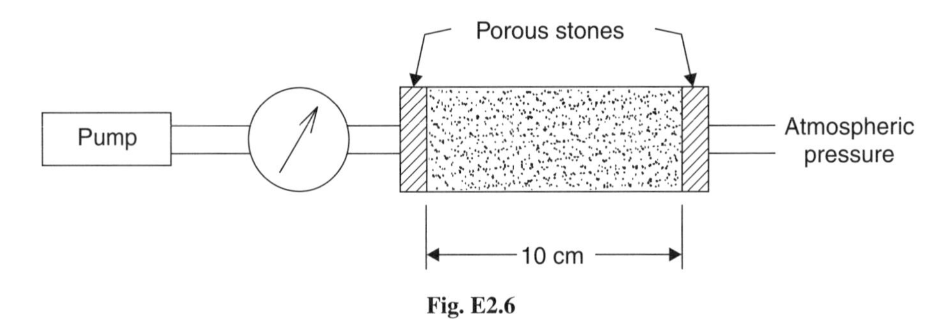

Fig. E2.6

SOLUTION: Assume that the permeability of the porous stones is k_p and the head difference over each stone is Δh_p; then the total head difference over the porous stone and the sample is

$$\Delta h = 2\Delta h_p + \Delta h_s \tag{E2.1}$$

where Δh_s is the head difference over the soil sample. According to Darcy's law, we have

$$q = k_p \frac{\Delta h_p}{L_p} A \tag{E2.2}$$

for the porous stones and

$$q = k_s \frac{\Delta h_s}{L_s} A \tag{E2.3}$$

for the soil sample. From Eq. (E2.2), we have

$$\Delta h_p = \frac{q}{A} \frac{L_p}{k_p} \tag{E2.4}$$

From Eq. (E2.3), we have

$$\Delta h_s = \frac{q}{A} \frac{L_s}{k_s} \tag{E2.5}$$

Substituting Eqs. (E2.4) and (E2.5) in (E2.1), we have

$$\frac{q}{A} \frac{L_s}{k_s} + 2\frac{q}{A} \frac{L_p}{k_p} = \Delta h \tag{E2.6}$$

which leads to

$$\begin{aligned} k_p &= \frac{2L_p}{\Delta h(A/q) - (L_s/k_s)} \\ &= \frac{2 \times 1 \text{ cm}}{20 \text{ cm} \times 50 \text{ cm}^2/100 \text{ mL/min} \times 1 \text{ mL}/1 \text{ cm}^3 - 10 \text{ cm}/1.67 \text{ cm/min}} \\ &= 0.5 \text{ cm/min} \end{aligned} \tag{E2.7}$$

So the permeability of the porous stone is 0.5 cm/min.

Example 2.7 What would be the water levels in the piezometers installed at either end of the soil sample of Example 2.6? (See Fig. E2.7.)

SOLUTION: From Example 2.6, Eq. (E2.4), we have the pressure difference over each of the porous stones expressed as

$$\Delta h_p = \frac{q}{A} \frac{L_p}{k_p} = \frac{100 \text{ mL/min}}{50 \text{ cm}^2} \times \frac{1 \text{ cm}^3}{1 \text{ mL}} \times \frac{1 \text{ cm}}{0.5 \text{ cm/min}} = 4 \text{ cm}$$

so the water level in the left piezometer is 20 cm − 4 cm = 16 cm, and the water level in the right piezometer is 4 cm.

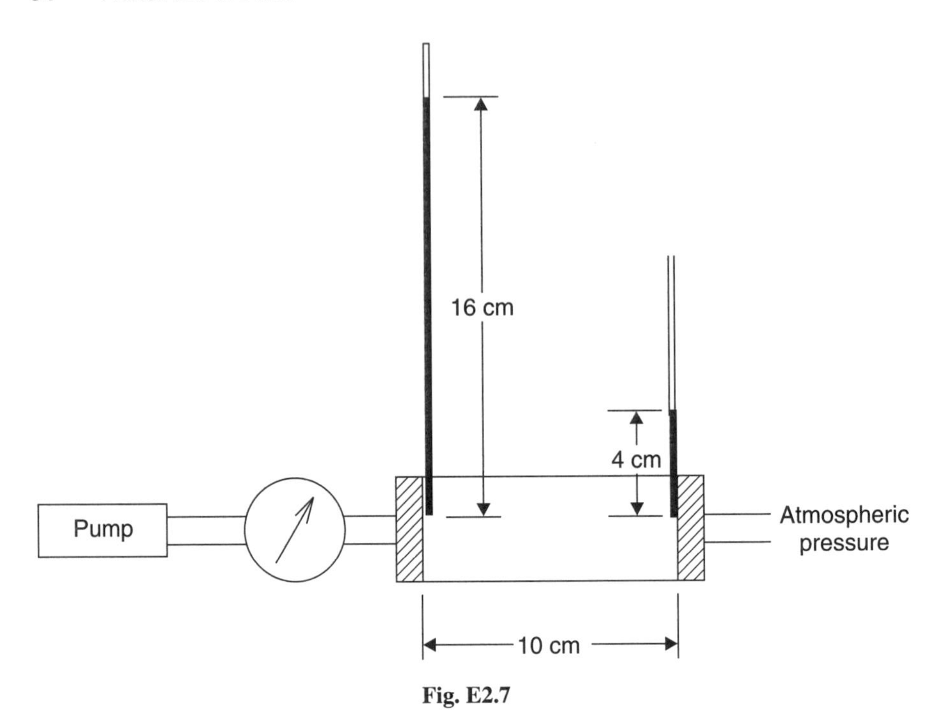

Fig. E2.7

PROBLEMS

2.1. Refer to the 3-ft-thick clay liner in Fig. P2.1. Determine the hydraulic gradient across the liner and the seepage through the liner.

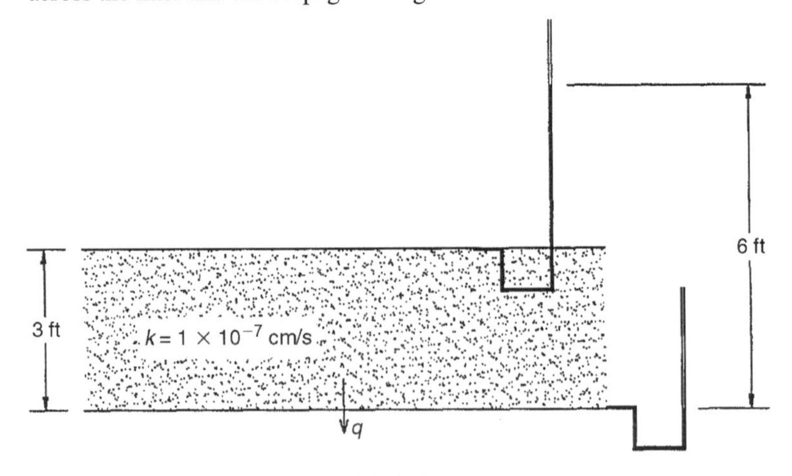

Fig. P2.1

2.2. For the two cases shown in Fig. P2.2, determine the pressure, elevation, and total heads at point A and at the two ends of the soil sample.

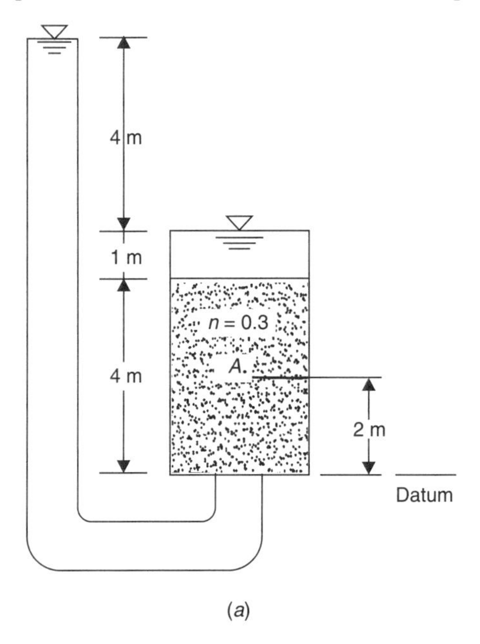

(*a*)

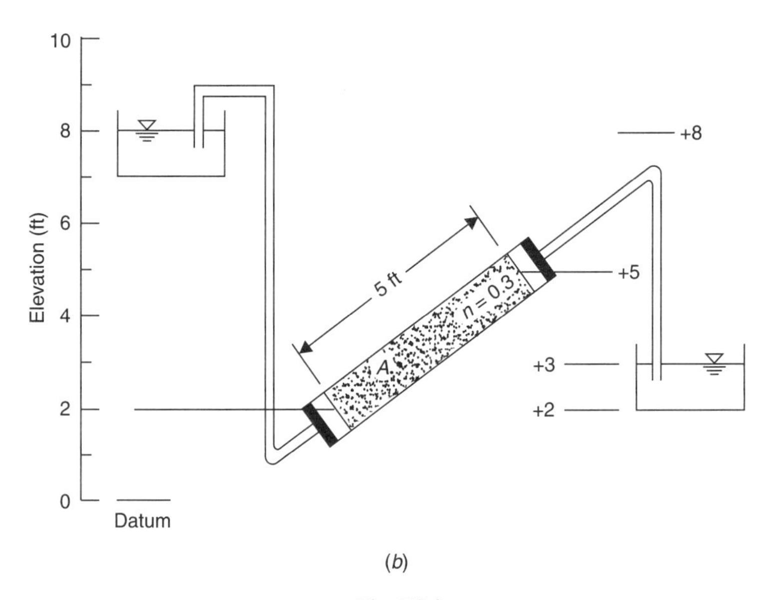

(*b*)

Fig. P2.2

2.3. For the two cases in Problem 2.2, determine the seepage and discharge velocities given $k = 5 \times 10^{-3}$ cm/s.

2.4. A desired goal in clay liner construction is to minimize the permeability so that seepage of waste liquid through the liner is minimal. Figure P2.4*a* and *b* show two options of using materials from different borrow pits. If $k_1 = 1 \times 10^{-7}$ cm/s and $k_2 = 1 \times 10^{-8}$ cm/s, $t = 3.0$ ft, $t_1 = t_2 = 1.5$ ft, which option would you prefer to minimize seepage through the liner? Justify your answer.

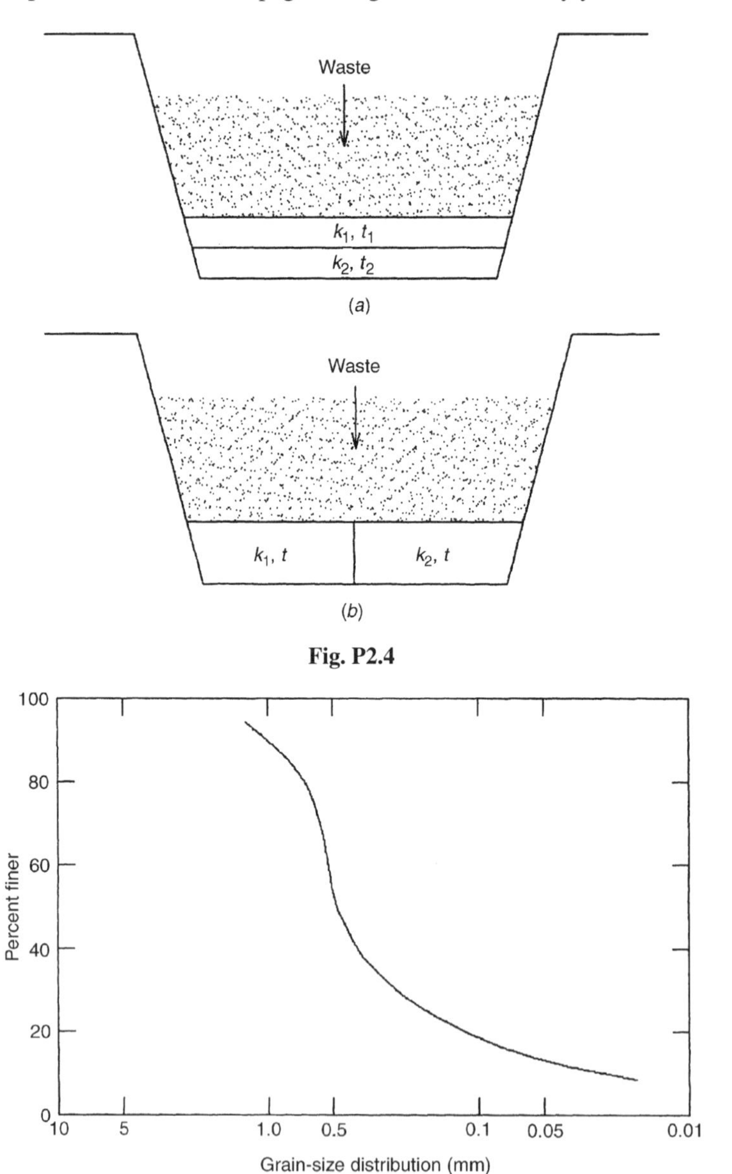

Fig. P2.4

Fig. P2.5

2.5. For a granular material with the grain-size distribution shown in Fig. P2.5, compare the estimates of hydraulic conductivity using all the relevant empirical methods discussed in Section 2.5. Porosity of the soil $n = 0.3$.

2.6. Determine the hydraulic conductivity from the following data obtained using a constant head test: $L = 30$ cm, $A = 105$ cm^2, $\Delta h = 40$ cm, water collected in 5 min = 800 mL. Also, determine the seepage velocity if porosity $n = 0.35$.

2.7. Can a variable head test be used to determine the hydraulic conductivity of the soil specimen in Problem 2.6? For a standpipe with cross-sectional area $a = 0.5$ cm^2, what would be the rate of drawdown of the water level in the pipe?

2.8. If in a variable head permeability test, the time interval is the same for head drops in the standpipe from h_1 to h_2 and from h_3 to h_4, what is the relationship between h_1, h_2, and h_3?

2.9. The soil specimen in Fig. P2.9 has three layers with different permeabilities, $k_1 = 1 \times 10^{-3}$ cm/s, $k_2 = 5 \times 10^{-3}$ cm/s, and $k_3 = 1 \times 10^{-4}$ cm/s. Determine (**a**) the flow rate through the specimen, and (**b**) total hydraulic heads at points *A, B, C,* and *D*.

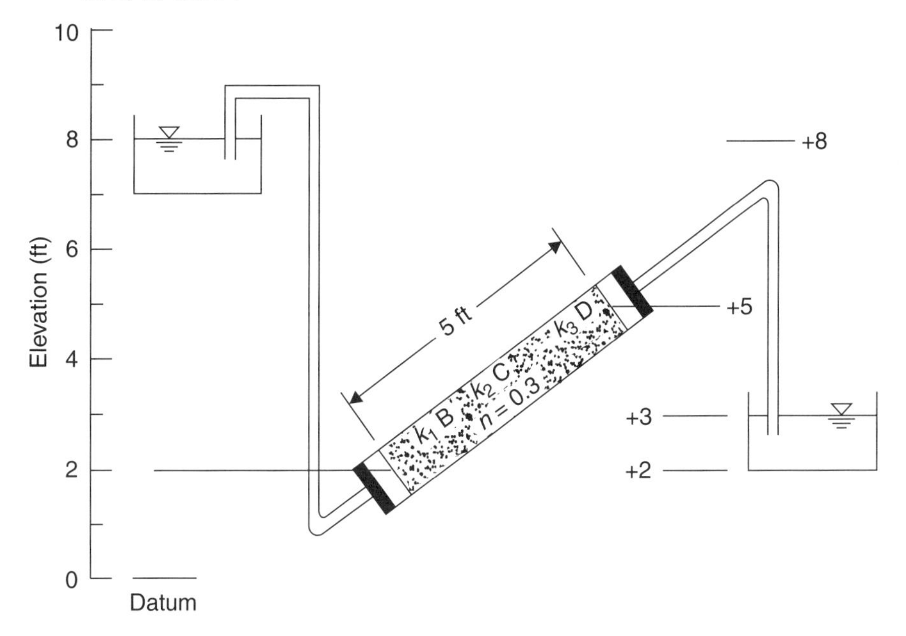

Fig. P2.9

2.10. For a well permeameter test in the field, it took 1.2 ft^3/hr to maintain a height (h_w) of 3.5 ft in the well. The radius of the well (r_w) is 0.2 ft, and the thickness of the unsaturated strata between the water level in the well and the groundwater table (T_u) is 4.5 ft. Determine the permeability of soils around the well if the test was conducted at 20°C.

CHAPTER 3

FLOW NETS

In the case of one-dimensional problems such as flow in a slender laboratory column, Darcy's law alone is sufficient to estimate the flow quantities when the hydraulic gradient and the geometrical dimensions of the soil sample are known. However, in a majority of the field cases, it is necessary to consider flow in two dimensions. An example is the flow taking place around a sheet pile installed to retain water (Fig. 3.1). Flow occurs from left to right of the structure as a result of head differences across the structure. It occurs in a two-dimensional (x–y) plane. Water flowing close to the structure travels a shorter distance (path ABC) than water traveling away from the structure (path DEF). The hydraulic head lost between points A and C and between D and F is the same although the travel distance is different. Thus in a two-dimensional domain, we should expect the gradients to be different. Application of Darcy's law alone is not sufficient to estimate flow quantities in such cases.

To tackle these problems, we need to combine Darcy's law with the mass/energy conservation principle. This could be accomplished using two different approaches. The first approach is a graphical method, which involves drawing flow nets. This method, given by Forchheimer (1930), is a simple method useful in many common problems of seepage. The second approach is to use either analytical or numerical solutions to the differential equation governing the seepage problem. Such solutions are useful for solving complicated flow domains where the graphical method becomes cumbersome. We take up the graphical method of flow nets in this chapter and defer the second approach to Chapter 4.

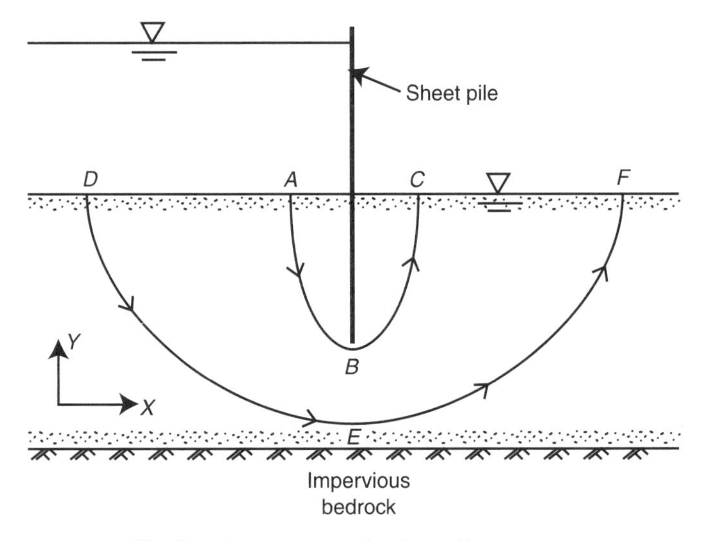

Fig. 3.1. Seepage around a sheet pile structure.

3.1 CONCEPT OF FLOW NETS

We start with the simple rectangular soil domain shown in Fig. 3.2*a* to illustrate the concept of flow nets. A hydraulic head difference equal to Δh is applied along the length of the domain. Flow is horizontal, so Δh is a result of the difference only in pressure heads, not in elevation heads. Considering the upstream boundary of the domain, the total hydraulic head is constant at all points on the line 0–0 and is equal to h_0. Similarly, along the downstream boundary of the domain 10–10, the total hydraulic head is constant at h_{10}. As water travels from the upstream to the downstream boundary, Δh is dissipated. When it arrives at the midpoint of travel, at section 5–5, we may expect that one-half of the total head difference is dissipated. Thus, at section 5–5,

$$h_5 = h_0 - \frac{\Delta h}{2} \tag{3.1}$$

Similarly, we can postulate that water loses one-tenth of the total head difference after it travels one-tenth of the way, two-tenths of the head difference after traveling two-tenths of the total length, and so on. Let us *choose* to identify 10 sections along the length of travel. As shown in Fig. 3.2*a*, each of these sections represents lines along which hydraulic head is the same. Hence, they are called *equipotential lines.* Thus,

$$h_1 = h_0 - \Delta h\left(\tfrac{1}{10}\right), \quad h_2 = h_0 - \Delta h\left(\tfrac{2}{10}\right), \quad \ldots, \quad h_9 = h_0 - \Delta h\left(\tfrac{9}{10}\right) \tag{3.2}$$

We can generalize this by saying that if the length of travel is divided into n_e segments, the head lost across each segment (whose length b is equal to L/n_e) is

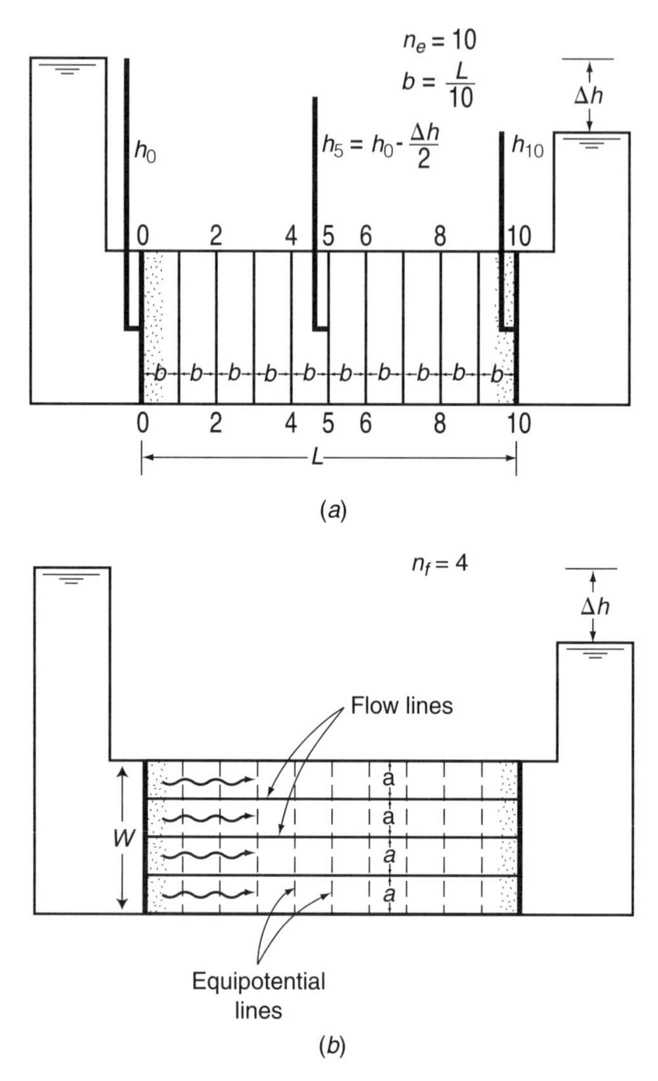

Fig. 3.2. Concept of flow nets: (*a*) equipotential lines in a rectangular flow domain; (*b*) completed flow net with flow lines and equipotential lines.

equal to $\Delta h/n_e$. Now consider the path that water takes as it travels through all these segments. In this relatively simple domain, we can postulate that water travels horizontally, along flow lines drawn perpendicular to the equipotential lines, losing its head continuously as it goes from the upstream to the downstream boundary. We can conceptualize the formation of horizontal flow channels in the soil. Again, let us *choose* to discretize the continuous flow using four channels, each of width a equal to $W/4$, where W is the total width of the domain (Fig. 3.2*b*). We consider the third dimension perpendicular to the paper as equal to 1 unit. We can now invoke Darcy's law to determine flow rate in each channel, q_c, as water travels from the upstream boundary to the first section, losing one-tenth of the total hydraulic head drop:

$$\begin{aligned} q_c = kiA &= k\frac{\Delta h/10}{b}(a \times 1) \\ &= k\frac{\Delta h}{10}\frac{a}{b} \end{aligned} \tag{3.3}$$

According to the mass conservation principle, the quantity of water leaving each successive section is the same in a single flow channel. Thus the flow rate given in Eq. (3.3) represents the flow rate along the entire length of the flow channel. Considering that we chose to divide the width of the domain into four channels, the total flow rate q through the domain is given by

$$q = 4q_c = k\,\Delta h\frac{4}{10}\frac{a}{b} \tag{3.4}$$

Equation (3.4) now gives us the flow rate in terms of the permeability, the total hydraulic head drop, and geometrical parameters of the two sets of lines that we chose to draw in the domain. Let us now generalize Eq. (3.4) for any choice of the number of equipotential drops and flow channels. Expressing these numbers as n_e and n_f, respectively, we get

$$q = k\,\Delta h\frac{n_f}{n_e}\frac{a}{b} \tag{3.5}$$

Since the drawing of the two sets of lines in Fig. 3.2*b* is a matter of choice, n_f n_e, a, and b may be varied; however, for all combinations of these four variables, the reader can easily verify that the flow rate determined using Eq. (3.5) will remain nearly the same. For this approach to be valid, n_f and n_e do not necessarily have to be integers. Although the flow net may contain rectangles where $a \neq b$, the process of drawing the net will be simplified greatly if squares are chosen ($a = b$). When this is done, Eq. (3.5) reduces to

$$q = k\,\Delta h\frac{n_f}{n_e} \tag{3.6}$$

The ratio n_f/n_e in Eq. (3.6) is often referred as the *shape factor*, as it represents the geometry of the flow net. We note that in the drawing of flow net in Fig. 3.2, the permeability of the soil does not play any role. Thus, the flow net in a soil domain is independent of the permeability of the soil. An exception is in the case of anisotropic soils, discussed later.

For the relatively simple rectangular domain chosen here, the student may verify that Eq. (3.4) is equivalent to Darcy's law where the width and length of the domain are written in terms of a and b, respectively. For this simple domain, used here only to illustrate the concept, a flow net is not needed to determine the flow rate. When one revisits the flow domain shown in Fig. 3.1, the use of flow nets will become readily obvious. For this domain, let us work out a pattern of flow channels from the upstream to the downstream end by drawing three flow lines (which result in four flow channels) as shown in Fig. 3.3*a*. In accordance with our example in Fig. 3.2, we will then proceed to draw another set of lines perpendicular to the flow lines ensuring that the elements enclosed by the two sets of lines are squares (Fig. 3.3*b*). Note that the squares thus formed may be curvilinear. Considerable time and effort are necessary,

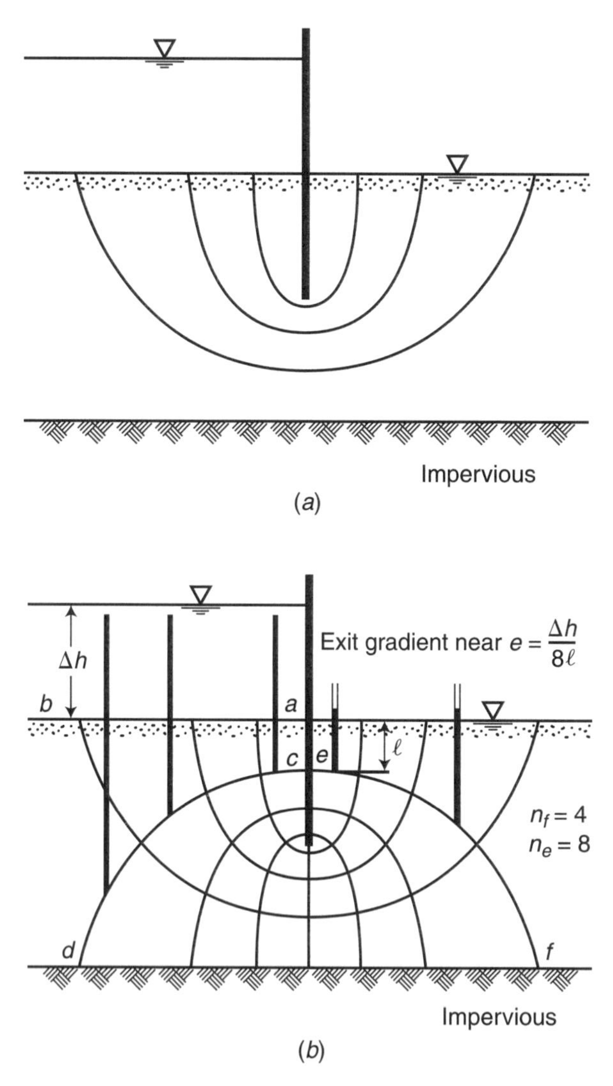

Fig. 3.3. Flow net for seepage around a sheet pile structure: (*a*) flow lines; (*b*) completed flow net with flow lines and equipotential lines.

especially in the beginning stages of learning to draw flow nets, to complete the flow net from Fig. 3.3*a* to Fig. 3.3*b*. Leaving to a later section a discussion of the guidelines that would help us in taking this leap, we see that the resulting number of equipotential drops is eight. The flow rate q under the structure can now be calculated readily using Eq. (3.6): $q = k\,\Delta h(4/8)$.

Apart from giving us the seepage rates, flow nets also indicate the hydraulic heads in the domain. In Fig. 3.3*b*, the total hydraulic head at any point on a given equipotential line is the same. Consider the equipotential line cd. The piezometric levels at the three points on this line indicate the pressure heads. If the impervious bottom boundary is used as the datum, elevation of the water levels in the piezometers with reference to the impervious datum represents the total hydraulic head along the potential line. Thus, the pressure heads at the three points are different, but the elevation heads at these points compensate for the differences in the pressure heads. The difference between the hydraulic heads on cd and ab is equal to one equipotential drop. Consider the equipotential line ef. The total hydraulic head at all points on this line equals the hydraulic head on line cd minus six equipotential drops. The pressure heads are considerably less on this line than on the line cd. Thus, the water pressures on the two sides of the structure are widely different. This has important implications on the stability of the structure, as discussed in later chapters. Another important consideration in the stability of water-retaining structures such as the sheet pile is the gradient with which water exits on the downstream side. The flow nets could be used to determine these exit gradients, as shown in Fig. 3.3*b*.

To sum up our conceptual understanding of flow nets, we note that when two sets of lines are drawn such that they are mutually perpendicular to each other, the geometrical parameters of the net formed in the flow domain could be used to determine the flow rate in the domain. Furthermore, the two sets of lines could be used to estimate the hydraulic heads and gradients at any given point in the domain. The mutual perpendicularity of the two sets of lines has a mathematical basis, as discussed in the next section. We will see at the end of that section that the graphical method discussed above is an outgrowth of the mathematical solution to equation governing flow of water in soils.

3.2 MATHEMATICAL BASIS FOR FLOW NETS

We seek a general expression that governs seepage in a flow domain of any given shape and under any given set of boundary conditions. As we do with similar processes in engineering, we invoke the universal law of conservation of mass and couple it with a cause-and-effect relationship. For the problem at hand, the mass conserved is water, and Darcy's law gives the cause-and-effect relationship between gradients and flow rates or velocities. The reader may observe parallelism between this and other processes in solid mechanics (where force equilibrium is coupled with Hooke's law), in heat conduction (where energy conservation is coupled with the law of thermal conductivity), and so on.

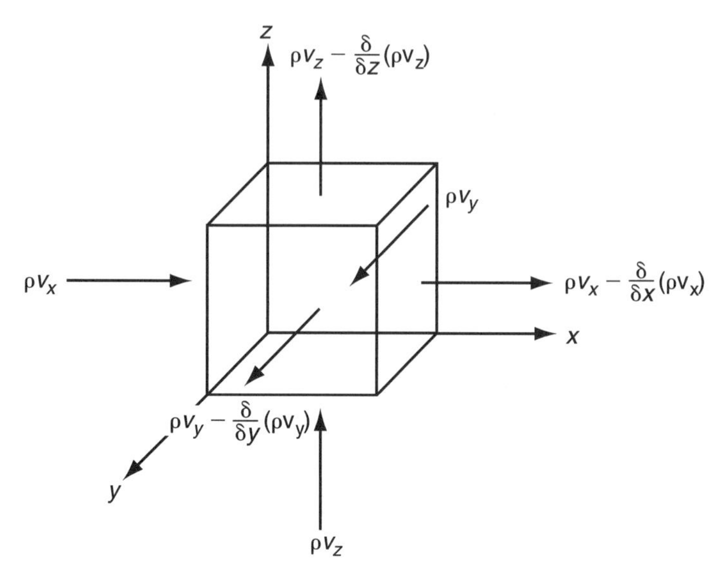

Fig. 3.4. Elemental control volume with seepage in three dimensions.

We apply mass conservation principle to the elemental control volume shown in Fig. 3.4. With reference to this volume, the principle dictates that

$$\begin{aligned}&(\text{rate of mass output}) - (\text{rate of mass input})\\ &\qquad = (\text{rate of change in storage in the control volume})\end{aligned} \tag{3.7}$$

When expressed mathematically in all three dimensions, the left-hand side of Eq. (3.7) becomes

$$\left[\rho v_x - \frac{\partial}{\partial x}(\rho v_x)\right] - \rho v_x + \left[\rho v_y - \frac{\partial}{\partial y}(\rho v_y)\right] - \rho v_y + \left[\rho v_z - \frac{\partial}{\partial z}(\rho v_z)\right] - \rho v_z$$

or

$$-\frac{\partial}{\partial x}(\rho v_x) - \frac{\partial}{\partial y}(\rho v_y) - \frac{\partial}{\partial z}(\rho v_z)$$

where ρ is the density of water and v_x, v_y, and v_z are the velocities of flow in the x, y, and z dimensions, respectively. Note that the mass flux of water entering or leaving the control volume is expressed as the product of density and velocity. In other words, the area of each of the six faces of the control volume is considered as a unit.

Storage in the elemental control volume changes due to either a change in density of the fluid over time, which causes expansion or contraction of the fluid, or a change in the pore volume of the element. The latter is possible when soil undergoes time-dependent compressibility. For fine-grained soils, this is well known to geotechnical engineers as consolidation. Accounting for these two factors on the right-hand side of Eq. (3.7), the mass conservation principle may be expressed as

$$-\frac{\partial}{\partial x}(\rho v_x) - \frac{\partial}{\partial y}(\rho v_y) - \frac{\partial}{\partial z}(\rho v_z) = \frac{\partial}{\partial t}(\rho n) \tag{3.8}$$

wherein the product of water density and porosity is used to indicate the mass of pore water in a cube of unit total volume. It is customary to express the rate of change in storage in terms of a parameter known as *specific storage*, S_s, which is defined as the volume of water that a unit volume of aquifer releases from storage under a unit decline in the hydraulic head. This parameter may be expressed as a function of compressibility of the soil skeleton and change in density of pore fluid. For details, the reader is referred to Freeze and Cherry (1979). Using specific storage, and assuming that the density of water does not vary with time, the mass conservation principle can be rewritten as

$$-\frac{\partial}{\partial x}(\rho v_x) - \frac{\partial}{\partial y}(\rho v_y) - \frac{\partial}{\partial z}(\rho v_z) = \rho S_s \frac{\partial h}{\partial t} \tag{3.9}$$

where h is the total hydraulic head. We now invoke Darcy's law and express velocity components in terms of hydraulic head h,

$$-\frac{\partial}{\partial x}\left(k_x \frac{\partial h}{\partial x}\right) - \frac{\partial}{\partial y}\left(k_y \frac{\partial h}{\partial y}\right) - \frac{\partial}{\partial z}\left(k_z \frac{\partial h}{\partial z}\right) = S_s \frac{\partial h}{\partial t} \tag{3.10}$$

where k_x, k_y, and k_z are permeabilities in the x, y, and z dimensions, respectively. ρ is eliminated in Eq. (3.10) because the variation of velocity components is much greater than that of ρ with respect to space coordinates. Assuming that the saturated medium is homogeneous and isotropic ($k_x = k_y = k_z = k$), Eq. (3.10) may be simplified as

$$\frac{\partial^2 h}{\partial x^2} + \frac{\partial^2 h}{\partial y^2} + \frac{\partial^2 h}{\partial z^2} = \frac{S_s}{k}\frac{\partial h}{\partial t} \tag{3.11}$$

Most groundwater flow modeling activities involve solving Eq. (3.11) in one form or other for a specific domain of interest. Under steady-state conditions, when there is no change in storage of the porous media, Eq. (3.11) reduces to the well-known *Laplace equation:*

$$\frac{\partial^2 h}{\partial x^2} + \frac{\partial^2 h}{\partial y^2} + \frac{\partial^2 h}{\partial z^2} = \nabla^2 h = 0 \tag{3.12}$$

which can now be taken as a mathematical law governing steady seepage in soils. This equation operates in accordance with the boundary conditions imposed on the specific domain under consideration. Let us revisit a one-dimensional case with the help of this equation, to understand how hydraulic head changes in a slender soil

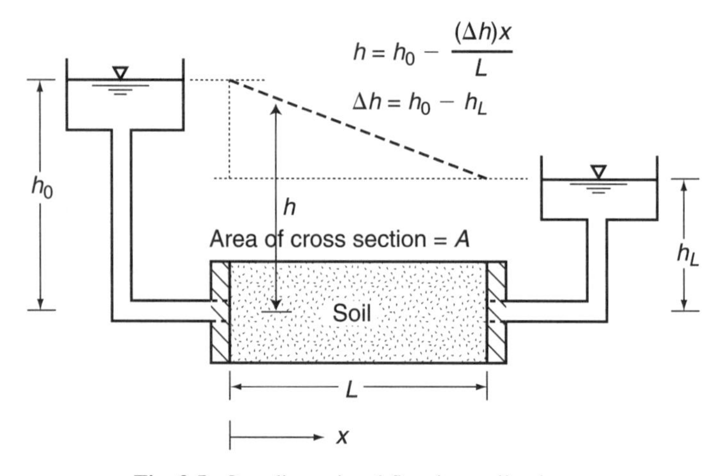

Fig. 3.5. One-dimensional flow in a soil column.

column subjected to a head difference, Δh (Fig. 3.5). The one-dimensional form of Eq. (3.12) is

$$\frac{\partial^2 h}{\partial x^2} = 0 \tag{3.13}$$

Integration of Eq. (3.13) twice gives us the general solution,

$$h(x) = C_1 x + C_2 \tag{3.14}$$

where C_1 and C_2 are integration constants. To evaluate these constants, we use the following simple boundary conditions:

$$h = \begin{cases} h_0 & \text{at } x = 0 \quad (3.15) \\ h_L & \text{at } x = L \quad (3.16) \end{cases}$$

where $h_0 > h_L$. When these conditions are used in Eq. (3.14), the solution becomes

$$h = h_0 - \frac{\Delta h\, x}{L} \tag{3.17}$$

which simply states that hydraulic head dissipates linearly (as shown in Fig. 3.5) in a homogeneous and isotropic soil medium, a fact that we hypothesized intuitively in our conceptual understanding of flow nets in Section 3.1.

For two-dimensional flow, Eq. (3.12) becomes

$$\frac{\partial^2 h}{\partial x^2} + \frac{\partial^2 h}{\partial y^2} = 0 \tag{3.18}$$

For this case, we seek two mathematical functions of spatial coordinates, $\phi(x,y)$ and $\psi(x,y)$, both of which satisfy Eq. (3.18). We consider $\phi(x,y)$ to be a velocity potential function whose derivative in a given direction indicates seepage velocity in that direction. Therefore, $\phi(x,y)$ represents the product of the permeability and hydraulic head, fulfilling the following conditions:

$$\frac{\partial \phi}{\partial x} = k\frac{\partial h}{\partial x} = v_x \tag{3.19}$$

and

$$\frac{\partial \phi}{\partial y} = k\frac{\partial h}{\partial y} = v_y \tag{3.20}$$

where $v_x$and v_y are the seepage velocities in the x and y directions, respectively. The function $\phi(x,y)$ represents a solution of the hydraulic head since it satisfies Eq. (3.18). Similarly, we find that there exists another function $\psi(x,y)$, given by

$$\frac{\partial \psi}{\partial x} = -\frac{\partial \phi}{\partial y} = -v_y \tag{3.21}$$

and

$$\frac{\partial \psi}{\partial y} = \frac{\partial \phi}{\partial x} = v_x \tag{3.22}$$

which also satisfies Eq. (3.18). Equations (3.21) and (3.22) are the Cauchy–Riemann equations, and the two functions $\phi(x,y)$ and $\psi(x,y)$ are said to be *conjugate harmonic,* which means that the two families of curves given by $\phi(x,y)$ = constant and $\psi(x,y)$ = constant represent mutually perpendicular trajectories. To understand the physical meaning of this second function, $\psi(x,y)$, consider a trajectory given by $\psi(x,y) = C_3$, a constant. Along this trajectory,

$$d\psi = \frac{\partial \psi}{\partial x}dx + \frac{\partial \psi}{\partial y}dy \tag{3.23}$$

Considering that $d\psi = 0$ along this trajectory, Eq. (3.23) can be reorganized as

$$\frac{dy}{dx} = -\frac{\partial \psi/\partial x}{\partial \psi/\partial y} = \frac{v_y}{v_x} \tag{3.24}$$

Equation (3.24) implies that the slope of the trajectory along which $\psi(x,y)$ = constant is in the direction of the resultant velocity in the domain. Hence this trajectory is a flow line or streamline, and the function $\psi(x,y)$ is referred as a *stream function.*

From the discussion above, we note that $\phi(x,y)$ and $\psi(x,y)$ satisfy the governing equation and represent two families of mutually perpendicular curves. The set of curves given by $\phi(x,y)$ = constant represent *equipotential lines*, and the curves given by $\psi(x,y)$ = constant represent *flow lines* or *streamlines*. Conversely, if two sets of curves could be drawn perpendicular to each other in a flow domain (of course, with due regard to the boundary conditions), the equation governing the seepage problem is solved graphically. This is the basis for drawing flow nets as described in Section 3.1.

3.3 SPECIAL CONDITIONS: ANISOTROPY AND LAYERED HETEROGENEITY

We assumed in our discussion on flow nets in the preceding two sections that the soil domain is isotropic and homogeneous. This is not always the case in reality. Stratification of soils, either in natural terrains or in engineered construction such as embankments, incorporates some degree of anisotropy or layered heterogeneity. The flow net concept can be extended for these soils, with some modifications. Consider the case where $k_x \neq k_y$. The governing equation for flow will then become

$$k_x \frac{\partial^2 h}{\partial x^2} + k_y \frac{\partial^2 h}{\partial y^2} = 0 \tag{3.25}$$

To bring it to the form shown in Eq. (3.18), we reorganize Eq. (3.25) as

$$\frac{\partial^2 h}{(k_y/k_x)\partial x^2} + \frac{\partial^2 h}{\partial y^2} = 0 \tag{3.26}$$

To make it identical to the Laplace equation (3.18), we transform the x axis to a new x' axis using $x' = x\sqrt{(k_y/k_x)}$ and write

$$\frac{\partial^2 h}{\partial x'^2} + \frac{\partial^2 h}{\partial y^2} = 0 \tag{3.27}$$

Thus, the problem of seepage in anisotropic soil reduces to an isotropic flow problem when the horizontal axis is transformed to account for the directional dependency of permeability. Once the flow domain is transformed to the $x'y$ plane, the flow net concept is valid. After the flow net is drawn on the transformed section, it can be projected back on to the original section by transforming the x' axis to the x axis. To calculate the flow rate from the shape factor of the flow net on the transformed section, an effective permeability $\bar{k} = \sqrt{k_x k_y}$ should be used. Thus,

$$q = \sqrt{k_x k_y}\, \Delta h \frac{n_f}{n_e} \tag{3.28}$$

To handle layered heterogeneity, we make sure that continuity of flow is maintained at the interface. This would mean that a flow channel approaching the interface will get either narrower or wider, depending on the permeability contrast. Thus, one should expect the flow lines to be deflected at the interface between two isotropic soils of different permeability. Consider a single flow channel at the interface shown in Fig. 3.6*a*, where square elements are used on the upstream side (left-hand side) of

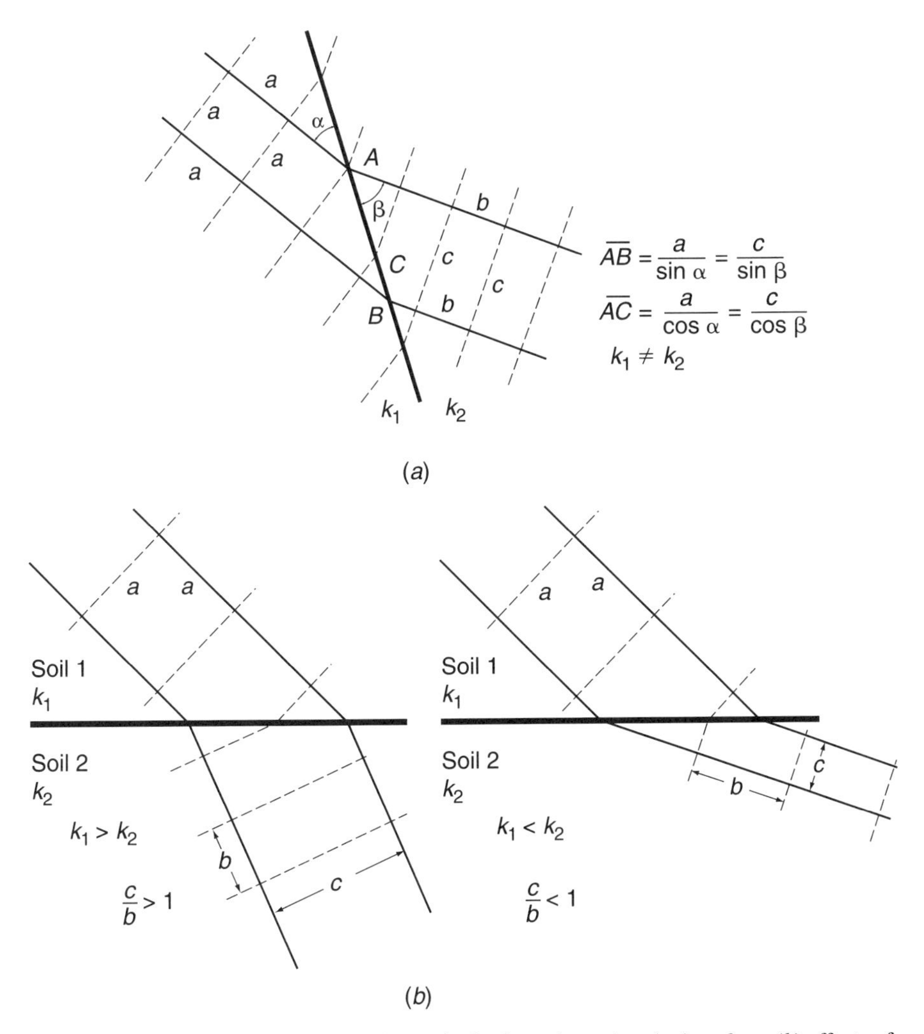

Fig. 3.6. Layered heterogeneity: (*a*) change in the flow channel at the interface; (*b*) effects of permeability contrast on flow channel changes at the interface.

the interface. If Δh represents the head drop between two consecutive equipotential lines, continuity of flow dictates that

$$q_c = k_1 a \frac{\Delta h}{a} = k_2 c \frac{\Delta h}{b} \tag{3.29}$$

which reduces to

$$\frac{c}{b} = \frac{k_1}{k_2} \tag{3.30}$$

Thus, the square elements on one side of the boundary can no longer remain square on the other side because of the permeability contrast. Equation (3.30) indicates that flow channels become wider when they approach material with a lower permeability, and narrower otherwise (Fig. 3.6*b*). This follows from intuition that the channel has to be wider to accommodate the same flow rate in a less conductive medium, and vice versa.

To complete the flow net on the downstream side of the interface, we also need the angles at which the flow lines will be distorted at the interface. From geometrical considerations (shown in Fig. 3.6*a*),

$$\frac{a}{\sin\alpha} = \frac{c}{\sin\beta} \tag{3.31}$$

and

$$\frac{a}{\cos\alpha} = \frac{b}{\cos\beta} \tag{3.32}$$

Combining Eqs. (3.30), (3.31), and (3.32) yields

$$\frac{c}{b} = \frac{\tan\beta}{\tan\alpha} = \frac{k_1}{k_2} \tag{3.33}$$

Equation (3.33) is referred as the *transfer condition,* which must be fulfilled when flow nets are drawn at interfaces between soils of different permeability. It was originally used by Forchheimer and is analogous to refraction of light rays crossing different media. Expressed in words, the condition states that the aspect ratio of the rectangular elements on the downstream side of the interface and the tangents of the intersecting angles with the interface are given by the ratio of permeabilities of the two soil types.

3.4 SPECIAL CONDITIONS: UNCONFINED FLOW

In the flow nets shown in Figs. 3.2 and 3.3, the flow is confined entirely within known boundaries. In contrast to these cases, flow occurring through an earth dam is unconfined (Fig. 3.7*a*). The upper boundary of the flow domain is not known a priori. This upper boundary, referred as the *phreatic line* or *line of seepage,* should be delineated during flow net construction. Approximate mathematical expressions are available to delineate the line of seepage, as discussed in Chapter 4. For instance, Kozeny (1931) solved the flow problem analytically in an open domain and concluded that under certain exit conditions, the flow lines (including the line of seepage) and the equipotential lines are confocal parabolas. We defer discussion of such mathematical solutions to Chapter 4 and pursue here a general graphical method to delineate the line of seepage.

Two conditions must be followed when drawing the line of seepage. First, the points of intersection of equipotential lines and the line of seepage represent the hydraulic heads along the equipotential lines. In a flow net containing square elements,

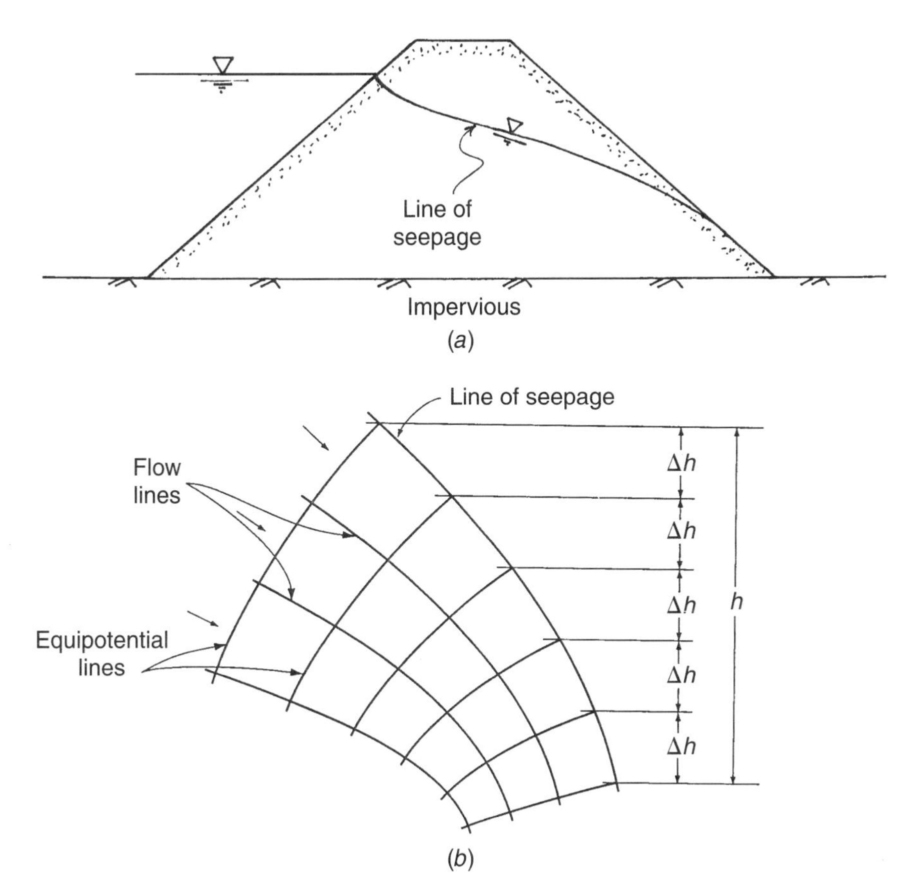

Fig. 3.7. Unconfined flow: (*a*) line of seepage in unconfined domains; (*b*) section of a flow net in the unconfined domains.

these points of intersection are equidistant in the vertical dimension (Fig. 3.7*b*). Second, transfer conditions must be followed at (1) the entrance point on the line of seepage (where water is entering into the dam from the reservoir), (2) point of interface when the line of seepage crosses two dissimilar soils in the domain, and (3) the discharge point where the line of seepage meets the downstream boundary. The transfer condition that must be followed when the line of seepage crosses dissimilar soils is the same as the one described in Section 3.3 for the layered heterogeneity. The conditions at the entrance and discharge points can be derived in the same way as the transfer condition between dissimilar soils. These conditions were developed and assembled systematically by Casagrande (1937) and are shown in Fig. 3.8. One should note that in the most common case of a downstream discharge face inclined less than or equal to 90°, the line of seepage must be tangent to that face at the discharge point.

As an example of how Casagrande derived these conditions, we consider the case of overhanging slopes (slopes inclined at angles greater than 90°). Limiting ourselves

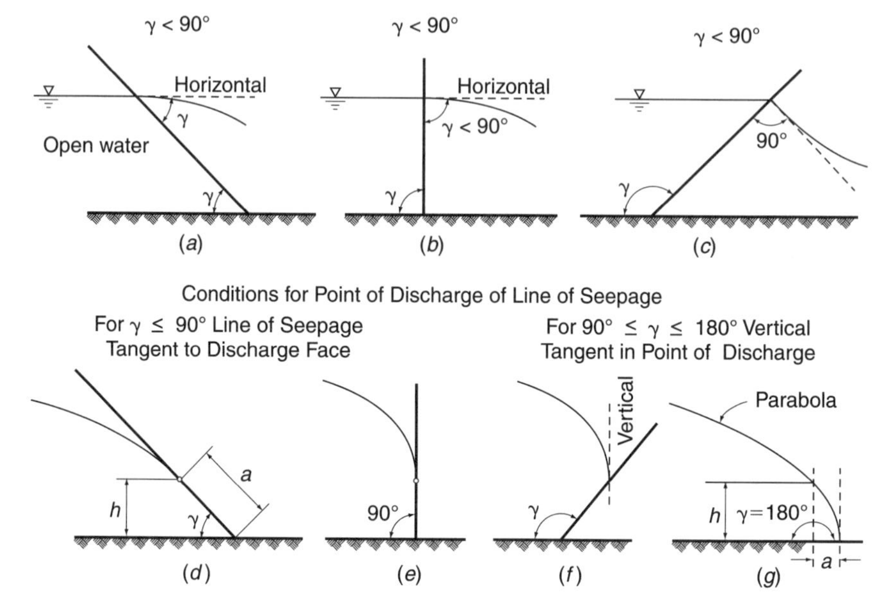

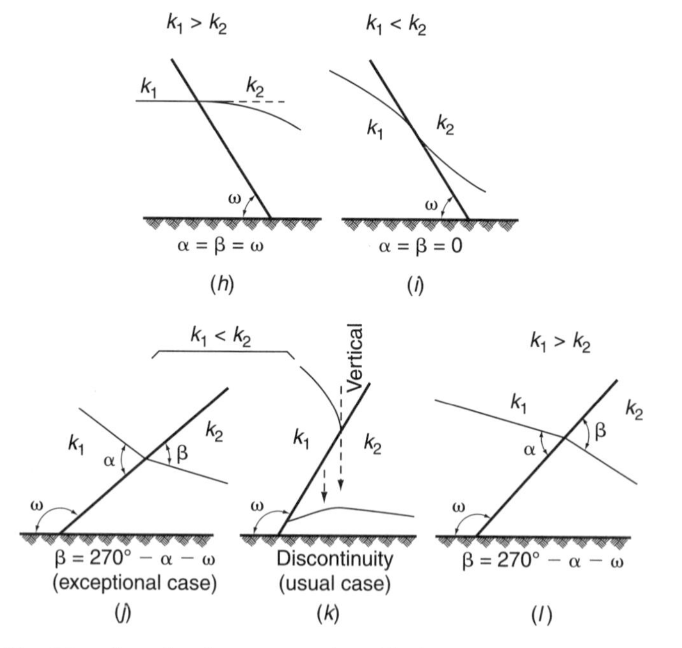

Fig. 3.8. Conditions for point of entrance, point of discharge, and deflection of line of seepage. (Adapted from Casagrande, 1937.)

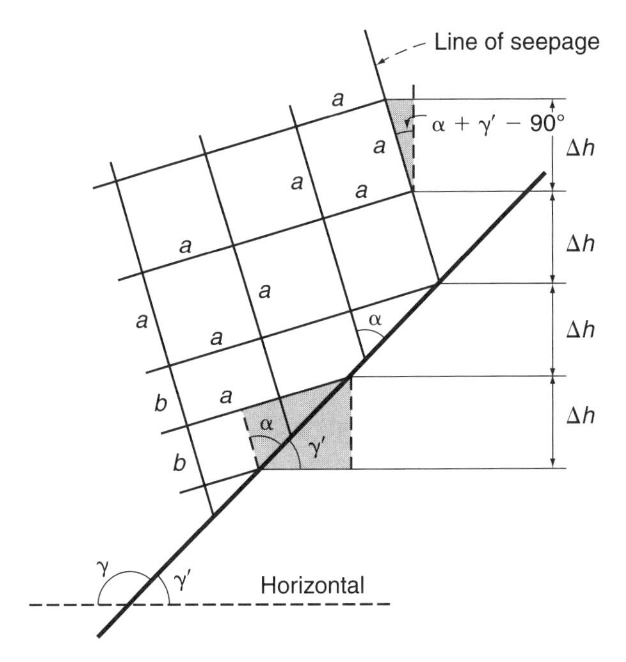

Fig. 3.9. Point of discharge of line of seepage for overhanging slopes. (Adapted from Casagrande, 1937.)

to a small portion of the flow net at the exit of the line of seepage (Fig. 3.9), let us arbitrarily fix the intersection angle α between the line of seepage and the face of the structure. We can readily see that this is not an accurate angle since the elements formed in the lower portion of the flow net are not squares. To determine the condition for α, we project the sides a and b in the shaded triangles to obtain expressions for a single equipotential drop, Δh,

$$\Delta h = \frac{b}{\cos\alpha}\sin\gamma = a\cos(\alpha + \gamma - 90°) \tag{3.34}$$

Since the condition $a = b$ must be fulfilled throughout the domain, the only solution for Eq. (3.34) will be $\alpha = 90° - \gamma$, which means that the line of seepage meets the overhanging slopes vertically.

3.5 GUIDELINES FOR FLOW NET CONSTRUCTION

Flow nets should be drawn keeping two fundamental rules in mind: (1) the flow lines and equipotential lines are mutually perpendicular; and (2) the elements enclosed by the two sets of lines are squares. These rules are deceptively simple, and flow net construction can be tedious and time consuming, particularly in the beginning stages. Added to these simple rules, the special conditions described in the preceding two sections must be met wherever appropriate. This should not, however, intimidate the student, as the end product can be truly rewarding for complicated flow domains. The student may also take consolation from the fact that computer software designed to give the same information may take just as long if one considers the time it takes to process the input and interpret the output.

A few guidelines are laid before us by experts in this field which if followed during flow net construction would greatly simplify the task. The flow net construction problems are classified as follows (in the order of increased difficulty):

1. Confined flow problems, where the boundaries are well defined
2. Unconfined flow problems, where the upper boundary or the line of seepage is not known a priori and the special conditions of Section 3.4 must be met
3. Flow problems, where the domain is anisotropic and/or heterogeneous and the special conditions of Section 3.3 must be met

The following general guidelines are applicable for all three classes of problems:

- The construction of the flow net should proceed from *outside* to *inside,* or from *generalities* to *details.* In other words, the overall nature of the flow net should always be kept in mind before zooming into a given area to work out fine details.
- The boundary conditions and transfer conditions should be obeyed. As discussed below, the boundary conditions will give us clues to draw the preliminary set of flow lines or equipotential lines for confined flow problems. The boundary conditions in many problems are such that the flow net is symmetrical (consider the flow net shown for the sheet pile structure in Fig. 3.3).
- Flow nets are better drawn on tracing papers. Cedergren advises us that the flow net could be drawn freehand on the reverse side before tracing it on the front. A suitable scale should be chosen to draw the section. Sections with large scales will lead to an undue burden in drawing flow nets. Sections with small scales, on the other hand, may obscure important details in complex shapes of flow domains.
- Casagrande advises us that four or five flow channels are usually sufficient for the first attempts at flow nets. Drawing too many flow lines may not increase the accuracy; it will increase the number of equipotential lines unnecessarily since square elements should be maintained. As shown in Fig. 3.10, additional flow lines could be added without much difficulty once the flow net is completed if more details are needed in any given area of the domain. Note that it is the ratio n_f/n_e that is of relevance in the flow rate estimation, not the absolute numbers of n_f or n_e.

In addition to these general guidelines, specific suggestions must be followed, as discussed below for the three different types of flow domains.

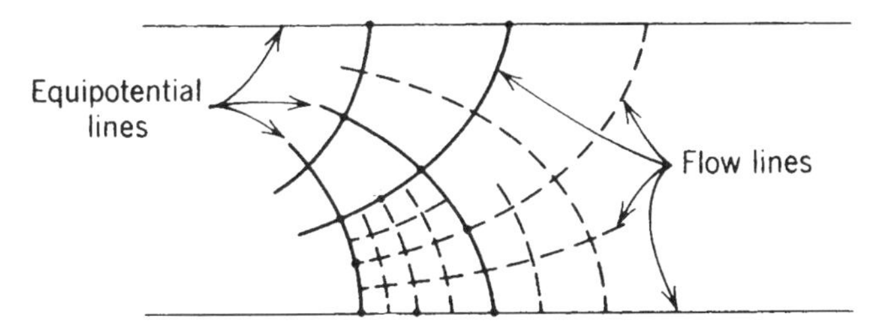

Fig. 3.10. Subdivision of flow channels and equipotential drops.

3.5.1 Confined Flow Domains

Confined flow domains are relatively simpler domains to draw flow nets. Several flow domains in this category yield symmetrical flow nets. Consider as an example the flow net for a flow domain beneath a concrete weir (Fig. 3.11). The boundaries for this problem are lines *AB* and *CD,* which are equipotential lines, and lines *BEFGHC* and *IJ,* which are flow lines (Fig. 3.11*a*). Because of the condition that flow lines and equipotential lines ought to be mutually perpendicular, we now know the directions in which the preliminary set of flow lines and equipotential lines should proceed from or meet the boundaries (Fig. 3.11*b*). At this step we have a choice to draw either equipotential lines or flow lines. Either set of lines could be drawn first. In the example

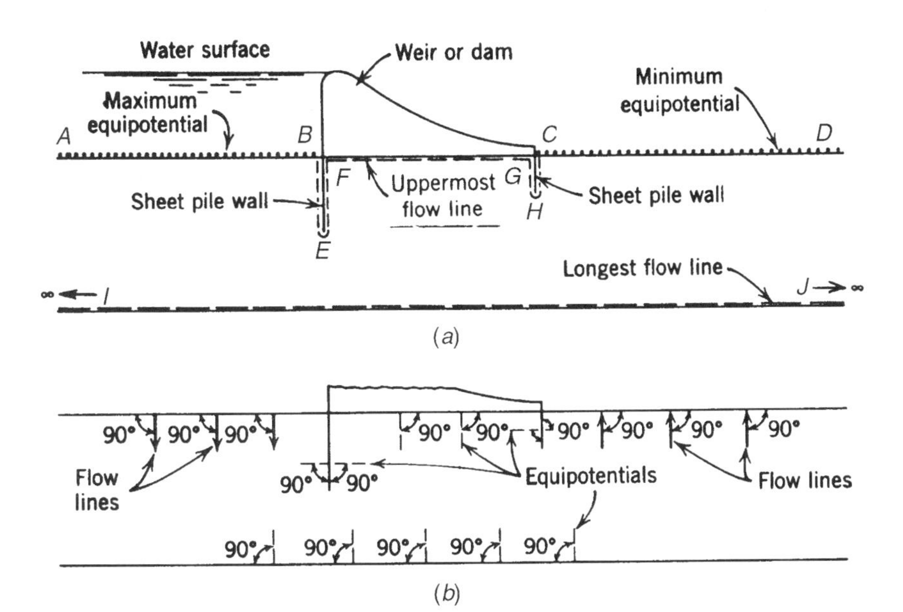

Fig. 3.11. Construction of a flow net for seepage beneath a concrete weir: (*a*) identify prefixed flow lines and equipotential lines; (*b*) look for prefixed starting directions of lines; (*continued*)

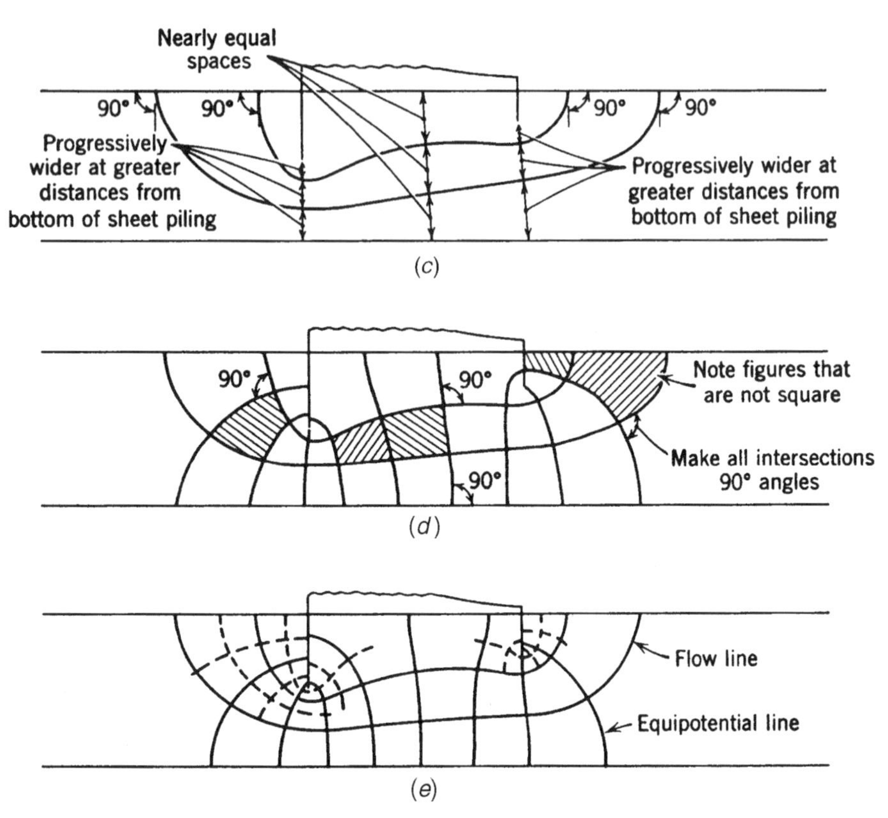

Fig. 3.11. (*Continued*) (*c*) draw a trial family of flow lines (or equipotentials) consistent with prefixed conditions; (*d*) keeping the lines drawn in (*c*), sketch the first trial flow net; make all lines intersect the other set of lines at 90° angles; (*e*) erase and redraw the lines until all figures are squares; subdivide as desired for detail and accuracy. (Adapted from Cedergren, 1989; copyright © John Wiley & Sons; reprinted with permission.)

shown here, flow lines are drawn first. Note that only two flow lines are chosen to be drawn (Fig. 3.11*c*). The trial family of flow lines is consistent with the condition that they should meet the equipotential boundaries (lines *AB* and *CD*) at right angles. We now proceed to draw the equipotential lines, ensuring as far as possible that the lines are perpendicular to the flow lines drawn originally (Fig. 3.11*d*). This is the step where we see at times that the two conditions of mutual perpendicularity and the square elements are not met at the same time–so the eraser becomes an important tool. After a few erasures and redrawings, the shaded elements could be made curvilinear squares. Very often, an entire flow line needs to be redrawn at this step. Once the number of flow lines is selected, it is a good practice not to change that number in the middle of flow net construction. As shown in Fig. 3.11*e*, additional flow lines and equipotential lines could be added with little difficulty once the overall shape of the flow net is known.

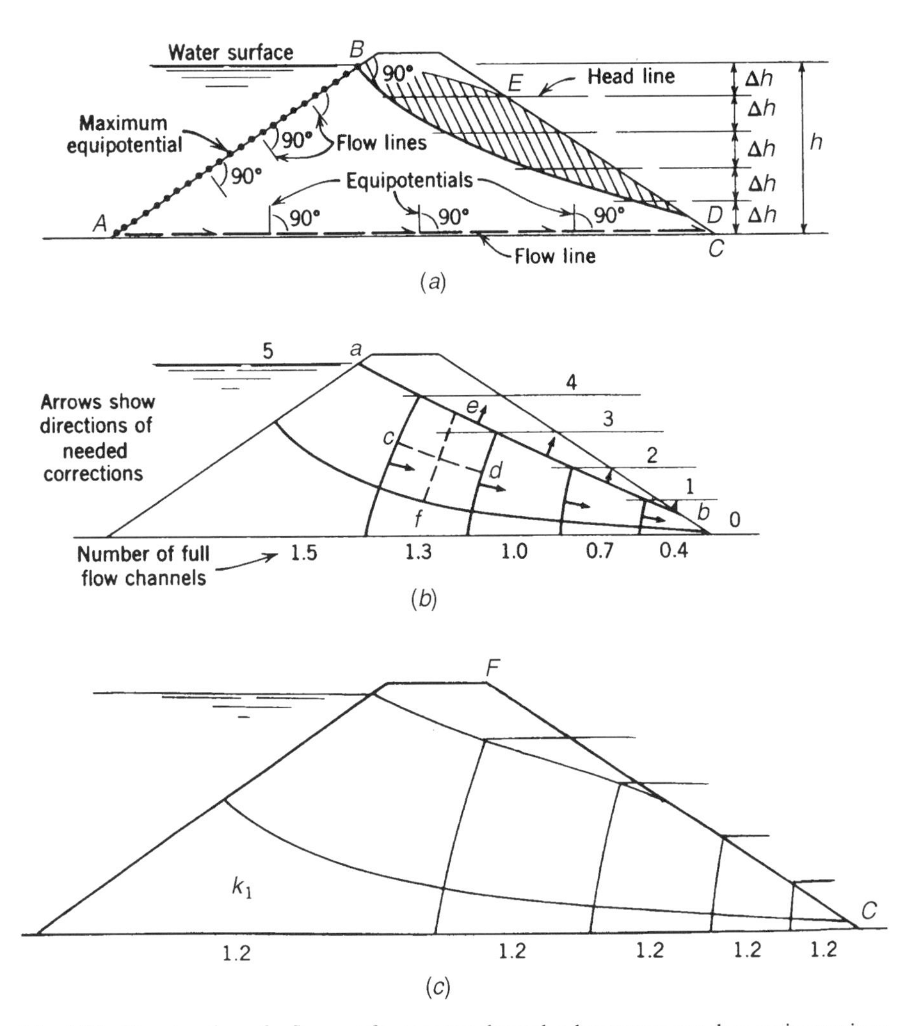

Fig. 3.12. Construction of a flow net for seepage through a homogeneous dam on impervious foundation: (*a*) known conditions; (*b*) trial saturation line and flow net; (*c*) final flow net. (Adapted from Cedergren, 1989; copyright © John Wiley & Sons; reprinted with permission.)

3.5.2 Unconfined Flow Domains

The *phreatic line* or *line of seepage* is unknown in unconfined flow domains. A trial line of seepage should be drawn first and it should gradually be refined, keeping the special conditions in Section 3.4 in mind. Consider flow in a homogeneous earth dam on impervious foundation as an example (Fig. 3.12). As shown in Fig. 3.12*a*, the phreatic line may lie anywhere in the shaded region. Consistent with our first condition, the points of intersection of equipotential lines and the line of seepage represent the hydraulic heads along the equipotential lines (Fig. 3.7*b*). If we choose five equipotential drops, we first mark "head" lines at equal vertical increments of Δh

and identify the points of intersection of the head lines with the tentative line of seepage. These become the points from which we proceed to draw equipotential lines in the domain. Consistent with the other two boundaries, we draw a set of equipotential lines. Keeping the requirement of square elements in mind, we then proceed to draw a flow line. However, as shown by the arrows in Fig. 3.12*b*, a number of adjustments are needed at this step. A good check at this point is to see if the total number of flow channels is the same throughout the domain. We note in Fig. 3.12*b* that the flow channels are changing from around 1.5 to 0.4. This requires us to adjust the line of seepage keeping the first condition in mind. The second set of conditions involving the entry and exit points of the line of seepage will help us at this step in altering the line of seepage. In our final flow net (Fig. 3.12*c*), the line of seepage exits tangential to the face of the structure as dictated by the conditions in Fig. 3.8, and the same number of flow channels, equal to 1.2, is maintained throughout our flow domain.

3.5.3 Anisotropic and Heterogeneous Domains

In accordance with the principles outlined in Section 3.3, flow nets in anisotropic domains require axis transformation. The new x' (horizontal axis) is given by $x' =$

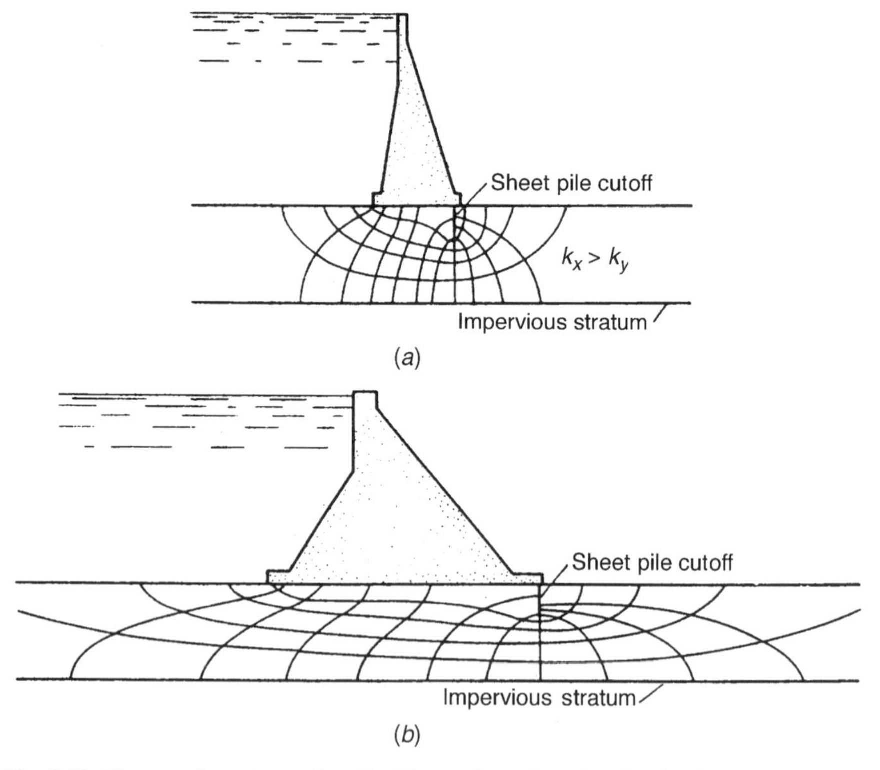

Fig. 3.13. Flow net for anisotropic soil: (*a*) transformed section for sketching the flow net; (*b*) retransformed section [flow net as sketched in (*a*) elongated to take into account greater permeability in the horizontal direction]. (Adapted from Spangler and Handy, 1973.)

$x\sqrt{k_y/k_x}$. Without changing the vertical scale, the section of the hydraulic structure is plotted using the transformed horizontal scale. The flow net in the transformed section is drawn in the same manner as in the case of isotropic soils described above. The flow rate is computed from the transformed section using Eq. (3.28). The flow net could be transformed back onto the natural section. However, the flow lines and equipotential lines transformed onto the original section do not generally intersect at right angles. It is also important to note that the hydraulic gradients in the domain can be determined only from the flow net on the true section. An example of flow net in anisotropic domains is shown in Fig. 3.13.

In drawing flow nets for heterogeneous or stratified domains, the transfer condition given by Eq. (3.33) must be met while continuing the flow lines across the interfaces. In all cases, the continuity of flow lines and equipotential lines must be maintained, although their direction may change abruptly. The number of flow channels must be constant throughout the flow net. Examples of flow net in heterogeneous domains are shown in Fig. 3.14.

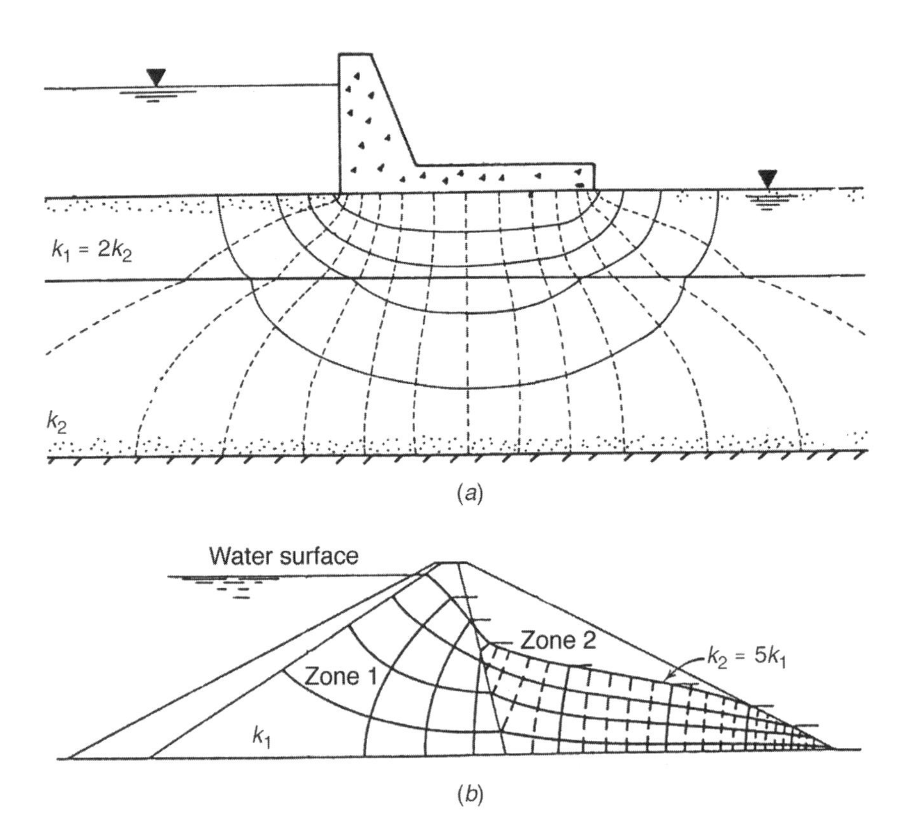

Fig. 3.14. Flow nets in the case of layered heterogeneity: (*a*) flow net under a dam with foundation layers of different permeability; (*b*) flow net for flow through a zoned dam with zones of different permeability. ((*b*) Adapted from Cedergren, 1989; copyright © John Wiley & Sons; reprinted with permission.)

3.6 MISCELLANEOUS FLOW NETS

Flow nets can be used to estimate flow rates and determine hydraulic heads in a variety of flow domains. Several types of seepage problems where flow nets become useful are shown in Figs. 3.15 to 3.21. The flow nets shown in these figures fall under four broad categories.

1. Flow nets used to map flow in and around geotechnical structures such as earth dams, retaining structures, and slopes are shown in Fig. 3.15. As discussed in detail in Chapter 6, seepage plays an important role in the stability of earthen structures. Collecting flow through drains to contain flow away from the downstream slope of earth dams (Fig. 3.15*a* and *b*) improves stability of slopes. Similarly, providing adequate drainage for water behind retaining structures (Fig. 3.15*c*) and in slopes (Fig. 3.15*d*) is important for the stability of these structures.
2. Flow nets are useful to map the flow in plan view and determine flow quantities toward wells and dewatered excavations as shown in Fig. 3.16. Given the proximity of the source of water, usually a river or a stream, the flow quantities toward the discharge point/area could be estimated in Fig. 3.16*a* and *b*. Flow domains mapped near an array of pumping and/or recharge wells (Fig.

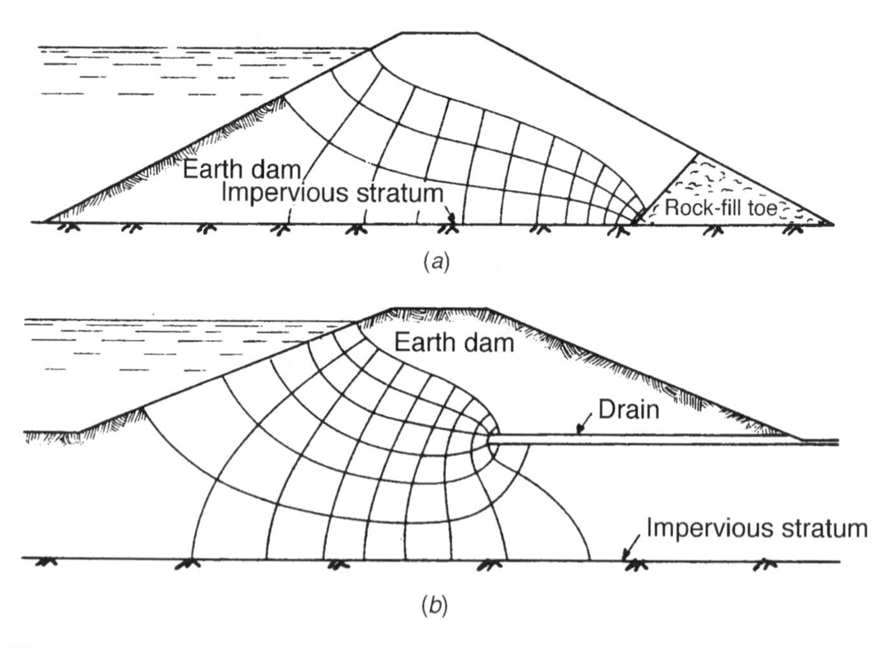

Fig. 3.15. Flow nets for seepage through and around geotechnical structures: (*a*) flow net through an earth dam provided with rock-fill toe; (*b*) flow net for an earth dam with a downstream drain; (*continued*)

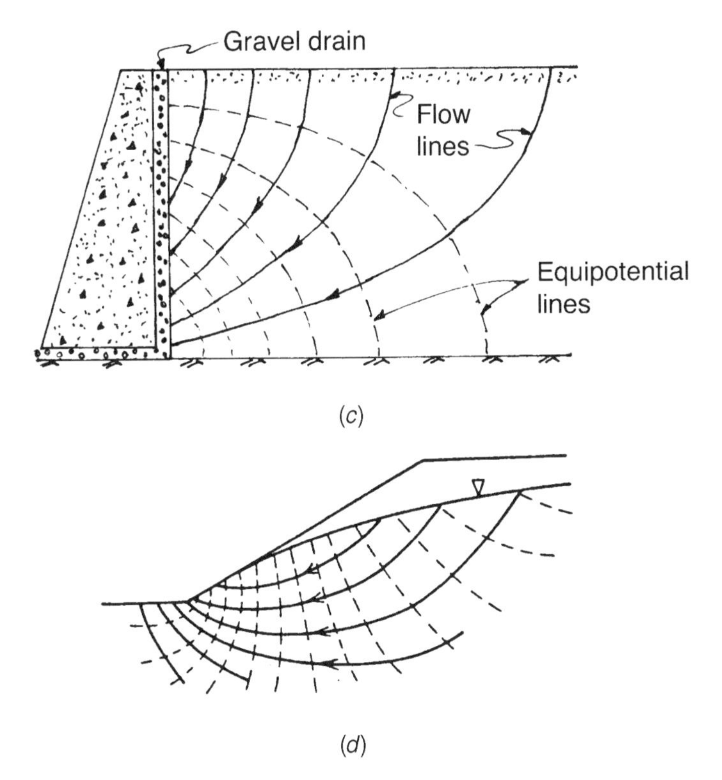

Fig. 3.15. (*Continued*) (*c*) flow net behind an earth retaining structure provided with a vertical drain; (*d*) flow net in a slope. [(*c*) Adapted from Lambe and Whitman, 1969; copyright © John Wiley & Sons; reprinted with permission; (*d*) adapted from Patton and Hendron, 1974.]

3.16*c* and *d*) are useful in geoenvironmental situations to determine the areas influenced by discharge and recharge operations.

3. As discussed later in Chapter 4, several analytical solutions exist for flow toward wells and slots located in confined and unconfined aquifers. Rarely is a flow net needed to determine flow rates from pumping wells. However, a study of flow nets around wells in confined and unconfined aquifers (Fig. 3.17) is useful to understand the extent of an aquifer affected by a single partly or fully penetrating well. The differences in flow nets between confined and unconfined aquifers, and between fully and partly penetrating wells, are worth pondering for a thorough grasp of flow net principles.
4. Flow nets can be drawn to map flow toward drains (Figs. 3.18 and 3.19) and trenches (Figs. 3.20 and 3.21) to estimate flow quantities. This is useful not only to estimate the design capacities of drains but also to determine the location and spacing of trenches and drains (see Fig. 3.19). The reader will readily see that flow around a deep tunnel in the subsurface is analogous to flow around the drain shown in Fig. 3.18*b*). Flow nets around trenches (Figs. 3.20 and 3.21) are useful in pavement drainage, discussed in Chapters 9 and 10.

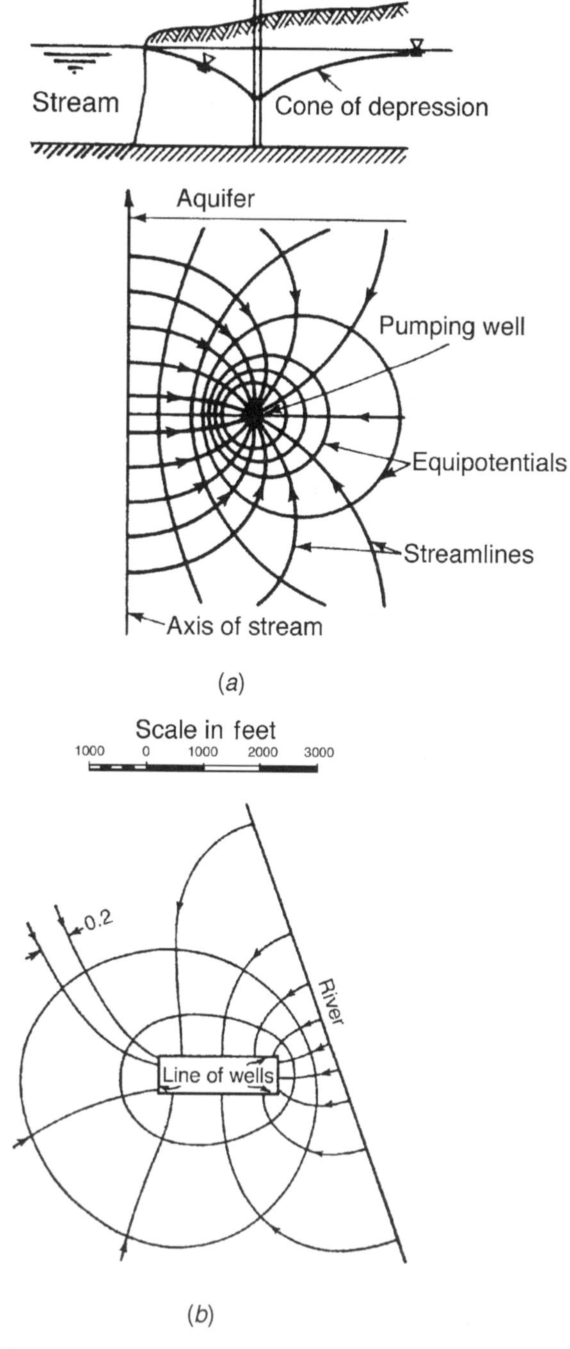

Fig. 3.16. Plan flow nets: (*a*) flow net for seepage toward a pumping well from a stream; (*b*) flow net for seepage toward a well system from a river; (*continued*)

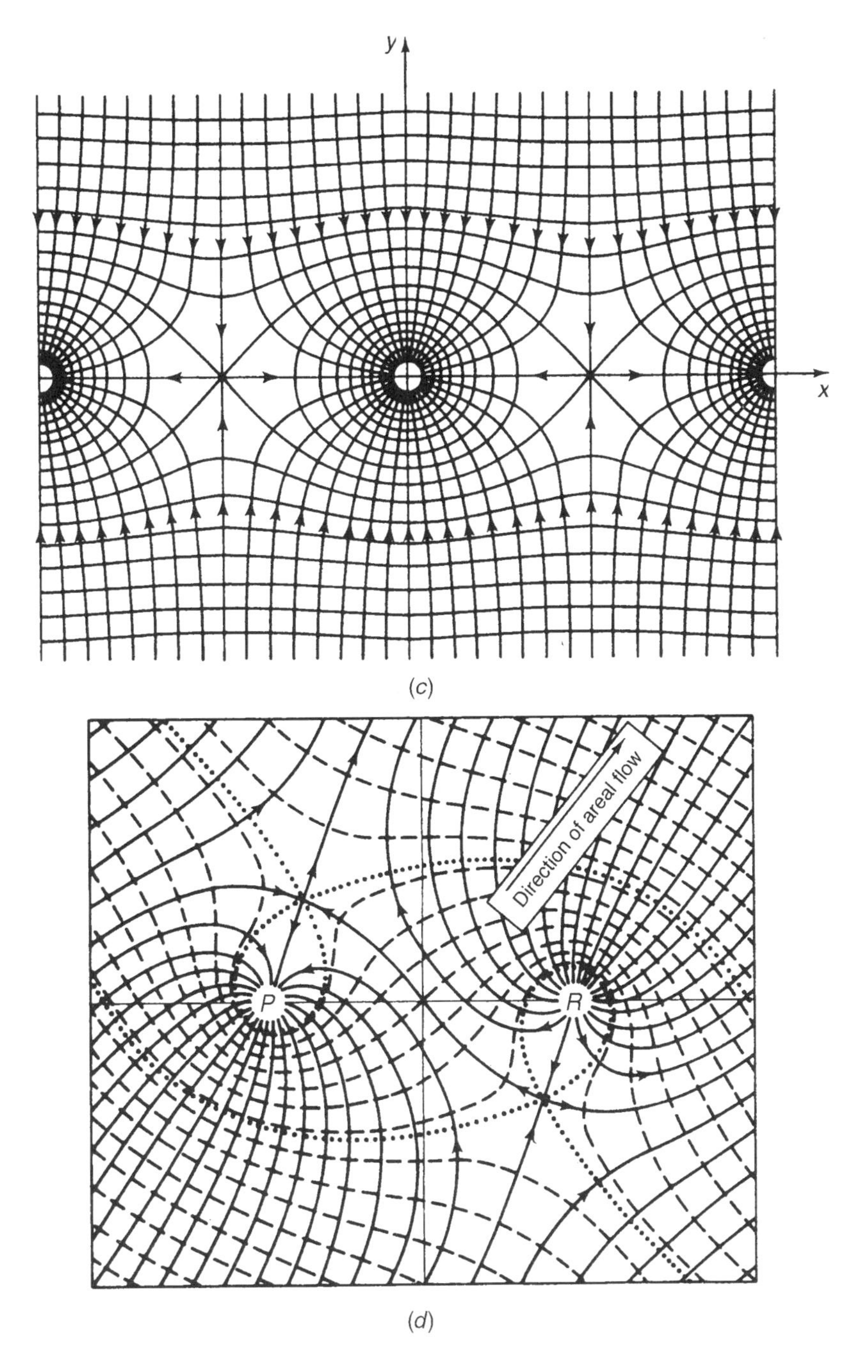

Fig. 3.16. (*Continued*) (*c*) flow lines and equipotential lines about an infinite array of wells; (*d*) flow net around a pair of pumping and recharging wells in uniform flow. [(*a, c*) Adapted from Bear, 1979; (*b*) adapted from Mansur and Kaufman, 1962; (*d*) adapted from Dacosta and Bennett, 1960.]

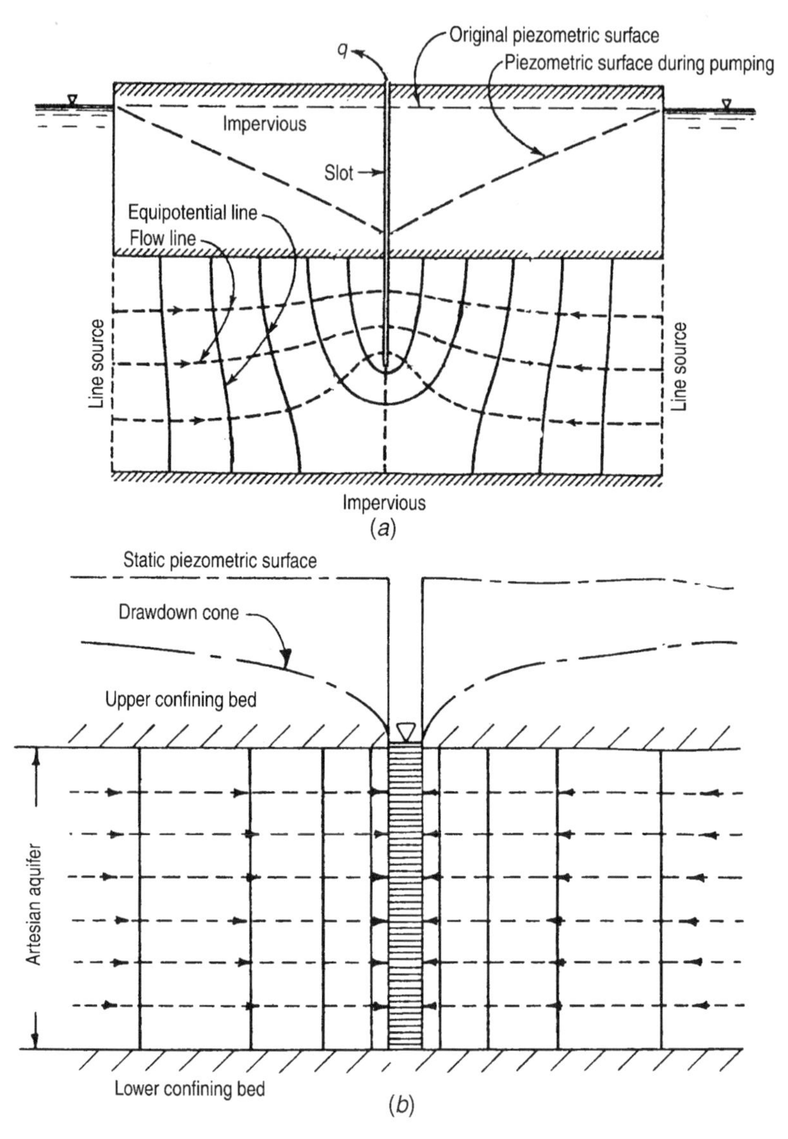

Fig. 3.17. Flow nets for seepage toward slots and wells in confined and unconfined aquifers: (*a*) flow net for artesian flow to a partially penetrating slot midway between two line sources of seepage, all of infinite length; (*b*) flow net for seepage toward a fully penetrating discharge well in an artesian aquifer; (*c*) flow net for seepage toward a fully penetrating discharge well in an unconfined aquifer; (*continued*)

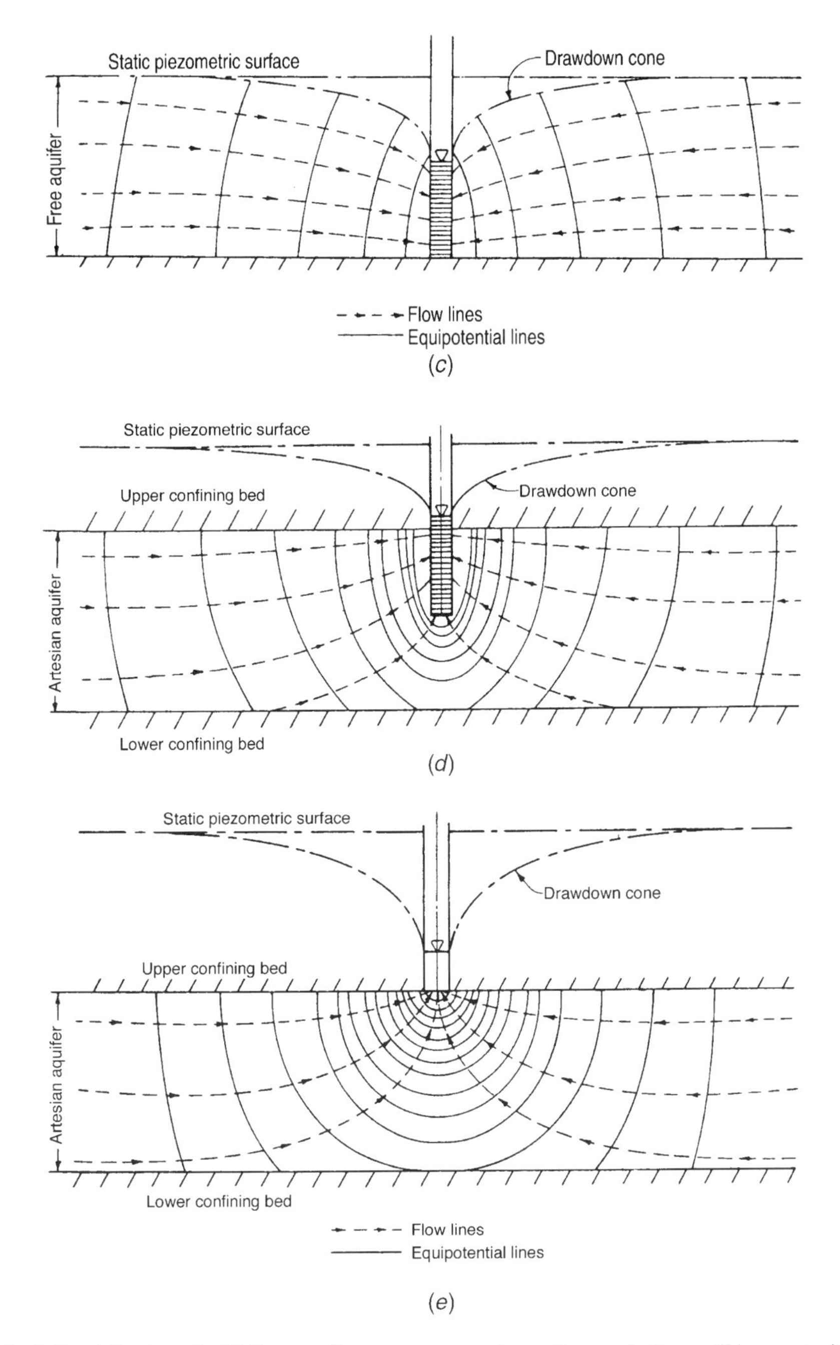

Fig. 3.17. (*Continued*) (*d*) flow net for seepage toward a partly penetrating well in an artesian aquifer; (*e*) flow net for seepage toward a discharge well resting on the top of an artesian aquifer. [(*a*) Adapted from Mansur and Kaufman, 1962; (*b–e*) adapted from U.S. Department of the Interior, 1977.]

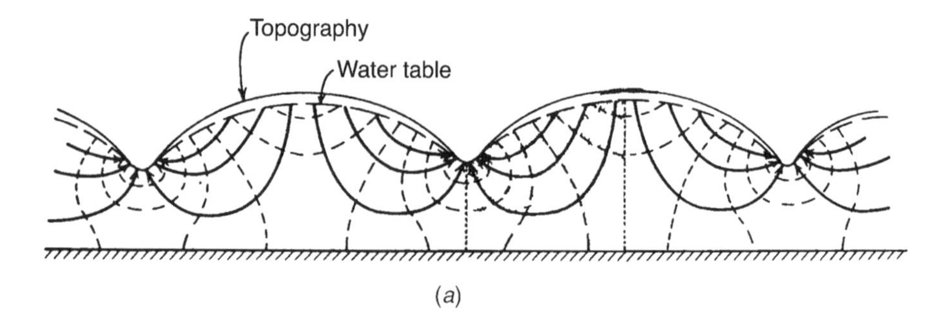

(a)

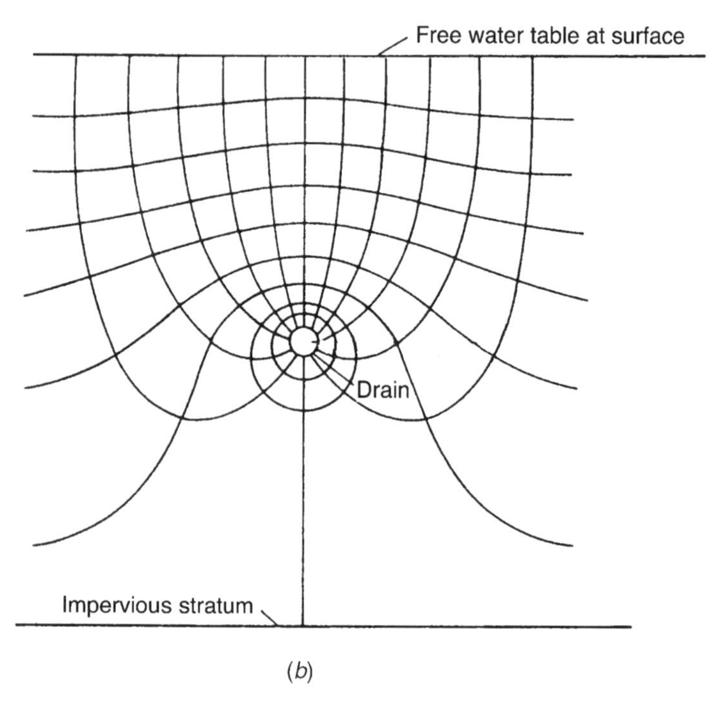

(b)

Fig. 3.18. Flow nets for seepage toward trenches and drains: (*a*) flow net for seepage toward trenches; (*b*) flow net for a subsurface drain or a tunnel. [(*a*) Adapted from Hubbert, 1940; (*b*) adapted from Spangler and Handy, 1973.]

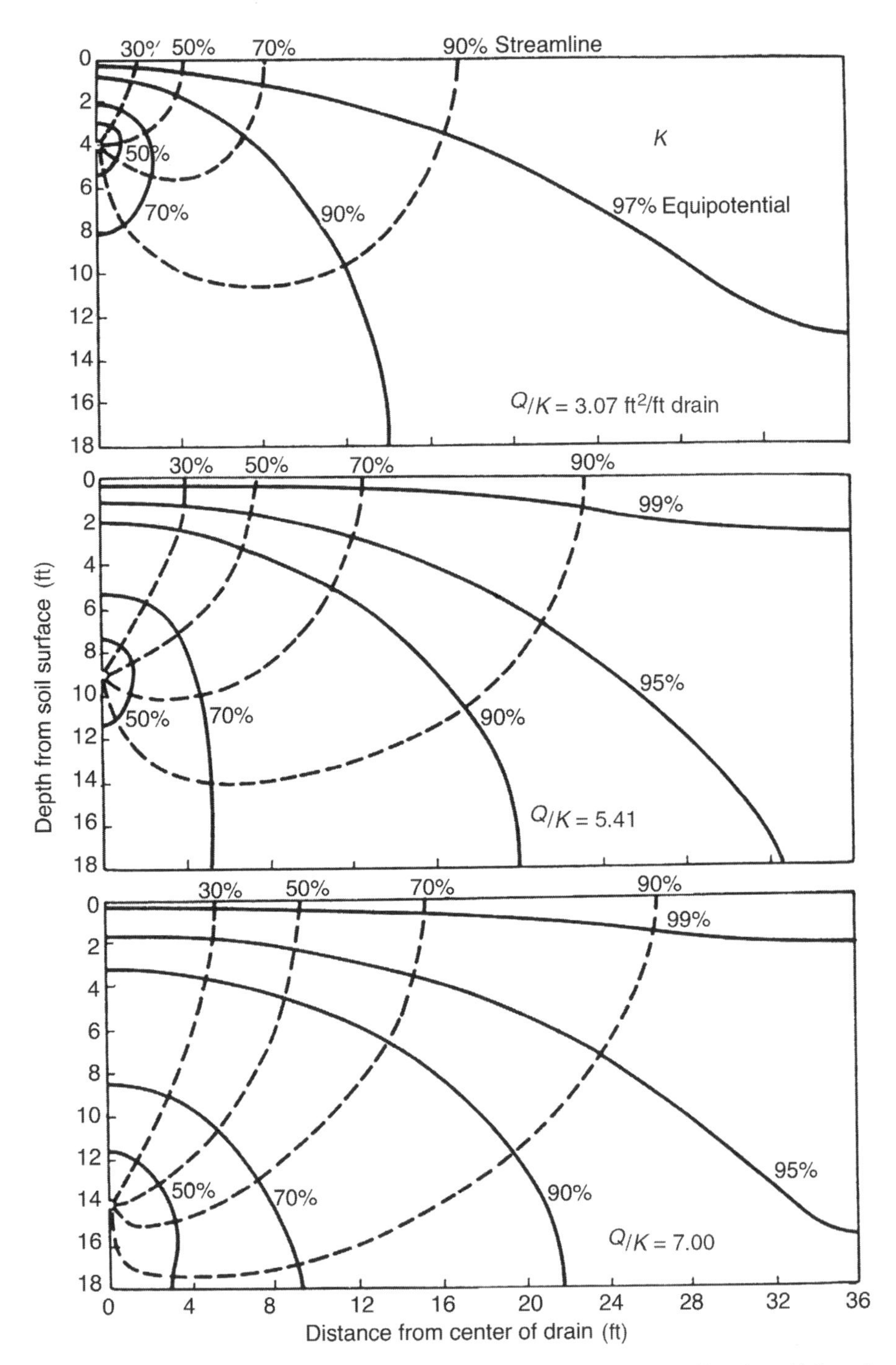

Fig. 3.19. Flow nets for drain lines at three different depths beneath the soil surface. (Adapted from Luthin, 1966.)

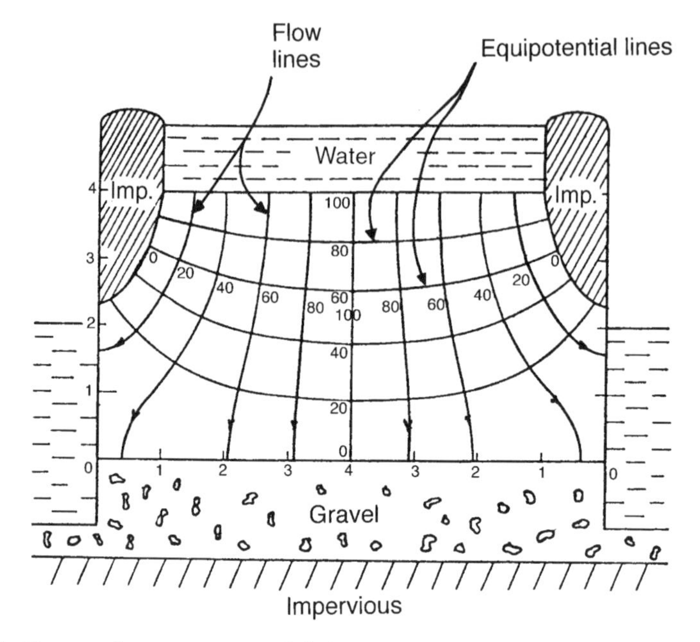

Fig. 3.20. Flow net for seepage toward drainage ditches in soil overlying gravel. (Adapted from Kirkham, 1960.)

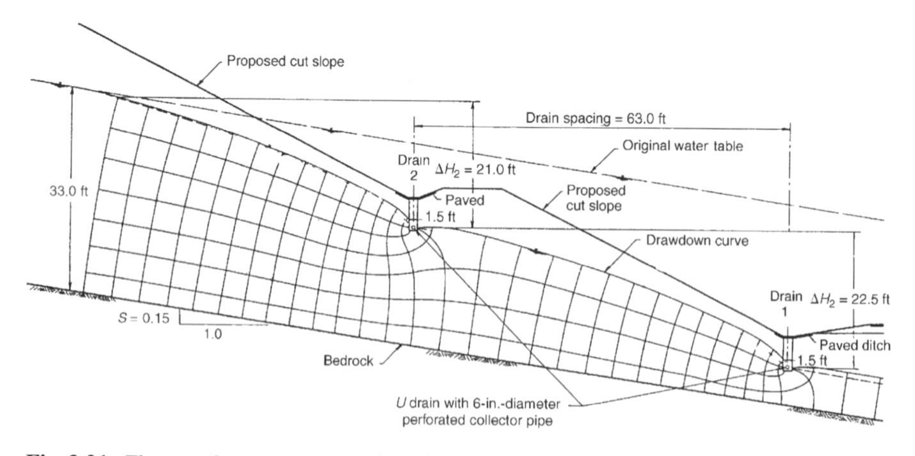

Fig. 3.21. Flow net for seepage toward two interceptor drains. (Adapted from Moulton, 1980.)

PROBLEMS

3.1. Determine the seepage rate under the dam shown in Fig. P3.1 using a flow net.

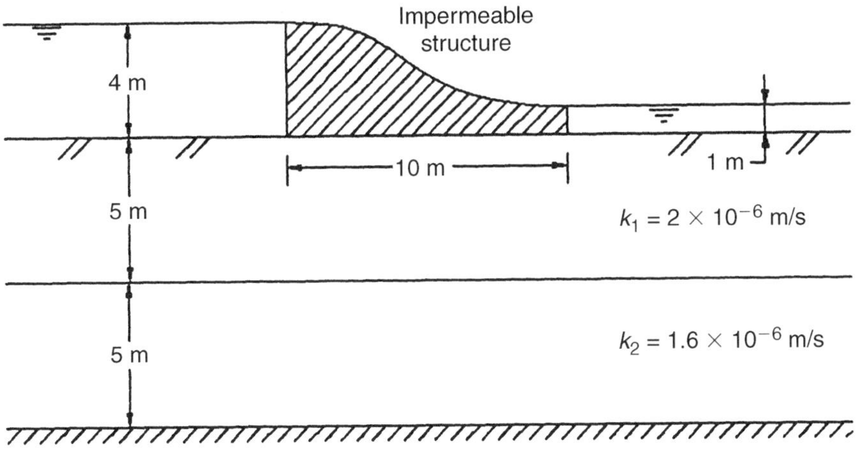

Fig. P3.1

3.2. Using the partial flow net shown in Fig. P3.2, calculate the elevation heads, pressure heads, and total heads at points *a, b, c, d, e, f, g, h, i,* and *j*.

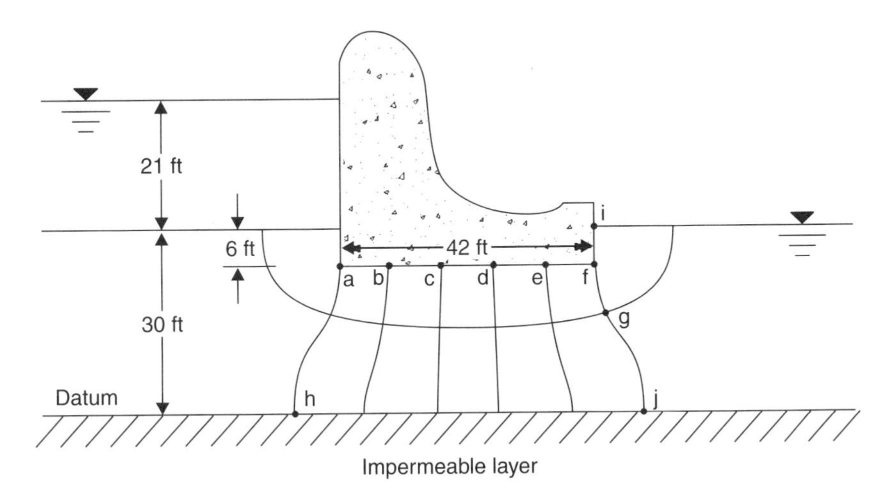

Fig. P3.2

3.3. Determine water levels in the piezometers shown in Fig. P3.3.

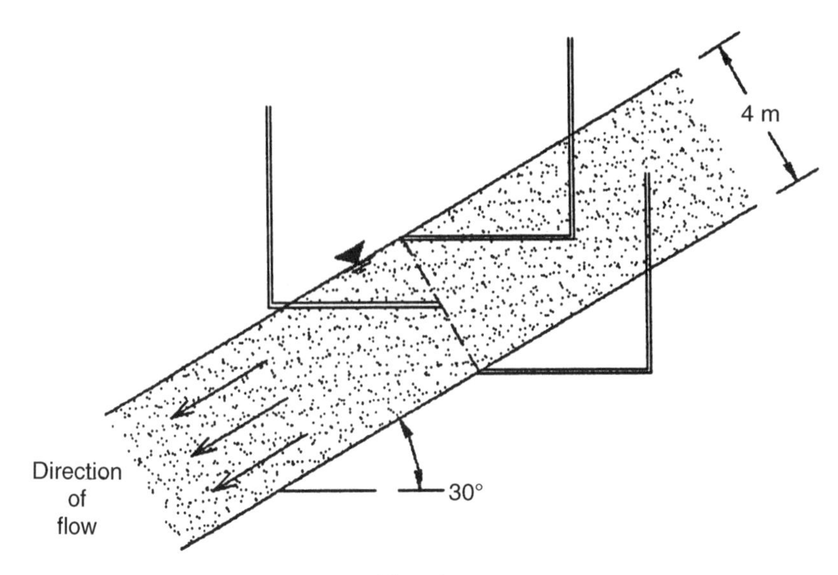

Fig. P3.3

3.4. Refer to Fig. P3.4. Draw flow nets for the following three cases:

(a) $D_1 = H$, $D_2 = H$

(b) $D_1 = 2H/3$, $D_2 = H/3$

(c) $D_1 = 3H/4$, $D_2 = H/4$

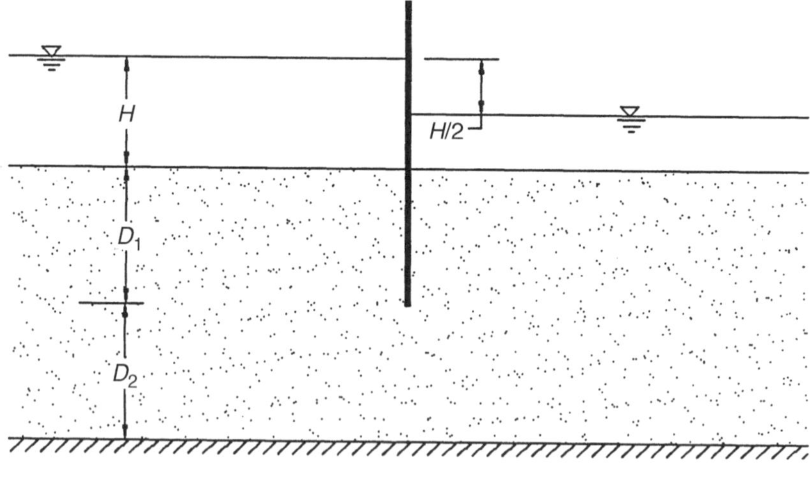

Fig. P3.4

3.5. Refer to the flow nets drawn in Problem 3.4, and calculate seepage rates per foot length of the sheet pile structure in terms of H. $k = 2 \times 10^{-3}$ cm/s. Also, determine exit gradients for the three different depths of the structure.

3.6. Draw flow nets in the transformed and real domains for the anisotropic flow domain in Fig. P3.6.

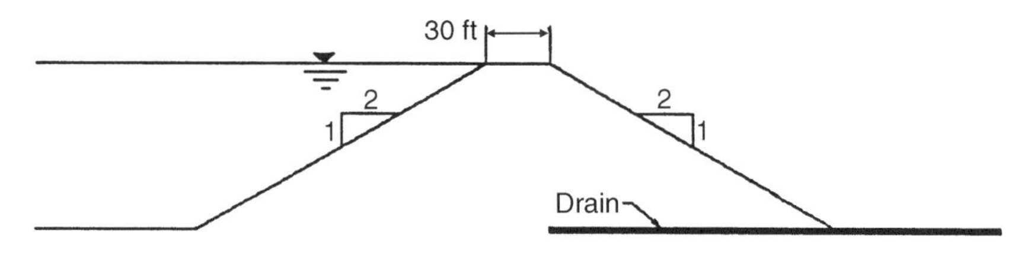

Homogeneous dam
and foundation

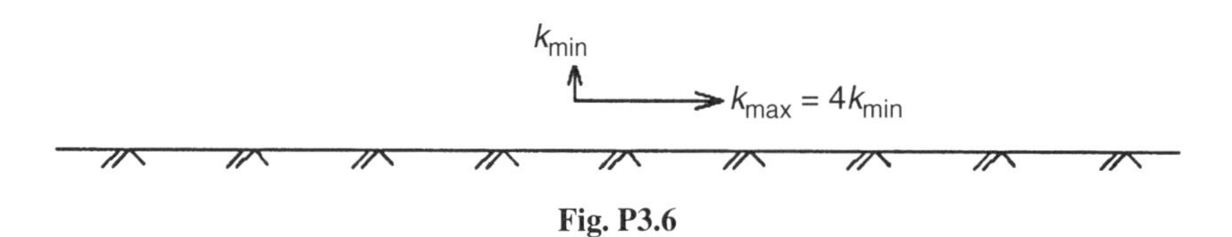

Fig. P3.6

3.7. Steady-state two-dimensional flow is occurring around the double-row sheet piles shown in Fig. P3.7. Draw a flow net and determine the rate of flow per meter length of wall. Also, determine the exit gradients near the interior of the sheet piles.

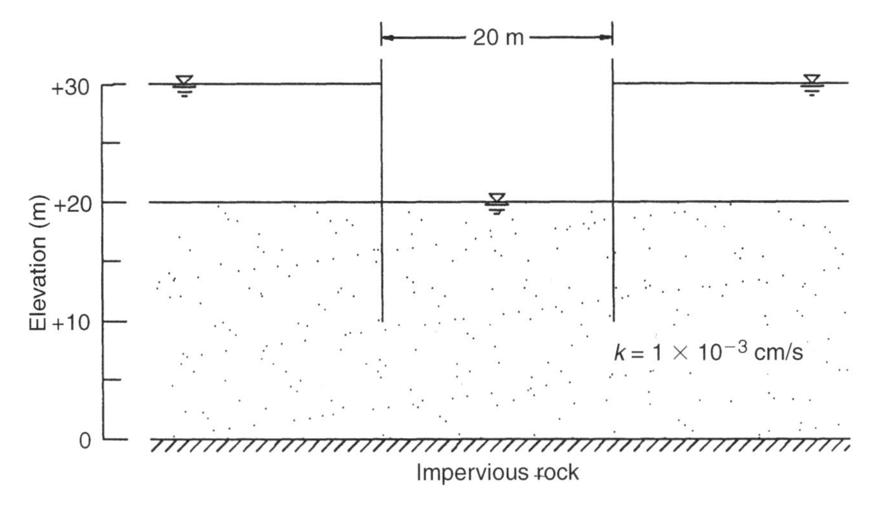

Fig. P3.7

3.8. As discussed in Chapter 9, seepage underneath hydraulic structures is controlled using either an impervious blanket on the upstream end of the structure or by placing a cutoff wall. Using Fig. P3.8, draw flow nets for the following cases:

(a) Only an impervious blanket is used upstream of length equal to H.
(b) Only a cutoff wall is used at the center of the structure with $D = H/2$.
(c) Both an impervious blanket and the cutoff wall are used in combination.

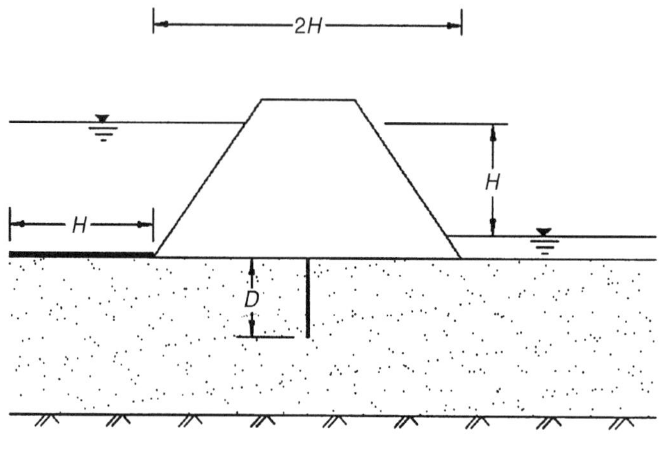

Fig. P3.8

3.9. Using the transformed domain in the flow nets in Fig. 3.13, determine $\sqrt{k_y/k_x}$.

3.10. Redraw the flow net in Fig. 3.14*b* for $k_2 = 10k_1$.

3.11. The plan flow net in Fig. 3.16*b* corresponds to a pervious confined aquifer 100 ft thick. If the difference in the hydraulic heads between the river and the slot bounded by the wells is 50 ft, determine the flow rate toward the slot. Use $k = 1 \times 10^{-3}$ cm/s.

3.12. Determine flow rates toward longitudinal trench drains in Fig. 3.21. Use $k = 1 \times 10^{-3}$ cm/s.

3.13. Consider the flow nets shown in Fig. P3.13. What should be the drainage capacities of the longitudinal drainage layer in Fig. P3.13*a* and the collector pipe in Fig. P3.13*b*?

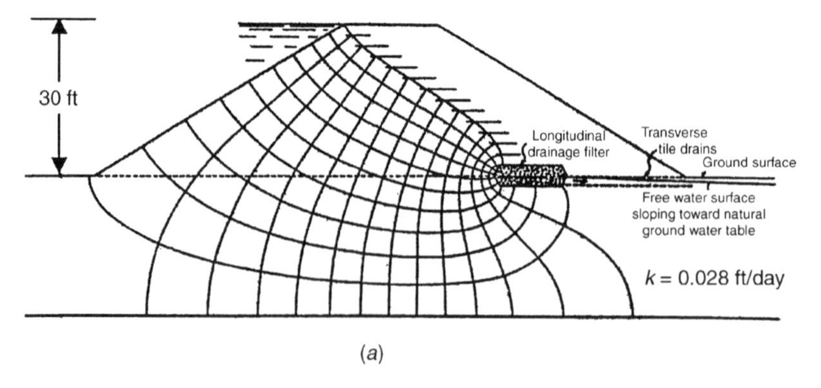

Fig. P3.13a. (Adapted from Casagrande, 1937.)

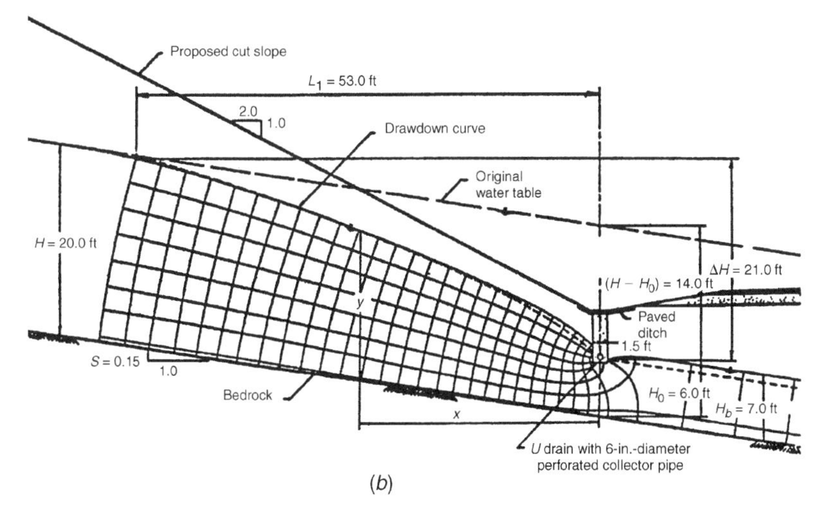

Fig. P3.13. (*Continued*)
(*b*) (Adapted from Moulton, 1980.)

CHAPTER 4

TWO-DIMENSIONAL FLOW: ANALYTICAL SOLUTIONS

Although the graphical solution of flow nets, discussed in Chapter 3, is suitable for a range of two-dimensional problems, there are several cases where simple closed-form solutions would prove to be desirable alternatives. The Laplace equation, which governs steady-state flow, has analytical solutions for flow domains with simple boundary conditions. In cases where boundary conditions and the flow domain are complicated, mathematical tools are available to transform the domain into equivalent simple domains for which closed-form solutions exist. We explore a number of such solutions in Chapters 4 and 5.

Many problems involving slots and discharge wells can be solved using Darcy's law and equation of continuity. Several others, particularly those with unconfined flow domains, can be simplified using the Dupuit–Forchheimer assumptions described below. Solutions for flow confined in complicated geometries, such as underneath a dam and around sheet piles, involve complex mathematics. We will find that for such confined problems, analytical solutions could be obtained using mathematical transformations. We defer this to Chapter 5 and proceed in this chapter with simpler solutions.

4.1 DUPUIT–FORCHHEIMER ASSUMPTIONS

Flow problems in unconfined domains can be greatly simplified with two assumptions concerning the phreatic surface, or line of seepage, in a two-dimensional domain. These assumptions, originally given by Dupuit in 1863 and later applied by Forchheimer in 1930s, can be stated as follows:

1. For small changes in the slope of line of seepage, the hydraulic head is independent of depth.
2. The hydraulic gradient causing flow is equal to the slope of the water table.

The first assumption means that the equipotential and flow lines are vertical and horizontal, respectively (Fig. 4.1*a* and *b*). There are no gradients in the vertical dimension. In other words,

$$h(x, y) = h(x) \tag{4.1}$$

Since the flow lines are horizontal, the hydraulic gradient, i, is equal to $\tan\theta$ (slope of the phreatic line), as specified in the second assumption (Fig. 4.1*c*):

$$i = \frac{\delta h}{\delta x} \tag{4.2}$$

The second assumption would thus mean using $\tan\theta$ instead of $\sin\theta$. The validity of these assumptions depends heavily on the magnitude of θ (i.e., the steepness of the phreatic line). Bear (1972) has shown that the error due to the second assumption is small when $i^2 \ll 1$. The corresponding condition in an anisotropic medium is

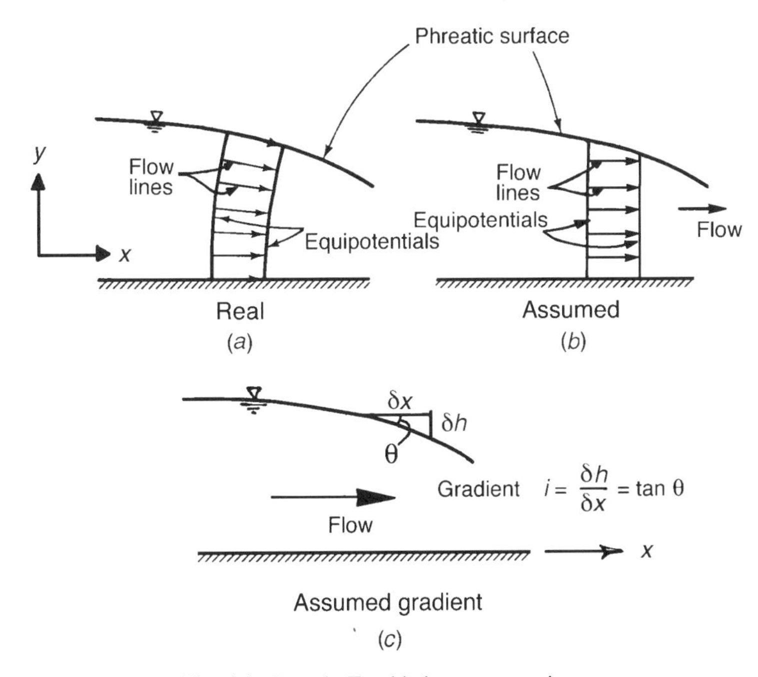

Fig. 4.1. Dupuit–Forchheimer assumptions.

$(k_x/k_y)i^2 \ll 1$, where k_x, k_y represent the permeability in the x and y directions, respectively.

4.2 SLOTS AND DITCHES

Flow through aquifers from an infinite line source on one end to an infinite slot or sink on the other end has one of the simplest mathematical solutions. Straightforward application of Darcy's law gives us the solution when flow occurs in a confined aquifer of constant thickness. In the case of unconfined flow, the Dupuit–Forchheimer assumptions described above come to our rescue in obtaining a mathematical solution. Details of these solutions are presented below.

4.2.1 Artesian (Confined) Flow

Consider flow through a confined pervious stratum of thickness D, shown schematically in Fig. 4.2. The hydraulic gradient causing flow is due to an unlimited line source on one end, with the water level maintained at a constant elevation in a slot on the other end. Consider that the line source and the slot extend infinitely in the lateral dimension. For a unit lateral dimension of the aquifer, *Darcy's law* may be expressed as

$$q = kiA = k\frac{dh}{dx}(D \cdot 1) \tag{4.3}$$

where q is the seepage rate through the aquifer, k the permeability of the aquifer, i the hydraulic gradient equal to dh/dx, and A the area of cross section for flow, equal to D for a unit lateral dimension. Rearranging the terms in Eq. (4.3) gives us

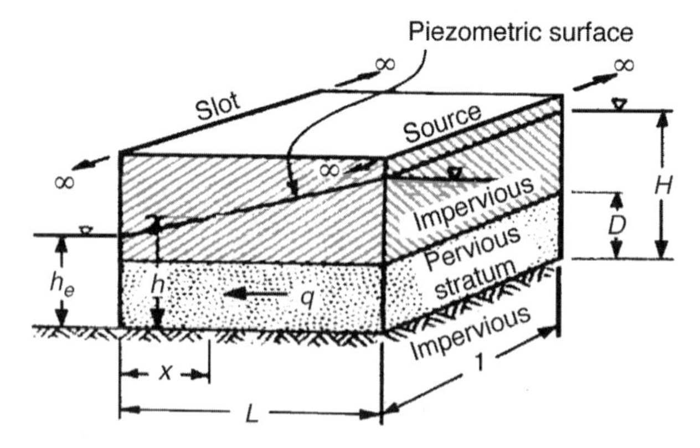

Fig. 4.2. Seepage from a line source to a fully penetrating slot through a confined aquifer.

$$dh = \frac{q}{kD} dx \tag{4.4}$$

Integration of Eq. (4.4) yields

$$h = \frac{qx}{kD} + C_1 \tag{4.5}$$

where C_1 is a constant of integration. Using the boundary conditions

$$h = \begin{cases} h_e & \text{at } x = 0 \quad (4.6) \\ H & \text{at } x = L \quad (4.7) \end{cases}$$

in Eq. (4.5), we get

$$h = \frac{qx}{kD} + h_e \tag{4.8}$$

and

$$q = \frac{kD}{L}(H - h_e) \tag{4.9}$$

Equations (4.9) and (4.8) express the seepage rate through the aquifer and variation of hydraulic head, respectively. Substitution of Eq. (4.9) in (4.8) gives an expression for hydraulic head independent of q:

$$h = H - \frac{L - x}{L}(H - h_e) \tag{4.10}$$

which gives us the variation of h in terms of x.

4.2.2 Gravity (Unconfined) Flow

In the case of an unconfined pervious stratum (Fig. 4.3), the area of cross section for flow varies in terms of x. We invoke Dupuit–Forchheimer's second assumption and express hydraulic gradient i as equal to dh/dx. Using Darcy's law gives us

$$q = kiA = k\frac{dh}{dx}(h \cdot 1) \tag{4.11}$$

Rearranging the terms in Eq. (4.11), we obtain

$$h\,dh = \frac{q}{k} dx \tag{4.12}$$

Integration of Eq. (4.12) yields

$$h^2 = \frac{2qx}{k} + C_2 \tag{4.13}$$

where C_2 is a constant of integration. Using the same boundary conditions as in the case of artesian flow [Eqs. (4.6) and (4.7)], we get

$$h^2 = \frac{2qx}{k} + h_e^2 \tag{4.14}$$

$$q = \frac{k}{2L}\left(H^2 - h_e^2\right) \tag{4.15}$$

and

$$h^2 = H^2 - \frac{L-x}{L}\left(H^2 - h_e^2\right) \tag{4.16}$$

Equation (4.15) is known as the *Dupuit equation for steady unconfined flow.* Unlike the case of artesian flow, the expression for hydraulic head variation represents a parabola [Eq. (4.16)], as shown in Fig. 4.3. Because of the boundary conditions used in its derivation, the parabola passes through the two points A and B at the two ends of the domain. This is, however, in violation of the conditions of the line of seepage shown in Fig. 3.8. When the downstream discharge face is at right angles, the line of seepage terminates at an elevation above the free water present outside the domain. Thus, a *seepage face* exists through which water trickles down into the free water body at the boundary. The water table elevations are not accurate near the upstream and downstream boundaries because of this seepage face. The errors may be significant when L/H and h_e/H are small. In such cases, the parabola given in Eq. (4.16) is modified as

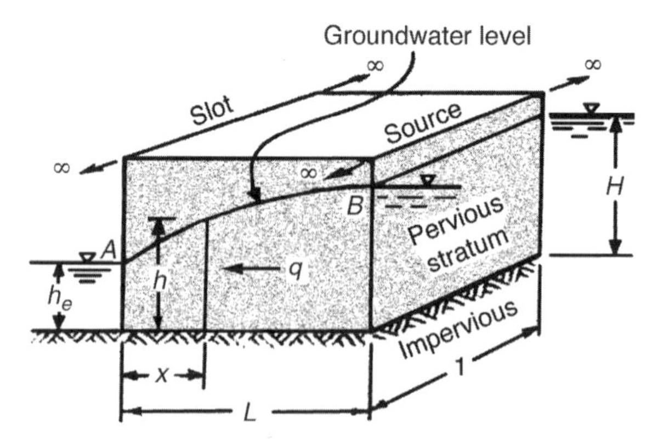

Fig. 4.3. Seepage from a line source to a fully penetrating slot through an unconfined aquifer.

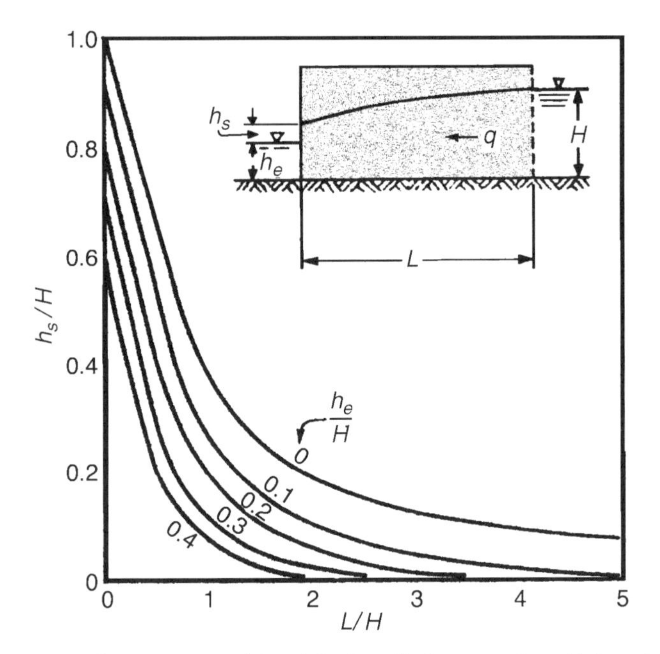

Fig. 4.4. Correction factor for the height of the free discharge surface. (Adapted from Chapman, 1956.)

$$h^2 = H^2 - \frac{L-x}{L}\left[H^2 - (h_e + h_s)^2\right] \tag{4.17}$$

where h_s is the height of the free discharge surface, obtained using Fig. 4.4.

Example 4.1 Determine the errors in hydraulic head due to ignoring the seepage face in gravity flow for the following conditions: (a) $H = 10$ m, $h_e = 5$ m, and $L = 20$ m, and (b) $H = 10$ m, $h_e = 1$ m, and $L = 10$ m.

SOLUTION: From Eq. (4.17), the error in hydraulic head due to ignoring the seepage face in gravity flow is

$$\Delta h = \sqrt{H^2 - \frac{L-x}{L}\left[H^2 - (h_e + h_s)^2\right]} - \sqrt{H^2 - \frac{L-x}{L}\left(H^2 - h_e^2\right)}$$

where h_s is the height of free discharge surface, obtained using Fig. 4.4.

(a) For $H = 10$ m, $h_e = 5$ m, and $L = 20$ m, $h_e/H = 0.5$ and $L/H = 2$. From Fig. 4.4, $h_s/H = 0$, so the error in hydraulic head due to ignoring the seepage face is zero.

(b) For $H = 10$ m, $h_e = 1$ m, and $L = 10$ m, $h_e/H = 0.1$ and $L/H = 1$. From Fig. 4.4, $h_s/H = 0.26$; the hydraulic head along L is plotted in Fig. E4.1 with and without the seepage face.

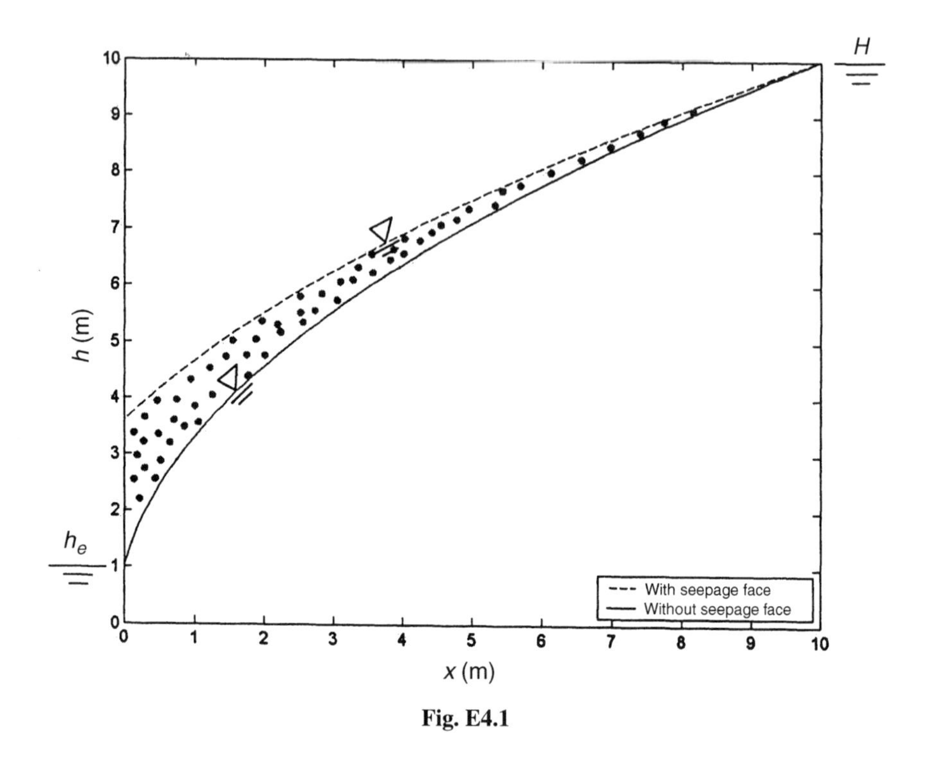

Fig. E4.1

4.2.3 Unconfined Flow in Stratified Aquifers

Consider the case of a two-layered aquifer system between two reservoirs (Fig. 4.5*a*). The phreatic surface is located in layer 2, and discharge takes place in both layers. The total discharge q may be expressed as

$$q = -k_1 a \frac{dh}{dx} - k_2 (h - a) \frac{dh}{dx} \tag{4.18}$$

Neglecting the seepage face at the downstream end, $h = h_e$ at $x = L$. Integrating from $x = 0$ (where $h = H$) to $x = L$, the reader can verify that Eq. (4.18) reduces to

$$q = \frac{k_2}{2L}(H - h_e)\left(H + h_e - 2a + 2a\frac{k_1}{k_2}\right) \tag{4.19}$$

or

$$q = k_1 a \frac{H - h_e}{L} + \frac{k_2}{2L}\left[(H - a)^2 - (h_e - a)^2\right] \tag{4.20}$$

Equation (4.20) implies that flow in a two-layered system may be expressed as the summation of two components: confined flow [Eq. (4.9)] and unconfined flow [Eq. (4.15)]. If the aquifer system consists of two vertical strata (Fig. 4.5*b*), we consider

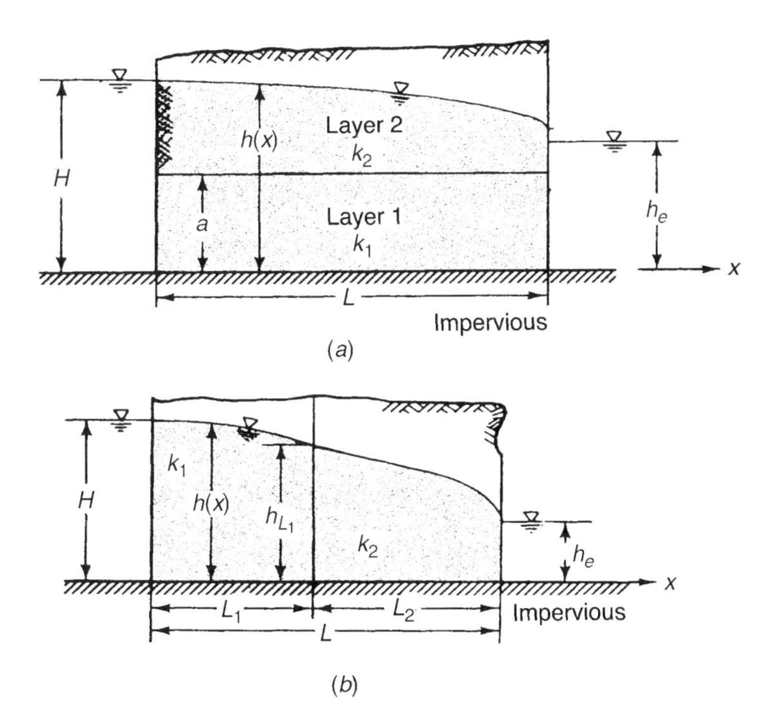

Fig. 4.5. Unconfined flow in stratified aquifers: (*a*) horizontal stratification; (*b*) vertical stratification. (Adapted from Bear, 1979; copyright © The McGraw-Hill Companies; reprinted with permission.)

the two regions separately in developing an equation for discharge. In the region $0 \leq x \leq L_1$,

$$q = -k_1 h \frac{dh}{dx} \tag{4.21}$$

Integrating Eq. (4.21) from $x = 0$ and $h = H$ to $x = L_1$ and $h = h_{L_1}$ yields

$$h_{L_1}^2 = H^2 - 2\frac{qL_1}{k_1} \tag{4.22}$$

For the region $L_1 \leq x \leq L$, we integrate Eq. (4.21) from $x = L_1, h = h_{L_1}$ to $x = L$, $h = h_e$, and replace k_1 by k_2. Thus,

$$h_{L_1}^2 = h_e^2 + \frac{2q(L - L_1)}{k_2} \tag{4.23}$$

Using Eqs. (4.22) and (4.23), we may express q as

$$q = \frac{H^2 - h_e^2}{2[L_2/k_2 + L_1/k_1]} \tag{4.24}$$

The shape of the phreatic surface may be obtained by integrating Eq. (4.21) from 0 to x. This will give us

$$h(x) = \begin{cases} \left[H^2 - \dfrac{H^2 - h_e^2}{k_1(L_1/k_1 + L_2/k_2)} x \right]^{1/2}, & 0 \leq x \leq L_1 \quad (4.25) \\ \left[h_e^2 + \dfrac{H^2 - h_e^2}{k_2(L_1/k_1 + L_2/k_2)} (L - x) \right]^{1/2}, & L_1 \leq x \leq L \quad (4.26) \end{cases}$$

4.2.4 Unconfined Flow with Infiltration

Figure 4.6 presents a general problem of two-dimensional flow with uniform infiltration i. We will obtain a solution to this problem assuming that the pervious soils of the unconfined aquifer are uniform and homogeneous with respect to permeability. The flow rate through a vertical section is given by

$$q = -kh\frac{dh}{dx} \tag{4.27}$$

However, q varies with respect to x because of the infiltration, i. Therefore,

$$\frac{dq}{dx} = i \tag{4.28}$$

or

$$q = ix + q_{at\ x=0} \tag{4.29}$$

Equating the two expressions for q, Eqs. (4.27) and (4.29), gives

$$-kh\frac{dh}{dx} = ix + q_{at\ x=0} \tag{4.30}$$

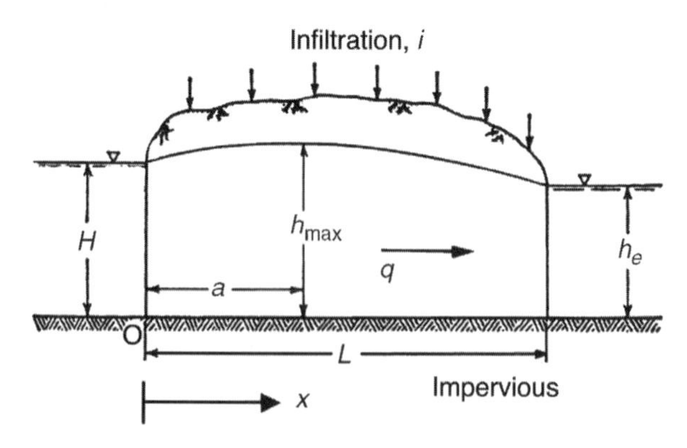

Fig. 4.6. Unconfined flow with infiltration.

Upon integration of Eq. (4.30) with the boundary conditions at $x = 0$ and $x = L$, we obtain

$$q_{at\ x=0} = \frac{k\left(H^2 - h_e^2\right)}{2L} - \frac{iL}{2} \tag{4.31}$$

Substitution of Eq. (4.31) in Eq. (4.29) enables us to express q through a vertical cross section in the domain as

$$q = \frac{k\left(H^2 - h_e^2\right)}{2L} - i\left(\frac{L}{2} - x\right) \tag{4.32}$$

Because of the asymmetry of the flow problem, the location where h is maximum is not at the center of the domain. At this location, called the *water divide*, $q = 0$; therefore, the distance a to the water divide can be obtained from Eq. (4.32) as

$$a = \frac{L}{2} - \frac{k}{i}\frac{H^2 - h_e^2}{2L} \tag{4.33}$$

Substituting Eq. (4.27) for the left-hand side of Eq. (4.32) and integrating, the variation of hydraulic head in the aquifer may be expressed as

$$h = \sqrt{H^2 - \frac{(H^2 - h_e^2)x}{L} + \frac{i}{k}(L - x)x} \tag{4.34}$$

The maximum head in the aquifer, h_{max}, is obtained by substituting a for x in Eq. (4.34).

4.2.5 Combined Artesian–Gravity Flow

When the downstream water level in the case of confined flow is kept very low, the piezometric surface may be lowered to a level below the top of the pervious stratum (Fig. 4.7). In this case, the artesian flow at the upstream end is combined with gravity

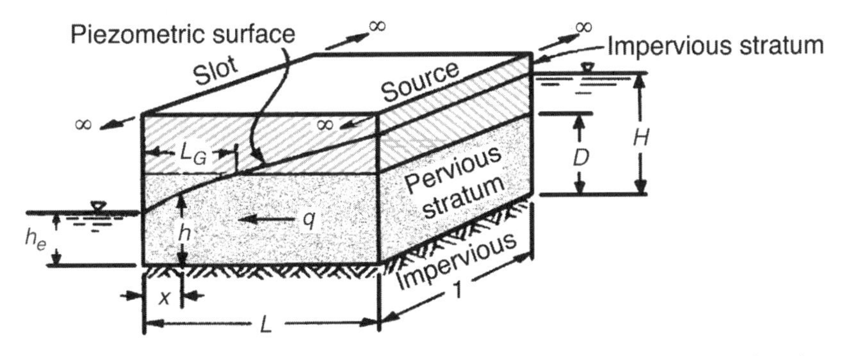

Fig. 4.7. Combined artesian–gravity flow from a line source to a fully penetrating slot.

flow at the downstream end. Let L_G denote the distance from the downstream slot to the point at which the flow changes from artesian to gravity. The discharge q_1 within the section $x = 0$ to L_G is given by Eq. (4.15) for unconfined flow. Similarly, the discharge q_2 within the section $x = L_G$ to L is given by Eq. (4.9) for confined flow. Thus,

$$q_1 = \frac{k}{2L_G}\left(D^2 - h_e^2\right) \tag{4.35}$$

and

$$q_2 = \frac{kD}{L - L_G}(H - D) \tag{4.36}$$

Since $q_1 = q_2$ at the section corresponding to $x = L_G$,

$$\frac{k}{2L_G}\left(D^2 - h_e^2\right) = \frac{kD}{L - L_G}(H - D) \tag{4.37}$$

or

$$L_G = \frac{L\left(D^2 - h_e^2\right)}{2DH - D^2 - h_e^2} \tag{4.38}$$

Substituting Eq. (4.38) for L_G in (4.35), discharge $q_1 = q_2 = q$ may be expressed as

$$q = k\frac{2DH - D^2 - h_e^2}{2L} \tag{4.39}$$

Using Eqs. (4.16) and (4.10), the hydraulic head in a combined artesian–gravity flow may be expressed as

$$h = \begin{cases} \sqrt{D^2 - \dfrac{L_G - x}{L_G}\left(D^2 - h_e^2\right)} & x \leq L_G \quad (4.40) \\[2ex] H - \dfrac{L - x}{L - L_G}(H - D) & x \geq L_G \quad (4.41) \end{cases}$$

To correct for vertical seepage face at the slot, h_e in Eq. (4.40) may be substituted by $h_e + h_s$, where h_s is given in Fig. 4.4. L and H in Fig. 4.4 must be substituted by L_G and D, respectively, to obtain h_s from Fig. 4.4.

Solutions are also available for a number of other special cases involving partly penetrating slots and multiple sources and slots. These are summarized in Figs. 4.8 to 4.10. Many of these solutions were obtained using empirical methods and experimental investigations. The solutions shown in Fig. 4.10, for instance, were based on experimental results from an electric analogy model for two-dimensional flow toward multiple trenches.

Artesian Flow

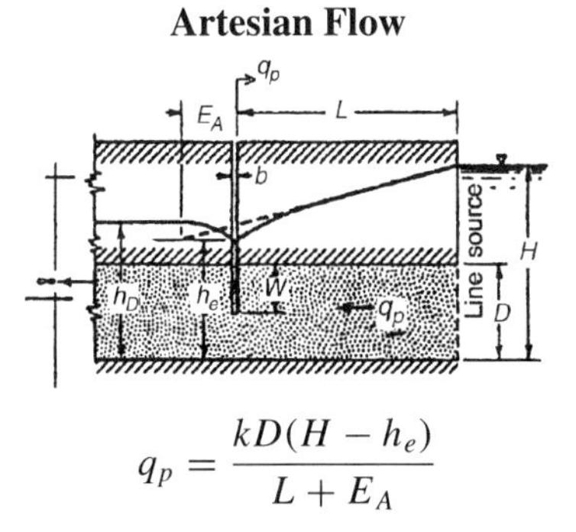

$$q_p = \frac{kD(H - h_e)}{L + E_A}$$

where E_A = extra-length factor, a function of *W/D* (see figure below)

$$h_D = \frac{E_A(H - h_e)}{L + E_A} + h_e$$

Gravity Flow

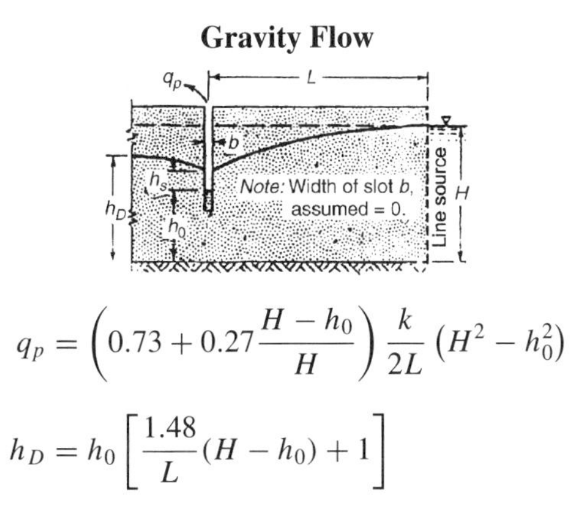

$$q_p = \left(0.73 + 0.27\frac{H - h_0}{H}\right)\frac{k}{2L}\left(H^2 - h_0^2\right)$$

$$h_D = h_0\left[\frac{1.48}{L}(H - h_0) + 1\right]$$

Note: Equations are applicable for $L/H \geq 3$.

Fig. 4.8. Solutions for flow toward partly penetrating slots. (Adapted from Mansur and Kaufman, 1962.)

Fully Penetrating Slots

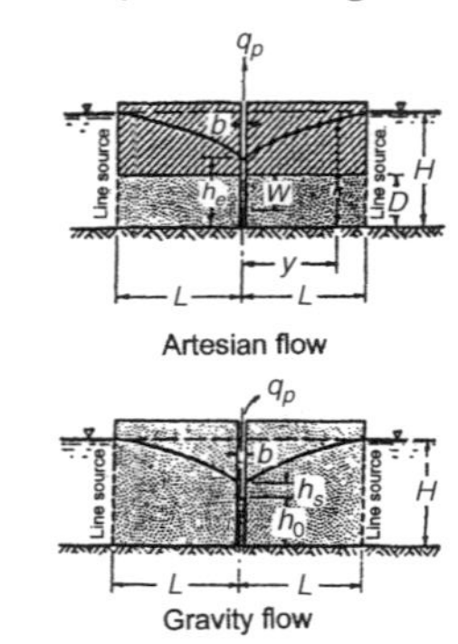

When slot is located midway between and parallel to line sources, q_p is twice the amount due to a single source.

Artesian flow:

$$q_p = \frac{2kD}{L}(H - h_e)$$

Gravity flow:

$$q_p = \frac{k}{L}(H^2 - h_0^2)$$

Artesian–gravity flow:

$$q_p = \frac{k\left(2DH - D^2 - h_0^2\right)}{L}$$

Artesian Flow
(partly penetrating)

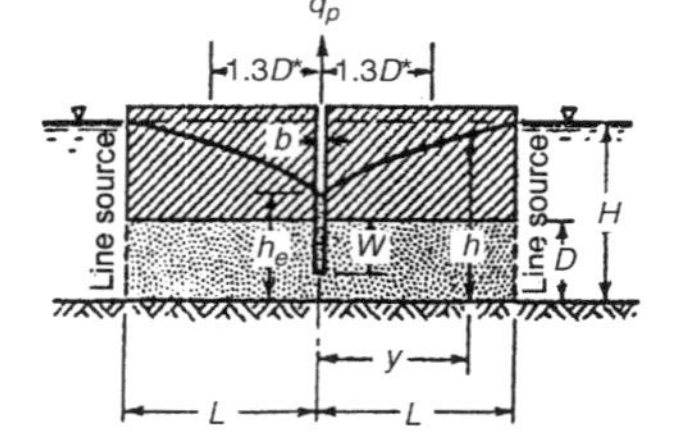

*Within distance (1.3*D*) the piezometric surface is nonlinear due to converging flow.

Note: Width of slot, *b* assumed = 0.

$$q_p = \frac{2kD(H - h_e)}{L + \lambda D}$$

λ = function of *W/D* (see figure below)

$$h = h_e + (H - h_e)\frac{y + \lambda D}{L + \lambda D} \quad (y > 1.3D)$$

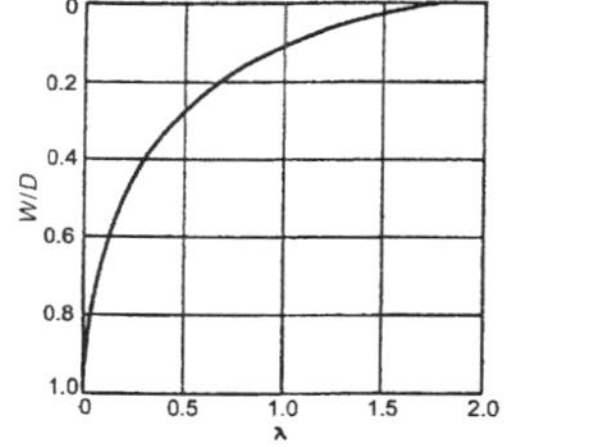

Gravity Flow
(partly penetrating)

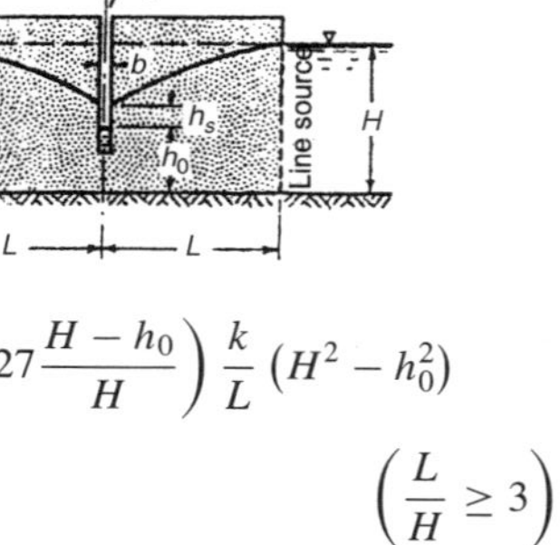

$$q_p = \left(0.73 + 0.27\frac{H - h_0}{H}\right)\frac{k}{L}(H^2 - h_0^2)$$

$$\left(\frac{L}{H} \geq 3\right)$$

Fig. 4.9. Solutions for flow toward fully and partly penetrating slots with two line sources. (Adapted from Mansur and Kaufman, 1962.)

Artesian Flow

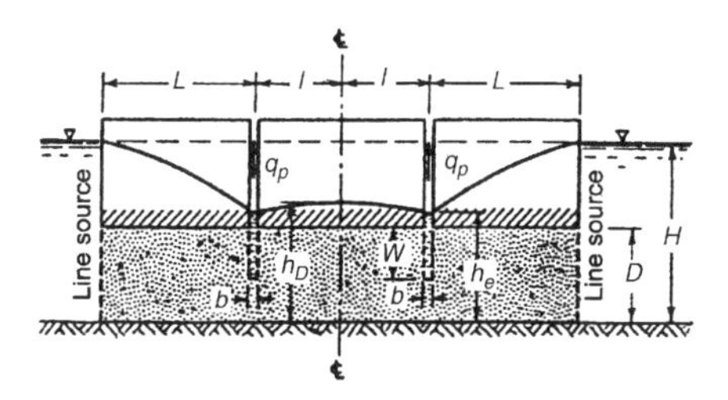

$$q_p = \frac{kD(H - h_e)}{L + E_A} \quad \text{(see Fig. 4.8 for } E_A\text{)}$$

$$h_D = \frac{E_A(H - h_e)}{L + E_A} + h_e \quad \text{(see Fig. 4.8 for } E_A\text{)}$$

Gravity Flow

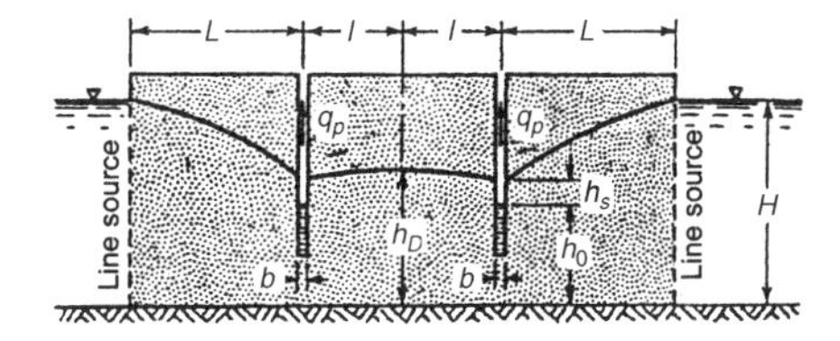

$$q_p = \left(0.73 + 0.27\frac{H - h_0}{H}\right)\frac{k}{2L}\left(H^2 - h_0^2\right); \left(\frac{L}{H} \geq 3\right)$$

$$h_D = h_0\left[\frac{C_1 C_2}{L}(H - h_0) + 1\right]$$

Use figures below to determine C_1 and C_2.

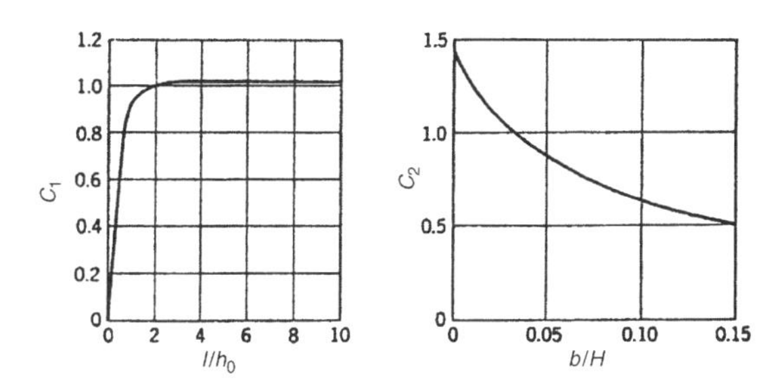

Fig. 4.10. Solutions for flow toward multiple slots. (Adapted from Mansur and Kaufman, 1962.)

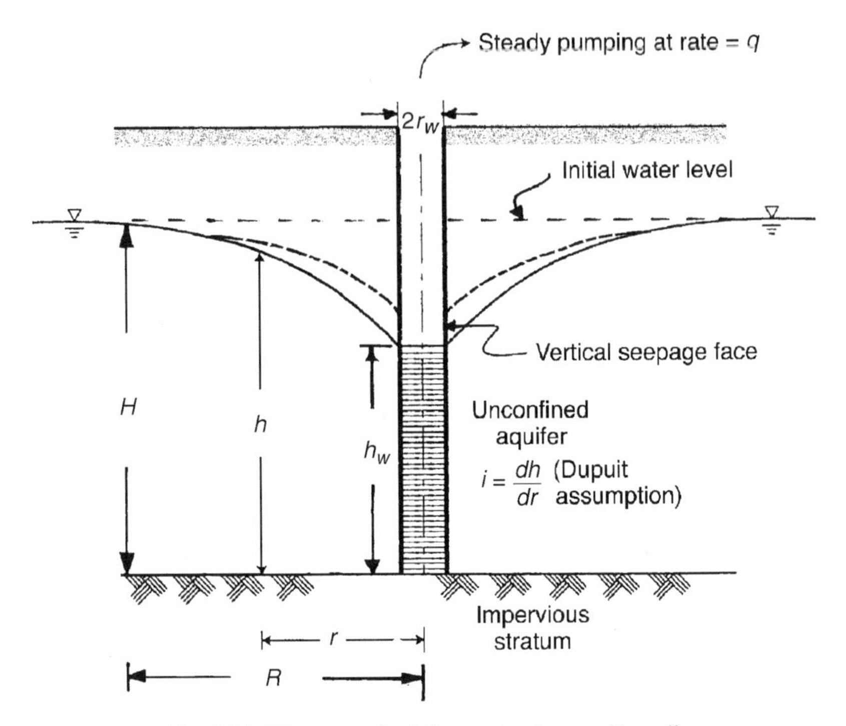

Fig. 4.11. Flow toward a fully penetrating gravity well.

4.3 DISCHARGE WELLS

Several engineering problems require solutions for groundwater levels when flow is occurring toward wells. The need commonly arises while dewatering an excavation for foundation construction purposes. As discussed in detail in Chapter 9, design of dewatering systems requires a thorough grasp of well hydraulics. In unconfined domains, there may be additional recharge due to infiltration of rainfall onto the phreatic surface, which must be accounted for in the solutions. The simplest problem is that of radial flow toward a well in an unconfined aquifer. The reader will note that the solution to this problem was used in Chapter 2 in indirect estimation of the hydraulic conductivity of soils (see Fig. 2.20). We revisit that problem to derive a simple equation commonly known as the *Dupuit–Forchheimer well discharge formula.*

Referring to Fig. 4.11, we use Dupuit's second assumption to express q through a cylindrical surface area at radius r as

$$q = k\frac{dh}{dr}(2\pi rh) \tag{4.42}$$

Integrating Eq. (4.42) between $h = h_w$ at $r = r_w$ and $h = H$ at $r = R$, we get

$$\frac{q}{2\pi k}\int_{r_w}^{R}\frac{dr}{r}=\int_{h_w}^{H}h\,dh \tag{4.43}$$

or

$$q=\frac{\pi k}{\ln(R/r_w)}\left(H^2-h_w^2\right) \tag{4.44}$$

Equation (4.44) is known as the *Dupuit–Forchheimer well discharge formula.* Here R is considered to be sufficiently large so that at $r = R$, the phreatic line is tangential to the original groundwater table. Hence R is known as the *radius of influence*. Several empirical expressions are available to determine R. Sichardt's empirical expression is commonly used:

$$R=C(H-h_w)\sqrt{k} \tag{4.45}$$

where R is the radius of influence (ft), C a dimensionless constant, ranging from 1.5 to 3; $H - h_w$ the drawdown at the well (ft), and k the coefficient of permeability (cm/s). This is described in more detail in Chapter 9 [see Eq. (9.27) and Fig. 9.44].

The upper boundary of integration can be changed to an arbitrary h at radius r to obtain

$$q=\frac{\pi k}{\ln(r/r_w)}\left(h^2-h_w^2\right) \tag{4.46}$$

Dividing Eq. (4.46) by (4.44), we get an equation that describes the shape of the phreatic line independent of discharge q:

$$h^2=H^2-\left(H^2-h_w^2\right)\frac{\ln(R/r)}{\ln(R/r_w)} \tag{4.47}$$

The well discharge formula suffers from the limitation described in Section 4.2: It does not consider the development of the seepage face at the well. The h predicted from Eq. (4.47) is generally considered to be valid for $r > H$. At small r, the discrepancies due to the development of the seepage face become significant. A number of investigators suggested corrections to this expression to account for the seepage face (Babbitt and Caldwell, 1948; Hansen, 1949; Boulton, 1951). Babbitt and Caldwell (1948) recommended the following expression to determine h when $r < H$:

$$h=H-\frac{C_x}{H}\frac{H^2-h_w^2}{\ln(R/r_w)}\ln\frac{R}{0.1H} \tag{4.48}$$

where C_x is a correction factor given as a function of r/R (Fig. 4.12).

Consider now the well shown in Fig. 4.13, fully penetrating a confined aquifer and pumping at a constant rate, q. A piezometric surface can be identified to represent

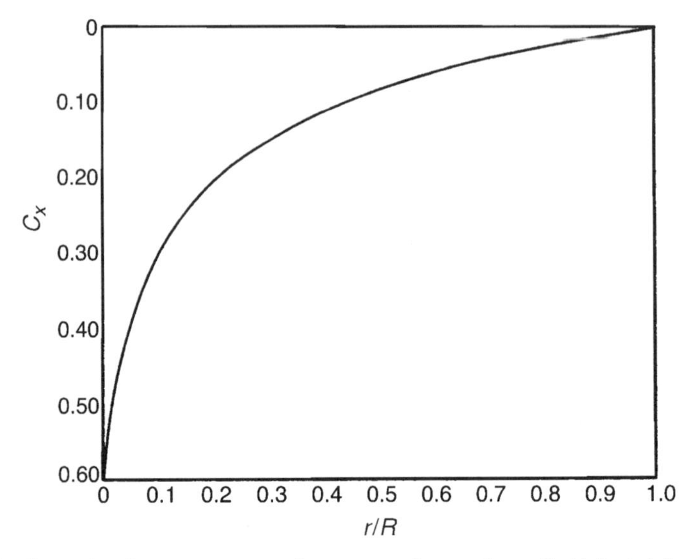

Fig. 4.12. Correction factor to account for seepage face at the well. (Adapted from Babbitt and Caldwell, 1948.)

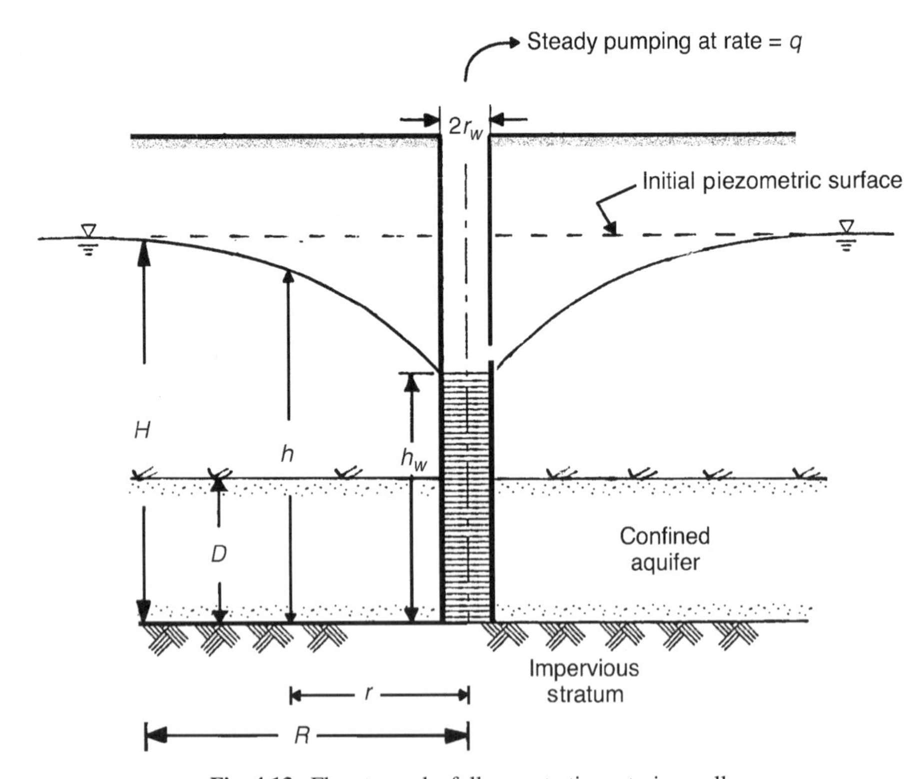

Fig. 4.13. Flow toward a fully penetrating artesian well.

the hydraulic heads in the confined aquifer, which shows a depression in the vicinity of the well and remains constant at distances far away from the well. We express q through a cylindrical surface area at radius r as

$$q = k\frac{dh}{dr}2\pi r D \tag{4.49}$$

Integrating Eq. (4.49) between $h = h_w$ at $r = r_w$ and $h = H$ at $r = R$, we get

$$\frac{q}{2\pi k}\int_{r_w}^{R}\frac{dr}{r} = D\int_{h_w}^{H} dh \tag{4.50}$$

or

$$q = \frac{2\pi k D}{\ln(R/r_w)}(H - h_w) \tag{4.51}$$

As in the case of wells in unconfined aquifers, the upper boundary of integration can be changed to an arbitrary h at radius r to obtain

$$q = \frac{2\pi k D}{\ln(r/r_w)}(h - h_w) \tag{4.52}$$

Dividing Eq. (4.52) by (4.51), we get an equation that describes the shape of the piezometric surface independent of q:

$$h = h_w + (H - h_w)\frac{\ln(r/r_w)}{\ln(R/r_w)} \tag{4.53}$$

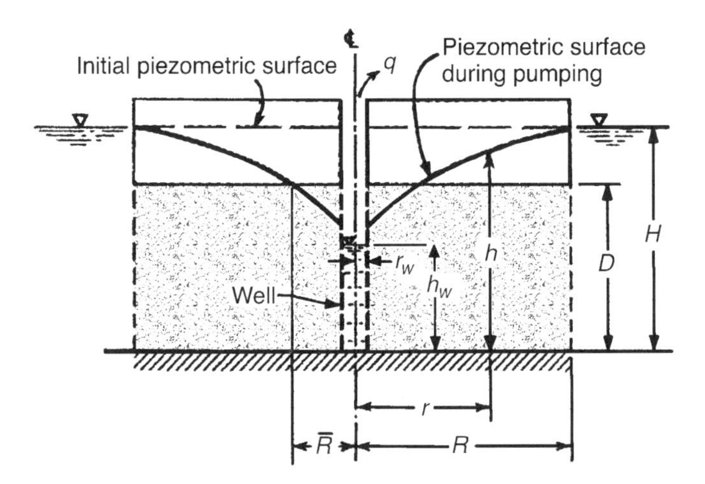

Fig. 4.14. Flow toward a fully penetrating artesian–gravity well.

Sometimes high pumping rates from wells in confined aquifers may lower the piezometric surface to a level below the top of the aquifers (Fig. 4.14). In this case, flow is unconfined in the vicinity of the well up to $\overline{R}$, and confined beyond $\overline{R}$. This is referred to as a combined artesian–gravity well. The discharge from such a well and the variation of hydraulic head at any distance r from the well can be obtained using the following equations developed by Muskat (1937):

$$q = \frac{\pi k \left(2DH - D^2 - h_w^2\right)}{\ln(R/r_w)} \tag{4.54}$$

and

$$h = \frac{H - D}{\ln(R/r_w)} \ln \frac{r}{r_w} + \sqrt{D^2 - \frac{D^2 - h_w^2}{\ln(R/r_w)} \ln \frac{R}{r}} \tag{4.55}$$

The distance $\overline{R}$ where flow changes from artesian to gravity is obtained using

$$\ln \overline{R} = \frac{(D^2 - h_w^2) \ln R + 2D(H - D) \ln r_w}{2DH - D^2 - h_w^2} \tag{4.56}$$

For wells partially penetrating confined and unconfined aquifers, empirical expressions are available for evaluating discharge and hydraulic head, as summarized in Fig. 4.15. Equations for multiple wells are discussed in Chapter 9 in the context of dewatering.

Artesian Well

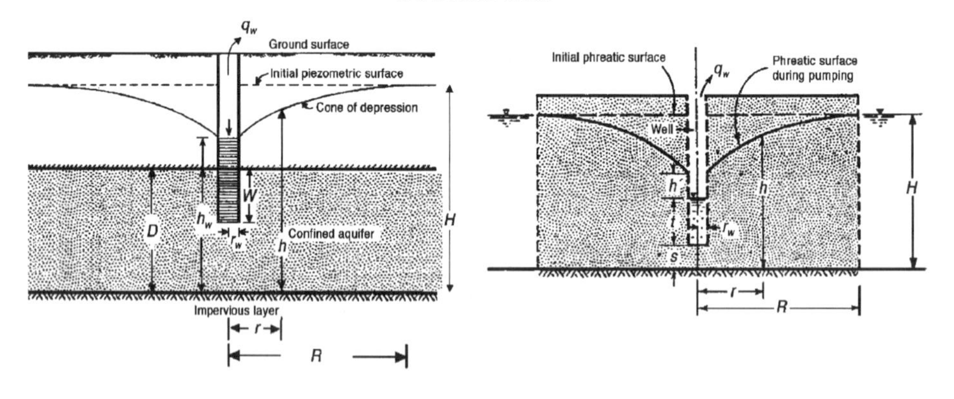

$$q_w = \frac{2\pi k D(H - h_w)G}{\ln(R/r_w)}$$

Fig. 4.15. Solutions for flow toward partly penetrating wells. (Adapted from Mansur and Kaufman, 1962.) *(continued)*

Artesian Well (*continued*)

$$G = \frac{\text{flow from partly penetrating well}}{\text{flow from fully penetrating well for the same drawdown}}$$

$$G = \frac{W}{D}\left(1 + 7\sqrt{\frac{r_w}{2W}}\cos\frac{\pi W/D}{2}\right) \qquad \text{(Kozeny)}$$

$$= \frac{\ln(R/r_w)}{\dfrac{D}{2W}\left[2\ln\dfrac{4D}{r_w} - \ln\dfrac{\Gamma(0.875(W/D))\Gamma(0.125(W/D))}{\Gamma(1-0.875(W/D))\Gamma(1-0.125(W/D))}\right] - \ln\dfrac{4D}{R}} \qquad \text{(Muskat)}$$

where Γ is the gamma function. The figure below shows G for a typical large diameter well ($r_w = 1$ ft) with a radius of influence of 1000 ft.

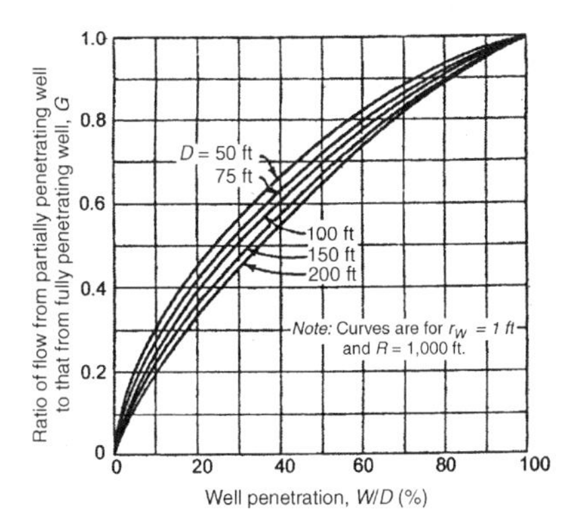

Drawdown for $r > D$, is approximately the same as that of a fully penetrating well (for the same well drawdown, $H - h_w$).

Gravity Well

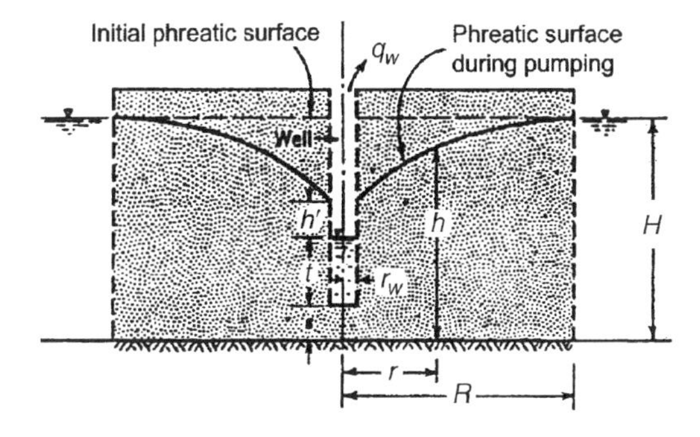

Fig. 4.15. (*Continued*)

Gravity Well *(continued)*

$$q_w = \frac{\pi k\left[(H-s)^2 - t^2\right]}{\ln(R/r_w)}\left[1 + \left(0.30 + \frac{10r_w}{H}\right)\sin\frac{1.8s}{H}\right]$$

For $r/h > 1.5$,

$$H^2 - h^2 = \frac{q_w}{\pi k}\ln\frac{R}{r}$$

For $r/h < 1.5$,

$$H - h = \frac{q_w p \ln(10R/H)}{\pi k H[1 - 0.8(s/H)^{1.5}]}$$

$$p = \begin{cases} 0.13\ln\dfrac{R}{r} & \text{for } 0.3 < \dfrac{r}{h} < 1.5 \\ \overline{C_x} + \Delta C & \text{for } \dfrac{r}{h} < 0.3 \end{cases}$$

$$\overline{C_x} = 0.13\ln\frac{R}{r} - 0.0123\ln^2\frac{R}{10r}$$

$$\Delta C = \frac{s}{h}\left[\left(\frac{1}{2.3}\ln\frac{R}{10r}\right)\left(1.2\frac{s}{H} - 0.48\right) + 0.113\ln\frac{2.4H}{R}\ln\frac{R}{34r}\right]$$

Fig. 4.15. *(Continued)*

4.4 SUBSURFACE DRAINS

Several analytical solutions are available to determine spacing of subsurface drains for specified rainfall/recharge rate and groundwater table conditions. These solutions are of practical interest in seepage control, dewatering, and irrigation operations. Dupuit–Forchheimer assumptions again allow us to simplify solutions in a number of these cases.

We start with the exact solution for spacing of drain pipes (Fig. 4.16) developed by Kirkham (1958). The pervious stratum in which the drains are located is assumed to be isotropic and homogeneous. The rainfall/recharge rate is assumed to be uniform and constant in time. Kirkham combined Darcy's laws and equation of continuity and solved the resulting partial differential equations with stream and potential functions. A steady-state drain spacing equation based on this theory was expressed as

$$h = \frac{Si}{k}\frac{1}{\pi}\frac{1}{(1-i/k)}\left[\ln\frac{\sin(\pi x/S)}{\sin(\pi r_d/S)} + \sum_{m=1}^{\infty}\frac{1}{m}\left(\cos\frac{2m\pi r_d}{S} - \cos\frac{2m\pi x}{S}\right)\left(\coth\frac{2m\pi d}{S} - 1\right)\right] \tag{4.57}$$

The maximum height h_{max} of water table above the drains corresponds to $x = S/2$. Thus,

$$h_{max} = \frac{Si}{k}\frac{1}{\pi}\frac{1}{(1-i/k)}\left[\ln\frac{S}{\pi r_d} + \sum_{m=1}^{\infty}\frac{1}{m}\left(\cos\frac{2m\pi r_d}{S} - \cos m\pi\right)\left(\coth\frac{2m\pi d}{S} - 1\right)\right] \tag{4.58}$$

A nomograph representing Eq. (4.58) is shown in Fig. 4.17 for use in practical applications.

Dupuit–Forchheimer assumptions allow us to obtain simpler solutions than the above with errors less than 5% when d and h_{max} are very small compared to S. Hooghoudt of Netherlands solved this problem in 1940 invoking Dupuit–Forchheimer assumptions. To present Hooghoudt's solution, we change the configuration of drains to parallel ditches (Fig. 4.18) and assume that the water table is in equilibrium with the uniform infiltration, i. Using Dupuit–Forchheimer assumptions, we assume a

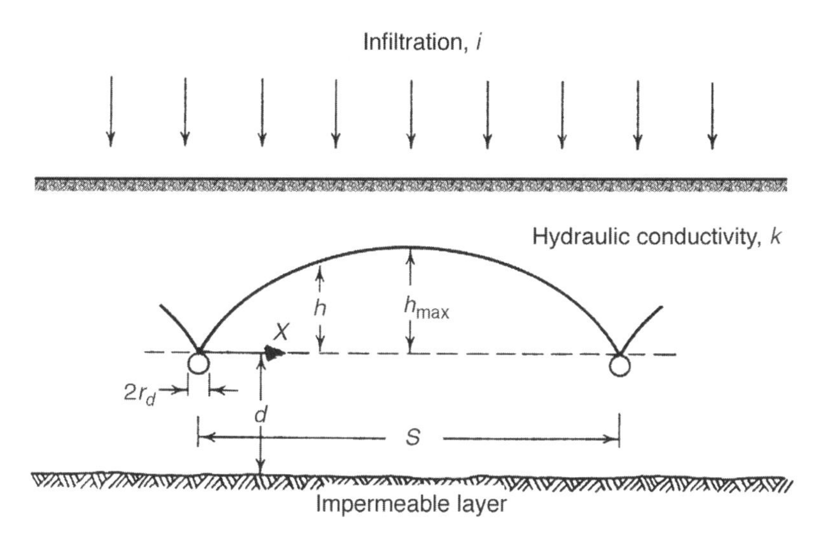

Fig. 4.16. Flow toward subsurface drains with steady infiltration.

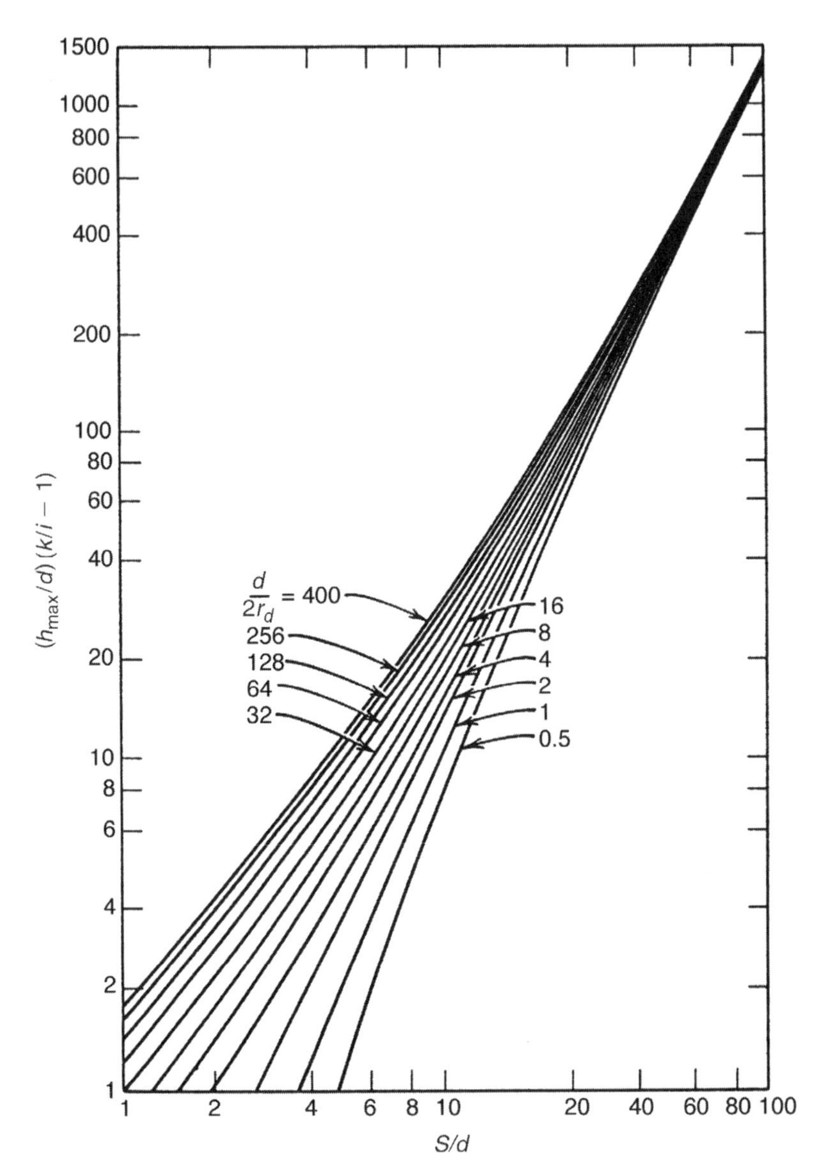

Fig. 4.17. Kirkham's potential theory solution for flow toward subsurface drains. (Adapted from Luthin, 1966.)

horizontal flow toward the drains. The radial flow that would occur in the presence of drains is ignored. Across a vertical plane located at x, the discharge is expressed using Darcy's law:

$$q = kh\frac{dh}{dx} \tag{4.59}$$

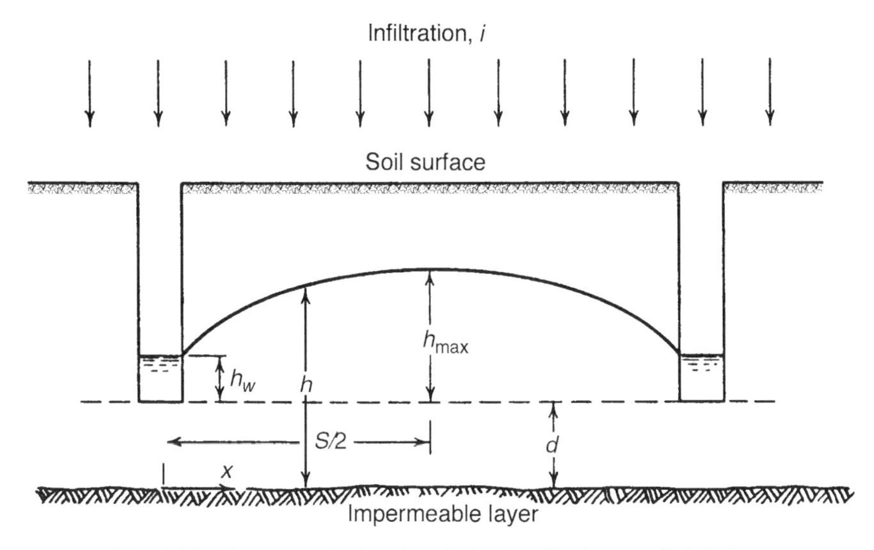

Fig. 4.18. Flow toward subsurface drains idealized as parallel ditches.

A second expression for q may be given in terms of infiltration i as

$$q = \left(\frac{S}{2} - x \right) i \tag{4.60}$$

Equating the discharge in Eqs. (4.59) and (4.60) gives us

$$\left(\frac{S}{2} - x \right) i = kh \frac{dh}{dx} \tag{4.61}$$

Integrating Eq. (4.61) from $x = 0$, $h = h_w + d$ to $x = S/2$, $h = h_{\max} + d$ yields

$$S = \left[\frac{4k}{i} \left(h_{\max}^2 - h_w^2 + 2dh_{\max} - 2dh_w \right) \right]^{1/2} \tag{4.62}$$

which is known as *Hooghoudt's equation for subsurface drains.* Two special cases of this equation are when the drain is empty ($h_w = 0$) and when the drains are fully penetrating up to the impermeable layer ($d = 0$). For these cases,

$$S = \begin{cases} \left[\dfrac{4kh_{\max}}{i} \left(2d + h_{\max} \right) \right]^{1/2} & h_w = 0 \qquad (4.63) \\ \left[\dfrac{4k}{i} \left(h_{\max}^2 - h_w^2 \right) \right]^{1/2} & d = 0 \qquad (4.64) \end{cases}$$

The shape of the water table is obtained by integrating Eq. (4.61) from the origin (0,0) to (x, h):

$$\frac{iSx}{2} - \frac{ix^2}{2} = \frac{kh^2}{2} \tag{4.65}$$

which reduces to the familiar form of the mathematical equation for ellipse upon transforming the origin to the midpoint between drains,

$$\frac{h^2}{S^2 i/4k} + \frac{x^2}{S^2/4} = 1 \tag{4.66}$$

where $S/2$ and $S/2\sqrt{i/k}$ represent semimajor and semiminor axes of the ellipse, respectively.

For a stratified soil with different permeabilities for the layers above and below the drains, k_1 and k_2, respectively, Hooghoudt's expression (4.63) for empty drains is modified to

$$S = \left(\frac{4k_1 h_{\max}^2}{i} + \frac{8k_2 d h_{\max}}{i}\right)^{1/2} \tag{4.67}$$

Hooghoudt's solution differs from the exact solutions of potential theory primarily because horizontal flow is assumed for the actual radial flow to drains. The discrepancy increases as d increases. To correct for this discrepancy, the two solutions were compared for various values of d, and charts for equivalent depths, d', were developed. The radius of the drains, which was not needed in the development of Hooghoudt's solution, was used to develop the charts shown in Figs. 4.19 and 4.20. For accuracy, the equivalent depths d' from the y axes of these figures should be substituted in Hooghoudt's solution expressed above. Other nomographs for drain-spacing solutions include those given by Visser (1954), shown in Figs. 4.21 and 4.22, and those by Toksoz and Kirkham (1971) for stratified soils, shown in Figs. 4.23 and 4.24.

When drains are placed over a sloping impermeable layer (usually for the purpose of lowering the groundwater table to stabilize slopes), the solutions given by Wooding and Chapman (1966) will be useful for design purposes. These solutions were obtained assuming that the flow lines are parallel to the sloping impermeable layer, referred to as the *extended form* of the Dupuit–Forchheimer assumption. Wooding and Chapman compared the approximate solutions obtained using this assumption with the exact solutions given by the hodograph method (Chapter 5). They observed that the values calculated for the slope of the phreatic surface at the highest point of the flow region and far downstream were in agreement with the exact solution. The drains were assumed to be placed either on the impermeable base (Fig. 4.25*a*) or at a height G above the base (Fig. 4.25*b*). The solutions obtained using the extended form of the Dupuit–Forchheimer assumption, are presented in Figs. 4.26 and 4.27. These solutions can be used to determine the required depth and spacing of drains. The parameter λ needed in Figs. 4.26 and 4.27 was defined as

$$\lambda = \frac{4i}{k(1 - i/k)^2 \tan^2 \alpha} \tag{4.68}$$

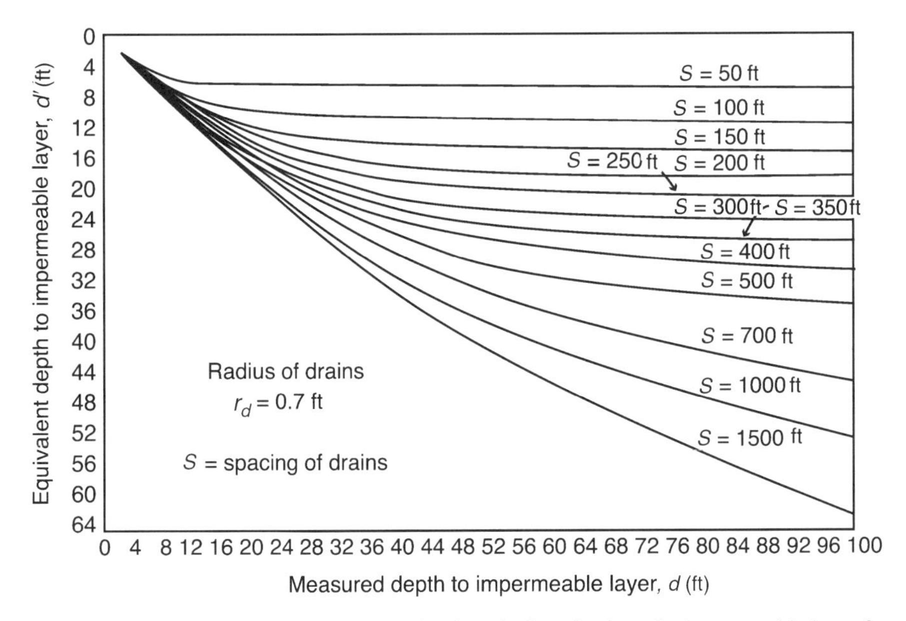

Fig. 4.19. Relationship between measured and equivalent depth to the impermeable layer for determining spacing between drains, $r = 0.7$ ft. (Adapted from Luthin, 1966.)

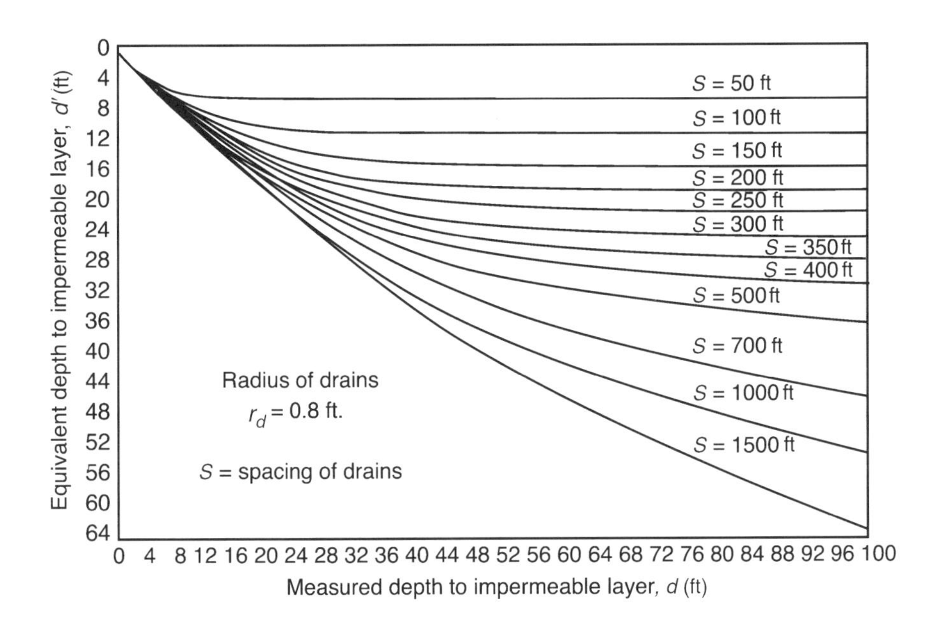

Fig. 4.20. Relationship between measured and equivalent depth to the impermeable layer for determining spacing between drains, $r = 0.8$ ft. (Adapted from Luthin, 1966.)

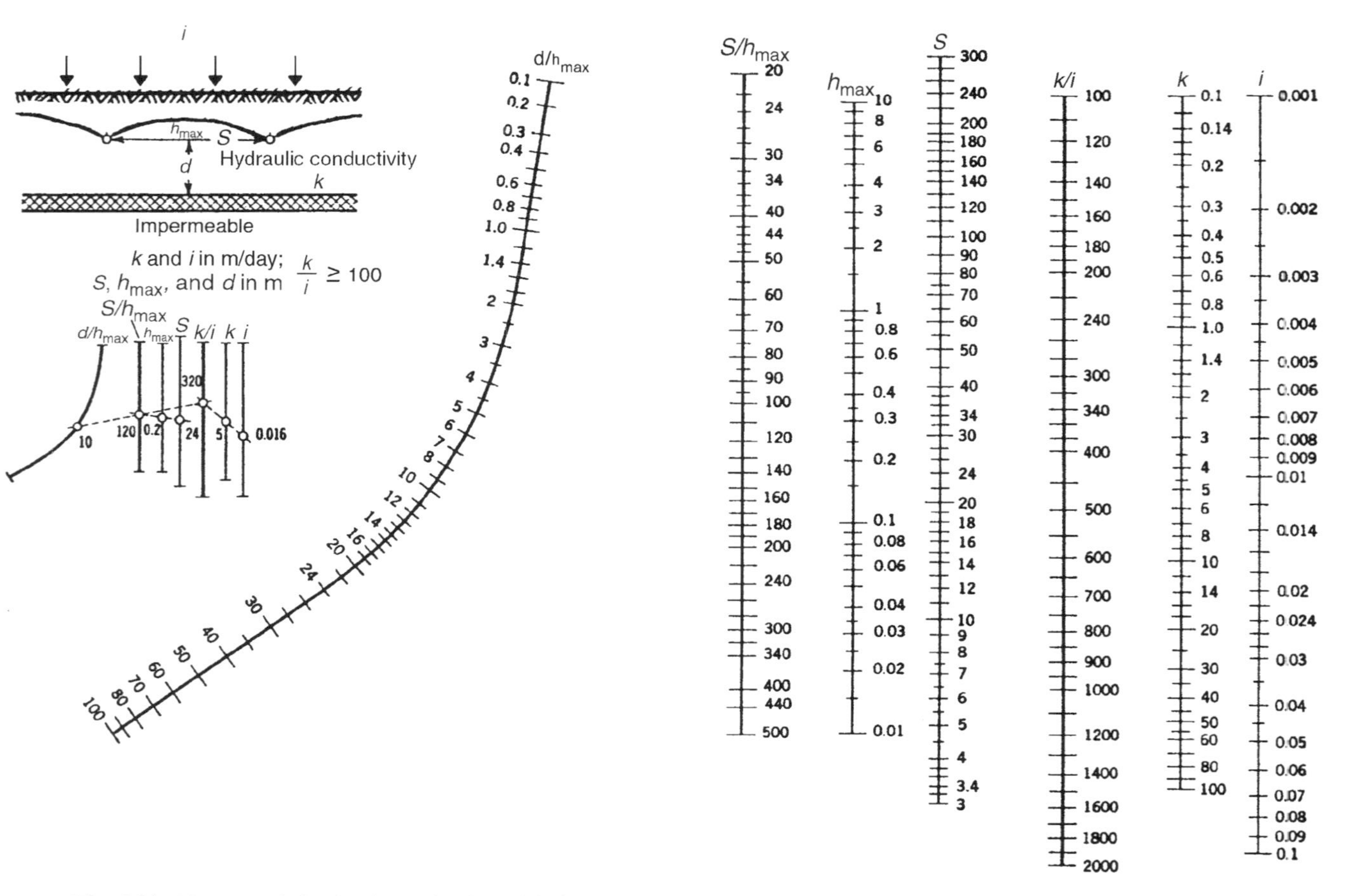

Fig. 4.21. Nomograph for the determination of drain spacings with $k/i \geq 100$. (Adapted from Visser, 1954; Luthin, 1966.)

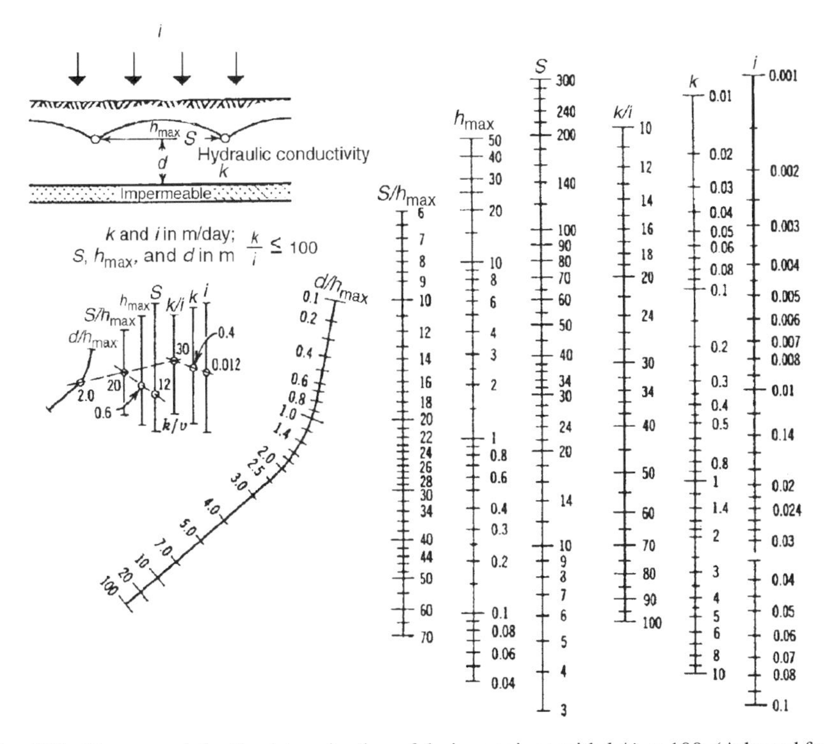

Fig. 4.22. Nomograph for the determination of drain spacings with $k/i \leq 100$. (Adapted from Visser, 1954; Luthin, 1966.)

Example 4.2 Compute the spacing required for drains if $h_{\max}$ were to be limited to 1 m given the following parameters: $i = 0.2$ cm/day, $r_d = 0.05$ m, $d = 1$ m, and $k = 2$ m/day. Use (a) Kirkham's potential theory solution, and (b) Hooghoudt's approximate solution.

SOLUTION: (a) *Kirkham's potential theory:*

$$\frac{d}{2r_d} = \frac{1 \text{ m}}{2 \times 0.05 \text{ m}} = 10$$

$$\frac{h_{\max}}{d}\left(\frac{k}{i} - 1\right) = \frac{1 \text{ m}}{1 \text{ m}}\left(\frac{2 \text{ m/day}}{0.2 \text{ cm/day}} \cdot \frac{100 \text{ cm}}{1 \text{ m}} - 1\right) = 999$$

so from Fig. 4.17, one finds that

$$\frac{S}{d} = 90$$

which indicates that the spacing required is $S = 90d = 90 \times 1 \text{ m} = 90$ m.

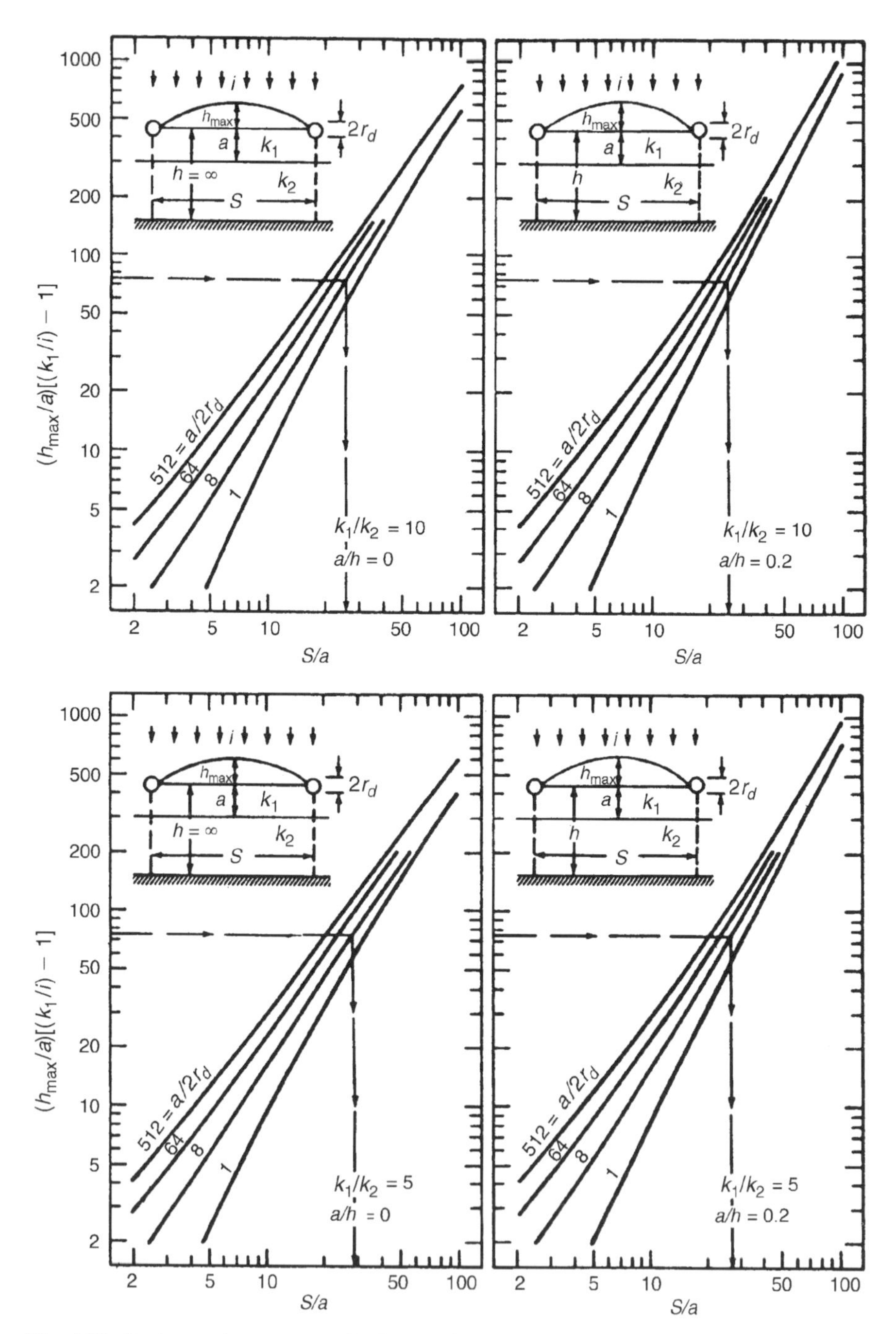

Fig. 4.23. Drain spacing nomographs for two layered soils with $k_1/k_2 > 1$. (Adapted from Toksoz and Kirkham, 1971.)

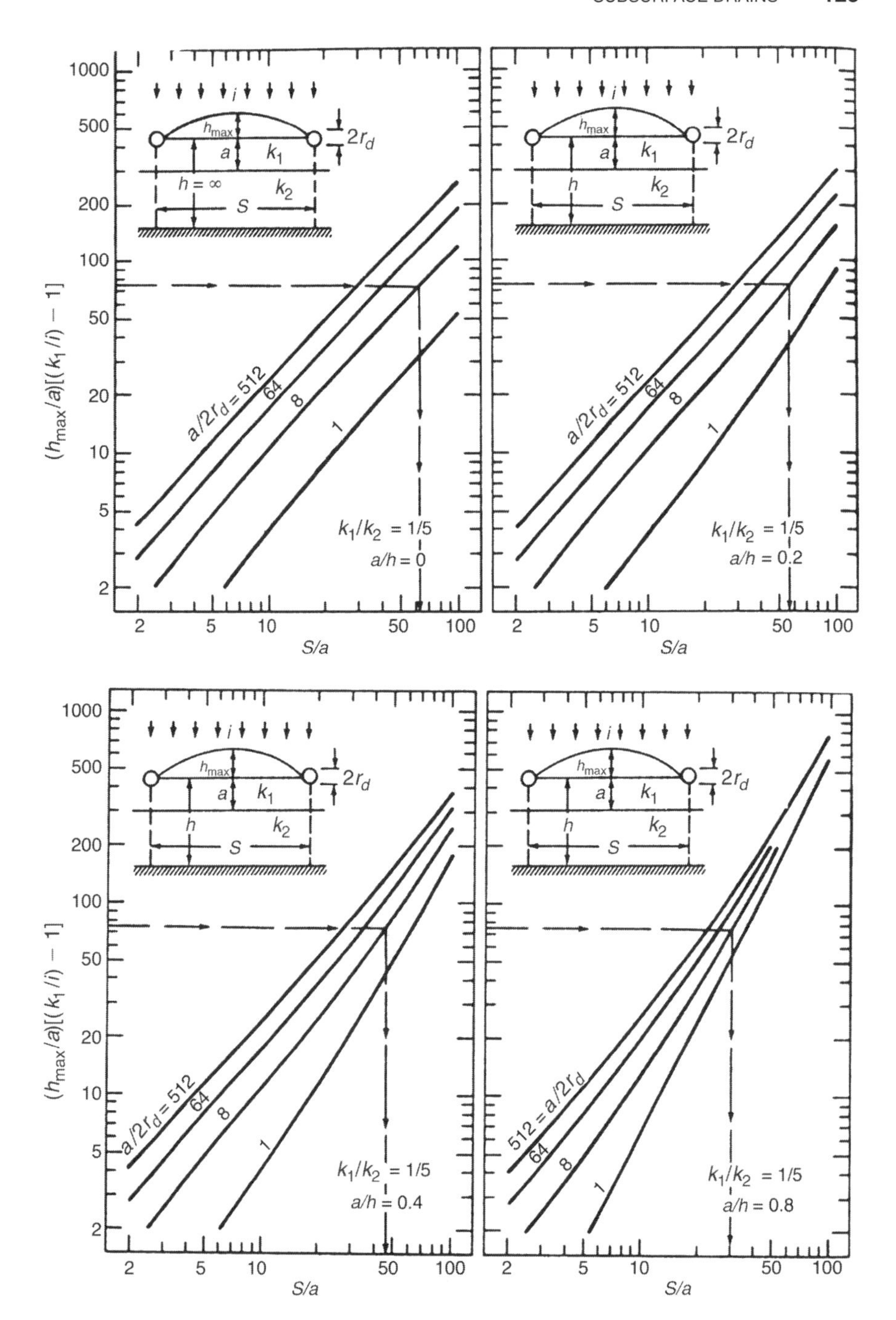

Fig. 4.24. Drain spacing nomographs for two layered soils with $k_1/k_2 < 1$. (Adapted from Toksoz and Kirkham, 1971.)

(b) *Hooghoudt's approximate solution:* From Eq. (4.63),

$$S = \left[\frac{4kh_{\max}}{i}(2d + h_{\max})\right]^{1/2}$$

$$= \left[\frac{4 \times 2 \text{ m/day}}{0.2 \text{ cm/day}} \cdot \frac{100 \text{ cm}}{1 \text{ m}} \cdot 1 \text{ m}(2 \times 1 \text{ m} + 1 \text{ m})\right]^{1/2}$$

$$\approx 109.5 \text{ m}$$

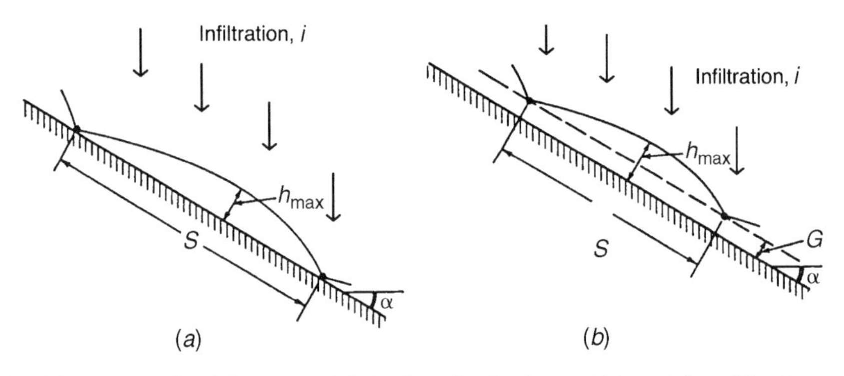

Fig. 4.25. Schematic of flow toward drains installed in slopes. (Adapted from Wooding and Chapman, 1966.)

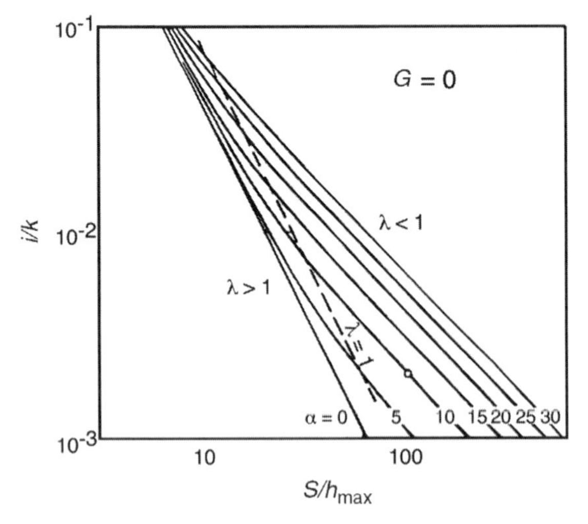

Fig. 4.26. Solutions for spacing of drains located on the impervious surface of slopes. (Adapted from Wooding and Chapman, 1966.)

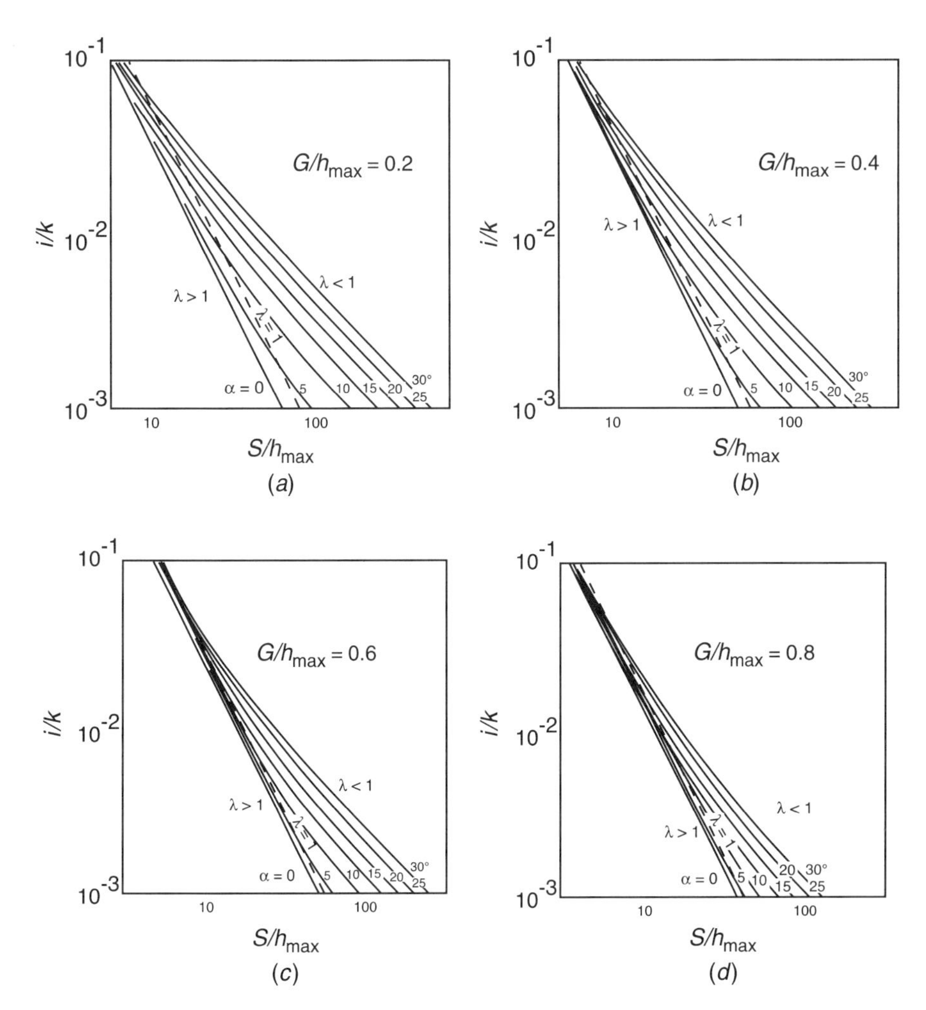

Fig. 4.27. Solutions for spacing of drains located above the impervious surface of slopes. (Adapted from Wooding and Chapman, 1966.)

4.5 SEEPAGE THROUGH EARTH DAMS

Flow through earth dams forms one of the special types of unconfined problems. Several investigators attempted to solve these problems mathematically. In this section we follow closely Casagrande's (1937) presentation of the various solution techniques. The Dupuit equation for steady unconfined flow [Eq. (4.15)] offers the simplest solution to calculate seepage quantities from the earth dam shown in Fig. 4.28. The equation represents a parabolic phreatic line. In the derivation of this equation, however, the reader will note that the boundary conditions at the entrance and exit ends of the line of seepage were not taken into account, as discussed in Section 4.2.

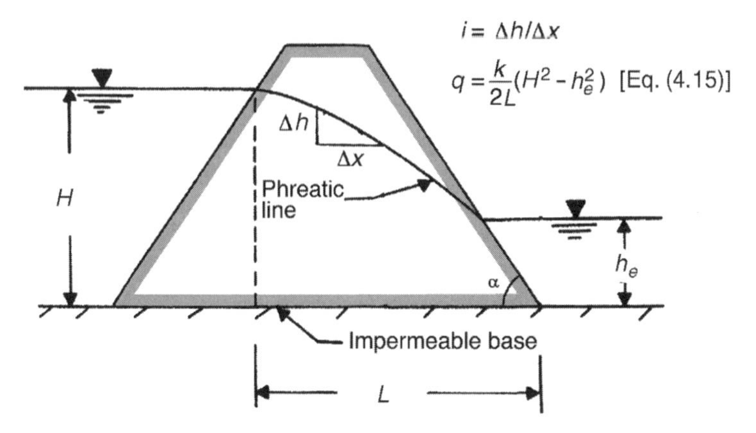

Fig. 4.28. Dupuit's solution for flow through earth dams.

For instance, when no tailwater is present (i.e. $h_e = 0$), Dupuit's parabola intersects the impervious base of the dam, which is inconsistent with the discharge condition that the line of seepage must be tangential to the discharge face (see Fig. 3.8). It is possible that Dupuit's parabola represents the solution with sufficient accuracy for small angles of α. As α increases, however, the errors due to the boundary conditions become significant. The solutions presented below will extend Dupuit's equation with due consideration to the boundary conditions.

4.5.1 Solution by Schaffernak and Iterson (Valid for $\alpha < 30°$)

Schaffernak (1917) and Iterson (1919) provide one of the earliest solutions accounting for the exit condition of the line of seepage. The line of seepage was considered to intersect the discharge face at a downstream slope distance equal to a (Fig. 4.29*a*). Dupuit's second assumption was still used (i.e., the hydraulic gradient is equal to the slope of the line of seepage). Therefore, the solution will be valid for relatively flat phreatic surfaces. Assuming no recharge from the top of the dam, the flow rate q through any vertical section of the dam can be expressed as equal to flow occurring through the vertical face *CD*. Thus,

$$q = kh\frac{dh}{dx} = k(a \sin \alpha)(\tan \alpha) \tag{4.69}$$

where $a \sin \alpha$ corresponds to $\overline{CD}$. With the boundary conditions $h = H$ for $x = L$ and $h = a \sin \alpha$ for $x = a \cos \alpha$, Eq. (4.69) could be integrated as

$$\int_{a \sin \alpha}^{H} h\, dh = \int_{a \cos \alpha}^{L} (a \sin \alpha)(\tan \alpha)\, dx \tag{4.70}$$

which leads to the following expression for the discharge face distance a on the downstream slope:

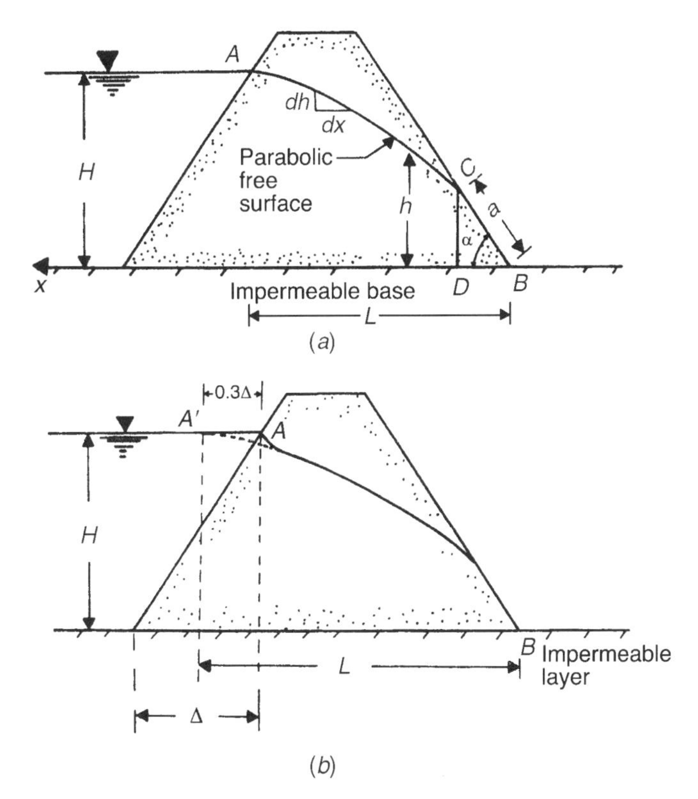

Fig. 4.29. Improved solution for flow through earth dams: (*a*) Shaffernak and Iterson's solution for $\alpha < 30°$; (*b*) Casagrande's correction for the origin of line of seepage.

$$a = \frac{L}{\cos\alpha} - \sqrt{\frac{L^2}{\cos^2\alpha} - \frac{H^2}{\sin^2\alpha}} \tag{4.71}$$

With a from Eq. (4.71), seepage through the dam could be evaluated using Eq. (4.69). Although the solution takes the exit condition of the line of seepage into account, the boundary condition at the entrance requires additional correction. Casagrande suggests that the start of the theoretical line of seepage should be moved to a point A' (Fig. 4.29*b*) chosen such that $AA' = 0.3\Delta$, where Δ is the horizontal projection of the upstream slope.

This solution becomes inaccurate at angles greater than $\alpha = 30°$. The deviations between Dupuit's gradient ($\tan\alpha$) used in Eq. (4.69) and the true gradient ($\sin\alpha$) for the line of seepage at the discharge point become significant for $\alpha > 30°$.

4.5.2 Casagrande's Solution for $\alpha > 30°$

Casagrande presented more accurate solutions when $\alpha > 30°$, considering the gradient along the line of seepage as equal to $\sin\alpha$. He expressed Darcy's law as

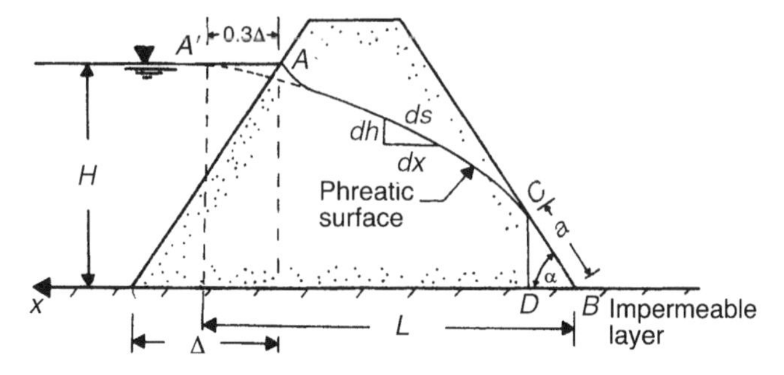

Fig. 4.30. Casagrande's solution for flow through earth dams when $\alpha > 30°$.

$$q = kh\frac{dh}{ds} = ka\sin^2\alpha \tag{4.72}$$

where s is measured along the line of seepage (Fig. 4.30). Two boundaries are now considered, one at $s = a$ and the other at $s = s_0$, where s_0 is the length of the curve $A'CB$. The vertical coordinates at the two boundaries are $h = a\sin\alpha$ and $h = H$, respectively. Integrating Eq. (4.72) between these two boundaries, we obtain

$$\int_{a\sin\alpha}^{H} h\,dh = \int_{a}^{s_0} (a\sin^2\alpha)\,ds \tag{4.73}$$

which is solved to express the length of the seepage face a as

$$a = s_0 - \sqrt{s_0^2 - \frac{H^2}{\sin^2\alpha}} \tag{4.74}$$

The discharge q through the dam can now be obtained substituting a from Eq. (4.74) in Eq. (4.72).

The length of the line of seepage, s_0, is needed to use this solution. Casagrande showed that s_0 can be approximated by the straight distance, $\overline{A'B} = \sqrt{H^2 + L^2}$ for $\alpha \leq 60°$. He suggests that if deviations up to 25% are permitted, the straight distance may be used for slopes up to 90°.

Casagrande also obtained approximate solutions for cases of overhanging discharge surfaces ($90° < \alpha < 180°$). He based these solutions on Kozeny's solution for $\alpha = 180°$. Hence we present Kozeny's solution first before discussing the solutions for intermediate α.

4.5.3 Kozeny's Solution for $\alpha = 180°$

$\alpha = 180°$ applies to the case of a horizontal discharge face such as that of a drainage blanket commonly provided in the downstream section of earth dams. Kozeny (1931) published a rigorous solution for flow over a horizontal impervious surface leading to

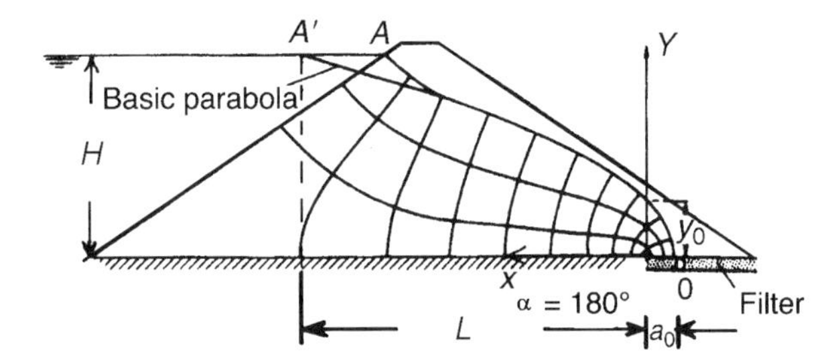

Fig. 4.31. Kozeny's solution for flow through earth dams when α = 180°.

a horizontal discharge face (Fig. 4.31). His solution was based on the theory of conformal transformation, which involves transforming the complicated flow domains into domains wherein flow lines and equipotential lines can easily be identified. As discussed in detail in Chapter 5, his theoretical solution suggests that the flow lines and equipotential lines shown in Fig. 4.31 could be represented using two families of confocal parabolas. The equation for the line of seepage is that of a parabola, expressed as

$$x = \frac{y^2 - y_0^2}{2y_0} \tag{4.75}$$

where x and y are the coordinates defined with the focus O (see Fig. 4.31) as the origin. The y_0 value can be obtained using the coordinates of one of the known points on the line of seepage in Eq. (4.75). Considering $y = H$ at $x = L$, y_0 can be expressed as

$$y_0 = \sqrt{L^2 + H^2} - L \tag{4.76}$$

From the basic properties of the parabola, the focal distance a_0 can be calculated using

$$a_0 = \frac{y_0}{2} \tag{4.77}$$

Seepage rate q can be obtained using the simple expression

$$q = ky_0 \tag{4.78}$$

wherein Eq. (4.76) may be substituted for y_0.

4.5.4 Casagrande's Solutions for 90° < α < 180°

For discharge surfaces corresponding to 90° < α < 180°, Casagrande compared solutions from hand-drawn flow nets with Kozeny's solution for α = 180°. Two such comparisons are shown in Fig. 4.32*a* and *b* for $\alpha = 90°$ and 135°, respectively. He noticed that the seepage face length a was consistently overpredicted by the Kozeny's parabola for the line of seepage. The difference Δa, plotted in Fig. 4.33, was found to have a systematic variation with α. The relationship, shown in Fig. 4.33, can be used to modify the basic Kozeny parabola and obtain a line of seepage.

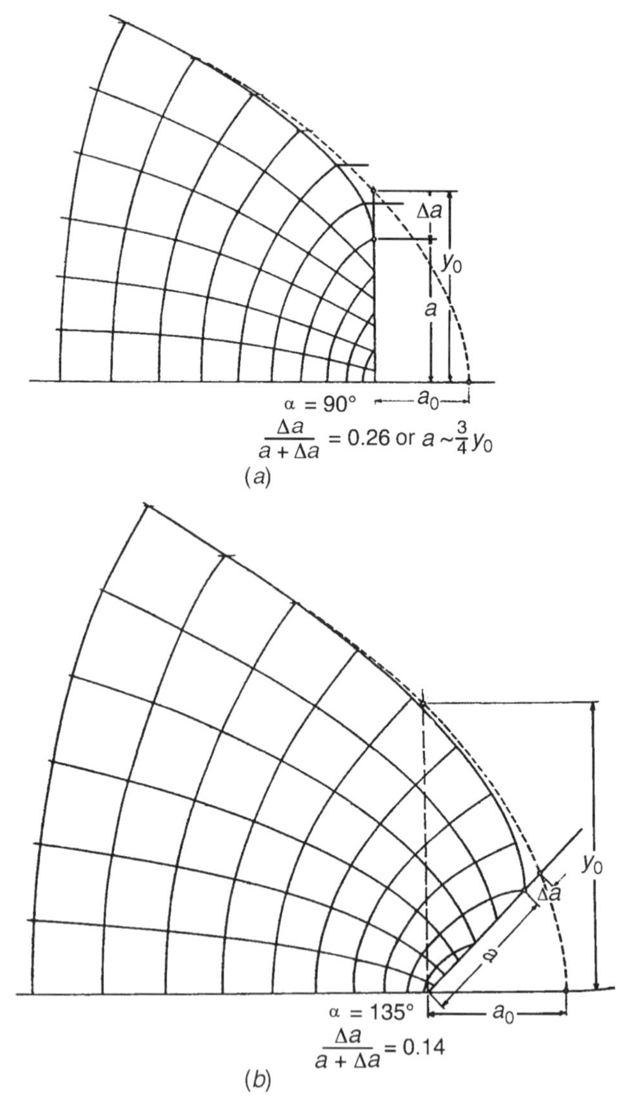

Fig. 4.32. Comparison of flow nets with Kozeny's solution: (*a*) $\alpha = 90°$; (*b*) $\alpha = 135°$. (Adapted from Casagrande, 1937; reprinted with permission of the New England Water Works Association, Holliston, Massachusetts.)

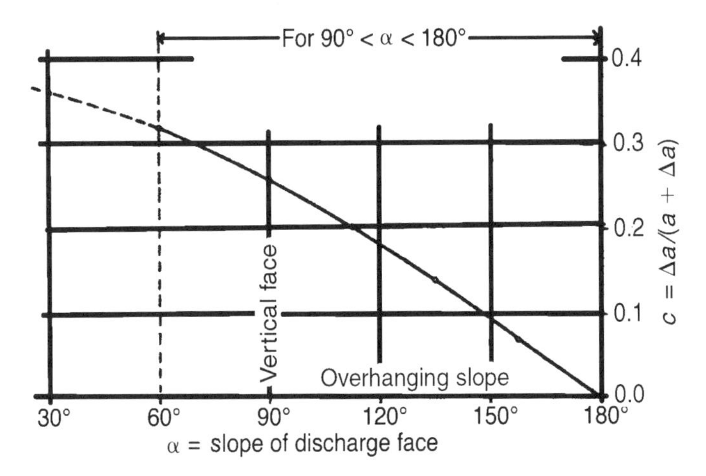

Fig. 4.33. Relationship between Δa and slope (α) of the discharge face. (Adapted from Casagrande, 1937; reprinted with permission of the New England Water Works Association, Holliston, Massachusetts.)

In summary, seepage through earth dams could be predicted with reasonable accuracy using either Eq. (4.78) or (4.72). Casagrande states that the two equations yield practically the same seepage rate for a range of α encountered in earth dams. Where tail water exists on the downstream side of dams, the seepage can be estimated by partitioning the dam into two sections at the tail water level. Seepage through the bottom section can be obtained using Darcy's law along an average flow length. Seepage through the upper section can be obtained using the solutions above and assuming the line of division between the two sections to be an impervious boundary. The total quantity of seepage is the sum of the quantities flowing through the two sections.

Example 4.3 Determine seepage through the earth dam cross section in Fig. 4.29*a* if $H = 30$ m, $L = 60$ m, and $\alpha = 30°$. Assume that the permeability $k = 1 \times 10^{-4}$ cm/s. No tail water is present. Use (a) Dupuit's, (b) Schaffernak's, and (c) Casagrande's methods.

SOLUTION: (a) *Dupuit's method:* Using Eq. (4.15), the discharge is estimated as

$$q = \frac{k}{2L}\left(H^2 - h_e^2\right)$$

$$= \frac{1 \times 10^{-4}\ \text{cm/s}}{2 \times 60\ \text{m}} \times \frac{1\ \text{m}}{100\ \text{cm}} \times \left[(30\ \text{m})^2 - 0\right]$$

$$= 7.5 \times 10^{-6}\ \text{m}^3\text{/s per meter length of the dam}$$

(b) *Schaffernak's method:* Using Eq. (4.69), the discharge is estimated as

$$q = k(a \sin \alpha)(\tan \alpha)$$

where a is evaluated using Eq. (4.71),

$$\begin{aligned} a &= \frac{L}{\cos \alpha} - \sqrt{\frac{L^2}{\cos^2 \alpha} - \frac{H^2}{\sin^2 \alpha}} \\ &= \frac{60 \text{ m}}{\cos 30^\circ} - \sqrt{\left(\frac{60 \text{ m}}{\cos 30^\circ}\right)^2 - \left(\frac{30 \text{ m}}{\sin 30^\circ}\right)^2} \\ &\approx 34.64 \text{ m} \end{aligned}$$

so the discharge

$$\begin{aligned} q &= k(a \sin \alpha)(\tan \alpha) \\ &= 1 \times 10^{-4} \text{ cm/s} \times \frac{1 \text{ m}}{100 \text{ cm}} \times 34.64 \text{ m} \times \sin 30^\circ \times \tan 30^\circ \\ &\approx 1 \times 10^{-5} \text{ m}^3\text{/s per meter length of the dam} \end{aligned}$$

(c) *Casagrande's method:* Using Eq. (4.72), the discharge is estimated as

$$q = ka \sin^2 \alpha$$

where a is evaluated using Eq. (4.74),

$$a = s_0 - \sqrt{s_0^2 - \frac{H^2}{\sin^2 \alpha}}$$

The value of s_0 is approximated as

$$s_0 \approx \sqrt{H^2 + L^2} = \sqrt{(30 \text{ m})^2 + (60 \text{ m})^2} \approx 67.08 \text{ m}$$

so

$$a = 67.08 \text{ m} - \sqrt{(67.08 \text{ m})^2 - \left(\frac{30 \text{ m}}{\sin 30^\circ}\right)^2} \approx 37.08 \text{ m}$$

and the discharge is

$$\begin{aligned} q &= 1 \times 10^{-4} \text{ cm/s} \times \frac{1 \text{ m}}{100 \text{ cm}} \times 37.08 \text{ m} \times \sin^2 30^\circ \\ &\approx 9.27 \times 10^{-6} \text{ m}^3\text{/s per meter length of the dam} \end{aligned}$$

PROBLEMS

4.1. Consider the confined aquifer of varying thickness shown in Fig. P4.1. Derive equations for the seepage rate through the aquifer and the hydraulic head distribution.

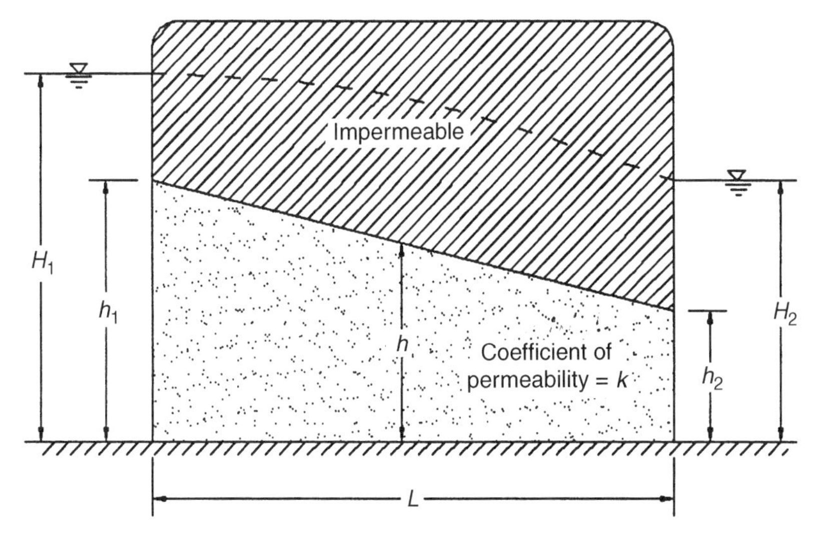

Fig. P4.1

4.2. Consider the drainage layer (Fig. P4.2) underneath a landfill. Derive an equation for the height of the leachate mound h in the drainage layer.

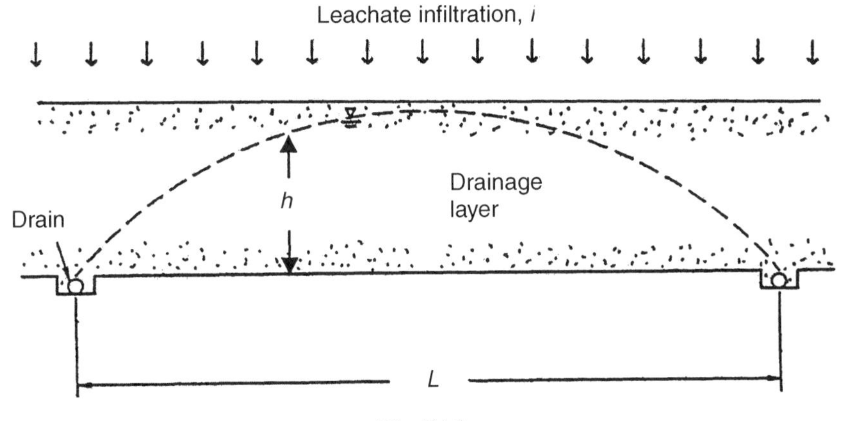

Fig. P4.2

4.3. Refer to Fig. P4.2. What should be the spacing of the drains if the mound were to be totally contained in a drainage layer of thickness 3 ft? Leachate flux from the landfill $i = 0.1$ ft/day, and the permeability of the drainage layer is 2 ft/day. What should be the discharge capacity of the drains?

4.4. A fully penetrating well is used to pump water from an unconfined aquifer at a rate of 100 gal/min. With reference to Fig. 4.11, $r_w = 0.3$ m, $h_w = 20$ m, and $H = 30$ m. The permeability of the aquifer soils $k = 0.005$ ft/min. Determine the steady pumping rate q and drawdown at a radius 15 m from the well. Use Sichardt's empirical equation, Eq. (4.45), to determine the radius of influence, R.

4.5. If the well in Problem 4.4 is partly penetrating with $s = 10$ m and $t = 10$ m (refer to Fig. 4.15), determine the pumping rate and drawdown at a radius 15 m from the well. All other parameters are as stated in Problem 4.4.

4.6. It is desired to stabilize a slope by providing drains, which limit the maximum height (h_{max}) of the water table to 1 m. Refer to Figs. 4.25 and 4.26, and determine the spacing required for the drains given $G = 0$, $\alpha = 25°$, and $i/k = 10^{-2}$.

4.7. Redo Problem 4.6 if the drains are placed above the impermeable surface such that $G = 0.6$ m.

4.8. How would you modify the solutions for flow through earth dams if the soil is anisotropic with respect to permeability?

4.9. Redo Example 4.3 assuming that the soils in the cross section are aniostropic with horizontal permeability $= 5 \times 10^{-6}$ cm/s and vertical permeability $= 1 \times 10^{-6}$ cm/s.

4.10. Redo Example 4.3 with a tailwater height of 10 m at the downstream end of the dam.

CHAPTER 5

ADVANCED SOLUTIONS FOR SEEPAGE: CONFORMAL MAPPING

From our discussion on analytical solutions for seepage in Chapter 4, it is clear that the solutions get complicated when the flow domain is irregular in shape or unknown altogether. In this chapter we seek mathematical functions that allow us to transform complicated domains where a seepage solution is needed into simpler domains where a solution could easily be found. Examples are (1) transformation of an unconfined flow domain through an earth dam into a simple rectangular flow domain, and (2) transformation of a confined flow domain underneath a hydraulic structure with cutoff walls into a rectangular domain or a half-space of a plane. Such transformations involve analytical functions of complex variables, so the reader would benefit by reviewing elements of complex variables prior to going through this chapter.

5.1 CONFORMAL MAPPING

We presented in Chapter 3 the governing equation for flow in soils, Eq. (3.18), as a basis for the development of flow nets. Two functions of spatial coordinates—a velocity potential function $\phi(x, y)$ and a stream function $\psi(x, y)$—were defined as the functions satisfying the governing Laplace equation. The properties of these functions were [Eqs. (3.19) to (3.22)]

$$\frac{\partial \phi}{\partial x} = v_x = k\frac{\partial h}{\partial x}, \qquad \frac{\partial \phi}{\partial y} = v_y = k\frac{\partial h}{\partial y} \tag{5.1}$$

and

$$\frac{\partial \psi}{\partial x} = -v_y, \qquad \frac{\partial \psi}{\partial y} = v_x \tag{5.2}$$

where v_x and v_y represent seepage velocities in the x and y directions, respectively. Comparison of Eqs. (5.1) and (5.2) indicates that

$$\frac{\partial \phi}{\partial x} = \frac{\partial \psi}{\partial y}, \qquad \frac{\partial \phi}{\partial y} = -\frac{\partial \psi}{\partial x} \tag{5.3}$$

known as the *Cauchy–Riemann conditions.* The two functions $\phi(x,y)$ and $\psi(x,y)$ are *conjugate harmonic,* meaning that the two families of curves given by $\phi(x,y) =$ constant and $\psi(x,y) =$ constant represent mutually perpendicular trajectories. For such functions, a complex potential w could be defined,

$$w = \phi + i\psi \tag{5.4}$$

such that it is an analytic function of the complex variable $z = x + iy$. This means that every point on z-space can be mapped or transformed onto a w-space using the analytic function. The flow region from the z-plane when transformed into the w-plane is known as a velocity hodograph. As seen in subsequent sections, such transformations allow us to determine flow and pressure head in complicated domains. Consider the simple transformation function

$$z = w^2 \tag{5.5}$$

Equating real and imaginary parts on both sides, we obtain

$$x = \phi^2 - \psi^2 \tag{5.6}$$

$$y = 2\phi\psi \tag{5.7}$$

These functions can now be used to map a rectangular grid in the w-plane shown in Fig. 5.1*a* onto the z-plane in Fig. 5.1*b*. For a line on the w-plane represented by

$$\psi = n = \text{constant} \tag{5.8}$$

Eqs. (5.6) and (5.7) can be solved to obtain a corresponding line on the z-plane,

$$x = \frac{y^2}{4n^2} - n^2 \tag{5.9}$$

Equation (5.9) represents a family of confocal parabolas as shown in Fig. 5.1*b*. Similarly, for a line on the w-plane represented by

$$\phi = m = \text{constant} \tag{5.10}$$

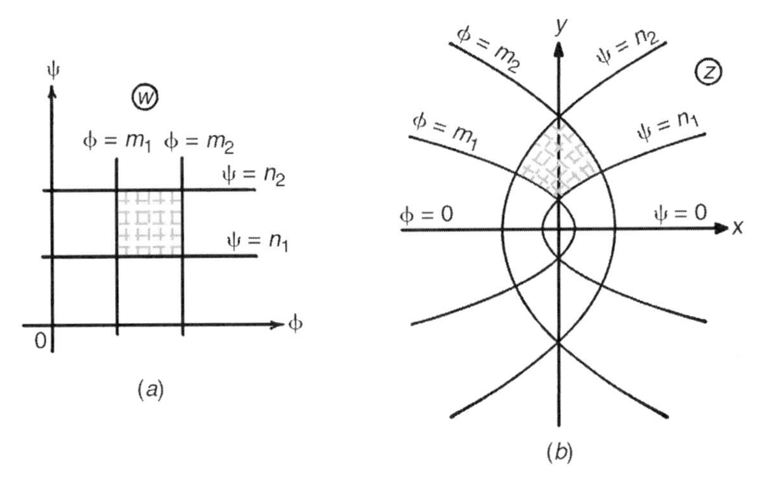

Fig. 5.1. Conformal mapping $(z = w^2)$: (*a*) w-plane; (*b*) z-plane.

the corresponding line on the z-plane can be expressed as

$$x = m^2 - \frac{y^2}{4m^2} \tag{5.11}$$

which represents another family of confocal parabolas conjugate with those of Eq. (5.9). Speaking in terms of flow nets, a simple rectangular grid on the w-plane can be transformed using the relatively simple expressions in Eqs. (5.9) and (5.11) to obtain the flow net represented by confocal parabolas.

5.2 SOLUTIONS USING ELEMENTARY TRANSFORMATION FUNCTIONS

Let us consider the Kozeny's solution for flow in earth dams (Fig. 5.2), which yields a parabola for the line of seepage, as discussed in Chapter 4 [Eq. (4.75)]. The solution could in fact be obtained using the transformation

$$z = Cw^2 \tag{5.12}$$

where C is a constant. The transformation equations therefore become

$$x = C\left(\phi^2 - \psi^2\right) \tag{5.13}$$

and

$$y = 2C\phi\psi \tag{5.14}$$

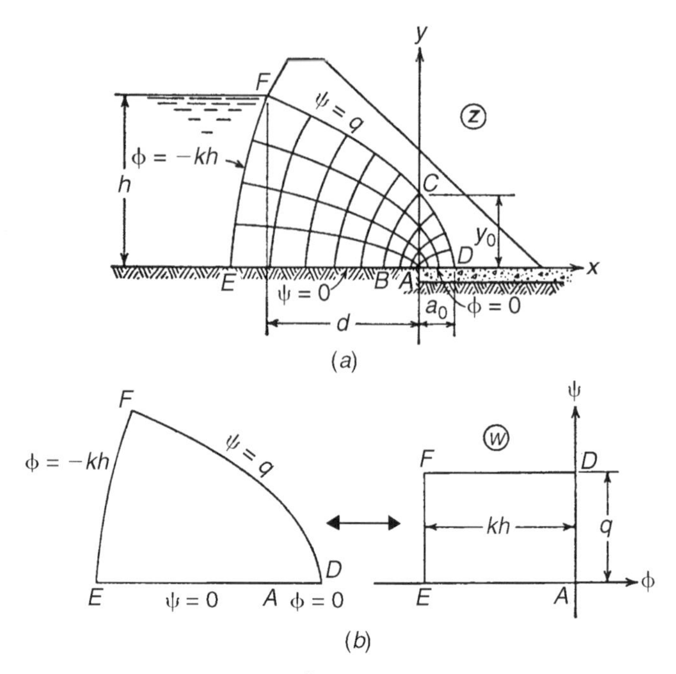

Fig. 5.2. Kozeny's solution using $z = Cw^2$: (*a*) earth dam on the z-plane; (*b*) transformation of the flow domain FDAE between the z and w planes.

To determine the constant C, we use the conditions along the free surface *FCD*, where

$$\psi = q \tag{5.15}$$

and

$$\phi = -ky \tag{5.16}$$

Note that in the z-plane, *FE* was taken to be a parabola, as opposed to the real upstream slope of the dam. This is in anticipation of our solution, which yields a parabola, and correction will have to be made to account for this discrepancy (compare flow nets in Figs. 4.31 and 5.2*a*). Substitution of the conditions along *FCD* in Eq. (5.14) yields

$$C = -\frac{1}{2kq} \tag{5.17}$$

The equation for the line of seepage, referred as *Kozeny's basic parabola,* may now be obtained by substituting Eqs. (5.15) to (5.17) in (5.13),

$$x = \frac{1}{2}\left(\frac{q}{k} - \frac{ky^2}{q}\right) \tag{5.18}$$

The focal distance a_0 can be determined by setting $y = 0$ in Eq. (5.18). Thus,

$$a_0 = \frac{q}{2k} \tag{5.19}$$

or

$$q = 2ka_0 \tag{5.20}$$

Similarly, the y-intercept of the line of seepage is obtained setting $x = 0$ in Eq. (5.18). Thus,

$$y_0 = \frac{q}{k} \tag{5.21}$$

Substituting y_0 for q/k in Eq. (5.18), the basic parabola can be expressed as

$$x = \frac{y_0^2 - y^2}{2y_0} \tag{5.22}$$

which is equivalent to Eq. (4.75). (Note that the positive directions of the x axes in Figs. 4.31 and 5.2 are opposite.)

As a second example, consider a simple flow domain underneath a flat-bottomed weir or dam (Fig. 5.3*a*). We will find a transformation scheme to map this onto a w-plane (Fig. 5.3*b*) and come up with analytical expressions for the streamlines, equipotential lines, and for the pressures developed at the base of the dam. This problem was solved using the complex velocity W (Polubarinova-Kochina, 1962), expressed as

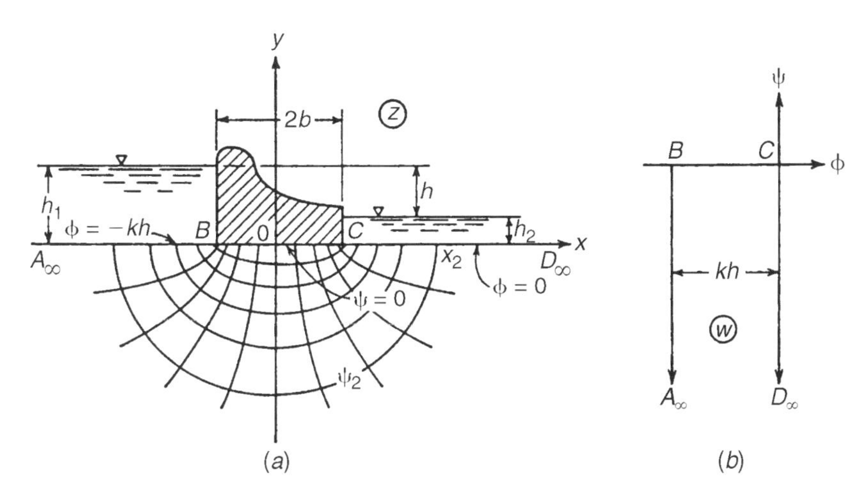

Fig. 5.3. Transformation of the flow domain underneath a flat-bottomed hydraulic structure: (*a*) structure in the z-plane; (*b*) ABCD transformed to the w-plane. (Adapted from Harr, 1962.)

$$W = \frac{dw}{dz} = \frac{\partial \phi}{\partial x} + \frac{i\,\partial \psi}{\partial x} = v_x - i v_y \tag{5.23}$$

W was expressed in terms of z to obtain the transformation function

$$W = v_x - i v_y = \frac{M}{\sqrt{b^2 - z^2}} \tag{5.24}$$

where M is the real constant and $2b$ is the width of the weir/dam. This function was chosen because it obeys the velocity conditions at the boundaries in the real flow domain. Note that according to Eq. (5.24), W is real along BC, where $-b < x < b$, and imaginary along AB and CD, where $|x| > b$. This is consistent with the flow velocities being horizontal along BC (i.e., $v_x \neq 0$, $v_y = 0$) and vertical along AB and CD (i.e., $v_x = 0$, $v_y \neq 0$). Integrating Eq. (5.24) with respect to z gives us

$$w = \phi + i\psi = M \sin^{-1}\frac{z}{b} + N \tag{5.25}$$

where N is a constant of integration. To determine the constants M and N, we use the conditions at points B and C.

At point B:

$$\psi = 0, \quad \phi = -kh, \quad w = -kh, \quad z = -b \tag{5.26}$$

At point C:

$$\psi = 0, \quad \phi = 0, \quad w = 0, \quad z = b \tag{5.27}$$

Note that $h = h_1 - h_2$ in Eq. (5.26). Substituting Eqs. (5.26) and (5.27) in (5.25), we get $M = kh/\pi$ and $N = -kh/2$. Thus,

$$w = \frac{kh}{\pi}\sin^{-1}\frac{z}{b} - \frac{kh}{2} = -\frac{kh}{\pi}\cos^{-1}\frac{z}{b} \tag{5.28}$$

or

$$z = b\cos\frac{\pi w}{kh} \tag{5.29}$$

Equating the real and imaginary components in Eq. (5.29) yields

$$x = b\cos\frac{\pi\phi}{kh}\cosh\frac{\pi\psi}{kh} \tag{5.30}$$

and

$$y = -b\sin\frac{\pi\phi}{kh}\sinh\frac{\pi\psi}{kh} \tag{5.31}$$

Solution of Eqs. (5.30) and (5.31) for constant values of ψ and ϕ, say ψ_n and ϕ_n, allows us to express the equations for streamlines and equipotential lines as

$$\frac{x^2}{b^2 \cosh^2\left(\pi\psi_n/kh\right)} + \frac{y^2}{b^2 \sinh^2\left(\pi\psi_n/kh\right)} = 1 \tag{5.32}$$

and

$$\frac{x^2}{b^2 \cos^2\left(\pi\phi_n/kh\right)} - \frac{y^2}{b^2 \sin^2\left(\pi\phi_n/kh\right)} = 1 \tag{5.33}$$

Equation (5.32) represents the streamlines as a series of confocal ellipses, and Eq. (5.33) represents the equipotential lines as a series of confocal hyperbolas. With the constant of integration M identified as equal to kh/π, Eq. (5.24) can be used to determine the velocity distribution along $ABCD$:

Along BC:

$$\text{horizontal velocity } v_x = \frac{kh}{\pi\sqrt{b^2 - x^2}} \tag{5.34}$$

Along AB and CD:

$$\text{vertical velocity } v_y = \mp\frac{kh}{\pi\sqrt{x^2 - b^2}} \tag{5.35}$$

where the minus sign applies to AB and the plus sign applies to CD. Equation (5.35) is particularly helpful in determining the exit gradients I_E along CD,

$$I_E = \frac{v_y}{k} = \frac{h}{\pi\sqrt{x^2 - b^2}} \tag{5.36}$$

It is possible to express velocity components v_x and v_y at any point (x, y) of the flow domain by separating real and imaginary components in Eq. (5.24). The reader can verify the following expressions:

$$v_x = \frac{kh}{\pi\sqrt{2}} \frac{\sqrt{\sqrt{(b^2 - x^2 + y^2)^2 + 4x^2y^2} + b^2 - x^2 + y^2}}{\sqrt{(b^2 - x^2 + y^2)^2 + 4x^2y^2}} \tag{5.37}$$

$$v_y = \mp\frac{kh}{\pi\sqrt{2}} \frac{\sqrt{\sqrt{(b^2 - x^2 + y^2)^2 + 4x^2y^2} - b^2 + x^2 - y^2}}{\sqrt{(b^2 - x^2 + y^2)^2 + 4x^2y^2}} \tag{5.38}$$

where the upper sign is used for $x < 0$ and the lower sign is used for $x > 0$. To determine the pressure along the foundation dam, we use Eq. (5.29). Considering that $\psi = 0$ along BC,

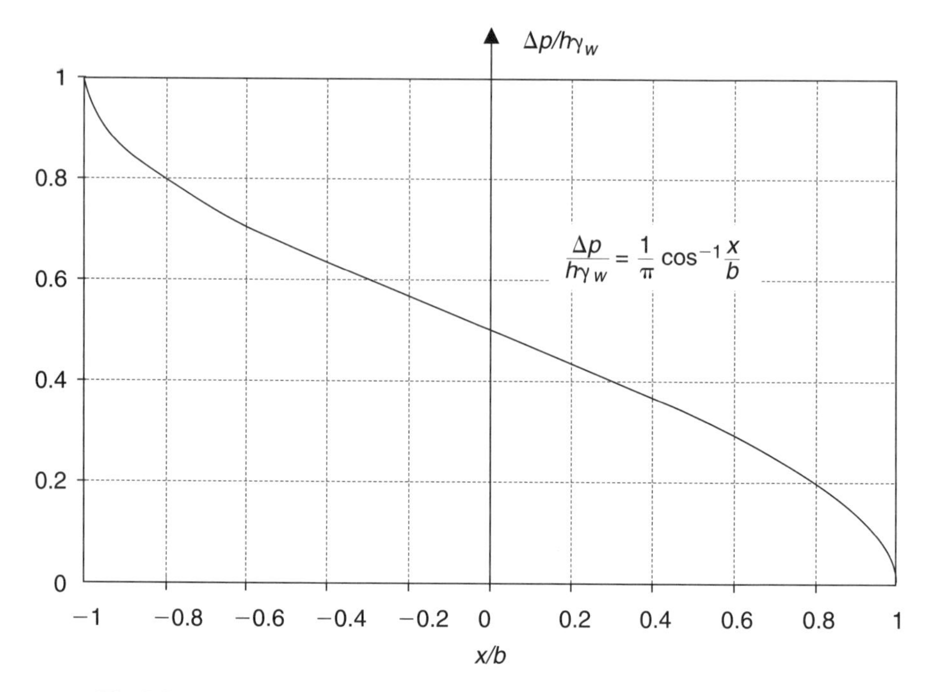

Fig. 5.4. Uplift pressure distribution along the base of the structure in Fig. 5.3.

$$x = b \cos \frac{\pi \phi}{kh} = b \cos \frac{\pi \, \Delta p}{\gamma_w h} \tag{5.39}$$

or

$$\frac{\Delta p}{\gamma_w} = \frac{h}{\pi} \cos^{-1} \frac{x}{b} \tag{5.40}$$

where Δp is the pressure along BC. Normalizing the pressure head $\Delta p/\gamma_w$ with the total differential head h, the distribution along BC may be plotted as shown in Fig. 5.4.

Example 5.1 For the hydraulic structure shown in Fig. 5.3, plot exit gradients on the downstream side as a function of x given $h = 10$ m and $b = 10$ m. How does the exit gradient distribution change if $b = 5$ m?

SOLUTION: The exit gradient I_E on the downstream side is estimated using Eq. (5.36),

$$I_E = \frac{h}{\pi \sqrt{x^2 - b^2}}$$

Figure E5.1 is the plot for $b = 10$ m and $b = 5$ m. The exit gradient becomes larger immediately after $x = b$, on the downstream side.

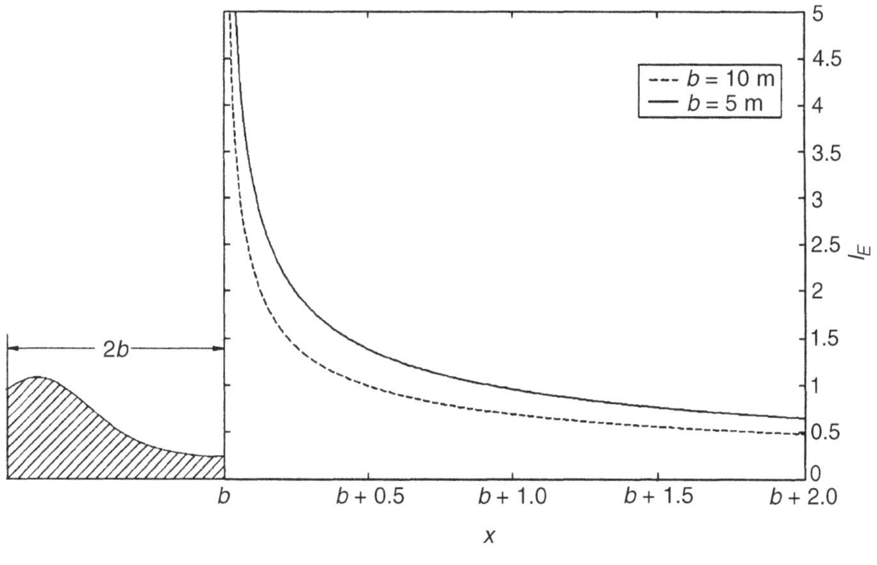

Fig. E5.1

5.3 SCHWARZ–CHRISTOFFEL TRANSFORMATION

Figure 5.3*a* represents a relatively simple case of flow domain with a flat-bottomed configuration of the dam. In more complicated flow domains, the equipotential and streamlines may not correspond to mathematical expressions of parabolas, ellipses, or hyperbolas. Pavlovsky considered a general set of problems where the flow region z is represented by a polygon. Such a polygon can be mapped onto the plane of the complex potential w using another simple polygon, generally a rectangle. An additional auxiliary complex variable space t was introduced. The upper or lower half of the t-plane could be used to map the z and w planes (Fig. 5.5). If the transformation functions are represented by F_1 and F_2,

$$z = F_1(t) \tag{5.41}$$

and

$$w = F_2(t) \tag{5.42}$$

all the elements of flow in the real flow domain could be determined using these functions. The complex velocity W, for instance, is given by

$$W = \frac{dw}{dz} = \frac{F_2'(t)}{F_1'(t)} \tag{5.43}$$

The conformal mapping of the polygon onto the half-plane of t-space is carried out using what is known as *Schwarz–Christoffel transformation.* To transform a polygon

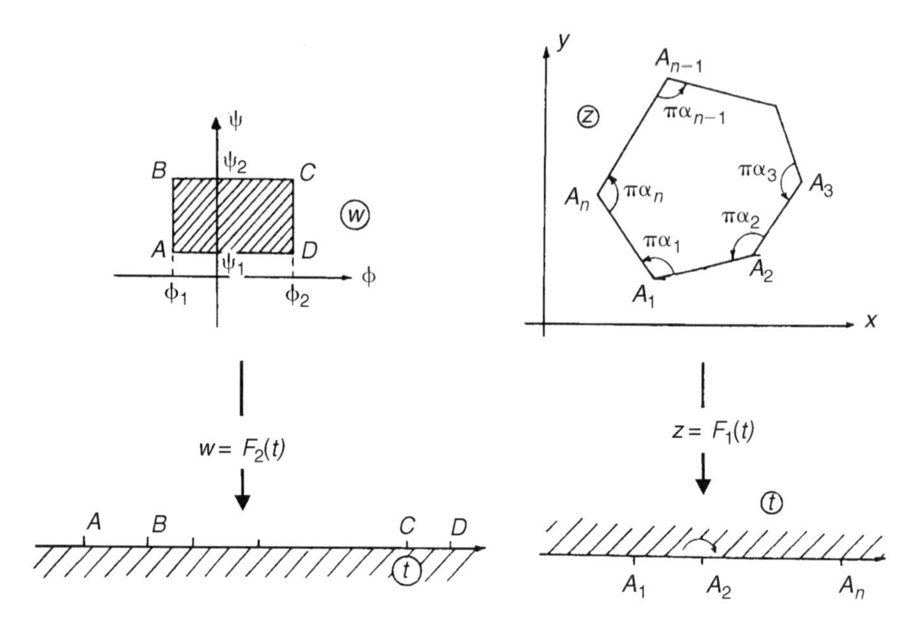

Fig. 5.5. Schwarz–Christoffel transformation between the z and w planes using the t-plane.

with vertices $A_1, A_2, \ldots, A_{n-1}, A_n$ from a z-plane to the upper half t-plane, the Schwarz–Christoffel transformation takes the form

$$z = M \int (t - a_1)^{\alpha_1 - 1} (t - a_2)^{\alpha_2 - 1} \cdots (t - a_n)^{\alpha_n - 1} \, dt + N \tag{5.44}$$

where $M, N, a_1, a_2, \ldots, a_n =$ constants, and $\alpha_1, \alpha_2 \ldots, \alpha_n$ represent the magnitudes of the interior angles of the polygon when multiplied by π. According to this formula, the vertices of the polygon map onto points of the real axis on t-space.

To demonstrate Schwarz–Christoffel transformation, we will discuss Pavlovsky's solution for the general problem of flow around a hydraulic structure on a semi-infinite porous medium shown in Fig. 5.6. The transformation function to map the bottom contour of the structure onto the t-plane was expressed as

$$z = \frac{d}{\pi m}\sqrt{t^2 - 1} - \frac{d}{\pi}\cosh^{-1} t + id \tag{5.45}$$

where d is as shown in Fig. 5.6a and m is obtained using

$$\frac{s\pi}{d} = \frac{\sqrt{1 - m^2}}{m} - \cos^{-1} m \tag{5.46}$$

which is plotted in Fig. 5.7. Using the transformation function in Eq. (5.45), the reader can verify that points A_∞, C, E, and G_∞ are mapped onto points $t = -\infty, -1, +1$, and $+\infty$ of the real axis of the t-plane, respectively. To transform the w-plane onto

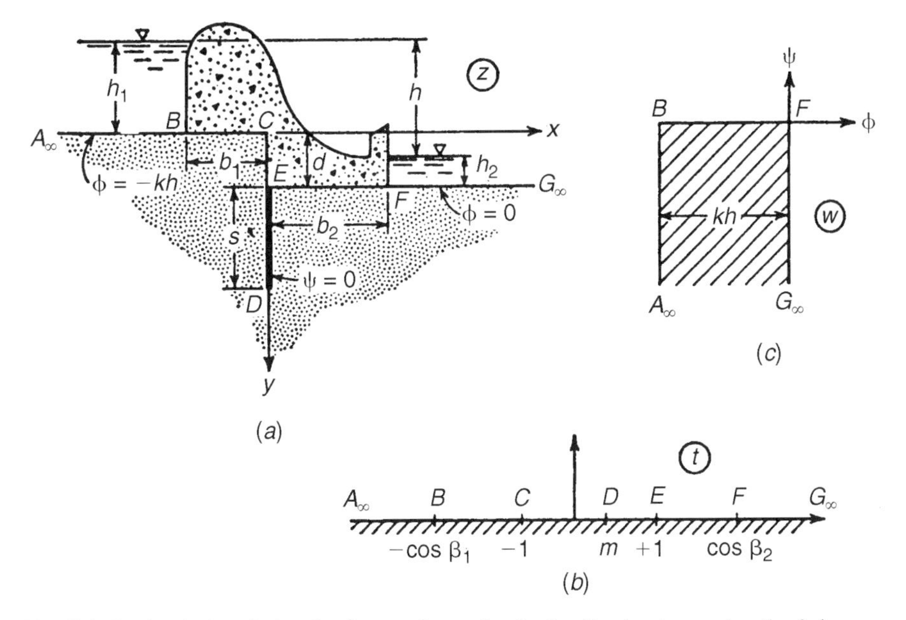

Fig. 5.6. Pavlovsky's solution for flow underneath a hydraulic structure using the Schwarz–Christoffel transformation: (*a*) structure in the z-plane; (*b*) transformation of the domain *ABCDEFG* onto the t-plane; (*c*) transformation of the domain on to the w-plane. (Adapted from Harr, 1962.)

the lower half of the t-plane, the following Schwarz–Christoffel transformation was used:

$$t = \lambda \cos \frac{w\pi}{kh} - \lambda_1 \tag{5.47}$$

where

$$\lambda = \frac{\cosh\beta_1 + \cosh\beta_2}{2} \tag{5.48}$$

and

$$\lambda_1 = \frac{\cosh\beta_1 - \cosh\beta_2}{2} \tag{5.49}$$

β_1 and β_2 in Eqs. (5.48) and (5.49) are obtained by solving

$$-\frac{b_1\pi m}{d} = \sinh\beta_1 + m\beta_1 \tag{5.50}$$

and

$$\frac{b_2\pi m}{d} = \sinh\beta_2 - m\beta_2 \tag{5.51}$$

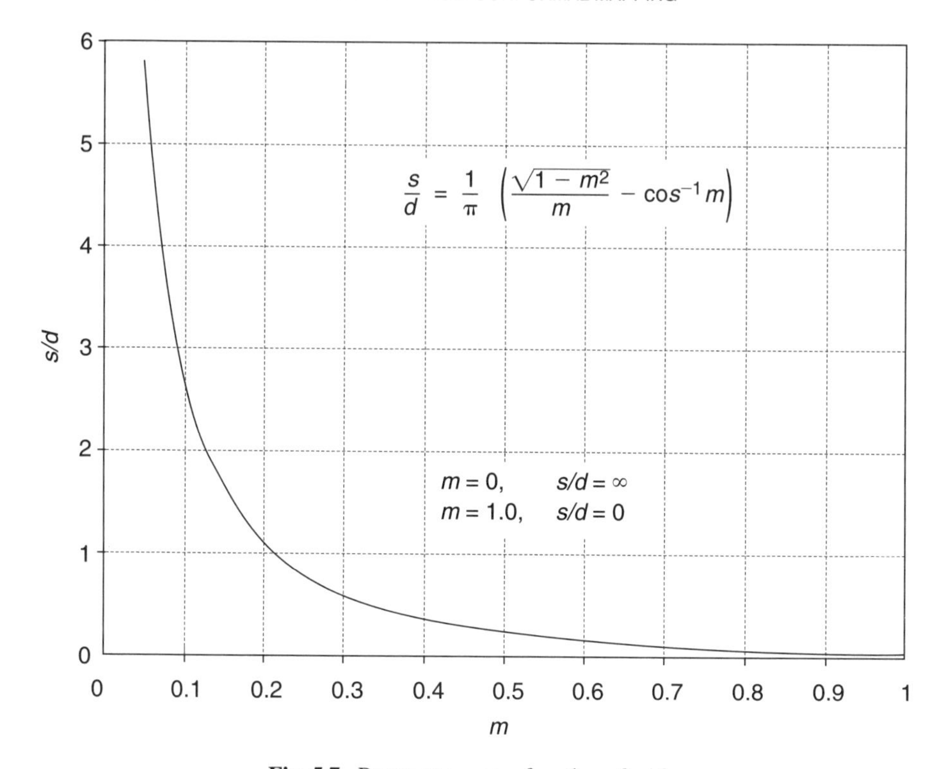

Fig. 5.7. Parameter m as a function of s/d.

where m is given by Eq. (5.46) or Fig. 5.7. β_1 and β_2 obtained using Eqs. (5.50) and (5.51) are such that the points B and F in the z-space are mapped to $-\cos\beta_1$ and $\cos\beta_2$, respectively, in the t-space. To obtain the function that transforms the z-plane to the w-plane, we substitute Eq. (5.47) for t in Eq. (5.45). Thus,

$$z = \frac{d}{\pi m}\left[\left(\lambda\cos\frac{\phi\pi}{kh} - \lambda_1\right)^2 - 1\right]^{1/2} - \frac{d}{\pi}\cosh^{-1}\left(\lambda\cos\frac{\phi\pi}{kh} - \lambda_1\right) + id \quad (5.52)$$

To determine the pressure distribution along the bottom of the structure, we note that

$$\phi = -k\left(\frac{p}{\gamma_w} - y\right) + C \quad (5.53)$$

At $y = 0$, $\phi = -kh$ and $p/\gamma_w = h_1$. Therefore,

$$C = -k\,(h - h_1) \quad (5.54)$$

Substituting Eq. (5.54) for C and using Eq. (5.47) to express ϕ in terms of t, the reader will verify that Eq. (5.53) can be written as

$$p = \gamma_w \left(\frac{h}{\pi} \cos^{-1} \frac{t + \lambda_1}{\lambda} + y - h + h_1 \right) \tag{5.55}$$

Using the coordinates of points C, D, and E, the pressure at those points may therefore be expressed as

$$p_{\text{at } C} = \gamma_w \left(\frac{h}{\pi} \cos^{-1} \frac{\lambda_1 - 1}{\lambda} - h + h_1 \right) \tag{5.56}$$

$$p_{\text{at } D} = \gamma_w \left(\frac{h}{\pi} \cos^{-1} \frac{\lambda_1 + m}{\lambda} + s + h_2 \right) \tag{5.57}$$

$$p_{\text{at } E} = \gamma_w \left(\frac{h}{\pi} \cos^{-1} \frac{\lambda_1 + 1}{\lambda} + h_2 \right) \tag{5.58}$$

Figure 5.8*a* and *b* illustrate two special cases of the problem described above, where $d = 0$. The reader may verify that the function mapping the z-plane to the t-plane could take the following simpler form in these cases:

$$z = s\sqrt{t^2 - 1} \tag{5.59}$$

According to this function, points B and F are mapped onto $t = -L_1$ and L_2, respectively, where

$$L_1 = \left[1 + \left(\frac{b_1}{s} \right)^2 \right]^{1/2} \tag{5.60}$$

and

$$L_2 = 1 \tag{5.61}$$

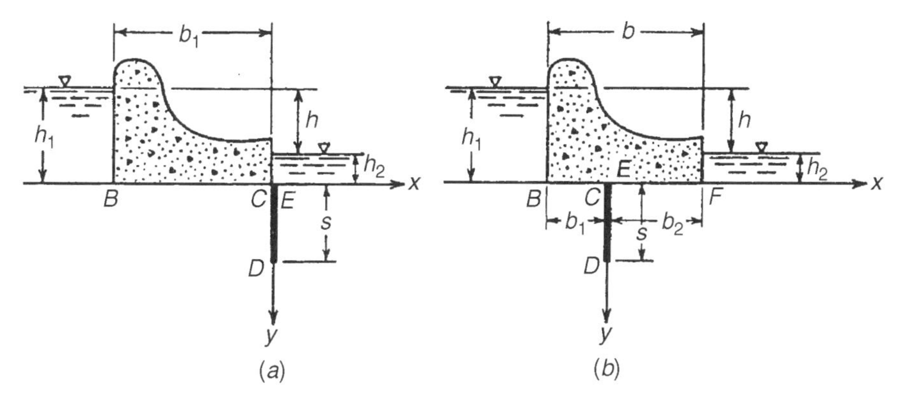

Fig. 5.8. Simpler configurations of flow underneath hydraulic structures: (*a*) cutoff at the downstream end; (*b*) intermediate cutoff. (Adapted from Harr, 1962.)

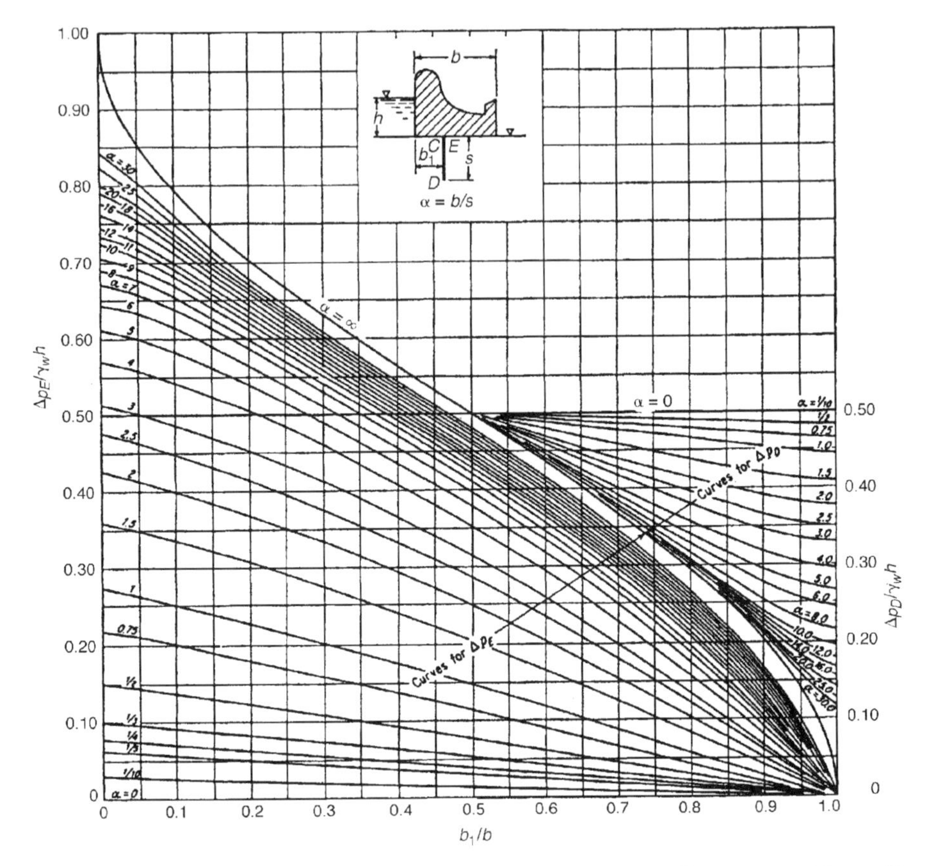

Fig. 5.9. Nomograms for determining pressures underneath a hydraulic structure with cutoff. (Adapted from Khosla et al., 1954.)

Leaving the complete exercise to the reader, we express the pressure distribution as

$$p = \begin{cases} \gamma_w \left(\dfrac{h}{\pi} \cos^{-1} \dfrac{\lambda_1 s \pm \sqrt{s^2 + x^2}}{\lambda s} + h_2 \right) & \text{for } y = 0 \qquad (5.62) \\[2ex] \gamma_w \left(\dfrac{h}{\pi} \cos^{-1} \dfrac{\lambda_1 s \pm \sqrt{s^2 - y^2}}{\lambda s} + y + h_2 \right) & \text{for } x = 0 \qquad (5.63) \end{cases}$$

For the case of the sheet pile at some intermediate point along the base of the structure (Fig. 5.8*b*), Khosla et al. (1954) developed a nomogram, shown in Fig. 5.9, which could be used directly to determine pressures at points D and E. To obtain pressure at point C, the curves for Δp_E are used with the x-axis changed to $1 - b_1/b$. Δp_C can then be expressed as

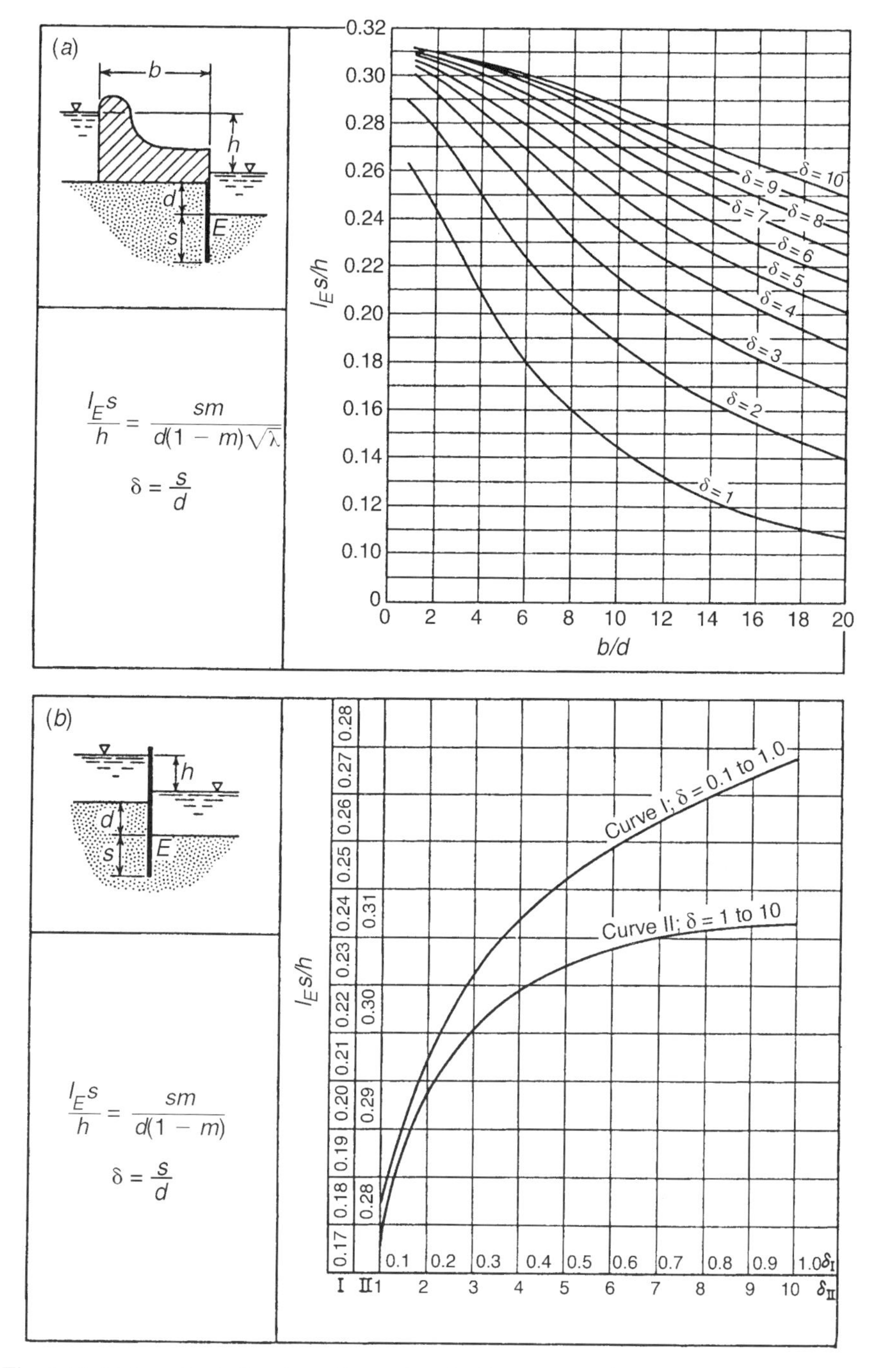

Fig. 5.10. Exit gradients for sheet piles with and without hydraulic structures. (Adapted from Khosla et al., 1954.) (*continued*)

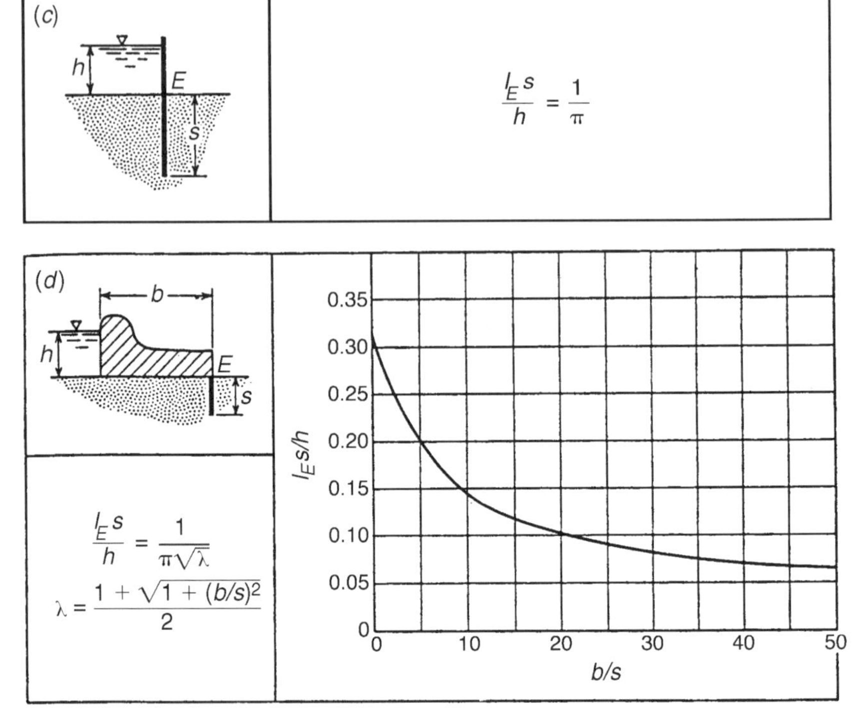

Fig. 5.10. (*Continued*)

$$\frac{\Delta p_C}{h\gamma_w} = 1 - \frac{\Delta p_E}{h\gamma_w} \tag{5.64}$$

Using the transformation functions $z = F_1(t)$ and $w = F_2(t)$, a simple expression could be derived for the exit gradient. In an isotropic flow regime, the gradient I may be expressed as

$$I = -\frac{dh}{ds} = \frac{1}{k}\frac{d\phi}{ds} = \frac{1}{k}\frac{d\phi}{dt}\frac{dt}{dz}\frac{dz}{ds} \tag{5.65}$$

where s is in the direction of flow line at the exit. $dz/ds = \cos\theta + i\sin\theta$, where θ is the angle between the flow line and the x-axis. Since the flow line at the exit intersects the downstream equipotential boundary at 90°, $\theta = 90°$, $dz/ds = i$. Along the flow line, ψ = constant; therefore, $d\phi/dt = dw/dt$. Thus, the exit gradient I_E may be simplified as

$$I_E = \frac{i}{k}\frac{dw}{dt}\frac{dt}{dz} \tag{5.66}$$

Given the two functions F_1 and F_2, it is possible to express I_E analytically. The expressions for I_E, along with figures to determine the same, are shown in Fig. 5.10 for a few configurations of hydraulic structures.

5.4 CONFINED FLOW IN LAYERS OF FINITE DEPTH

Polubarinova-Kochina (1962) and Harr (1962) discuss transformation functions useful to map flow domains in layers of finite depth. Directing advanced students to these references, we proceed by cataloging some solutions to common problems.

1. *Hydraulic structure without a sheet pile on a layer of finite depth.* The discharge q and pressure distribution along the base are as shown in Figs. 5.11 and 5.12, respectively.
2. *Hydraulic structure with sheet pile(s) on a layer of finite depth.* For symmetrical and nonsymmetrical locations of the sheet pile, the discharge could be determined using Figs. 5.13 and 5.14, respectively. A comparison of the discharges for various locations of the sheet pile shows that the discharge is maximum when the sheet pile is at the center; however, the variation with position is small (Fig. 5.15). The pressure drop across the sheet pile as a percentage of total pressure drop across the structure is shown in Fig. 5.16 for two specific locations of sheet pile: symmetric and at the upstream end or heel of the structure. The figure shows the relative effects of depth of embedment of sheet pile s and the thickness of the layer, T. It is seen that larger pressure drops could be accomplished with smaller depths of embedment when T is small.

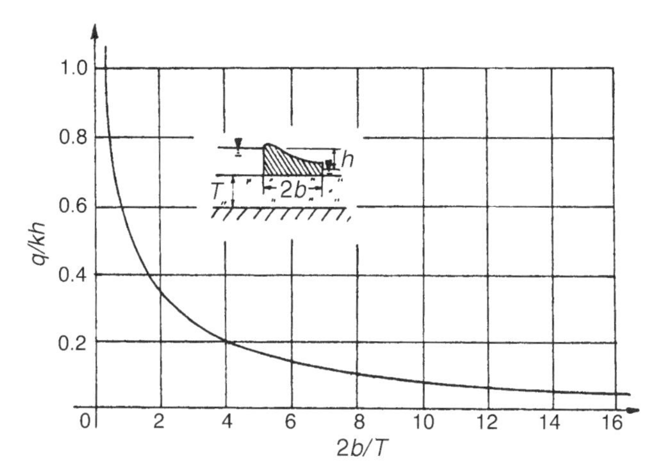

Fig. 5.11. Solution for seepage underneath a flat-bottomed dam or weir resting on a layer of finite depth. [Adapted from Polubarinova-Kochina, 1962; copyright © 1962 (renewed 1990) Princeton University Press; reprinted with permission.]

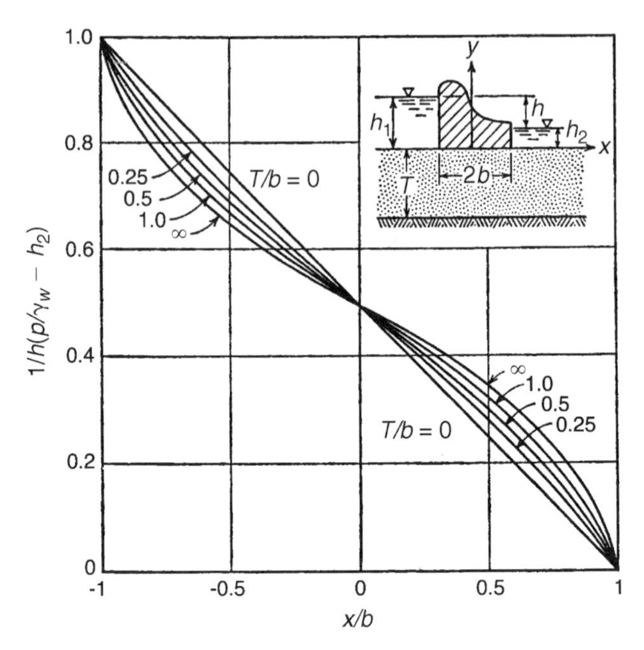

Fig. 5.12. Uplift pressure distribution along the base of a flat-bottomed dam or weir resting on a layer of finite depth. [Adapted from Polubarinova-Kochina, 1962; copyright © 1962 (renewed 1990) Princeton University Press; reprinted with permission.]

3. *Double-wall sheet pile cofferdam.* For a cofferdam consisting of two rows of sheet piles (Fig. 5.17*a*), the discharge q is given as a function of modulus m in Fig. 5.17*b*. The modulus m needed in Fig. 5.17*b* could be determined using Fig. 5.18 for specific values of s, b, and d. It is crucial to determine exit gradients for these structures, since piping is a common mode of failure. The exit gradients are shown in Fig. 5.19 in terms of modulus m and d/b.
4. *Single sheet pile embedded in a two-layer system.* The discharge around a sheet pile is plotted in Fig. 5.20 in terms of s/T and ε. ε is a dimensionless parameter expressed as

$$\tan \pi\varepsilon = \sqrt{\frac{k_2}{k_1}} \tag{5.67}$$

According to Eq. (5.67), ε ranges between 0 and $\frac{1}{2}$ when the ratio of permeabilities varies from 0 to ∞. The exit gradients at the downstream end of the sheet pile are shown in Fig. 5.21, again in terms of s/T and ε.
5. *Flat-bottomed structure on a two-layered system.* Plots showing discharge q in terms of $2b/T$ and ε are shown in Fig. 5.22.

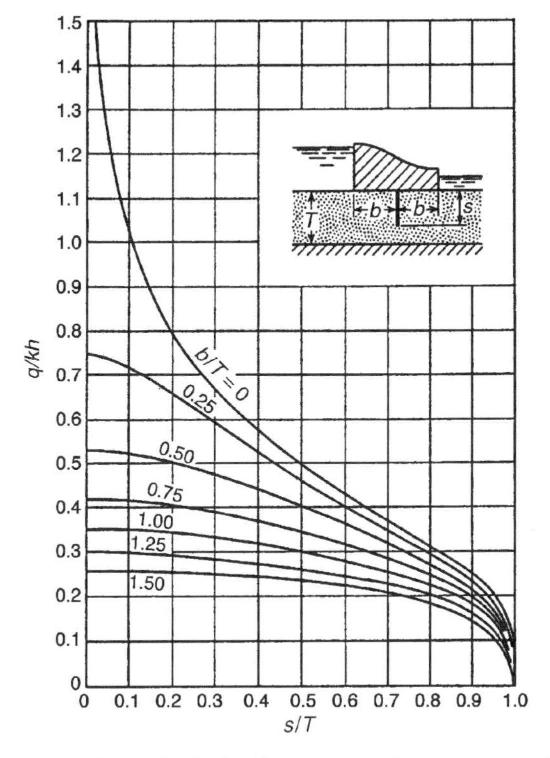

Fig. 5.13. Seepage rate underneath a hydraulic structure with a symmetrically located cutoff. (Adapted from Polubarinova-Kochina, 1952.)

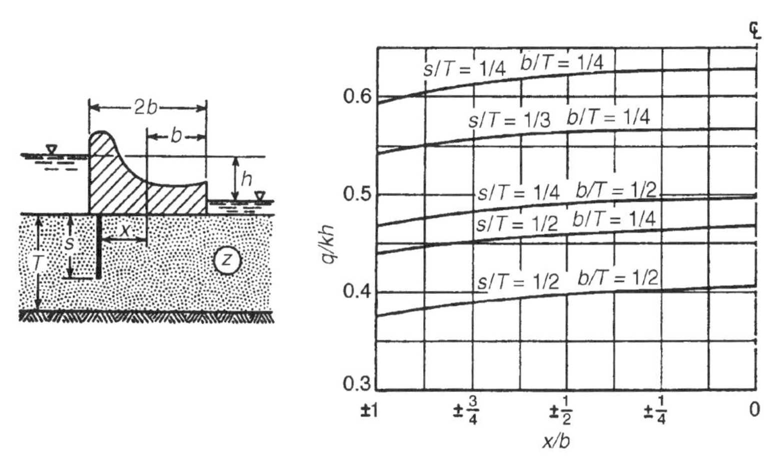

Fig. 5.14. Seepage rate underneath a hydraulic structure for any general location of cutoff. (Adapted from Harr, 1962.)

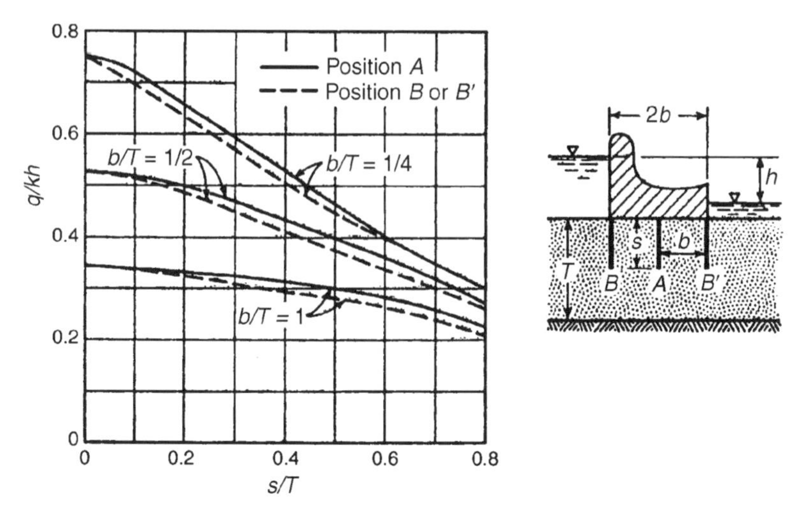

Fig. 5.15. Comparison of seepage for various locations of cutoff. (Adapted from Harr, 1962.)

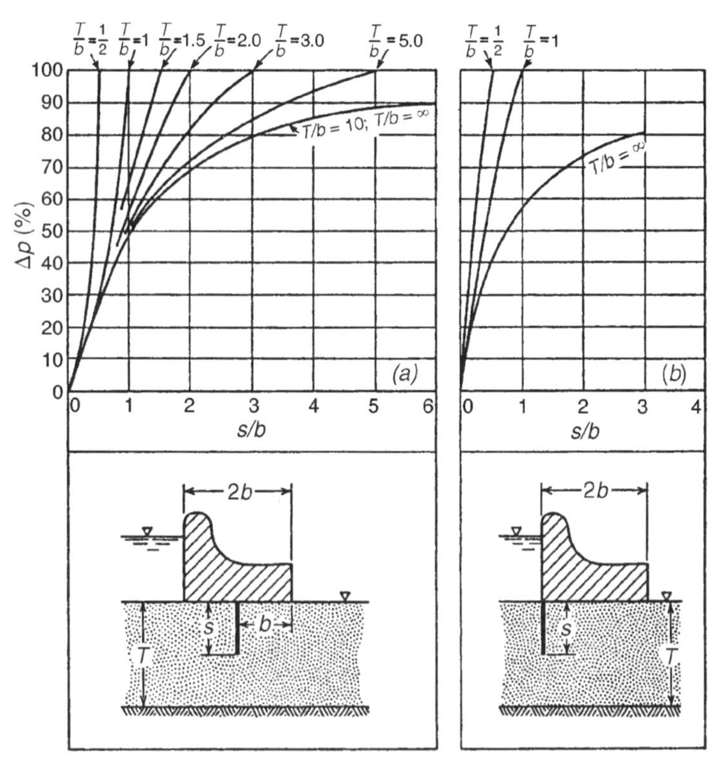

Fig. 5.16. Pressure drop across the sheet pile as a percentage of total pressure drop across the structure. (Adapted from Pavlovsky, 1956.)

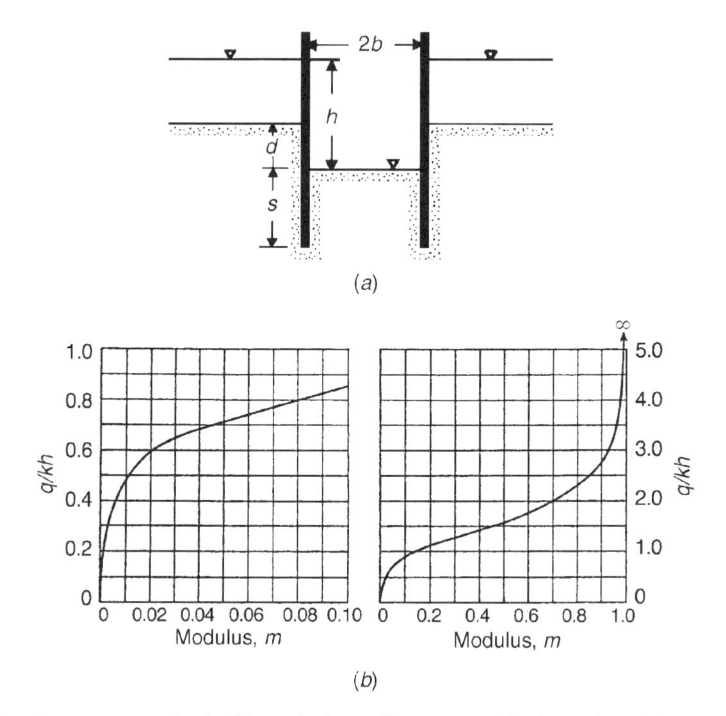

Fig. 5.17. Seepage around a double-wall sheet pile system: (*a*) schematic of the structure; (*b*) seepage as a function of modulus m. (Adapted from Harr, 1962.)

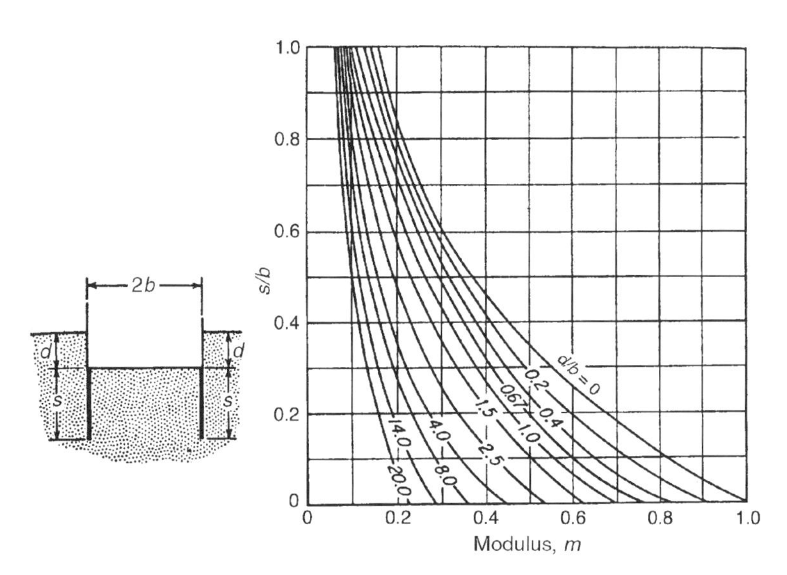

Fig. 5.18. Modulus m versus s/b for a double-walled sheet pile system. (Adapted from Harr, 1962.)

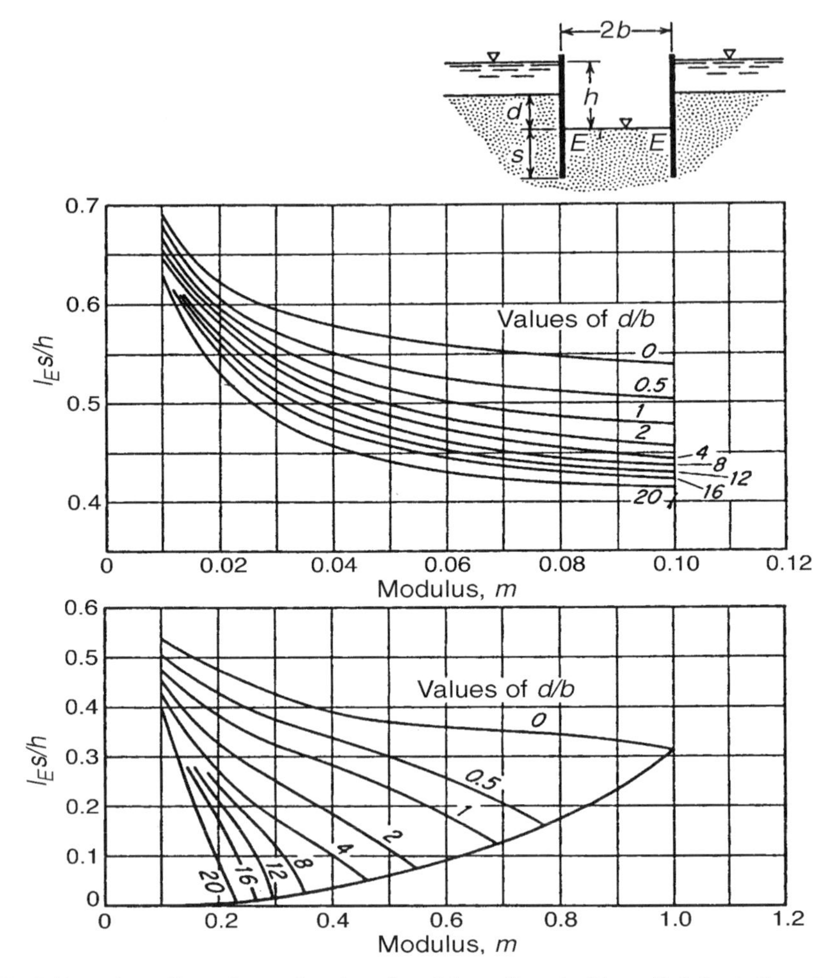

Fig. 5.19. Exit gradients I_E as a function of modulus m for a double-walled sheet pile system. (Adapted from Harr, 1962.)

5.5 METHOD OF FRAGMENTS

In the case of a general flow domain with multiple cutoff walls and several dissimilar flow elements, an approximate method known as the *method of fragments* (Pavlovsky, 1956) may be used. As the name implies, this method involves dividing the flow domain into a number of fragments (Fig. 5.23). The nearly vertical equipotential lines at the cutoff walls may be used to partition the flow domain in Fig. 5.23 into four segments. We express discharge q in the mth segment as

$$q = \frac{kh_m}{\phi_m} \qquad m = 1, 2, \ldots, n \tag{5.68}$$

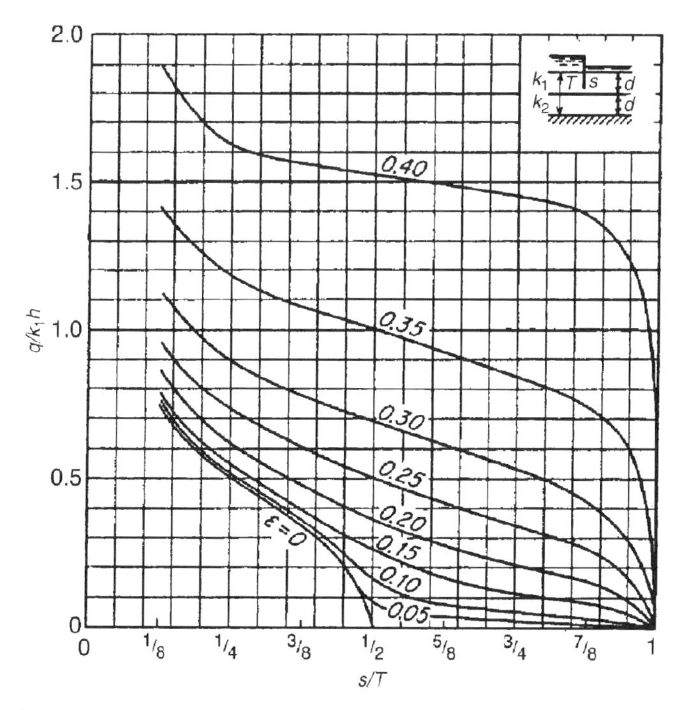

Fig. 5.20. Seepage around a sheet pile embedded in a two-layer system. (Adapted from Polubarinova-Kochina, 1952.)

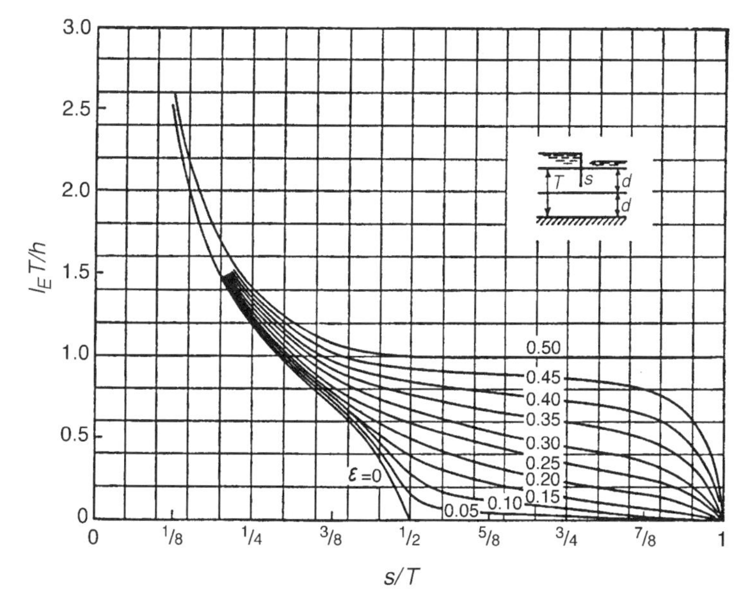

Fig. 5.21. Exit gradients for a sheet pile embedded in a two-layer system. (Adapted from Polubarinova-Kochina, 1952.)

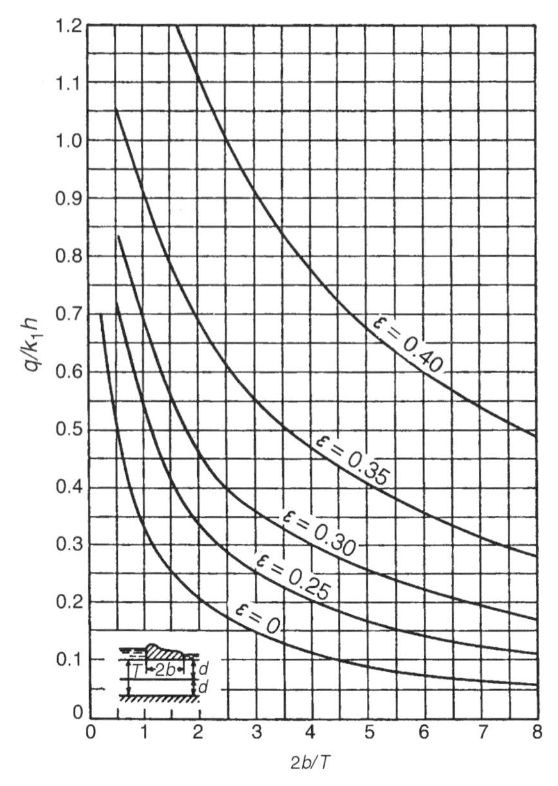

Fig. 5.22. Seepage underneath a flat-bottomed dam or weir resting on a two-layered system. (Adapted from Polubarinova-Kochina, 1952.)

where h_m is the head loss through the mth fragment and ϕ_m is a dimensionless form factor. This is an expression similar to Eq. (3.6) where a shape factor is used instead of ϕ_m. Since the discharge through all fragments is the same,

$$q = \frac{kh_1}{\phi_1} = \frac{kh_2}{\phi_2} = \cdots = \frac{kh_n}{\phi_n} \tag{5.69}$$

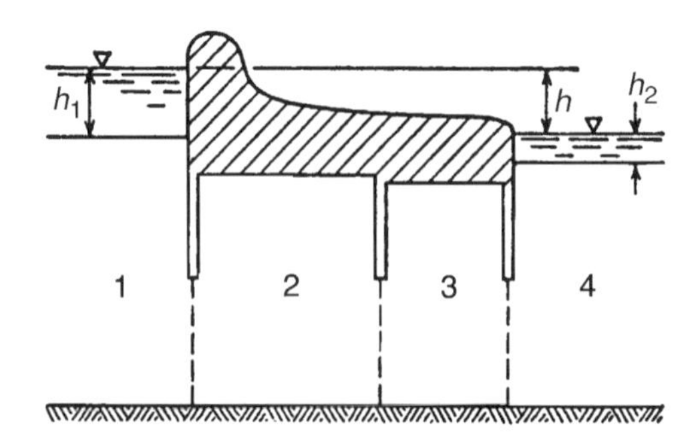

Fig. 5.23. Division of a flow domain into various fragments. (Adapted from Harr, 1962.)

and

$$q = k\frac{\sum h_m}{\sum_{m=1}^{n} \phi_m} = \frac{kh}{\sum_{m=1}^{n} \phi_m} \tag{5.70}$$

Using Eqs. (5.69) and (5.70), the head loss in the mth fragment can be expressed as

$$h_m = \frac{h\phi_m}{\sum_{m=1}^{n} \phi_m} \tag{5.71}$$

The problem then reduces to identifying form factors for each type of flow fragment. Based on analytical solutions similar to those presented earlier, Pavlovsky cataloged the form factors for six different types of fragments. These are shown in Table 5.1.

TABLE 5.1 Summary of Fragment Types and Form Factors

Fragment Type	Illustration	Form Factor Φ^a
I	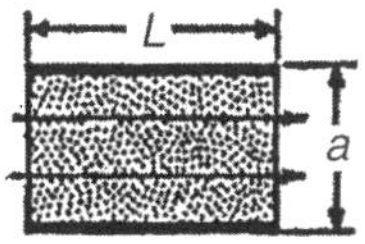	$\Phi = \frac{L}{a}$
II		$\Phi = \frac{K}{K'}$; $m = \sin\frac{\pi s}{2T}$ $I_E = \frac{h\pi}{2KTm}$
III	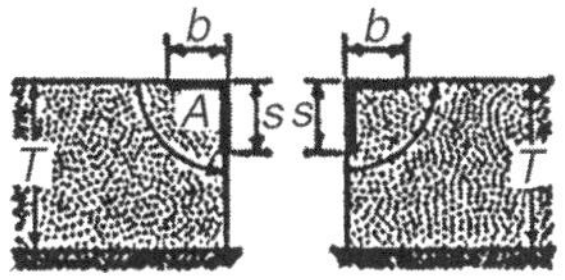	$\Phi = \frac{K}{K'}$ $m = \cos\frac{\pi s}{2T}\sqrt{\tanh^2\frac{\pi b}{2T} + \tan^2\frac{\pi s}{2T}}$
IV	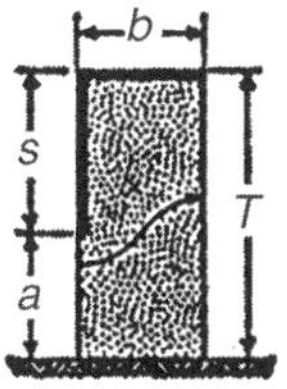	$s \geq b$: $\Phi = \ln\left(1 + \frac{b}{a}\right)$ $b \geq s$: $\Phi = \ln\left(1 + \frac{s}{a}\right) + \frac{b - s}{T}$
	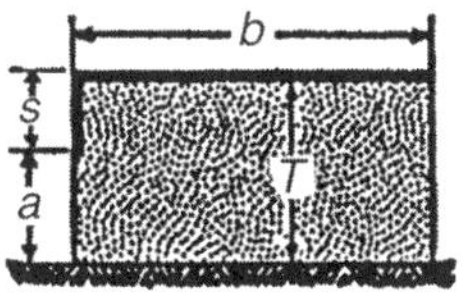	

(*continued*)

TABLE 5.1 Summary of Fragment Types and Form Factors (*Continued*)

Fragment Type	Illustration	Form Factor Φ^a
V		$L \leq 2s$: $\Phi = 2\ln\left(1 + \frac{L}{2a}\right)$ $L \geq 2s$: $\Phi = 2\ln\left(1 + \frac{s}{a}\right) + \frac{L - 2s}{T}$
VI		$L > s' + s''$: $\Phi = \ln\left[\left(1 + \frac{s'}{a'}\right)\left(1 + \frac{s''}{a''}\right)\right] + \frac{L - (s' + s'')}{T}$ $L = s' + s''$: $\Phi = \ln\left[\left(1 + \frac{s'}{a'}\right)\left(1 + \frac{s''}{a''}\right)\right]$ $L < s' + s''$: $\Phi = \ln\left[\left(1 + \frac{b'}{a'}\right)\left(1 + \frac{b''}{a''}\right)\right]$ where $b' = \frac{L + (s' - s'')}{2}$ $b'' = \frac{L - (s' - s'')}{2}$

Source: Adapted from Harr (1962).

[a] h is the head loss through a fragment.

TABLE 5.2 Ratios of Elliptic Integrals of the First Kind for Use in Table 5.1

m^2	K/K'	m^2	K/K'	m^2	K/K'	m^2	K/K'
0.000	0.000	0.21	0.745	0.51	1.009	0.81	1.378
0.001	0.325	0.22	0.754	0.52	1.018	0.82	1.397
0.002	0.349	0.23	0.763	0.53	1.028	0.83	1.417
0.003	0.366	0.24	0.773	0.54	1.037	0.84	1.439
0.004	0.379	0.25	0.782	0.55	1.047	0.85	1.461
0.005	0.389	0.26	0.791	0.56	1.057	0.86	1.485
0.006	0.398	0.27	0.800	0.57	1.066	0.87	1.510
0.007	0.406	0.28	0.808	0.58	1.076	0.88	1.537
0.008	0.413	0.29	0.817	0.59	1.087	0.89	1.567
0.009	0.420	0.30	0.826	0.60	1.097	0.90	1.599

(*continued*)

TABLE 5.2 Ratios of Elliptic Integrals of the First Kind *(Continued)*

m^2	K/K'	m^2	K/K'	m^2	K/K'	m^2	K/K'
0.01	0.426	0.31	0.834	0.61	1.107	0.991	2.381
0.02	0.471	0.32	0.843	0.62	1.118	0.992	2.418
0.03	0.502	0.33	0.852	0.63	1.129	0.993	2.461
0.04	0.526	0.34	0.860	0.64	1.140	0.994	2.510
0.05	0.547	0.35	0.869	0.65	1.151	0.995	2.568
0.06	0.565	0.36	0.877	0.66	1.162	0.996	2.639
0.07	0.582	0.37	0.886	0.67	1.174	0.997	2.731
0.08	0.598	0.38	0.895	0.68	1.186	0.998	2.860
0.09	0.612	0.39	0.903	0.69	1.198	0.999	3.081
0.10	0.625	0.40	0.911	0.70	1.211	0.9991	3.115
0.11	0.638	0.41	0.920	0.71	1.224	0.9992	3.152
0.12	0.650	0.42	0.929	0.72	1.237	0.9993	3.195
0.13	0.662	0.43	0.938	0.73	1.251	0.9994	3.244
0.14	0.674	0.44	0.946	0.74	1.265	0.9995	3.302
0.15	0.684	0.45	0.955	0.75	1.279	0.9996	3.373
0.16	0.695	0.46	0.964	0.76	1.294	0.9997	3.465
0.17	0.706	0.47	0.973	0.77	1.310	0.9998	3.594
0.18	0.716	0.48	0.982	0.78	1.326	0.9999	3.814
0.19	0.726	0.49	0.991	0.79	1.343	1	infinity
0.20	0.735	0.50	1.000	0.80	1.360		

The ratios of elliptic integrals of the first kind needed to determine form factors are given in Table 5.2.

Example 5.2 Use the method of fragments to determine seepage for the hydraulic structure with downstream cutoff shown in Fig. E5.2.

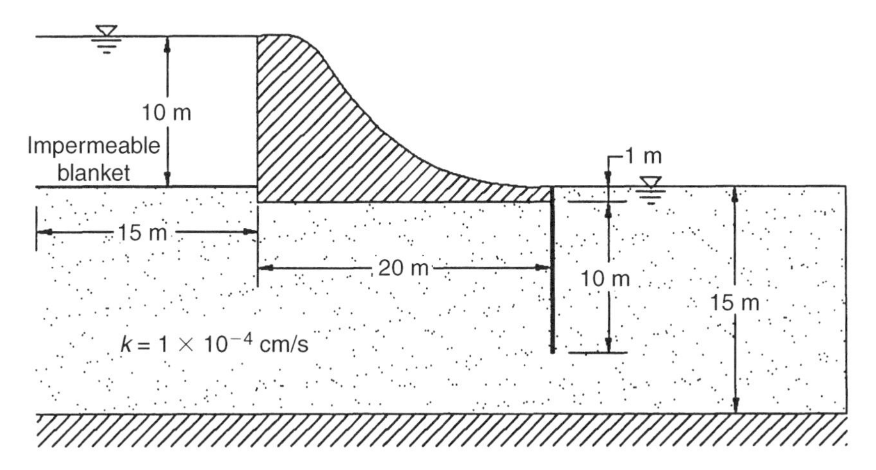

Fig. E5.2

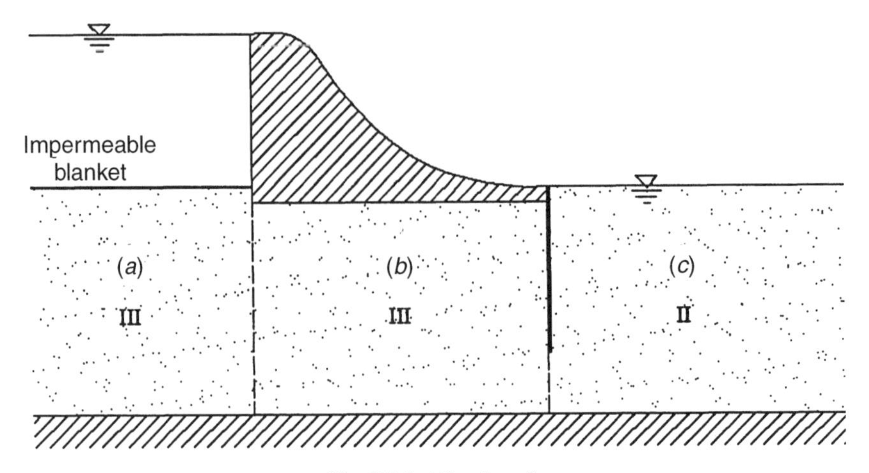

Fig. E5.2 (*Continued*)

SOLUTION: (a) Fragment type III, $b = 15$ m, $s = 1$ m, $T = 15$ m,

$$
\begin{aligned}
m &= \cos\frac{\pi s}{2T}\sqrt{\tanh^2\frac{\pi b}{2T} + \tan^2\frac{\pi s}{2T}} \\
&= \cos\frac{\pi \times 1\text{ m}}{2 \times 15\text{ m}}\sqrt{\tanh^2\frac{\pi \times 15\text{ m}}{2 \times 15\text{ m}} + \tan^2\frac{\pi \times 1\text{ m}}{2 \times 15\text{ m}}} \\
&= 0.918
\end{aligned}
$$

so $m^2 = 0.843$. From Table 5.2, for $m^2 = 0.843$, $K/K' = 1.445$. From Table 5.1, $\Phi = 1.445$.

(b) Fragment type III, $b = 20$ m, $s = 10$ m, $T = 14$ m,

$$
\begin{aligned}
m &= \cos\frac{\pi s}{2T}\sqrt{\tanh^2\frac{\pi b}{2T} + \tan^2\frac{\pi s}{2T}} \\
&= \cos\frac{\pi \times 10\text{ m}}{2 \times 14\text{ m}}\sqrt{\tanh^2\frac{\pi \times 20\text{ m}}{2 \times 14\text{ m}} + \tan^2\frac{\pi \times 10\text{ m}}{2 \times 14\text{ m}}} \\
&= 0.996
\end{aligned}
$$

so $m^2 = 0.992$. From Table 5.2, for $m^2 = 0.992$, $K/K' = 2.42$. From Table 5.1, $\Phi = 2.42$.

(c) Fragment type II, $s = 11$ m, $T = 15$ m,

$$
m = \sin\frac{\pi s}{2T} = \sin\frac{\pi \times 11\text{ m}}{2 \times 15\text{ m}} = 0.914
$$

so $m^2 = 0.835$. From Table 5.2, for $m^2 = 0.835$, $K/K' = 1.428$. From Table 5.1, $\Phi = 1.428$.

So the seepage is

$$q = \frac{kh}{\sum_{n=1}^{3} \Phi_n} = \frac{10^{-4}\ \text{cm/s} \times 10\ \text{m}}{1.445 + 2.42 + 1.428} \times \frac{100\ \text{cm}}{1\ \text{m}} \approx 1.9 \times 10^{-2}\ \text{cm}^2/\text{s}$$

PROBLEMS

5.1. Refer to Problem 3.1. Use the solutions discussed in Section 5.4 to determine the seepage rate underneath the structure.

5.2. Refer to Problem 3.4. Determine the seepage rate using the solutions discussed in Section 5.4 for all three cases. Use $H = 10$ m and $k = 2 \times 10^{-3}$ cm/s.

5.3. Refer to Problem 3.7. Use the solutions discussed in Section 5.4 to determine the seepage rate q and exit gradients.

5.4. Refer to Fig. 5.8*a*. Map the pressure distribution along the base of the structure and along the sheet pile given the following parameters: $b_1 = 15$ m, $s = 5$ m, $h_1 = 8$ m, and $h_2 = 3$ m.

5.5. Refer to Fig. 5.8*b*. Map the pressure distribution along the base of the structure and along the sheet pile given $b_1 = 5$ m, $b_2 = 10$ m, $s = 5$ m, $h_1 = 8$ m, and $h_2 = 3$ m.

5.6. For a single sheet pile with $h = 10$ m (refer to Fig. 5.10*c*), plot the exit gradient as a function of depth of embedment (s) of the pile.

5.7. For the sheet pile shown in Fig. P5.7, determine the seepage using the method of fragments.

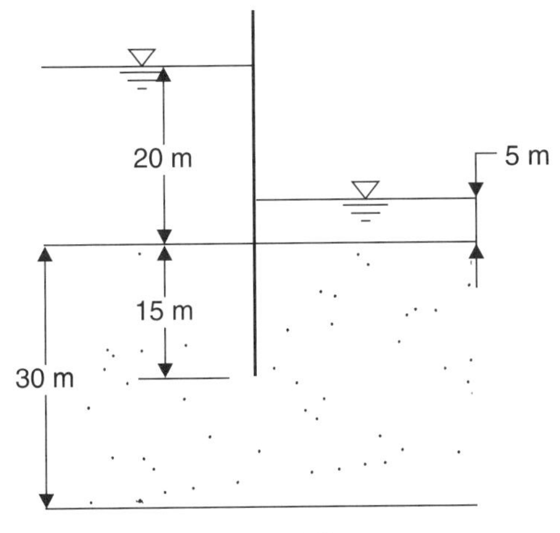

Fig. P5.7

5.8. Refer to Problem 3.8. Use the method of fragments to determine the seepage for all three cases.

CHAPTER 6

SEEPAGE FORCES

Water loses energy as it travels in soils from locations of high energy to those of low energy. The *principle of conservation of energy,* which states that energy can be neither gained nor lost, implies that the energy lost by the seepage water is transferred to the soil skeleton. This transferred energy acts as a force on the soil skeleton through frictional drag in the direction of flow. We will see later in this chapter that the seepage force thus created in soils is related directly to the hydraulic gradient causing flow.

Often, the magnitude of the seepage forces is significant enough to disturb force balance and destabilize soils. If the seepage forces are not counteracted by other structural forces, failure may take place, such as in the case of collapse of retaining structures, earth dam failures, landslides, and so on. A key component in the design of geotechnical infrastructure is to manipulate or control the seepage forces by providing adequate drainage measures. In this chapter we quantify the seepage forces and study their role in geotechnical stability. We restrict ourselves to a general discussion of the role of seepage in geotechnical stability, leaving the specifics of individual geotechnical structures to students trained in advanced soil mechanics concepts.

6.1 EFFECTIVE STRESS PRINCIPLE

To understand the role of seepage forces in geotechnical stability, we must note that any loads applied on soils are borne by both the soil skeleton (consisting of soil solids) and pore water. The *effective stress principle* enables us to partition the total stresses on a soil system into the individual stresses carried by soil skeleton and pore water.

Consider the saturated soil mass shown in Fig. 6.1*a* and the state of stress on a fictitious plane *AA*. The total stress at this elevation, σ, is given by

$$\sigma = H_w \gamma_w + H_s \gamma_{sat} \tag{6.1}$$

where γ_w is the unit weight of water, γ_{sat} the saturated unit weight of soil, and H_w and H_s are as shown in Fig. 6.1*a*. The first term on the right-hand side of Eq. (6.1) is the stress due to standing water on soil mass, and the second term is due to the weight of the saturated soil mass. Alternatively, σ could be expressed in terms of the submerged unit weight of soil:

$$\sigma = H \gamma_w + H_s \gamma_{sub} \tag{6.2}$$

where H is the total height shown in Fig. 6.1*a* and γ_{sub} is the submerged unit weight of soil. Considering $\gamma_{sat} = \gamma_{sub} + \gamma_w$, the expressions (6.1) and (6.2) are equivalent.

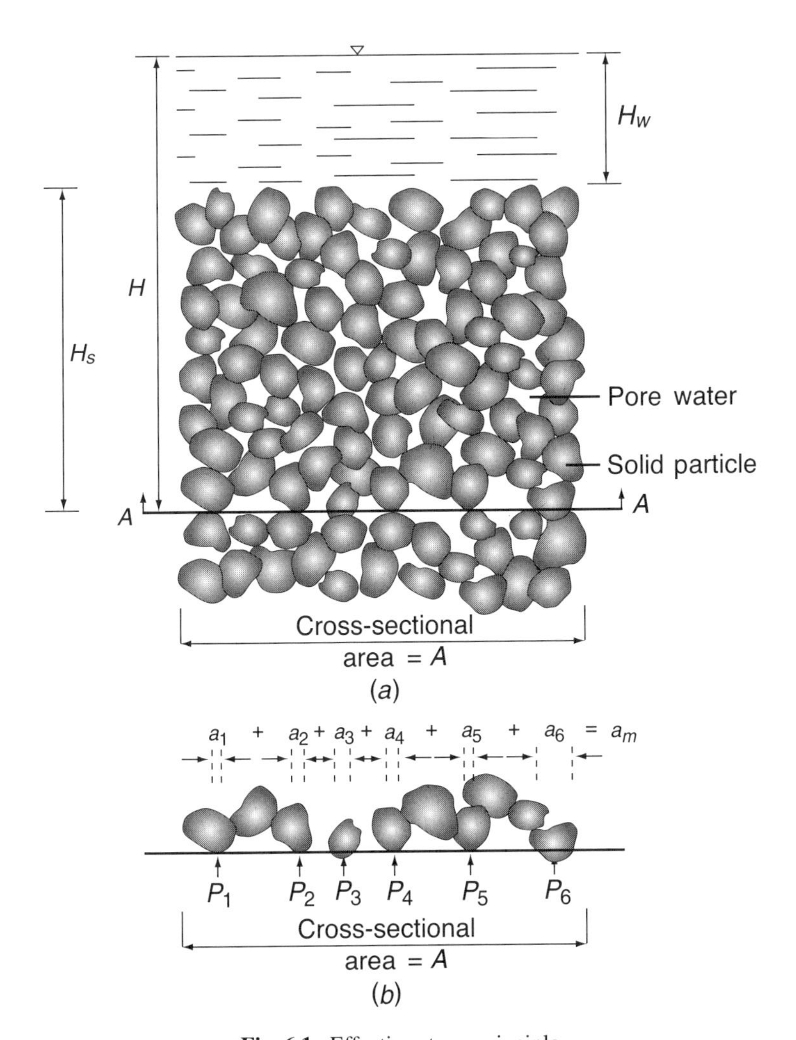

Fig. 6.1. Effective stress principle.

The stress carried by pore water on the plane AA, under hydrostatic conditions, is simply

$$u = H\gamma_w \qquad (6.3)$$

where u is the pore-water pressure, also referred to as neutral stress. Comparison of Eqs. (6.2) and (6.3) indicates that $H_s\gamma_{\text{sub}}$ is the remainder of the total stress, which is borne by solid grains of the soil skeleton. In general terms, one can state that the total stress in a soil system is the summation of two components: pore-water pressure or neutral stress (u) and intergranular pressure or effective stress, σ', or

$$\sigma = \sigma' + u \qquad (6.4)$$

Although Eq. (6.4) is conceptually simple when viewed in terms of stresses, one must note the difficulty involved in determining the individual forces. The pore water pressure u on the plane AA acts not on the entire cross-sectional area A of the plane but over an area equal to $A - a_m$, where A is the total cross-sectional area of the sample and a_m is the total intergranular contact area (see Fig. 6.1*b*). Similarly, the stress carried by solid particles acts only on the area a_m. In reality, the contribution of the soil skeleton to total stress is the summation of the vertical components of the numerous granule-to-granule contact loads P_1, P_2, P_3, ... shown in Fig. 6.1*b*. In expressing the stresses as in Eq. (6.4), we are therefore neglecting a_m in determining neutral stress u and expressing *effective stress* σ' as the force per unit area carried by the soil skeleton. The effective stress, determined by subtracting u from σ, is a key parameter in assessing the stability of the geotechnical infrastructure. As seen in the following sections, pore-water pressure may increase drastically under flow conditions, although the total stress σ may remain constant. This leads to a reduction in effective stress, which may mean partial or total loss of solid particle contacts in the physical sense, causing quicksand conditions and other instabilities.

Effective stress also influences the shear strength of soils. Trusting that the reader will refer to geotechnical engineering textbooks for details, we quickly note here that the shear strength of cohesionless soils is proportional to the normal stress on the plane of shear failure,

$$\tau = \sigma' \tan \phi \qquad (6.5)$$

where τ is the shear strength, σ' the normal effective stress on the failure plane, and ϕ the angle of internal friction of the soil. Many instabilities in geotechnical infrastructure occur due to lack of adequate shear strength. The key to enhancing stability in the presence of water is therefore to maximize the effective stresses and increase shear strength while minimizing the shear stresses driving failure. This is a fine balancing act, as we will see in Section 6.4.

6.2 STRESSES IN SOILS WITH SEEPAGE

Let us reconsider the saturated soil mass in Fig. 6.1 and determine the stresses under three possible conditions: no seepage (Fig. 6.2*a*), downward seepage (Fig. 6.2*b*), and

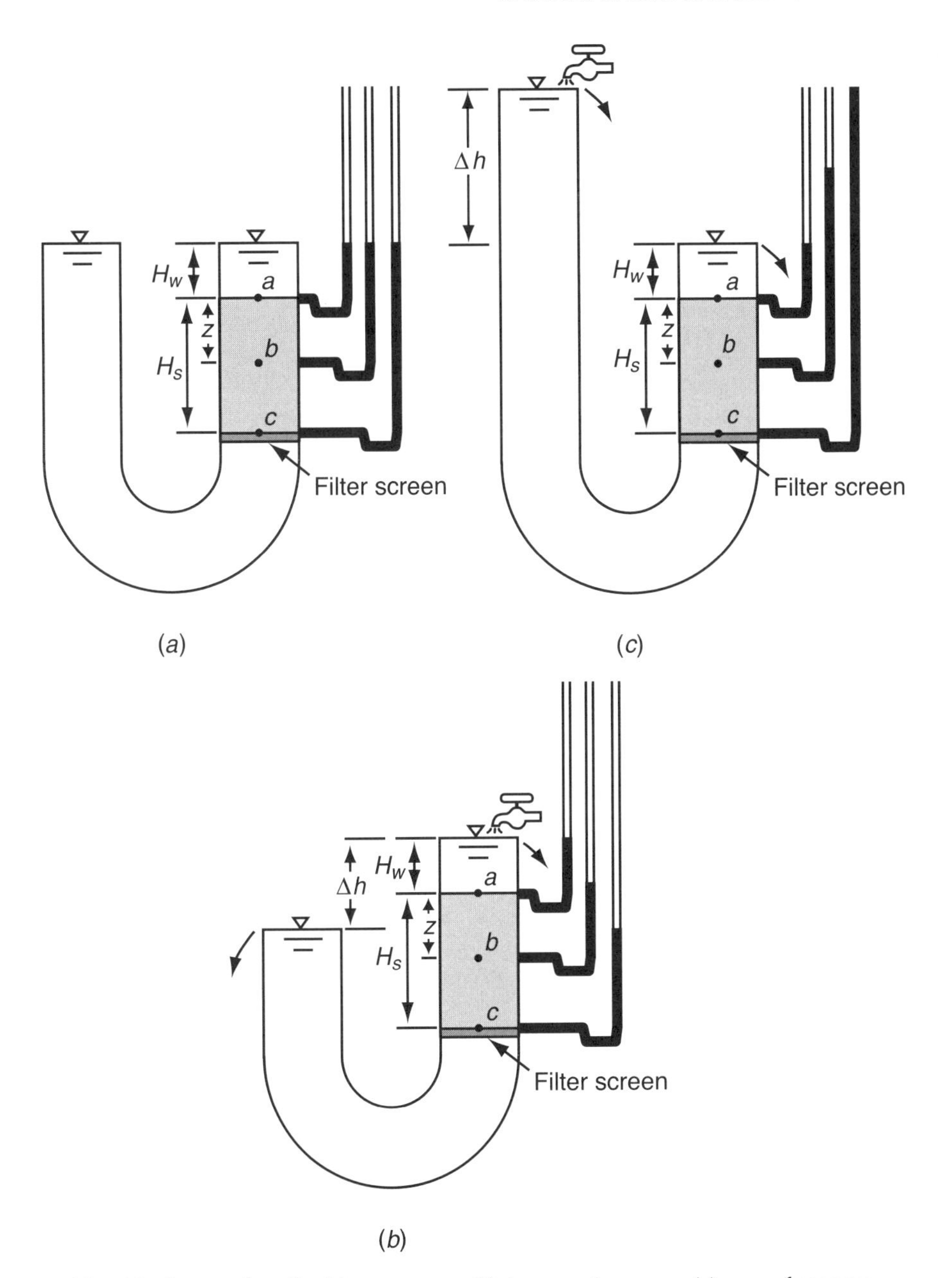

Fig. 6.2. Stresses in soils: (*a*) no seepage, (*b*) downward seepage; (*c*) upward seepage.

upward seepage (Fig. 6.2*c*). As seen in Fig. 6.2*a*, the piezometric levels at the three points *a*, *b*, and *c* in the soil remain at the same elevation in the absence of seepage. The direction of seepage controls the trends in piezometric levels, as shown in Fig. 6.2*b* and *c*. The total stress on any plane is due to the load caused by saturated soil mass and water above that plane. Therefore, it does not vary among the three cases. For the three points *a*, *b* and *c*, σ may be expressed as

$$\sigma_a = H_w \gamma_w \tag{6.6a}$$

$$\sigma_b = H_w \gamma_w + z\gamma_{\text{sat}} \tag{6.6b}$$

and

$$\sigma_c = H_w \gamma_w + H_s \gamma_{\text{sat}} \tag{6.6c}$$

The pore-water pressures or neutral stresses differ among the three cases. In the absence of seepage (Fig. 6.2*a*),

$$u_a = H_w \gamma_w \tag{6.7a}$$

$$u_b = (H_w + z)\gamma_w \tag{6.7b}$$

and

$$u_c = (H_w + H_s)\gamma_w \tag{6.7c}$$

Let Δh denote the difference between pressure heads at the top and bottom of the soil sample, which causes seepage in the cases shown in Fig. 6.2*b* and *c*. The hydraulic gradient i in both cases is therefore equal to $\Delta h/H_s$. Considering that the change in pressure head along the length of the soil sample is linear, the hydraulic head at point b in the case of downward seepage (Fig. 6.2*b*) may be expressed as

$$h_b = H_w + z - iz \tag{6.8a}$$

and in the case of upward seepage (Fig. 6.2*c*) as

$$h_b = H_w + z + iz \tag{6.8b}$$

The pore-water pressures in the presence of seepage are then given as follows:

Downward seepage (Fig. 6.2*b*):

$$u_a = H_w \gamma_w \tag{6.9a}$$

$$u_b = (H_w + z - iz)\gamma_w \tag{6.9b}$$

$$u_c = (H_w + H_s - \Delta h)\gamma_w \tag{6.9c}$$

Upward seepage (Fig. 6.2*c*):

$$u_a = H_w \gamma_w \tag{6.10a}$$

$$u_b = (H_w + z + iz)\gamma_w \tag{6.10b}$$

$$u_c = (H_w + H_s + \Delta h)\gamma_w \tag{6.10c}$$

The effective stresses may now be expressed in terms of the total stresses [Eq. (6.6)] and pore-water pressures [Eqs. (6.7), (6.9), and (6.10)] using the effective stress principle, $\sigma' = \sigma - u$. The expressions are listed below for the three cases.

No seepage (Fig. 6.2*a*):

$$\sigma'_a = 0 \tag{6.11a}$$

$$\sigma'_b = z(\gamma_{sat} - \gamma_w) = z\gamma_{sub} \tag{6.11b}$$

$$\sigma'_c = H_s(\gamma_{sat} - \gamma_w) = H_s\gamma_{sub} \tag{6.11c}$$

Downward seepage (Fig. 6.2*b*):

$$\sigma'_a = 0 \tag{6.12a}$$

$$\sigma'_b = z\gamma_{sub} + iz\gamma_w \tag{6.12b}$$

$$\sigma'_c = H_s\gamma_{sub} + \Delta h\gamma_w \tag{6.12c}$$

Upward seepage (Fig. 6.2*c*):

$$\sigma'_a = 0 \tag{6.13a}$$

$$\sigma'_b = z\gamma_{sub} - iz\gamma_w \tag{6.13b}$$

$$\sigma'_c = H_s\gamma_{sub} - \Delta h\gamma_w \tag{6.13c}$$

The effect of seepage in soils is therefore to increase effective stresses in the case of downward seepage and to decrease them in the case of upward seepage.

6.3 SEEPAGE FORCE AND CRITICAL HYDRAULIC GRADIENTS

Consider the portion of the saturated soil mass between the elevations of points a and b in Fig. 6.2. The forces acting on this soil portion are as shown in Fig. 6.3. In the presence of seepage, the body forces are supplemented or counteracted by a force due to seepage, which may be expressed as [see Eqs. (6.12*b*) and (6.13*b*)]

$$F_s = iz\gamma_w A \tag{6.14}$$

where A is the area of cross section of the soil sample. Considering that the volume of soil equals zA, the seepage force per unit volume will reduce to a simple expression,

$$f_s = \frac{iz\gamma_w A}{zA} = i\gamma_w \tag{6.15}$$

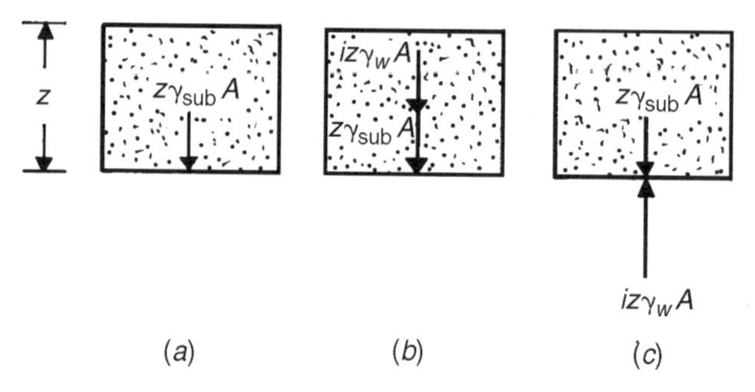

Fig. 6.3. Body forces and seepage forces on a soil element: (*a*) no seepage; (*b*) downward seepage; (*c*) upward seepage.

In other words, the seepage force per unit volume of soil is proportional to the hydraulic gradient, with the constant of proportionality being equal to the unit weight of water. This expression for seepage force becomes useful in determining the stability of soils in slopes and in downstream soil wedges at sheet pile installations, as discussed in Section 6.4.

When the hydraulic gradients are high, it is possible that the seepage stresses may equal or exceed the stresses due to the submerged weight of soil. This presents a unique problem in the case of upward seepage (Fig. 6.3*c*). When effective stresses become zero, the interparticle contacts are lost and soil begins to behave like a liquid, a condition often referred as *quicksand* or *boiling*. For this condition to occur,

$$\sigma' = z\gamma_{\text{sub}} - iz\gamma_w = 0 \tag{6.16}$$

The gradient satisfying Eq. (6.16), known as the *critical gradient* i_{cr}, is given by

$$i_{\text{cr}} = \frac{\gamma_{\text{sub}}}{\gamma_w} \tag{6.17}$$

The reader will verify from definitions of soil void ratio e and specific gravity G_s that the submerged unit weight of soils may be expressed as

$$\gamma_{\text{sub}} = \frac{(G_s - 1)\gamma_w}{1 + e} \tag{6.18}$$

Substituting Eq. (6.18) in (6.17) yields

$$i_{\text{cr}} = \frac{G_s - 1}{1 + e} \tag{6.19}$$

For void ratios of sand ranging from 0.5 to 1.0, the range of i_{cr} corresponds to 1.12 to 0.84, for a specific gravity of 2.68 for sands.

It is important to note that seepage forces act in the direction of flow, independent of the orientation of soil mass and regardless of gravity. For the same orientation of the soil mass, seepage stresses reverse their direction if the flow direction is reversed (Fig. 6.4).

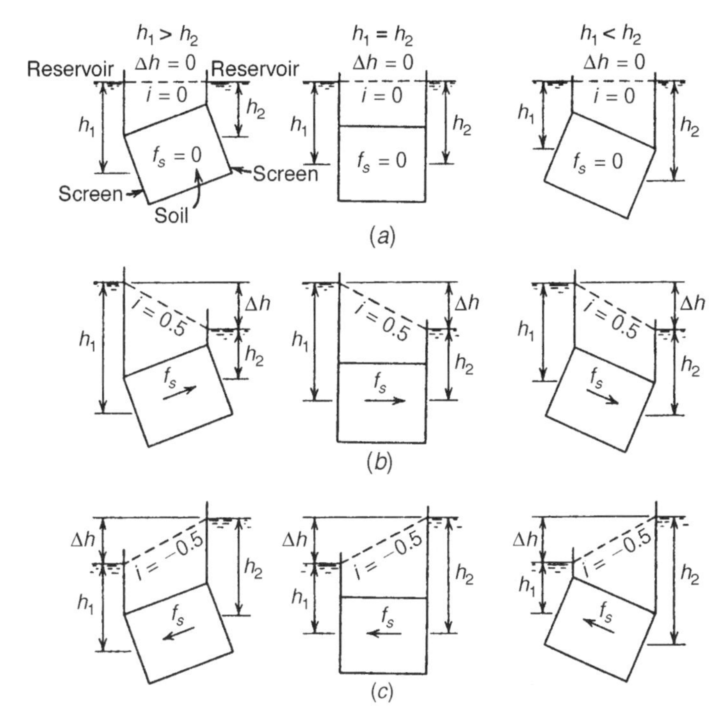

Fig. 6.4. Direction of seepage forces relative to the orientation of a soil sample. (Adapted from Cedergren, 1977.)

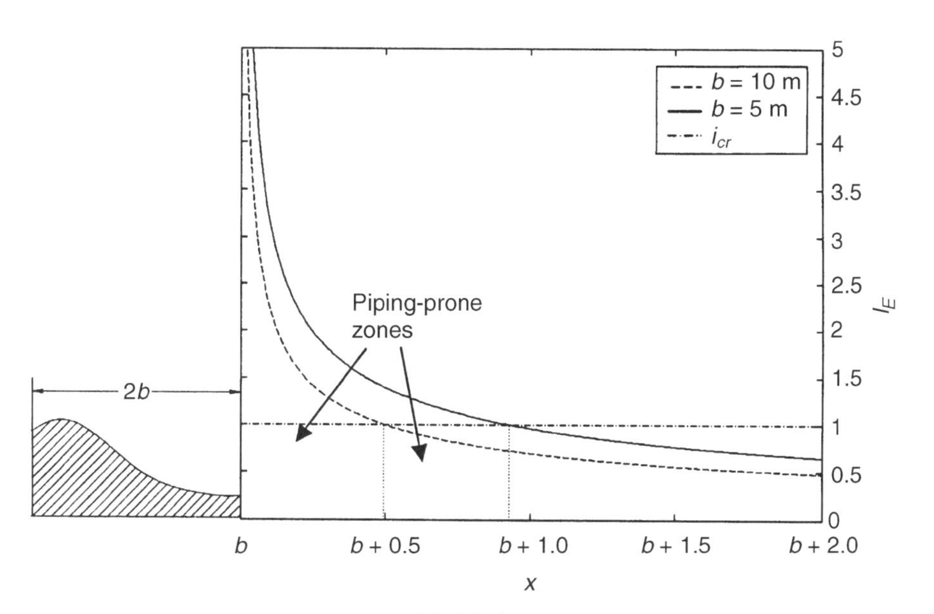

Fig. E6.1

Example 6.1 Exit gradients were plotted for a hydraulic structure in our solution to Example 5.1. Revisit that solution and show how these gradients compare with the critical gradient if the sandy soils at the structure have a void ratio of 0.7 and a specific gravity of 2.70.

SOLUTION: The critical gradient is estimated using Eq. (6.19):

$$i_{\text{cr}} = \frac{G_s - 1}{1 + e} = \frac{2.7 - 1.0}{1 + 0.7} = 1.0$$

From Fig. E6.1, the exit gradient is larger than the critical gradient within zones of 0.49 m and 0.93 m for $b = 10$ m and 5 m, respectively.

6.4 ROLE OF SEEPAGE IN STRUCTURAL STABILITY

Because seepage through soils is associated with forces, any geotechnical structure subjected to seepage through or underneath must be evaluated for stability with due consideration to seepage forces. In this section we analyze a range of geotechnical systems that are prone to structural instability due to seepage.

6.4.1 Stability of Hydraulic Structures

There are two important aspects of seepage that influence the stability of hydraulic structures: internal and exit gradients, and uplift pressures along the bottom of the structures. The exit gradients must be compared with critical hydraulic gradients to ensure the safety of downstream soils. This is one of the motivations for deriving analytical expressions for exit gradients in Chapter 5. As shown in Fig. 6.5*a*, the exit gradients at the downstream may sometimes exceed i_{cr}, initiating boiling, or quicksand conditions. The soil is liquefied if the exit gradient is greater than i_{cr}. This results in further increase in hydraulic gradient because of the shortening of seepage path through soil media. An opening or a pipe is soon developed which progresses upstream as more and more soil is liquefied and scooped out of the pipe (Fig. 6.5*b*). This process, referred to as *piping,* is a well-known dam collapse/failure mechanism. It is possible for the pipe to reach the upstream water, in which case the pipe may be enlarged further, causing catastrophic failure. Generally, cohesionless soils (fine sands and silts) are prone to piping failures. Cohesive soils have strong interparticle bonds, which resist particle separation and washing. The exception is in the case of dispersive clays, where interparticle bonds are weak and are easily influenced by slight changes in the chemistry of the eroding fluid.

Cedergren (1989) presented an interesting analysis of how the position of a cutoff influences the magnitude of exit gradients and the degree of safety against piping. Flow nets for three different foundation configurations of a hydraulic structure are shown in Fig. 6.6. When a sheet pile cutoff is located under the upstream edge of the structure, the exit gradients downstream are very high (Fig. 6.6*a*). Locating the

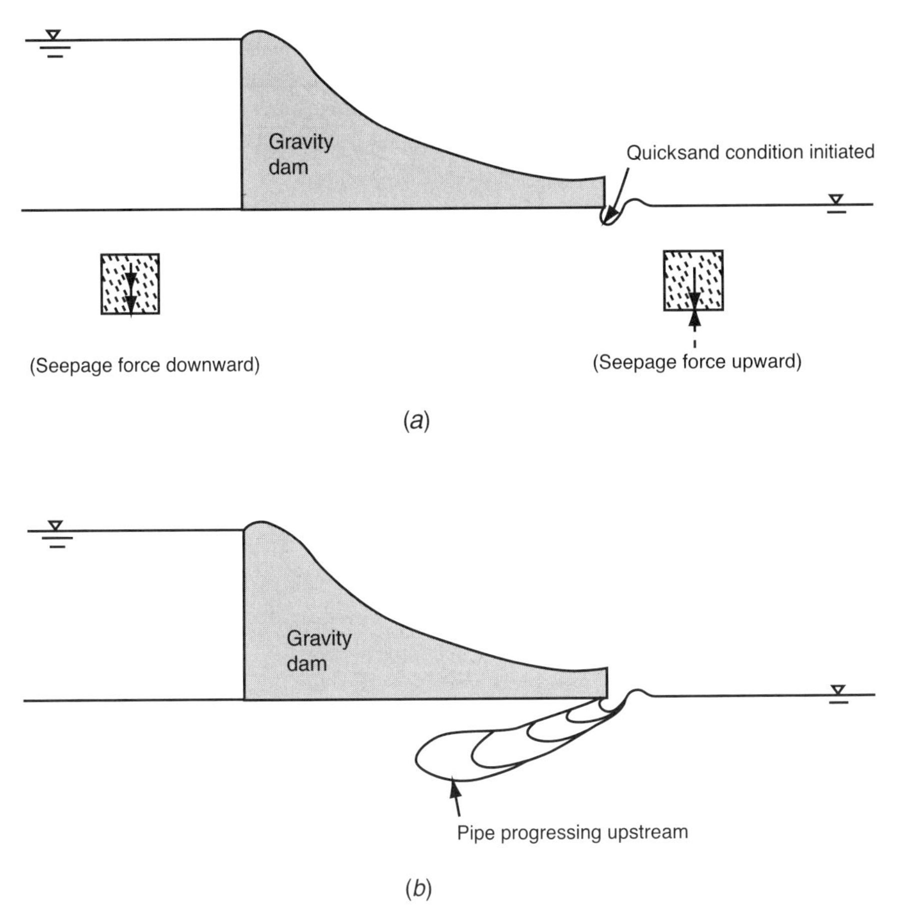

Fig. 6.5. Stability of hydraulic structures: (*a*) initiation of quick sand conditions at the downstream; (*b*) progressing pipe.

cutoff under the downstream edge of the structure (Fig. 6.6*b*) maintains an identical but transposed flow net with the same seepage quantities. However, the exit gradients in this case are drastically reduced. Using a cutoff under the upstream edge and a drain at the downstream edge (Fig. 6.6*c*) slightly increases seepage quantities (15% in this case) but provides safe exit gradients.

Piping may also be initiated in the interior of an earth dam. Flow occurring at high internal gradients may cause internal erosion. Although filters provided adjacent to the core on the downstream side generally protect the cores against loss of material, it is advisable to keep the internal gradients low. As shown in Fig. 6.7, the seepage gradients through cores in zoned dams are controlled by the size and shape of impervious cores. The exit gradients at cores may vary from less than unity for wide cores to more than 10 for dams with narrow inclined cores. Depending on the type of material used for cores, the material may be eroded into the filters, initiating a pipe at

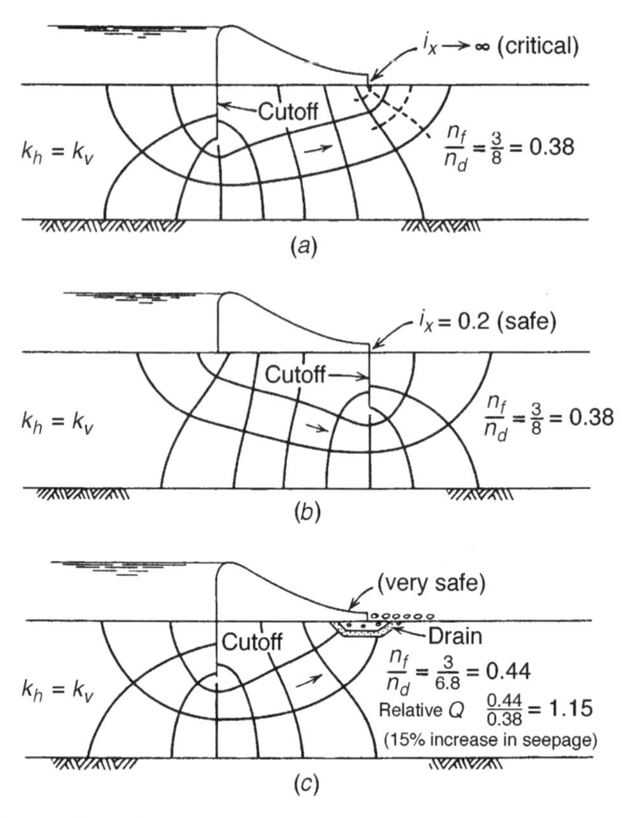

Fig. 6.6. Influence of cutoff location on exit gradients. (Adapted from Cedergren, 1989; copyright © John Wiley & Sons; reprinted with permission.)

the interface of the core and the filter (Fig. 6.8). Even in the absence of high hydraulic gradients, it is advisable to assume the presence of concentrated leaks at the interface of core and filter materials. Differential settlements of dams and poor construction practices may cause these leaks. When sufficient gradients exist, these concentrated leaks may progress backward toward the upstream, causing flow instabilities through the filter and leading eventually to breach of the dam.

In addition to the internal and exit gradients, the uplift pressure distribution along the base of the hydraulic structure also plays a crucial role in the stability of the structure. As shown in Fig. 6.9 for a gravity dam on a pervious foundation, the lateral pressure due to water in the reservoir, the weight of the structure, and the uplift pressure distribution are the three major forces controlling the force and moment equilibrium of the structure. The uplift pressure generally decreases from upstream to downstream. It can be determined using either flow nets (discussed in Chapter 3) or analytical solutions (Chapter 5). When sheet pile cutoffs are used, their location controls the distribution of pressure (see Fig. 6.6). The magnitude and location of the

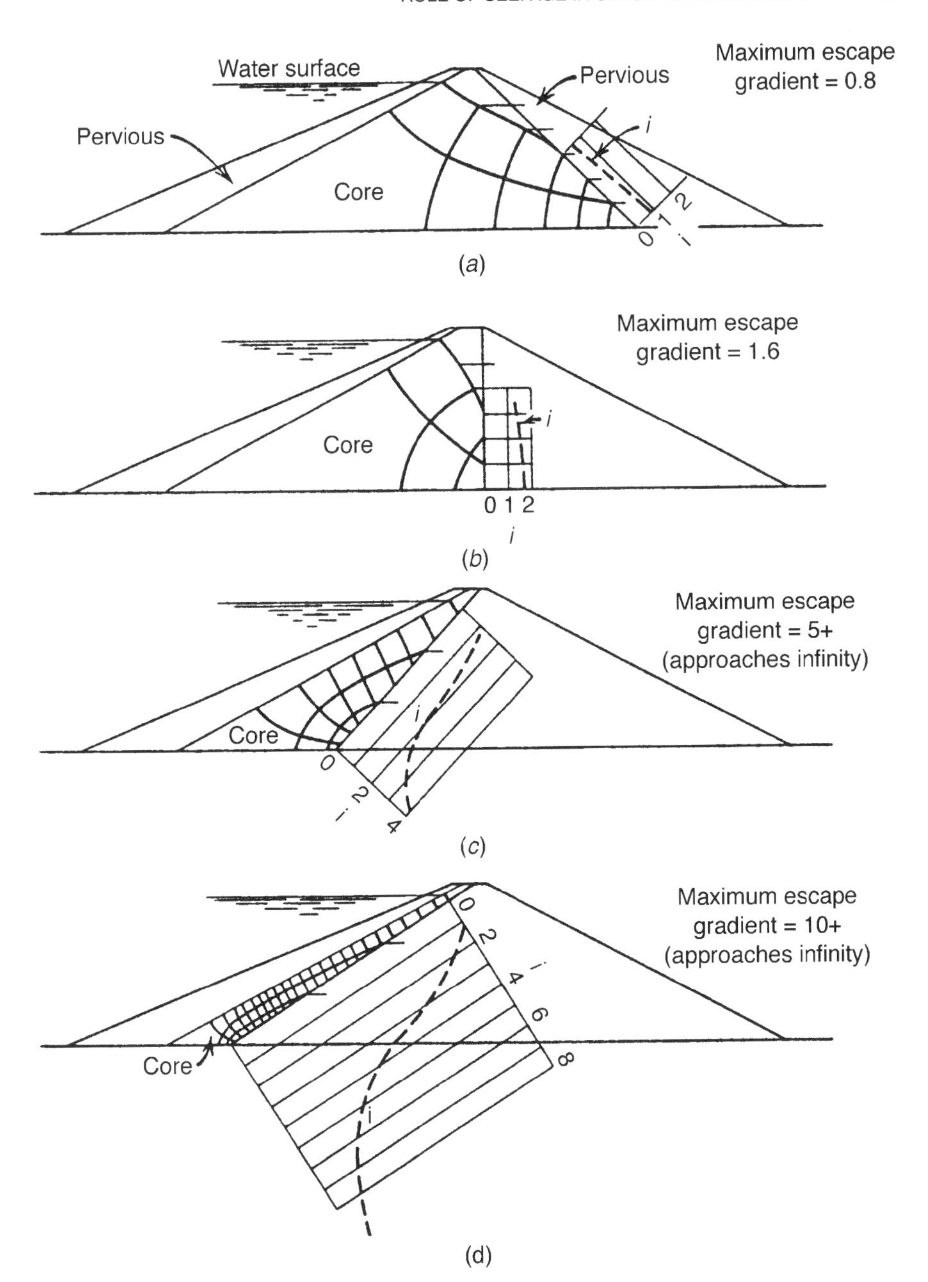

Fig. 6.7. Gradients at the downstream end of the core. (Adapted from Cedergren, 1989; copyright © John Wiley & Sons; reprinted with permission.)

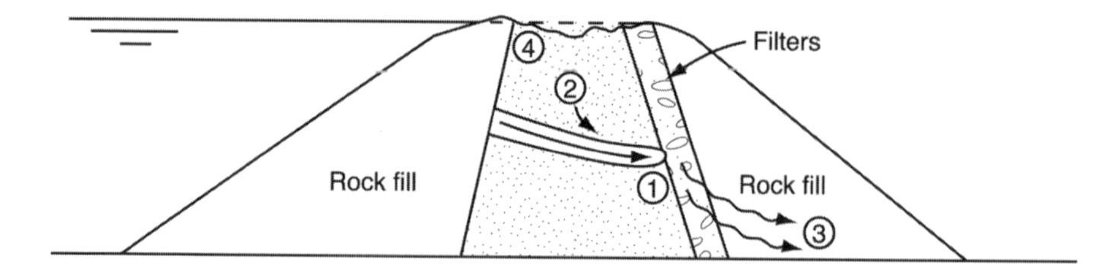

Fig. 6.8. Failure of earth dam due to internal erosion and piping.

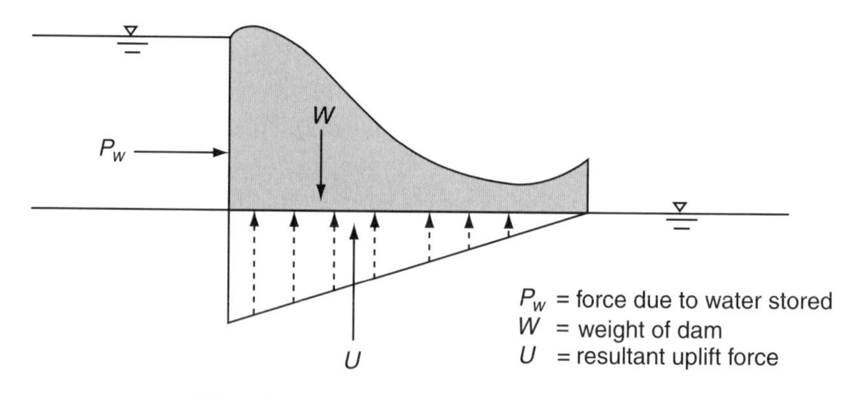

Fig. 6.9. Force balance on a hydraulic structure.

resulting uplift force is determined by calculating the area and location of the center of gravity of the pressure distribution diagram.

6.4.2 Heaving and Piping at Sheet Pile Structures

When driven into the ground, interlocking steel sheets reduce seepage toward the site where seepage protection is needed. This is particularly suitable when coarse-grained soils are present or in stratified soils with alternating fine-grained and pervious layers where horizontal permeability greatly exceeds vertical. Seepage quantities into excavations can be estimated using flow nets (Chapter 3) or advanced analytical solutions (Chapter 5). An important consideration in using cutoff walls for reduction of seepage is the exit gradients. Seepage on the downstream side of sheet piles may produce piping in dense sands or heaving in loose sands (Fig. 6.10). The rising of a mass of

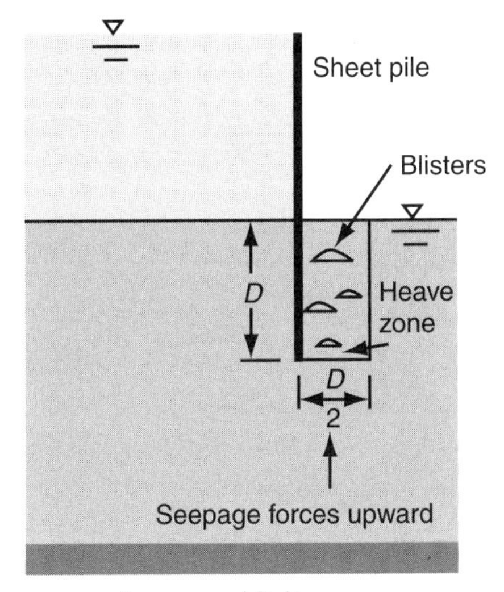

Fig. 6.10. Development of heaving at the downstream side of sheet pile structures.

soil as a result of large seepage forces is referred to as *heaving*. During the heaving process, soil expands locally with increased void ratio, often forming water blisters. The blisters gradually rise to the top as soil particles at the roof of the blisters collapse. Terzaghi (1922) stated that heaving generally occurs within a distance equal to half of the sheet pile embedment depth. A comparison of the magnitudes of the upward seepage force and the downward body force due to the submerged weight of soil in this zone allows us to determine the stability of the soil. A factor of safety against heaving could be defined as

$$\text{FS} = \frac{W'}{F_s} = \frac{\frac{1}{2}\left(D^2\right)\gamma_{\text{sub}}}{\frac{1}{2}\left(D^2\right)i_{\text{av}}\gamma_w} \tag{6.20}$$

or

$$\text{FS} = \frac{\gamma_{\text{sub}}}{i_{\text{av}}\gamma_w} \tag{6.21}$$

where W' is the submerged weight of soil in the heave zone and i_{av} is the average hydraulic gradient in the heave zone.

To improve the factor of safety against heaving, an inverted filter of weight W may be placed over the soil wedge (Fig. 6.11). This increases the factor of safety to

$$\text{FS} = \frac{W' + W}{F_s} \tag{6.22}$$

The filter must, of course, be designed such that its particles are coarse enough to permit upward seepage of water but fine enough to prevent penetration of soil particles into its voids. Filter design is discussed in detail in Chapter 7.

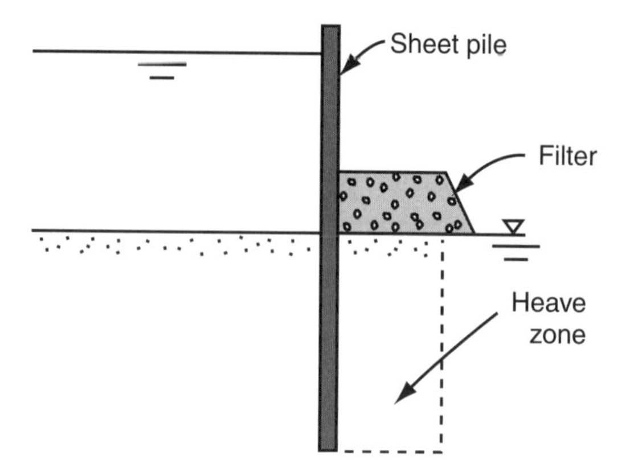

Fig. 6.11. Use of an inverted filter to protect soils from heaving.

To prevent piping or heaving problems, the sheet pile must be provided adequate penetration depth below the bottom of excavation. The penetration required for cutoff walls in sands of infinite and finite depth can be determined using Fig. 6.12. For specified values of w, H_w, and H_1, the depth of penetration D can be determined for a given factor of safety. Generally, a safety factor of 1.5 to 2 is desired against piping or heaving.

In the case of stratified soils, the exit seepage gradients will be different from the case of homogeneous soils in Fig. 6.12. The presence of coarse sands below fine sands makes flow in the fine sands nearly vertical and generally increases seepage gradients in the fine layer compared to the homogeneous cross section of Fig. 6.12. Similarly, the presence of a fine layer below coarse sands constricts flow beneath the cutoff wall and generally decreases seepage gradients in the coarse layer. Guidelines for protecting sheet piles against piping and heaving in stratified soils are given in Fig. 6.13.

6.4.3 Slope Stability

Stability assessment of slopes is one of the most common tasks in geotechnical engineering practice. Failures due to instability of slopes range from landslides and avalanches in the case of natural slopes to slumping of soils in finite slopes (such as on a highway embankment or a dike). The failures are generally due to sliding on a critical surface where the driving shear forces are greater than the shear strength. The critical surface could be the interface between topsoils and intact bedrock in the case of infinite slopes (Fig. 6.14*a* and *b*) or a cylindrical surface in finite slopes (Fig. 6.14*c* and *d*). The driving shear forces are components of the self-weight of the soil mass in the downward direction of the slope. The shear strength on the critical surface resisting the driving forces consists of two components, cohesive and frictional resistance, given by

$$\tau = c + \sigma' \tan \phi \tag{6.23}$$

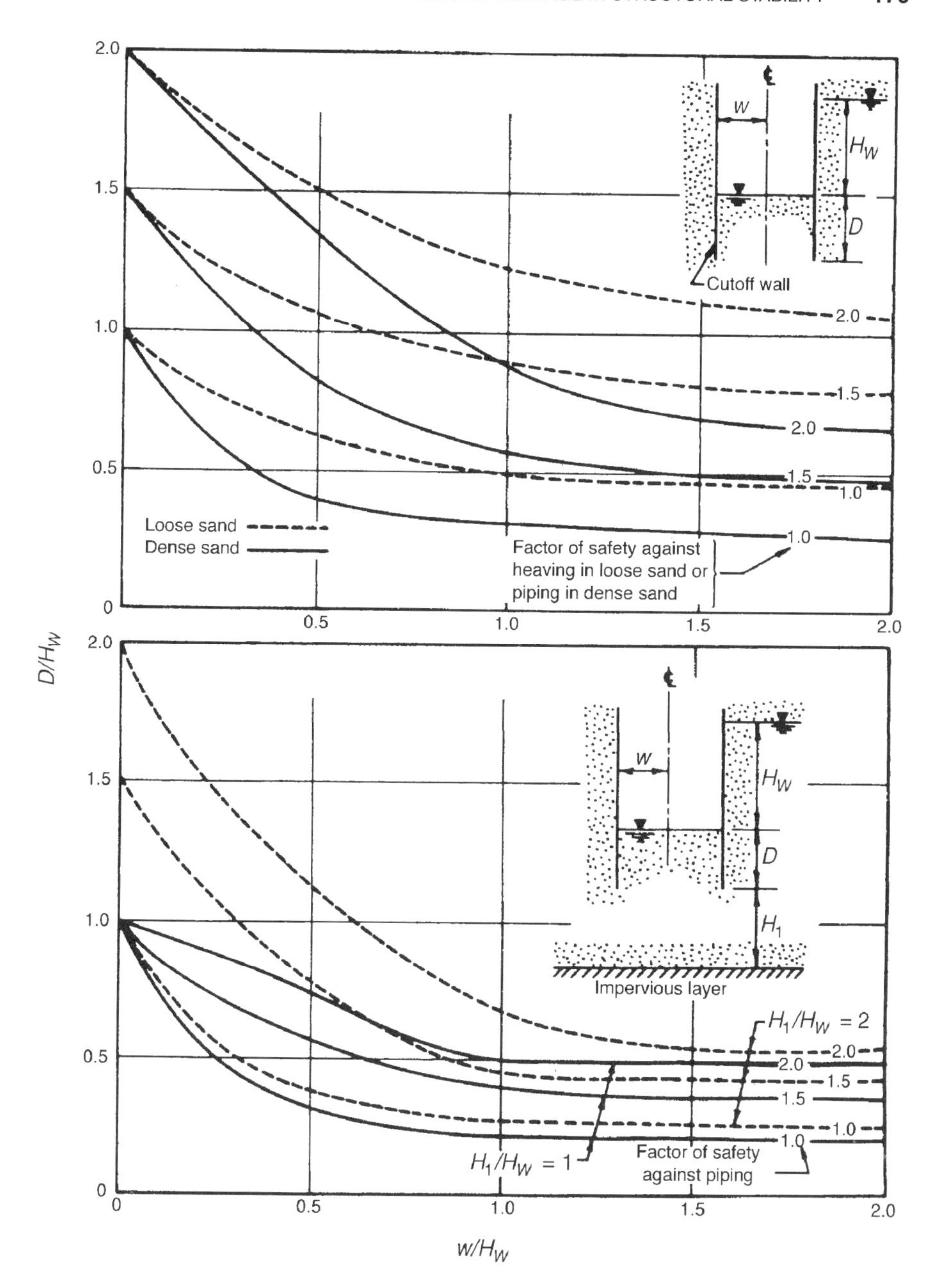

Fig. 6.12. Penetration of cutoff wall to prevent piping or heaving in isotropic sands. (Adapted from U.S. Department of the Navy, Naval Facilities Engineering Command, 1982.)

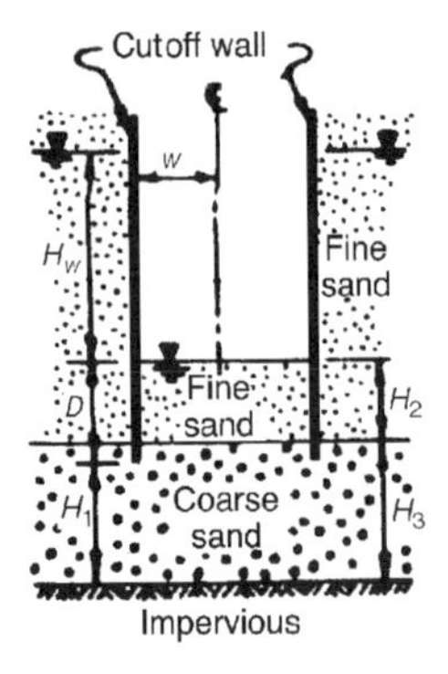

Coarse sand underlying fine sand

If top of coarse layer is at a depth below cutoff wall bottom greater than width of excavation, safety factors of Fig. 6.12 for infinite depth apply.

If top of coarse layer is at a depth below cutoff wall bottom less than width of excavation, the uplift pressures are greater than for the homogeneous cross section. If permeability of coarse layer is more than 10 times that of fine layer, failure head (H_w) = thickness of fine layer (H_2).

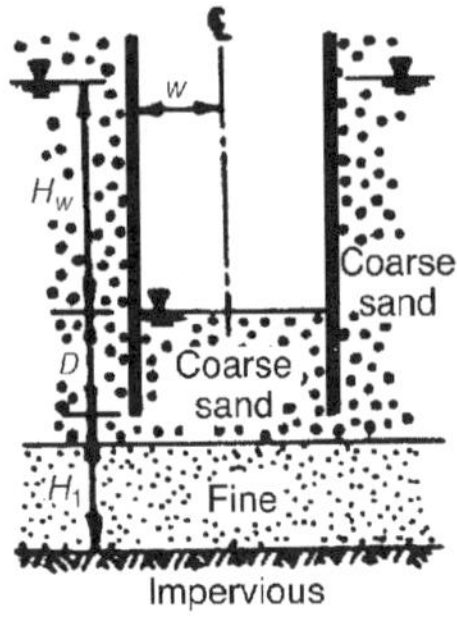

Fine sand underlying coarse sand

If top of fine layer lies below cutoff wall bottom, safety factors are intermediate between those for an impermeable boundary at top or bottom of the fine layer (Fig. 6.12).

If top of the fine layer lies above cutoff wall bottom, the safety factors of Fig. 6.12 are somewhat conservative for penetration required.

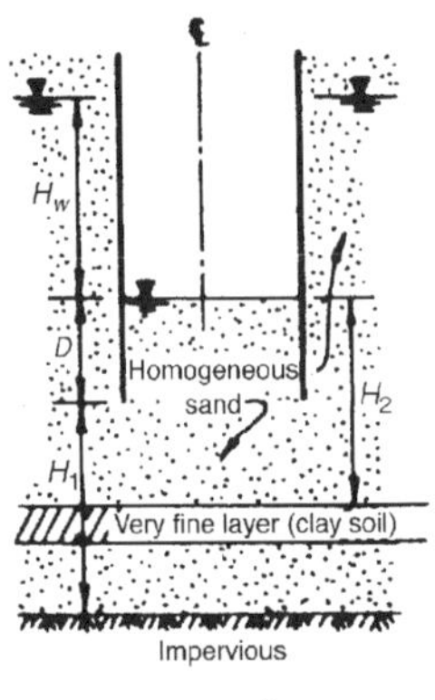

Fine layer in homogeneous sand stratum

If the top of fine layer is at a depth greater than width of excavation below cutoff wall bottom, safety factors of Fig. 6.12 apply, assuming impervious base at top of fine layer.

If top of fine layer is at depth less than width of excavation below cutoff wall tips, pressure relief is required so that unbalanced head below fine layer does not exceed height of soil above base of layer.

If fine layer lies above subgrade of excavation, final condition is safer than homogeneous case, but dangerous condition may arise during excavation above the fine layer and pressure relief is required as in the preceding case.

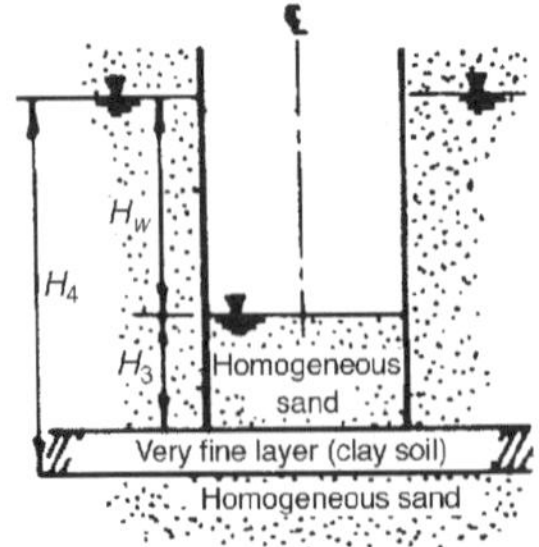

To avoid bottom heave, $\gamma_T \times H_3$ should be greater than $\gamma_w \times H_4$.

γ_T = total unit weight of the soil

γ_w = unit weight of water

Fig. 6.13. Penetration of cutoff wall to prevent piping or heaving in stratified sands. (Adapted from U.S. Department of the Navy, Naval Facilities Engineering Command, 1982.)

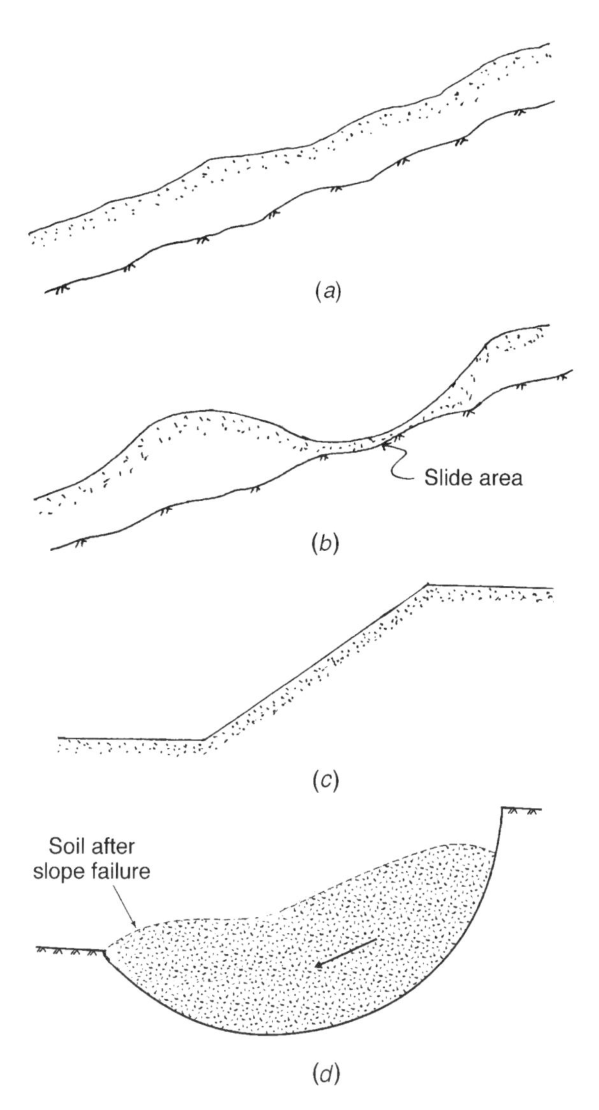

Fig. 6.14. Slope stability: (*a*) intact infinite slope; (*b*) failed infinite slope; (*c*) intact finite slope; (*d*) failed finite slope.

Equation (6.23) is an extension of Eq. (6.5) with the cohesive component c added. Note that the normal effective stress (σ') in Eq. (6.23) is also derived from the self-weight of the soil. Thus, as shown schematically in Fig. 6.15, one component of self-weight of the soil mass (T) tends to drive failure, and the other component (N) contributes to shear strength (τ) resisting failure. The key to stabilizing slopes is therefore to maximize N and minimize T. A factor of safety FS against slope failure is defined as the ratio of the sum of resisting forces τ on the entire failure surface to the sum of driving forces T from all the soil elements in the failure zone,

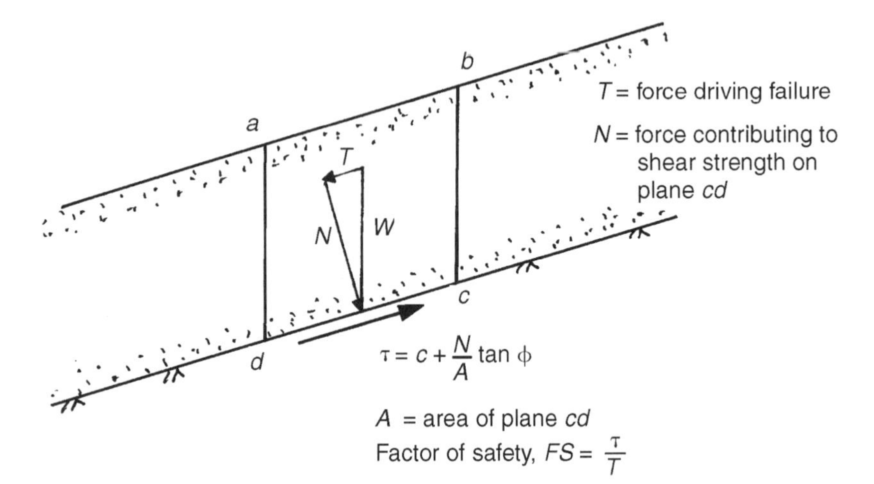

Fig. 6.15. Force balance of an element on an infinite slope.

$$\mathrm{FS} = \frac{\Sigma\,\tau}{\Sigma\,T} \tag{6.24}$$

In the case of curved failure surfaces, the driving forces (which have different directions) cannot be added arithmetically; therefore, the factor of safety is defined in terms of moments about a point, such as the center of the cylindrical failure surface (Fig. 6.16). Thus,

$$\mathrm{FS} = \frac{\text{Moments of } \tau}{\text{Moments of } T} \tag{6.25}$$

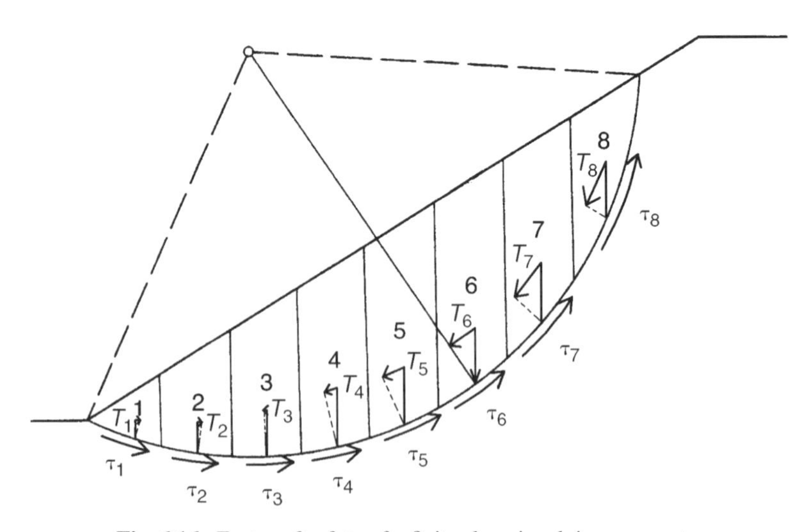

Fig. 6.16. Factor of safety of a finite slope involving moments.

Seepage in soils affects stability primarily in two ways:

1. It produces pore-water pressures, which reduce effective stresses (σ') and lower shear strength τ in Eq. (6.23).
2. It produces seepage forces, which may increase the driving forces or overturning moments.

Seepage may also affect slope stability by altering the structure of soils and thus reducing or eliminating cohesive strength c; however, this falls in the purview of the physicochemical behavior of soils and is beyond the scope of this book. To illustrate the effect of seepage, consider the stability of an infinite slope under three conditions:

1. The slope is dry (Fig. 6.17*a*).
2. The slope is saturated to the level of the ground surface with seepage parallel to slope (Fig. 6.17*b*).
3. The slope is saturated to the level of the ground surface, and the presence of a highly pervious gravel layer at some depth forces the seepage to be vertical (Fig. 6.17*c*).

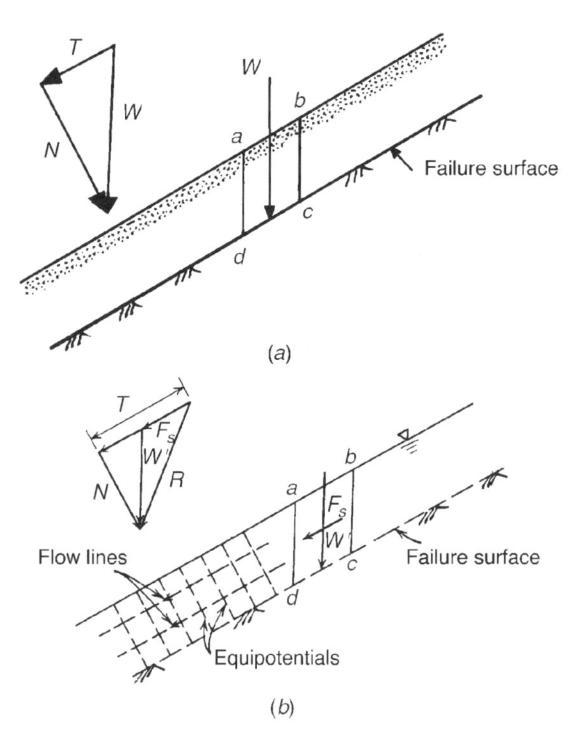

Fig. 6.17. Force balance of an element on an infinite slope: (*a*) under dry conditions; (*b*) seepage parallel to slope; (*continued*)

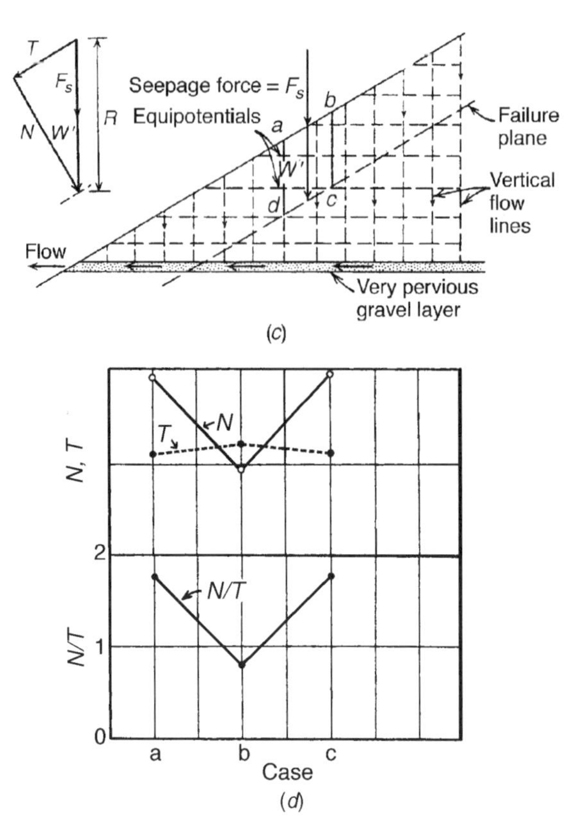

Fig. 6.17. (*Continued*) (*c*) vertically downward seepage; (*d*) magnitudes of N and T, and the ratio N/T, for the three cases. (Adapted from Cedergren, 1989; copyright © John Wiley & Sons; reprinted with permission.)

As shown in the force diagram for Fig. 6.17*a*, the self-weight of the slope is the only contributing force for normal and shear forces on the failure plane. For a slope with seepage parallel to the failure plane (Fig. 6.17*b*), the vertical body force is reduced to the submerged weight of soil (W') as a result of buoyancy. In addition, the presence of seepage force F_s tilts the resultant force R in the downward direction of slope. This increases the shear component T and reduces the normal component N. When the seepage is vertical (Fig. 6.17*c*), the seepage force F_s adds to the submerged weight of soil, thus increasing the normal component N. Depending on the magnitude of seepage force, the changes in N and T could destabilize the slope under conditions in Fig. 6.17*b*, and could enhance the stability of the slope under conditions in Fig. 6.17*c*. The factor, N/T could be used as an index of relative stability of these slopes, as shown in Fig. 6.17*d*. This illustration justifies using horizontal gravel drains in slopes as in Fig. 6.17*c*, to maintain stability.

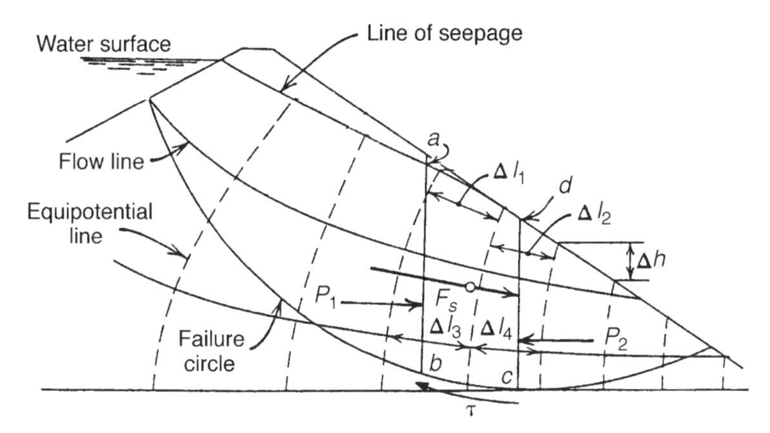

Fig. 6.18. Force balance of an element *abcd* in a finite slope. (Adapted from Cedergren, 1989; copyright © John Wiley & Sons; reprinted with permission.)

The stability of finite slopes can be studied in a similar manner. However, the analysis is more complicated in the case of finite slopes because the components of N and T vary in their directions within the failure region and a single element cannot be used for stability analysis as done above. Consider the element *abcd* in the finite slope shown in Fig. 6.18. The stability of the element is governed by its self-weight (which is almost submerged in this case), the seepage force F_s in the direction of flow, the lateral forces on the element, P_1 and P_2, and the reaction from soil at the bottom of the element τ. The flow gradient is different at different regions in the element, since the flow distances for a single equipotential drop vary within the element ($\Delta l_1 \neq \Delta l_2 \neq \Delta l_3 \neq \Delta l_4$). To determine the magnitude of the seepage force, an average gradient may be used. The position and direction of F_s may also be determined approximately based on the center of gravity of the flow region in the element and the mean direction of flow in the element. When the gradients are nonuniform in the element, the position of F_s should be shifted slightly toward the higher gradients.

Stability of finite slopes is also affected by the rate at which seepage occurs. Taylor (1948) provided one of the first comprehensive analyses on how water varies the loading conditions on a slope. As shown in Fig. 6.19, four conditions of loading were identified: dry conditions (Fig. 6.19*a*), complete submergence of slope (Fig. 6.19*b*), sudden drawdown conditions (Fig. 6.19*c*), and steady seepage subsequent to drawdown (Fig. 6.19*d*). The force polygons for each of these cases are also shown in Fig. 6.19. The force S in these figures represents the shear strength available on the failure plane, and F_t represents the force required for equilibrium. The two forces may not necessarily be equal. Note that the force F_n represents the resultant normal force on the failure plane, which contributes to shear strength on the plane. Under dry conditions (Fig. 6.19*a*), the dry weight W of the wedge

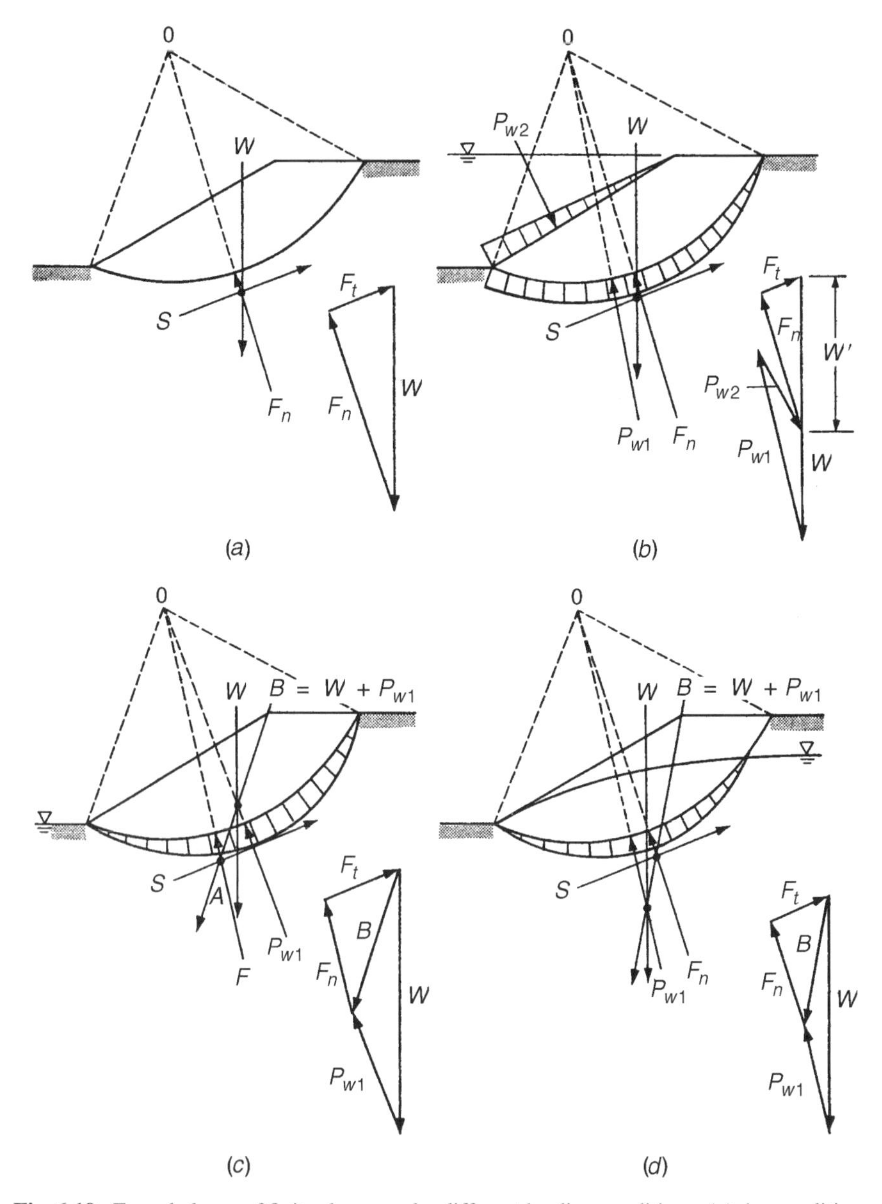

Fig. 6.19. Force balance of finite slopes under different loading conditions: (*a*) dry conditions; (*b*) submerged conditions; (*c*) sudden drawdown conditions; (*d*) steady seepage conditions. (Adapted from Taylor, 1948; Wu, 1981.)

alone contributes to F_n and F_t. The presence of water under submerged conditions (Fig. 6.19*b*) leads to hydrostatic forces P_{w1} and P_{w2}, the combined effect of which is to reduce W to the submerged weight of the wedge, W'. This results in a smaller F_t needed for equilibrium, making this case less severe. In contrast, sudden drawdown conditions (Fig. 6.19*c*) eliminate P_{w2}. P_{w1} remains, although with its magnitude and location modified. This leads to an undesirable direction for the resulting force B. The force F_t needed for equilibrium in this case is generally much greater than in the case of submerged conditions. Due to steady seepage occurring in the slope sometime after drawdown (Fig. 6.19*d*), P_{w1} changes its magnitude and direction, leading to a B generally more favorable than in the case of sudden drawdown. The F_t needed for equilibrium in this case is less than in the case of sudden drawdown. Thus, in terms of slope stability, the conditions represented in Fig. 6.19*b* and *d* are generally less severe than those in Fig. 6.19*a* and *b*.

Slopes that are marginally stable can be stabilized by providing adequate drainage. One of the simplest ways (simple in concept, not necessarily in practice) is to move the flow domain away from the slope into the interior of the slope. Consider the downstream slope in Fig. 6.20*a*. The phreatic line touches the downstream slope, and the bottom portion of the slope is subjected to flow gradients. In addition to causing slope instability, the gradients may also be high enough to cause soil erosion. Moving the flow domain away from the slope by providing a drainage trench at the downstream toe (Fig. 6.20*b*) will help in preventing erosion and in stabilizing the

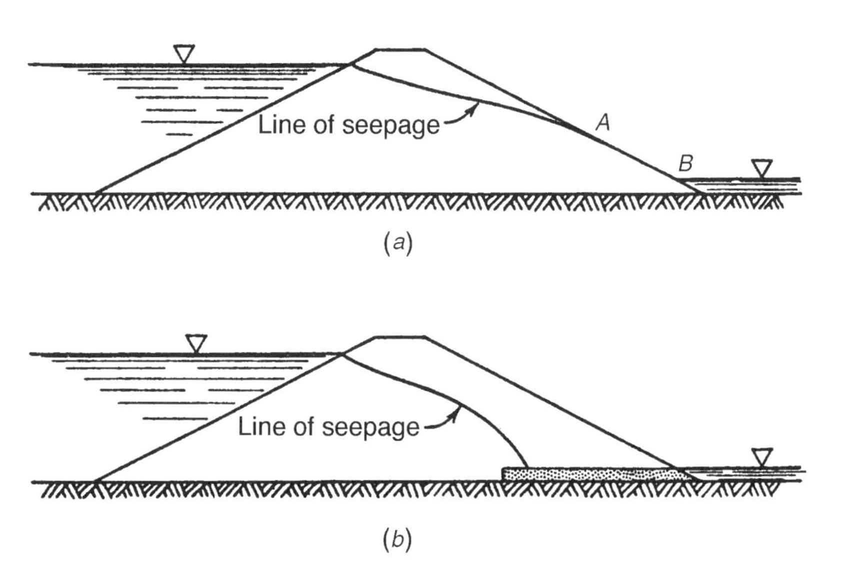

Fig. 6.20. Stabilization of slopes: (*a*) line of seepage touching the downstream face; (*b*) provision of a drainage trench at the downstream toe.

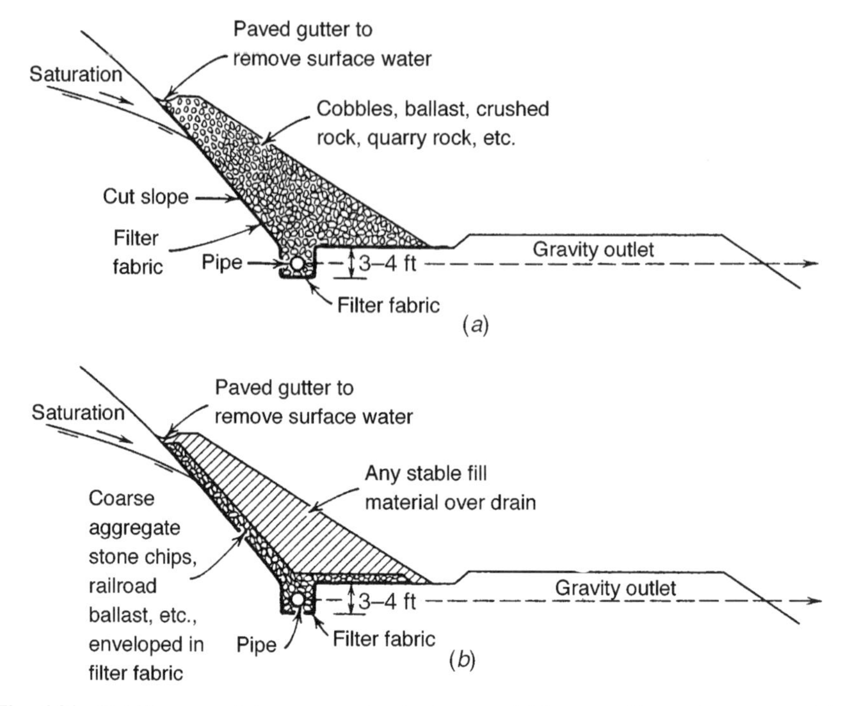

Fig. 6.21. Stabilization of slopes using toe buttresses: (*a*) buttress fill consisting of cobbles, ballast, crushed rock, quarry rock, and so on; (*b*) buttress fill consisting of any stable earthen material. (Adapted from Cedergren, 1989; copyright © John Wiley & Sons; reprinted with permission.)

slope. Toe buttresses (Fig. 6.21) can also be used to stabilize slopes with seepage. The buttresses may consist of coarse-grained filter material or any stable fill material. In either case, filters must be provided in between the slope and the buttress to permit movement of water and prevent soil particle movement. Note that the weight of these buttresses is a force contributing to the stability of soil elements at the toe of the slope.

Another common practice is to use horizontal drains to cause the flow direction to be vertical and to provide a safe outlet for water (Fig. 6.22*a* and *b*). These drains usually consist of small-diameter wells drilled horizontally, or slightly inclined, to intercept seepage. Sometimes, vertical drains are also used in conjunction with horizontal drains to intercept groundwater flowing laterally from the surrounding regions (Fig. 6.22*c*). The drains have to be designed such that their discharge capacity is sufficient to remove water that reaches them. Cedergren (1977) provides case studies of horizontal and vertical drains used to stabilize troublesome highway slopes.

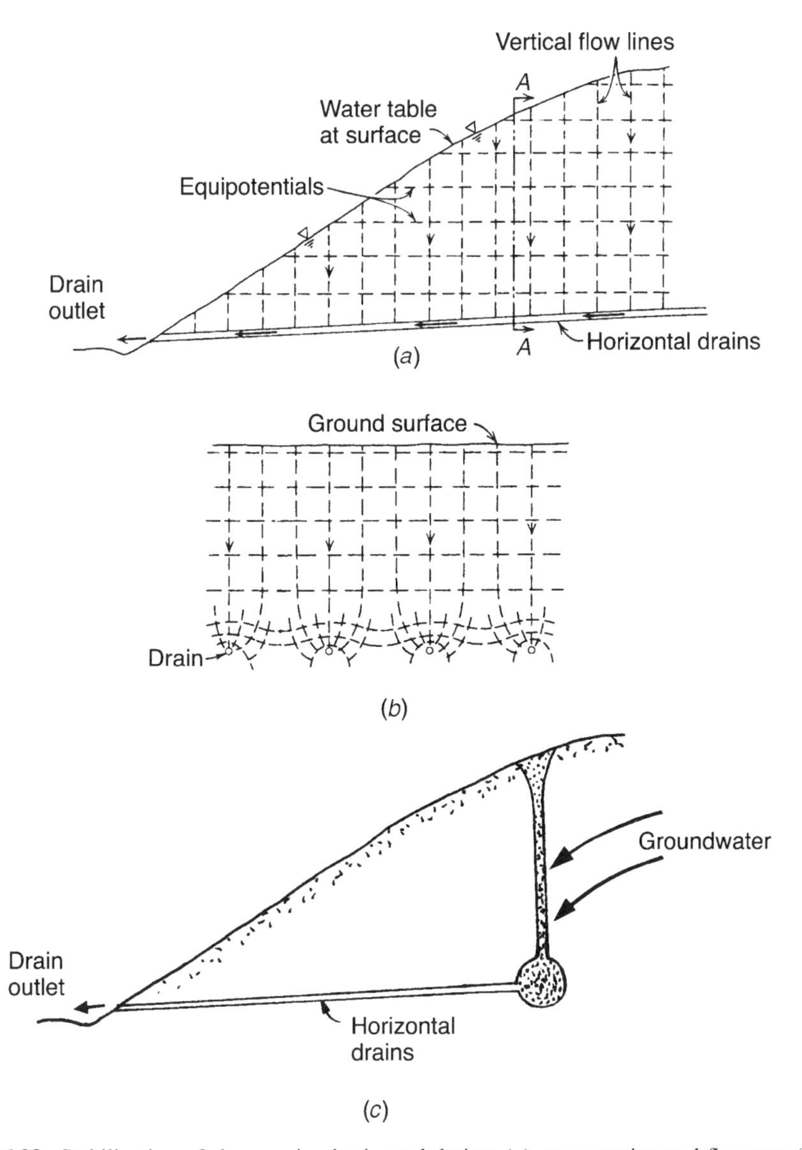

Fig. 6.22. Stabilization of slopes using horizontal drains: (*a*) cross section and flow net; (*b*) longitudinal flow net at section *A*–*A*; (*c*) groundwater interception using vertical drainage wells. (Adapted from Cedergren, 1989; copyright © John Wiley & Sons; reprinted with permission.)

Example 6.2 Geotechnical investigations at a hill slope revealed a 5-m layer of clayey sand overlying a rock stratum. The properties of the topsoil are: dry density $\gamma_{dry} = 17$ kN/m^3, saturated density $\gamma_{sat} = 20$ kN/m^3, cohesion $c = 2$ kN/m^2, coefficient of internal friction $\phi = 25°$. Calculate the factor of safety for a hill slope for the three conditions depicted in Fig. E6.2: (a) the slope is dry; (b) the slope is

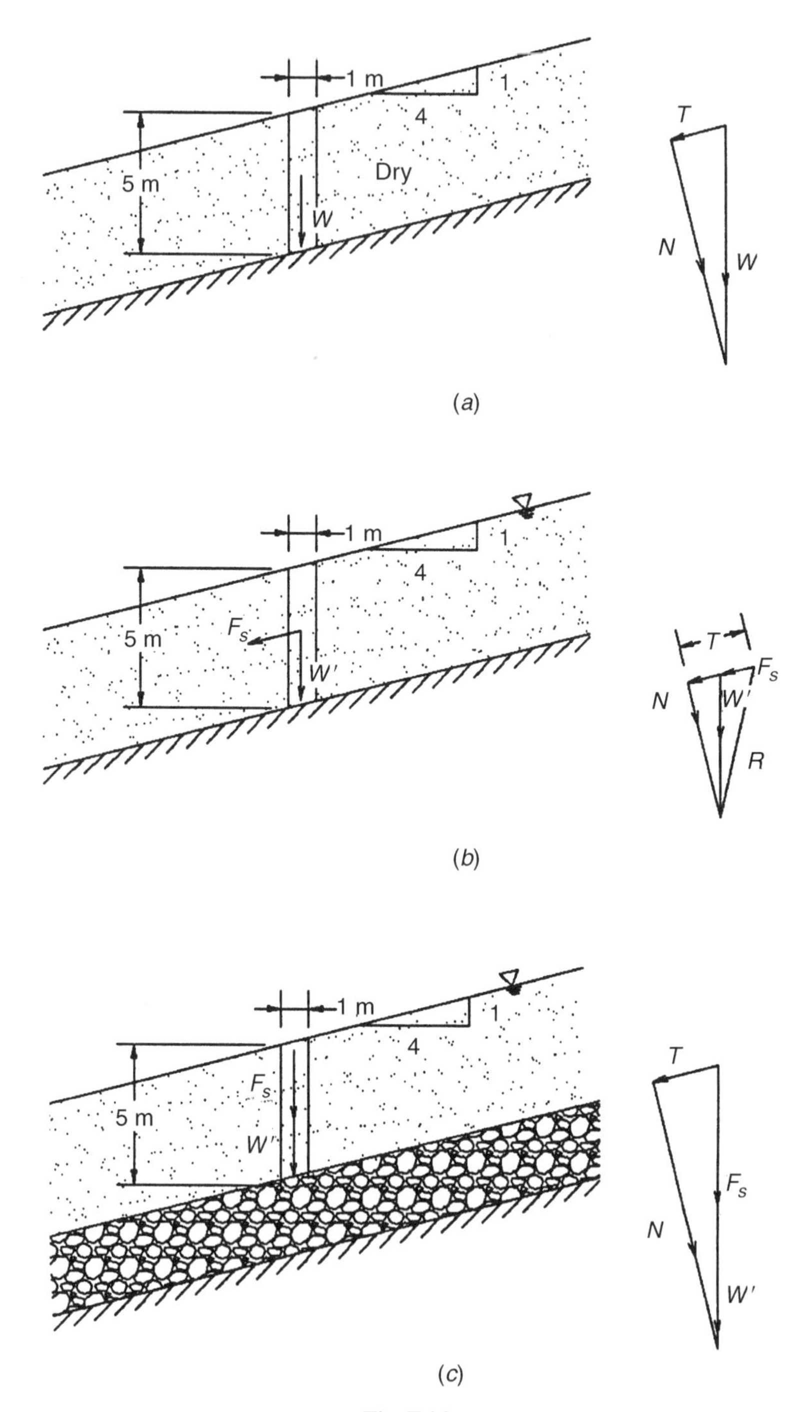

Fig. E6.2

completely saturated with seepage occurring parallel to slope; and (c) a gravel layer exists at the bottom, causing vertical seepage in the slope.

Given data: Slope angle = 1(V): 4(H); $c = 2$ kN/m^2; $\phi = 25°$; $\gamma_{sat} = 20$ kN/m^3; $\gamma_{dry} = 17$ kN/m^3; assume that the width and length of the slice = 1 m.

SOLUTION: (a) *Dry slope:* Weight of the slice $W = 5 \times 1 \times 1 \times 17 = 85$ kN. From the force diagram, $T = 21$ kN; $N = 82.5$ kN. Factor of safety = $(2 \times 1 + 82.5 \times \tan 25°)/21 = 1.93$.

(b) *Saturated slope with parallel seepage:* Submerged weight of the slice $W' = 5 \times 1 \times 1 \times (20 - 9.81) = 50.95$ kN. Hydraulic gradient $i = 0.25$. Seepage force $F_s = i\gamma_w \times$ volume of the element $= 0.25 \times 9.81 \times 5 \times 1 \times 1 = 12.26$ kN. From the force diagram, $T = 27.5$ kN; $N = 50$ kN. Factor of safety = $(2 \times 1 + 50 \times \tan 25°)/27.5 = 0.92$.

(c) *Saturated slope with vertical seepage:* Submerged weight of the slice $W' = 5 \times 1 \times 1 \times (20 - 9.81) = 50.95$ kN. Hydraulic gradient $i = 1$. Seepage force $F_s = i\gamma_w \times$ volume of the element $= 1 \times 9.81 \times 5 \times 1 \times 1 = 49.05$ kN. From the force diagram, $T = 23.75$ kN; $N = 97.5$ kN. Factor of safety = $(2 \times 1 + 97.5 \times \tan 25°)/23.75 = 2.00$.

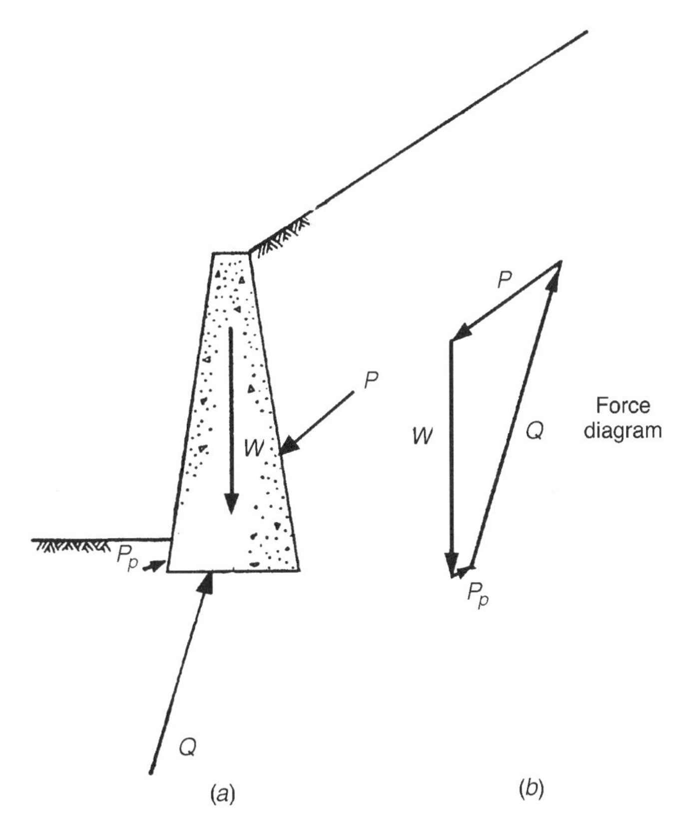

Fig. 6.23. Force balance of an earth-retaining structure.

6.4.4 Stability of Retaining Structures

Retaining structures are walls constructed with concrete or reinforced concrete for the purpose of supporting steep slopes or soils at high elevation. The force balance on these structures is shown in Fig. 6.23. The lateral force due to soil on the retaining wall is resisted by the self-weight of the wall and the reaction from the foundation soil. Water behind the wall adds to the lateral force P. The lateral force due to water is generally significant enough to warrant a drainage provision. To drain the water out, weep holes at regular horizontal and vertical increments are made in the wall by embedding pipes (Fig. 6.24*a*). The weep holes are protected from soil erosion using filter materials/fabric in the interior. Other types of drainage include continuous back drains along the entire back of the wall (Fig. 6.24*b*) and inclined drainage blankets (Fig. 6.24*c*). Note that the inclined drainage blankets cause the seepage forces to be vertically downward, thus protecting the retaining structure from these forces.

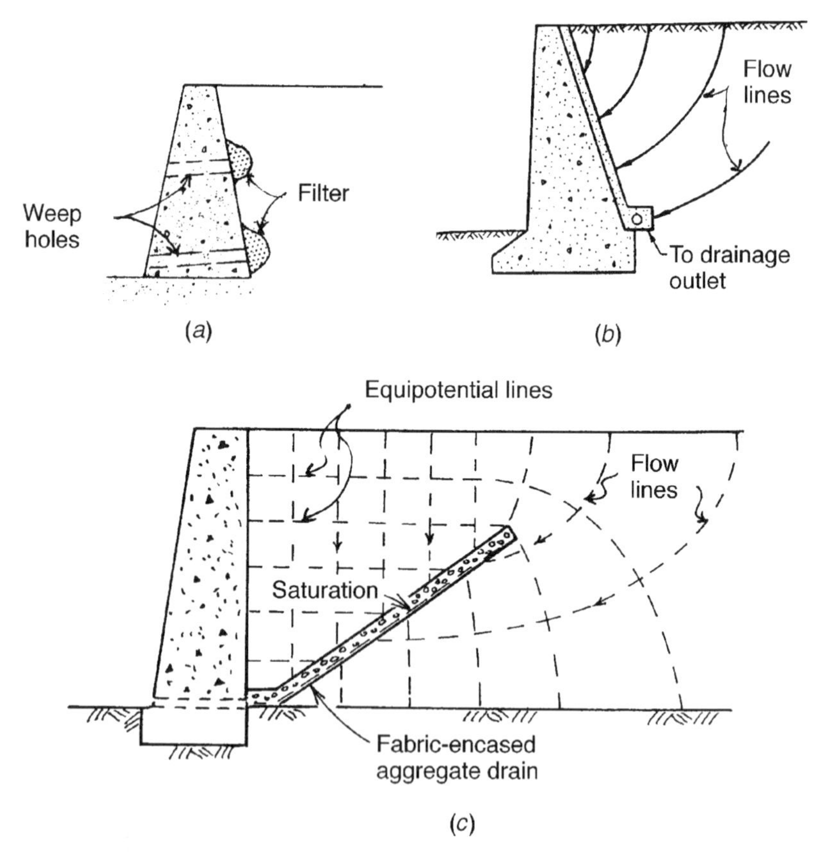

Fig. 6.24. Drainage provisions for retaining structures: (*a*) weep holes; (*b*) continuous back drains; (*c*) inclined drainage blankets.

PROBLEMS

6.1. Determine the uplift pressure distribution and the magnitude and location of the resulting force for the three cases in Fig. 6.6. The width of the structure at the base is 30 m, and the upstream head is 20 m.

6.2. Refer to Problem 3.7. This problem was also solved in Chapter 5 (Problem 5.3), and seepage rate and exit gradients were determined. Compute the factor of safety against heaving if the soils are characterized as loose sands with a void ratio of 0.8 and a specific gravity of 2.68.

6.3. For a single sheet pile, determine the factor of safety against heaving as a function of the depth of embedment of the pile. The upstream head is 10 m. The soils are loose sands with void ratio $e = 0.6$ and specific gravity $G_s = 2.65$ (*Hint:* Use the solution to Problem 5.6.)

6.4. Using the solution plotted in Fig. 6.12, determine the depth of embedment D of a sheet pile cutoff system if a factor of safety of 2 is needed against piping. Use $H_w = 10$ m, $w = 10$ m, and H_1 (depth from the bottom of the cutoff to the impervious layer) $= 10$ m.

6.5. Rework Example 6.2 if the gradient of the slope is 1 vertical to 3 horizontal. All other parameters are as in Example 6.2.

6.6. Estimate the magnitude and location of the seepage force F_s acting on the element *abcd* in Fig. 6.18 if the head difference between upstream and downstream reservoirs is 50 m, and the flow net is drawn to the scale 1 cm $=$ 10 m.

CHAPTER 7

SOIL FILTERS AND DRAINAGE LAYERS

Use of filters is so common in our daily life that its definition is found even in nontechnical sources. The *American Heritage Dictionary* (3rd edition; Boston: Houghton Mifflin) defines *filter* as "a porous material through which a liquid or gas is passed in order to separate the fluid from suspended particulate matter" of the filtered medium. The same source gives another definition relevant in a different context: "any of various electric, electronic, acoustic, or optical devices used to reject signals, vibrations, or radiations of certain frequencies while passing others." The parallel meaning in both of these definitions is that a filter material screens a filtered medium. The reader may readily see the filter applications in our daily life, ranging from filters in coffeemakers to cigarette filters and air-conditioning filters.

The definition is pretty much the same in our context of seepage in soils. Filters are needed in a number of contexts to protect the solid particles from getting transported while allowing water to exit. Examples of filter applications are shown in Fig. 7.1. In groundwater wells, for instance, protective filters are necessary so that seepage water from the natural soil flows through the well screen without carrying solid particles with it (Fig. 7.1*a*). Groundwater, which may carry colloid-sized clay particles, is filtered through fine and coarse filters before it enters the well screen. Perhaps a more dramatic necessity for filters is in the area of earth dams and embankments, where filters are necessary to prevent migration of particles eroded from within the core and prevent failures due to subsidence and collapse (Fig. 7.1*b*). In pavements and geotechnical structures, water must be drained away from the protected soils to avoid high seepage pressures. Figure 7.1*c* shows a typical longitudinal interceptor drain used for this purpose. Such drains consist of perforated or slotted pipes, which are

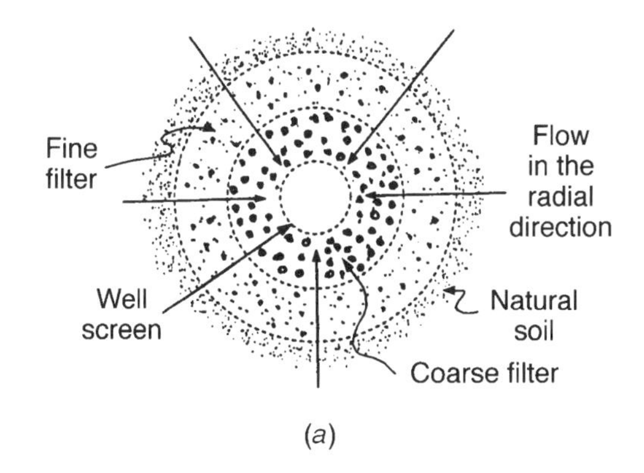

(*a*)

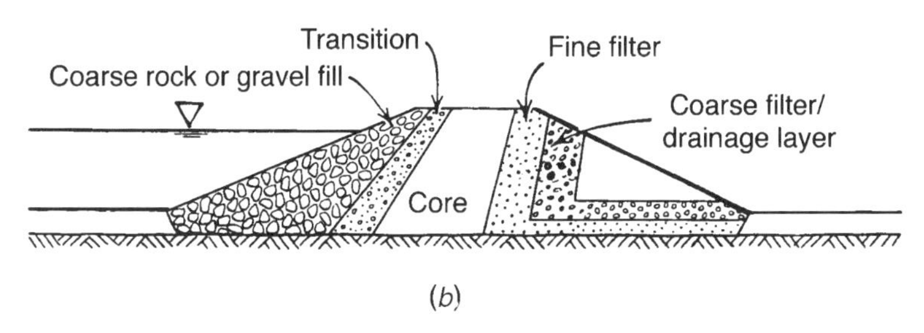

(*b*)

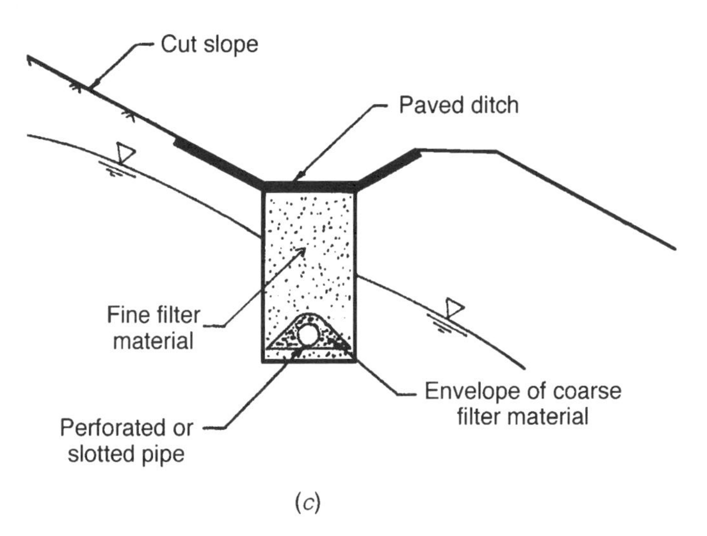

(*c*)

Fig. 7.1. Examples of filters and drains.

enveloped in filter materials to prevent natural soil particles from clogging the pipes. As shown in all of these examples, fine and coarse filter materials are often used in combination for sequential filtration.

In this chapter we learn how to design filters for diverse types of filtered soils. In that attempt we will find it necessary to review each situation separately and identify the criteria that are most suited. This is where we need to distinguish between filters and drainage layers. *Drainage layers* serve the additional purpose of transmitting known or predictable quantities of water in addition to serving as filters. For instance, a filter placed adjacent to the core of an earth dam (Fig. 7.1b) not only must protect the core from being eroded but must provide adequate transmitting capacities to drain water away from the dam. Otherwise, the excess water will surely find its path of least resistance and will be uncontrolled in its effects on the overall stability of the structure.

7.1 SOIL FILTER REQUIREMENTS

The U.S. Department of the Interior (1955) states that a protective filter has to satisfy four main requirements:

1. The filter material should be more pervious than the base material in order that no hydraulic pressure will build up to disrupt the filter and adjacent structures.
2. The voids of the in-place filter material must be small enough to prevent base material particles from penetrating the filter, which may cause clogging and failure of the protective filter system.
3. The layer of the protective material must be sufficiently thick to provide a good distribution of all particle sizes throughout the filter and to provide adequate insulation for the base material where frost action is involved.
4. Filter material particles must be prevented from movement into the drainage pipes using sufficiently small slot openings or perforations; additional coarser filter zone may be necessary.

Of the four above, the first two are major requirements and are apparently conflicting. For the filter material to be pervious, it should contain sufficiently large particles, which are associated with large voids. However, the presence of large voids may conflict with the second requirement. This conflict is the reason why design of filters typically involves checking for two separate criteria, as discussed later. The third requirement focuses on the thickness of the filter, and the fourth requirement addresses those scenarios where a drain pipe or well screen is installed with filter material around it.

The International Commission on Large Dams (ICOLD, 1994) expanded on these requirements further to give the characteristics of an ideal filter or filter zone as follows:

- Filter materials should not segregate during processing, handling, placing, spreading, or compaction. This requires the gradation of the material to be sufficiently uniform.
- Filters should not change in gradation; they should not degrade or break down during processing, handling, placing, spreading, and/or compaction or degrade with time due to freeze–thaw or seepage flow. Breakdown of the material may cause changes in pore structure and/or permeability, thus affecting the first two requirements of the filters.
- Filter materials should not have apparent or real cohesion or the ability to cement as a result of chemical, physical, or biological action. A cohesionless material will have less tendency to crack, even though cracking may have damaged an adjacent core zone.
- Filter materials should be internally stable, so that the finer fraction of the filter itself may not migrate and penetrate the coarser fraction. Internal stability is a problem with broadly graded filter materials, which have a tendency to segregate.
- Filters placed adjacent to earth dam cores must have the ability to control and seal a concentrated leak through the core. This is usually ensured with the help of a no erosion filter test, discussed in Section 7.5.2.

7.2 BASIC FILTER DESIGN CRITERIA

Several criteria have been in use to design filters since the early part of the twentieth century. All of these criteria were developed primarily to fulfill the two primary functions of filters: particle retention function and permeability function. The basic criteria used to fulfill these two functions are as follows.

7.2.1 Retention Criterion

The voids created by filter material particles should be small enough to hold the majority of base soil particles and prevent their migration into filters. Based on spherical particle geometry, one can show that for a small sphere to be held intact in the pores of three large spheres, the diameter of the larger spheres must be 6.5 times the diameter of the smaller sphere (Fig. 7.2*a*). However, base soils and filter materials are far from being uniform spheres. It is generally established that if the pore spaces in filters are small enough to hold the 85% size (D_{85}) of adjacent soils in place, the finer soil particles will also be held in place (Fig. 7.2*b*). Terzaghi (1922) suggested one of the earlier criteria following this logic:

$$\frac{D_{15}}{d_{85}} < 4 \tag{7.1}$$

where D_{15} is the filter particle size corresponding to 15% finer and d_{85} is the particle size in base/protected soil corresponding to 85% finer.

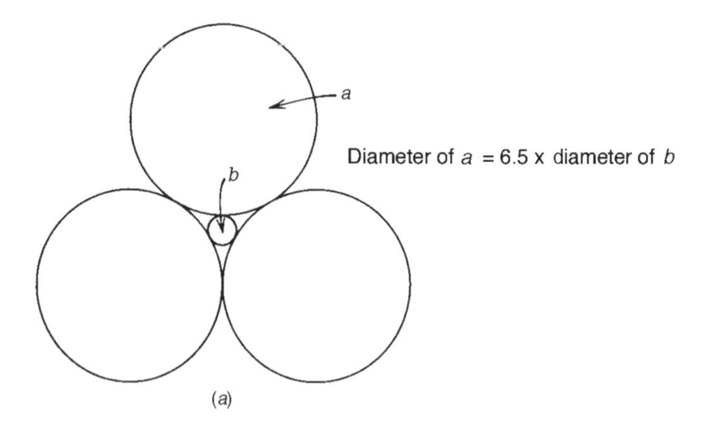

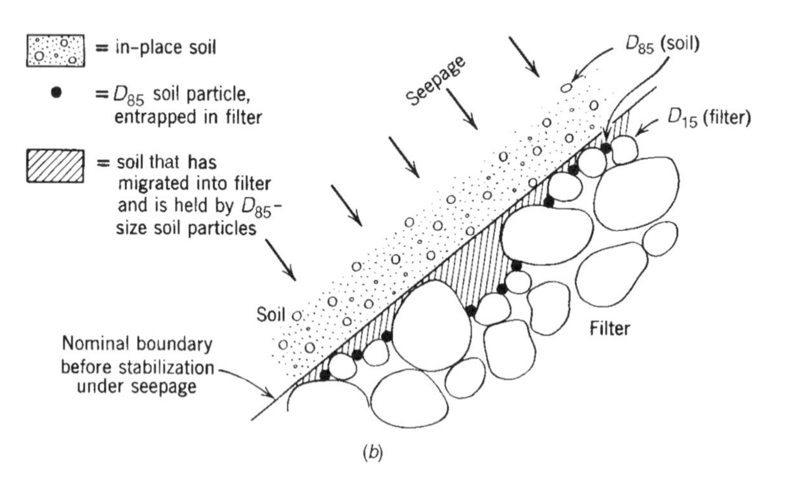

Fig. 7.2. Retention of base soil particles by filters. (Adapted from Cedergren, 1989; copyright © John Wiley & Sons; reprinted with permission.)

7.2.2 Permeability Criterion

To enable seepage from adjacent base soils without buildup of excess pore pressure, the pore sizes (therefore, the particle sizes) of the filter material should be sufficiently large. Permeability ratios (i.e., ratios of permeability of filter material to base material) of 25 to 100 are often used to achieve the permeability function of the filter. Since permeability is related to particle sizes, a criterion relating particle sizes of filter and base soils may again be used as the permeability criterion. One of the earlier expressions suggested by Terzaghi (1922) was

$$\frac{D_{15}}{d_{15}} > 4 \tag{7.2}$$

where D_{15} is the filter particle size corresponding to 15% finer and d_{15} is the size of particles in base/protected soil corresponding to 15% finer. Terzaghi developed these

TABLE 7.1 Summary of Early Filter Design Criteria

Investigators	Base Material	Filter Material	Criteria Developed
Terzaghi (1922)	Uncertain whether criteria based on experiments or conservative reasoning		$\frac{D_{15}}{d_{85}} < 4 < \frac{D_{15}}{d_{15}}$
Bertram (1939)	Uniform quartz and Ottawa sands	Uniform quartz and Ottawa sands	$\frac{D_{15}}{d_{85}} < 6$, $\frac{D_{15}}{d_{15}} < 9$
Newton and Hurley (1940)	Well-graded gravelly sand	Natural bank gravels, finer sizes screened out successively, fairly uniform filters	$\frac{D_{15}}{d_{15}} < 32$, $\frac{D_{15}}{d_{50}} < 15$
Waterways Experiment Station (1941, 1948)	Random material types; fine to coarse sands	Random types, including natural pit-run gravels	$\frac{D_{15}}{d_{85}} < 4, < 20$ $\frac{D_{50}}{d_{50}} < 25$, $\frac{D_{15}}{d_{85}} < 5$ Gradation of filters should be more or less parallel to base; filter should be well graded
Office, Chief of Engineers, Corps of Engineers	All types	Concrete sand and coarse aggregate generally recommended	$\frac{D_{15}}{d_{85}} < 5$, $\frac{D_{15}}{d_{15}} < 5$
U.S. Bureau of Reclamation, 1947 (for canal slopes and drains under structures; discontinued in 1955)	Artificially blended materials of various ranges, including uniform material	Artificially blended uniform filters; artificially blended well-graded filters	$\frac{D_{50}}{d_{50}} > 5, < 10$ $\frac{D_{50}}{d_{50}} > 12, < 58$ $\frac{D_{15}}{d_{15}} > 12, < 40$
Providence District, Corps of Engineers, 1942	All types	Certain general types recommended	Filter design curve, C_u of base vs. $\frac{D_{15}}{d_{15}}$

Source: Adapted from U.S. Army Corps of Engineers (1948).

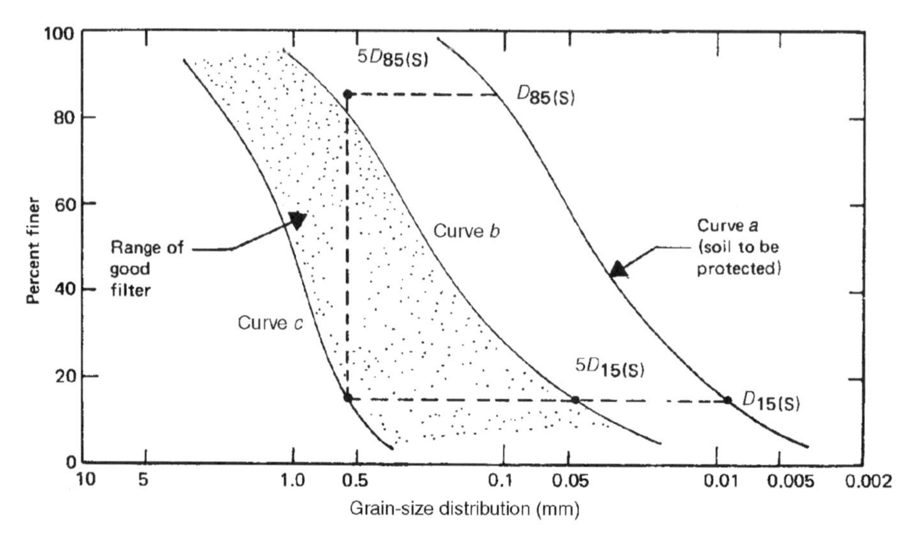

Fig. 7.3. Design of soil filters using U.S. Army Corps of Engineers' filter criteria. (Adapted from Das, 1983; copyright © The McGraw-Hill Companies, reprinted with permission.)

criteria in connection with the design of rockfill dams in North Africa which were resting on uniform sands. These criteria were found to be generally successful in filter design for earthfill dams, although several agencies and investigators proposed similar criteria. Table 7.1 summarizes the criteria proposed by several investigators and agencies for different base and filter materials. One can use the criteria and construct the grain size distribution curve of filter required, as shown in Fig. 7.3. Curve *a* corresponds to the base soil to be protected using the filter, and curves *b* and *c* show the band within which the grain size distribution must be chosen for the filter. The criteria demonstrated in Fig. 7.3 are those proposed by the U.S. Army Corps of Engineers (see Table 7.1).

7.3 EXTENDED FILTER CRITERIA

The basic criteria discussed in Section 7.2 were not intended to ensure internal stability of filters. Furthermore, the criteria were developed and validated for a few specific, mostly cohesionless base soils. Based on extensive experimental work by Sherard (1984a,b) and Sherard and Dunnigan (1989), the U.S. Soil Conservation Service and the U.S. Bureau of Reclamation begun adopting a more extensive set of criteria to design soil filters. These criteria are shown in Table 7.2. For the purpose of developing the criteria, the base/protected soils were divided into four categories. The graphical representation of the four categories of the base soils is shown in Fig. 7.4 along with the criteria. Using these criteria, the Natural Resources Conservation Service of the U.S. Department of Agriculture proposed a 12-step filter design method. Segregation

TABLE 7.2 Criteria for Filters

Base Soil Category	Base Soil Description, and Percent Finer Than No. 200 Sieve[a]	Filter Criteria[b]
1	Fine silts and clays; more than 85% finer	$D_{15} \leq 9 \times d_{85}$[c]
2	Sands, silts, clays, and silty and clayey sands; 40 to 85% finer	$D_{15} \leq 0.7$ mm
3	Silt and clayey sands and gravels; 15 to 39% finer	$D_{15} \leq \frac{40 - A}{40 - 15}(4 \times d_{85} - 0.7 \text{ mm}) + 0.7 \text{ mm}$[d]
4	Sands and gravels; less than 15% finer	$D_{15} \leq 4 \times d_{85}$[e]

Source: Adapted from U.S. Department of Agriculture, Soil Conservation Service, 1986; Bureau of Reclamation, 1987.

[a]Category designation for soil containing particles larger than the No. 4 (4.75 mm) sieve determined from a gradation curve of the base soil which has been adjusted to 100% passing the No. 4 (4.75 mm) sieve.

[b]Filters are to have a maximum particle size of 75 mm (3 in.) and a maximum of 5% passing the No. 200 (0.075 mm) sieve with the plasticity index (PI) of the fines equal to zero.

[c]When $9 \times d_{85}$is less than 0.2 mm, use 0.2 mm.

[d]A = percent passing the No. 200 (0.075 mm) sieve after any regrading. When $4 \times d_{85}$ is less than 0.7 mm, use 0.7 mm.

[e]In category 4, the d_{85} may be determined from the original gradation curve of the base soil without adjustments for particles larger than 4.75 mm.

and possible instability of the filter were also considered in this method in addition to particle retention and permeability functions. The objective of the design method, outlined below, is to construct the filter gradation curve for a given base/protected soil.

Step 1: Plot the gradation curve (grain-size distribution) of the base soil material. A sufficient number of samples should be used to define the range of grain sizes for the base soil.

Step 2: Proceed to step 4 if the base soil contains no gravel (material larger than a No. 4 sieve).

Step 3: Prepare adjusted gradation curves for base soils that have particles larger than No. 4 (4.75 mm) sieve size. The procedure to adjust the curves is described below.

a. Obtain a correction factor by dividing 100 by the percent passing the No. 4 (4.75 mm) sieve.
b. Multiply the percentage passing each sieve size of the base soil smaller than the No. 4 (4.75 mm) sieve by the correction factor determined above.

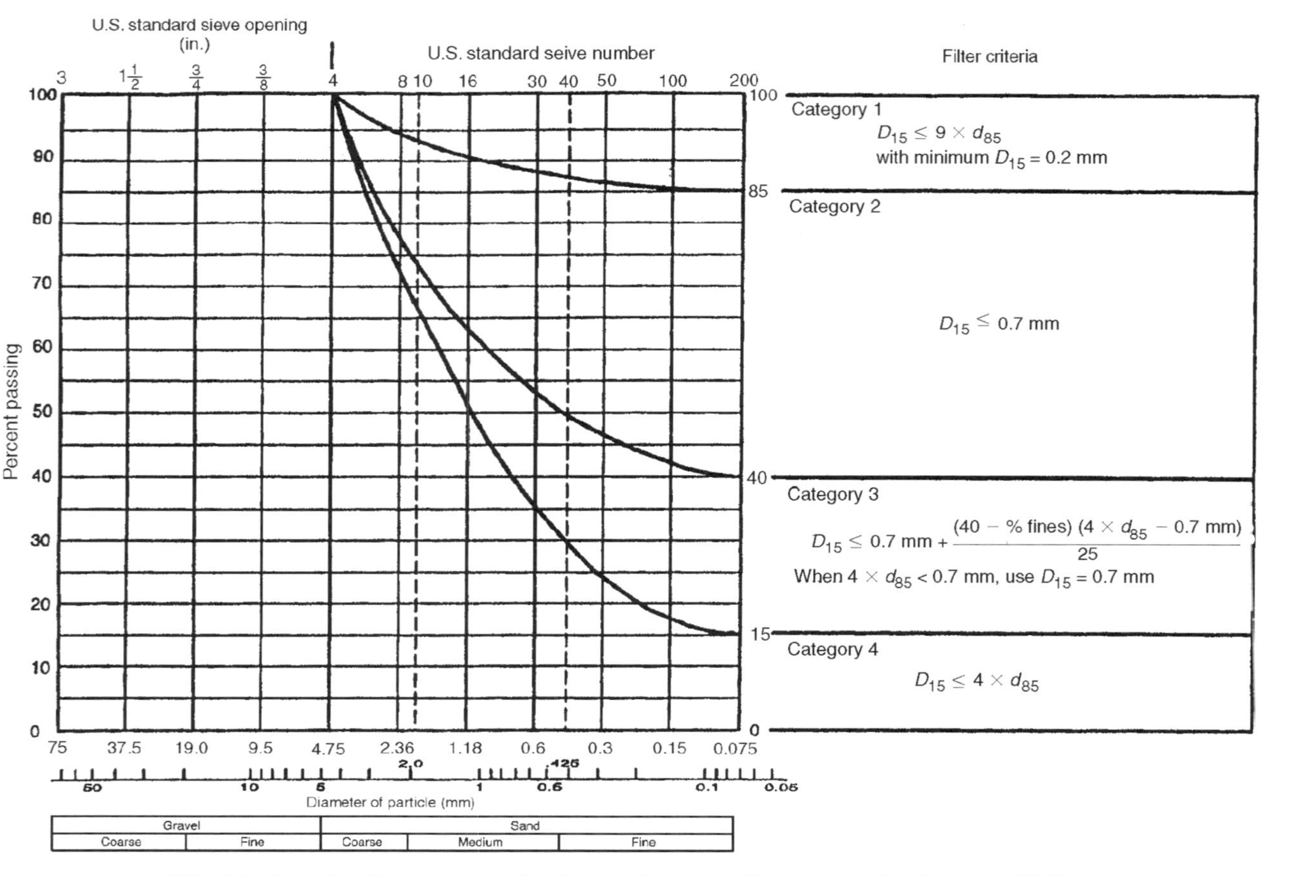

Fig. 7.4. Extended filter criteria by the National Resources Conservation Service of the USDA.

c. Plot these adjusted percentages to obtain a new gradation curve.
d. Use the adjusted curve to determine the percentage passing the No. 200 (0.075 mm) sieve in step 4.

Step 4: Use Table 7.2 or Fig. 7.4 to categorize the base soil according to the percentage passing the No. 200 (0.075 mm) sieve.

Step 5: To satisfy filtration requirements, determine the maximum allowable D_{15} size for the filter in accordance with Table 7.2 or Fig. 7.4. The maximum D_{15} may be adjusted for certain noncritical uses of filters where significant hydraulic gradients are not predicted, such as bedding beneath riprap and concrete slabs. For fine-clay-base soils that have d_{85} sizes between 0.03 and 0.1 mm, a maximum D_{15} value of ≤ 0.5 mm is still conservative. For fine-grained silt that has a low sand content, plotting below the *A* line, a maximum D_{15} of 0.3 mm may be used.

Step 6: To satisfy the permeability criterion, the D_{15} size for the filter should be greater than or equal to four times the d_{15} value of the base soil before regrading. A minimum $D_{15}$of 0.1 mm is recommended.

Step 7: The width of the allowable filter design band must be kept relatively narrow to prevent the use of possibly gap-graded filters. Adjust the maximum and minimum D_{15} sizes for the filter band determined in steps 5 and 6 so that the ratio is 5 or less at any given percentage passing of 60 or less. The criteria are summarized in Table 7.3. The following additional requirements should be followed to decrease the probability of using a gap-graded filter.

First, calculate the ratio of the maximum D_{15} to the minimum D_{15} sizes determined in steps 5 and 6. If this ratio is greater than 5, adjust the values of these control points so that the ratio of the maximum D_{15} to the minimum D_{15} is no greater than 5. If the ratio is 5 or less, no adjustments are necessary. Label the maximum D_{15} size as control point 1 and the minimum D_{15} size as control point 2. Proceed to step 8.

The decision on where to locate the final D_{15} sizes within the range established with previous criteria should be based on one of the following considerations:

TABLE 7.3 Other Filter Design Criteria

Design Element	Criteria
To prevent gap-graded filters	The width of the designed filter band should be such that the ratio of the maximum diameter to the minimum diameter at any given percent passing value $\leq 60\%$ is ≤ 5.
Filter band limits	Coarse and fine limits of a filter band should each have a coefficient of uniformity of 6 or less.

a. Locate the design filter band at the maximum D_{15} side of the range if the filter will be required to transmit large quantities of water (serve as a drain as well as a filter). With the maximum D_{15} size as the control point, establish a new minimum D_{15} size by dividing the maximum D_{15} size by 5, and locate a new minimum D_{15} size. Label the maximum D_{15} size control point 1 and the minimum D_{15} size control point 2.

b. Locate the band at the minimum D_{15} side of the range if it is probable that there are finer base materials than those sampled and filtering is the most important function of the zone. With the minimum D_{15} size as the control point, establish a new maximum D_{15} size by multiplying the minimum D_{15} size by 5, and locate a new maximum D_{15} size. Label the maximum D_{15} size as control point 1 and the minimum D_{15} size as control point 2.

c. The most important consideration may be to locate the maximum and minimum D_{15} sizes within the acceptable range of sizes determined in steps 5 and 6, so that a standard gradation available from a commercial source or other gradations from a natural source near the site would fall within the limits. Locate a new maximum D_{15} and minimum D_{15} within the permissible range to coincide with the readily available material. Ensure that the ratio of these sizes is 5 or less. Label the maximum D_{15} size as control point 1 and the minimum D_{15} size as control point 2.

Step 8: Adjust the limits of the design filter band so that the coarse and fine sides have a coefficient of uniformity of 6 or less. The width of the filter band should be such that the ratio of maximum to minimum diameters is less than or equal to 5 for all percent passing values of 60 or less. Calculate a maximum D_{10} value equal to the maximum D_{15} size divided by 1.2. This factor of 1.2 is based on the assumption that the slope of the line connecting D_{15} and D_{10} corresponds to a coefficient of uniformity of about 6. Calculate the maximum permissible D_{60} size by multiplying the maximum D_{10} value by 6. Label this as control point 3. Determine the minimum allowable D_{60} size for the fine side of the band by dividing the determined maximum D_{60} size by 5. Label this as control point 4.

Step 9: For all base soil categories, a minimum D_5 of 0.075 mm should be used. The D_{100} of the filter should be less than or equal to 3 in. (75 mm). Label these control points as 5 and 6, respectively. The filter material finer than 0.425 mm must be nonplastic.

Step 10: To minimize segregation during construction, the relationship between the maximum D_{90} and the minimum D_{10} of the filter is important. Calculate a preliminary minimum D_{10} size by dividing the minimum D_{15} size by 1.2. This factor of 1.2 is based on the assumption that the slope of the line connecting D_{15} and D_{10} corresponds to a coefficient of uniformity of about 6. Determine the maximum D_{90} using Table 7.4. Label this as control point 7. Sand filters that have a D_{90} less than about 20

TABLE 7.4 Segregation Criteria for All Base Soil Categories

If D_{10} Is (mm):	Then Maximum D_{90} Is (mm):
< 0.5	20
0.5–1.0	25
1.0–2.0	30
2.0–5.0	40
5.0–10	50
> 10	60

mm generally do not require special adjustments for the broadness of the filter band. For coarser filters and gravel zones that serve both as filters and drains, the ratio of D_{90}/D_{10} should decrease rapidly with increasing D_{10} sizes.

Step 11: Connect control points 4, 2, and 5 to form a partial design for the fine side of the filter band. Connect control points 6, 7, 3, and 1 to form a design for the coarse side of the filter band. This results in a preliminary design for a filter band. Complete the design by extrapolating the coarse and fine curves to the 100% finer value. For purposes of writing specifications, select appropriate sieves and corresponding percent finer values that best reconstruct the design band and tabulate the values.

Step 12: Filters adjacent to perforated pipe must be designed to have a D_{85} size greater than or equal to the perforation size. For critical structure drains where rapid gradient reversal (surging) is probable, it is recommended that the D_{15} size of the material surrounding the pipe be no smaller than the perforation size.

Examples 7.1–7.4 To understand the use of extended criteria, the reader should work out the problems of Figs. E7.1 to E7.4 corresponding to the four base soil categories. These examples are adapted from the *National Engineering Handbook* (Part 633) of the National Resources Conservation Service (NRCS). The control points for the design filter bands are shown at the bottom of the figures.

7.4 DRAINAGE CRITERIA

In addition to fulfilling the filter criteria, soils used as drainage layers must have adequate discharge capacities. Given the amounts of water that need to be discharged, one can use Darcy's law and estimate the permeability needed for the drainage layer for a given thickness of the layer. Conversely, the thickness needed for the layer can be estimated given its permeability. Thus,

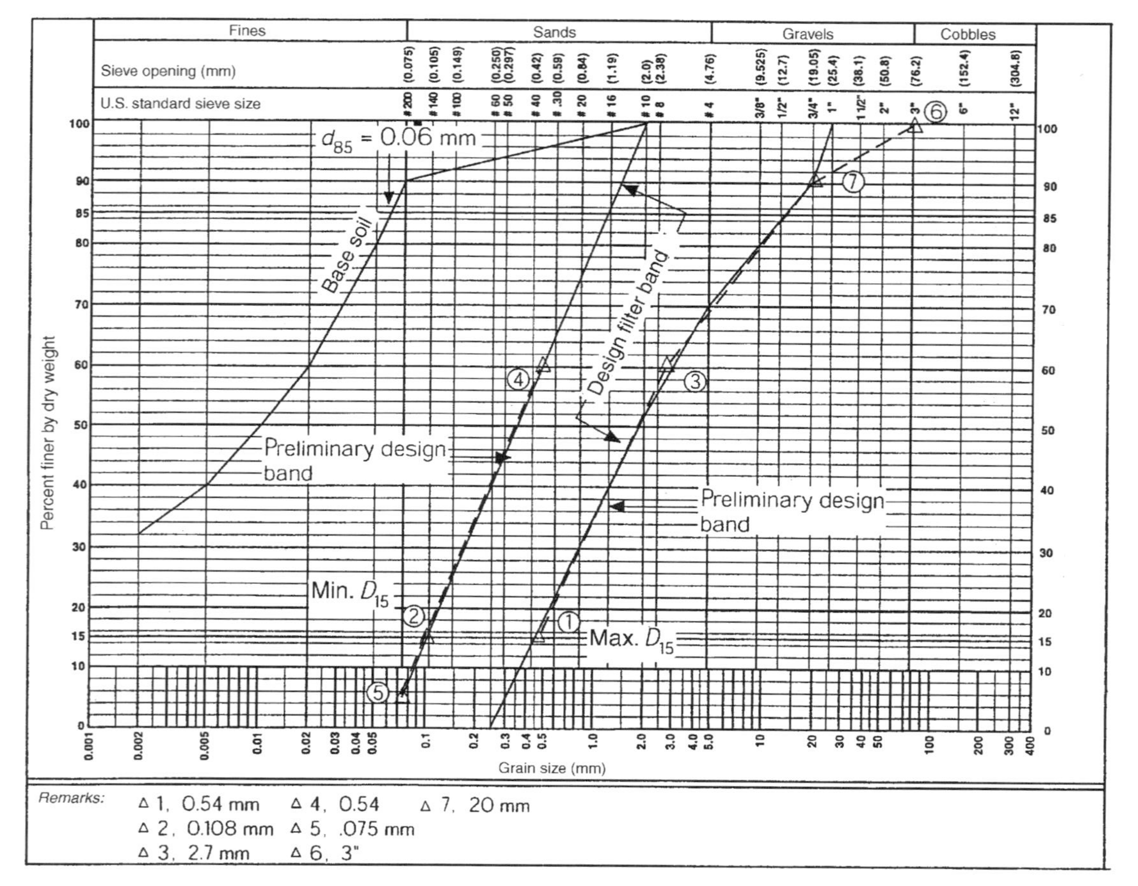

Fines
Sands
Gravels
Cobbles
Sieve opening (mm)
U.S. standard sieve size
Percent finer by dry weight
Grain size (mm)
d_{85} = 0.06 mm
Base soil
Design filter band
Preliminary design band
Preliminary design band
Min. D_{15}
Max. D_{15}
Remarks:
Δ 1, 0.54 mm
Δ 4, 0.54
Δ 7, 20 mm
Δ 2, 0.108 mm
Δ 5, .075 mm
Δ 3, 2.7 mm
Δ 6, 3"

Fig. E7.1

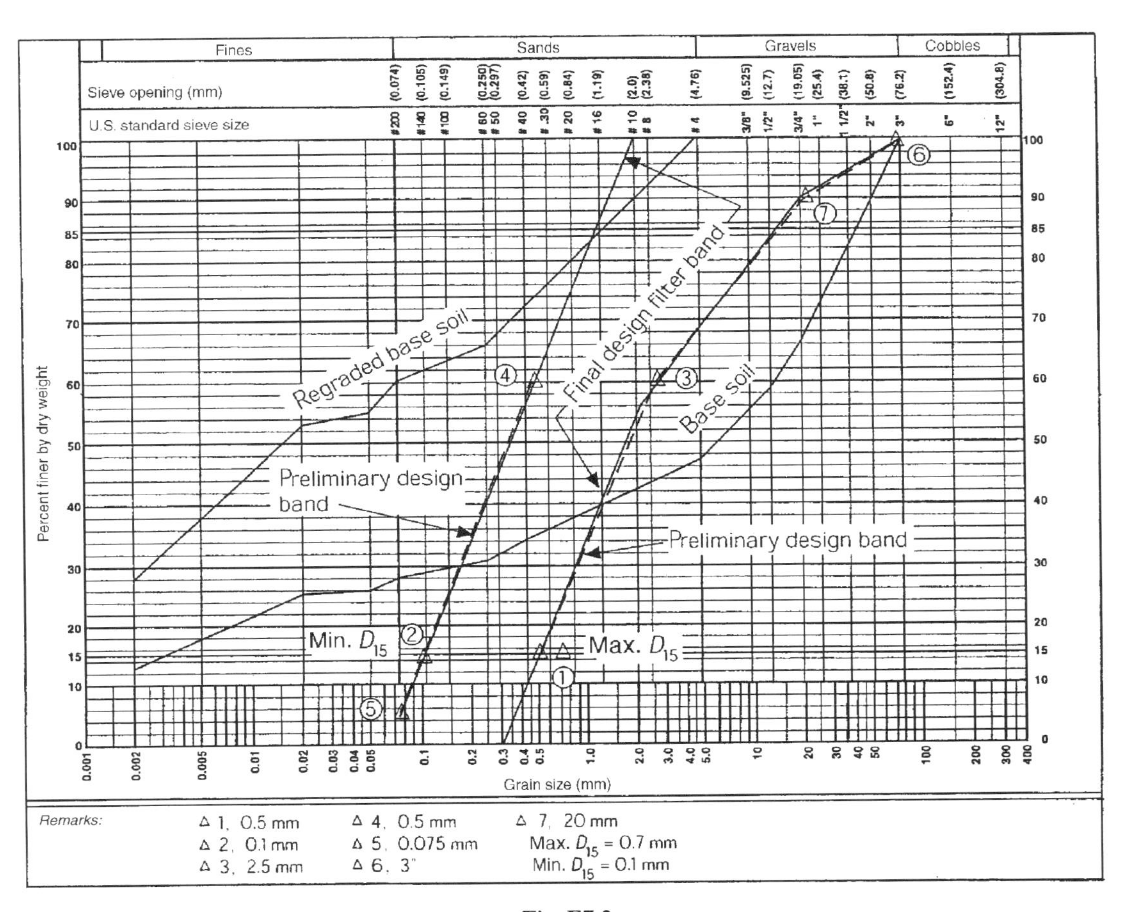
Fines
Sands
Gravels
Cobbles
Sieve opening (mm)
U.S. standard sieve size
Percent finer by dry weight
Grain size (mm)
Regraded base soil
Final design filter band
Base soil
Preliminary design band
Preliminary design band
Min. D_{15}
Max. D_{15}
Remarks:
Δ 1, 0.5 mm
Δ 2, 0.1 mm
Δ 3, 2.5 mm
Δ 4, 0.5 mm
Δ 5, 0.075 mm
Δ 6, 3"
Δ 7, 20 mm
Max. D_{15} = 0.7 mm
Min. D_{15} = 0.1 mm

Fig. E7.2

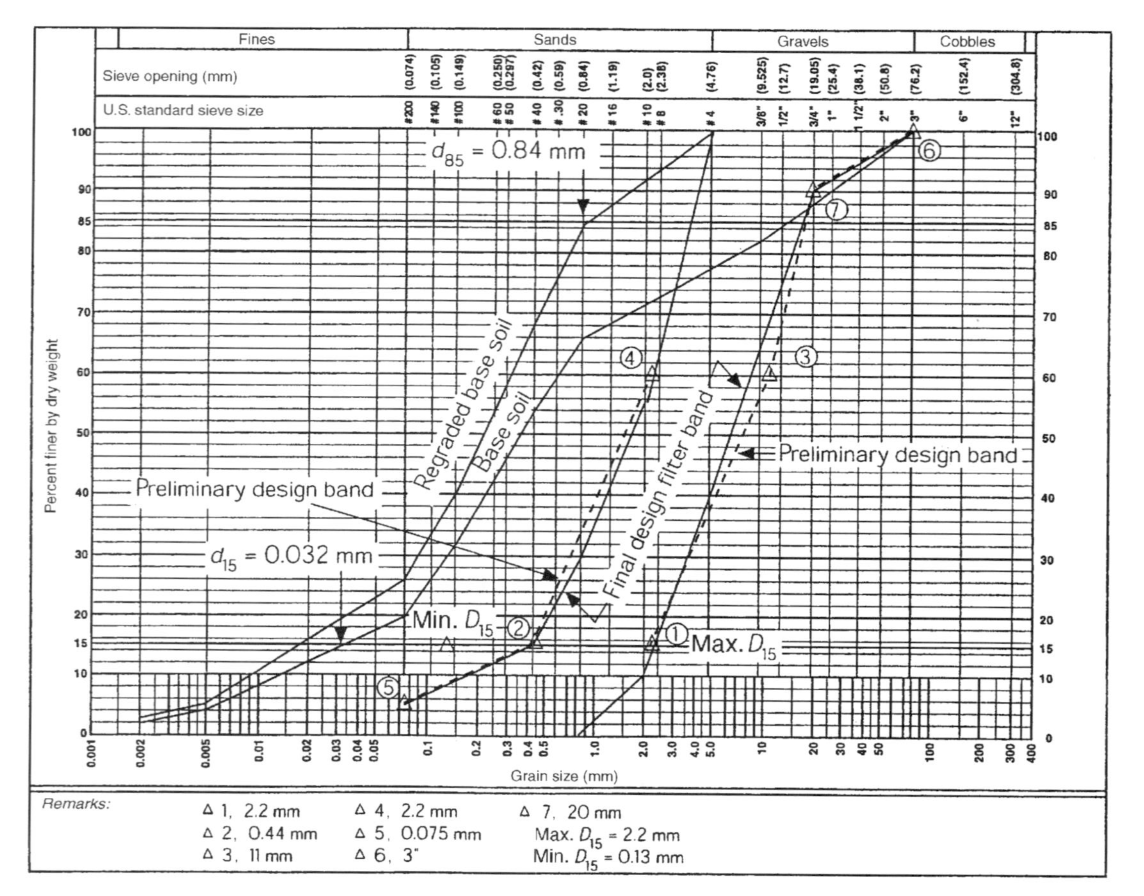

Fines
Sands
Gravels
Cobbles
Sieve opening (mm)
U.S. standard sieve size
Percent finer by dry weight
Grain size (mm)
d85 = 0.84 mm
d15 = 0.032 mm
Preliminary design band
Regraded base soil
Base soil
Final design filter band
Preliminary design band
Min. D15
Max. D15
Remarks:
Δ 1, 2.2 mm
Δ 2, 0.44 mm
Δ 3, 11 mm
Δ 4, 2.2 mm
Δ 5, 0.075 mm
Δ 6, 3"
Δ 7, 20 mm
Max. D15 = 2.2 mm
Min. D15 = 0.13 mm

Fig. E7.3

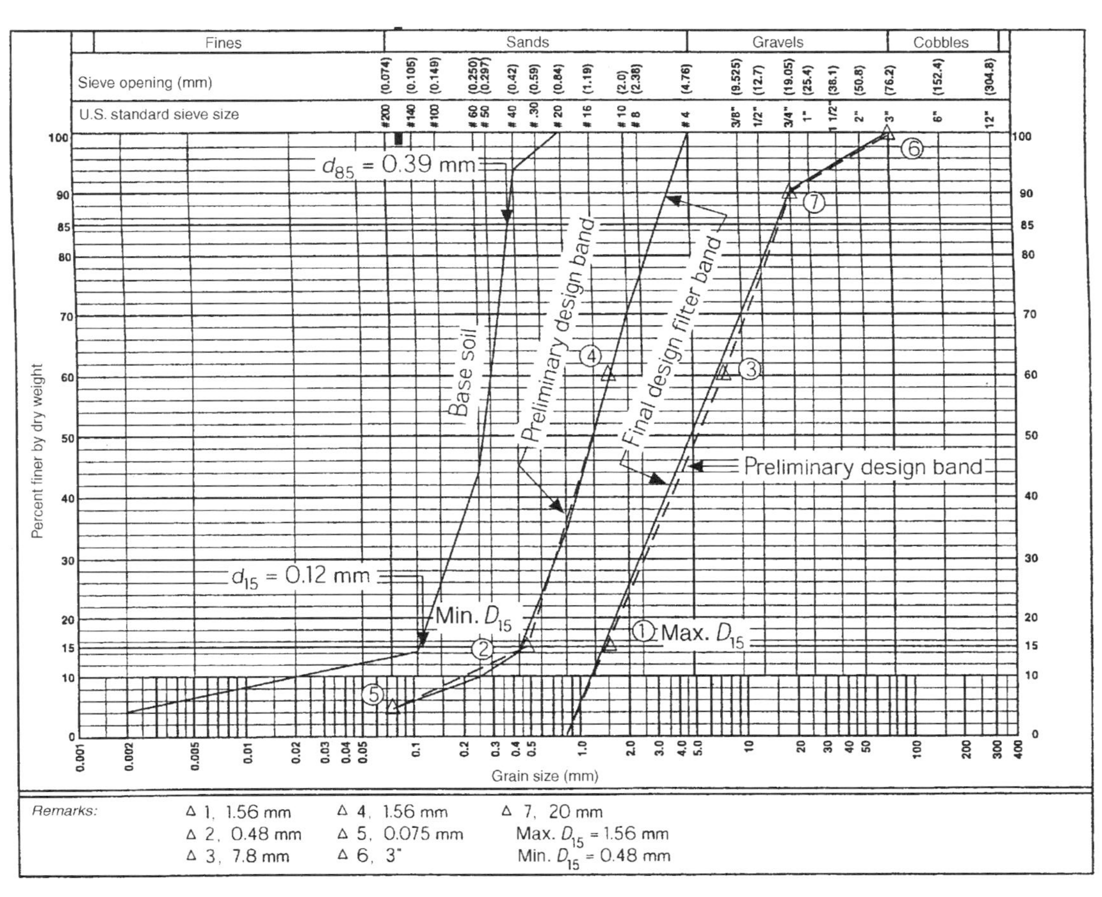
Fines
Sands
Gravels
Cobbles
Sieve opening (mm)
U.S. standard sieve size
Percent finer by dry weight
Grain size (mm)
d_{85} = 0.39 mm
d_{15} = 0.12 mm
Base soil
Preliminary design band
Final design filter band
Preliminary design band
Min. D_{15}
Max. D_{15}
Remarks:
Δ 1, 1.56 mm
Δ 2, 0.48 mm
Δ 3, 7.8 mm
Δ 4, 1.56 mm
Δ 5, 0.075 mm
Δ 6, 3"
Δ 7, 20 mm
Max. D_{15} = 1.56 mm
Min. D_{15} = 0.48 mm

Fig. E7.4

$$k = \frac{Q}{iA} \tag{7.3}$$

or

$$A = \frac{Q}{ki} \tag{7.4}$$

where k is the hydraulic conductivity of the drainage layer, Q the total discharge capacity needed in a given structure, i the hydraulic gradient, and A the area of cross section through which flow is occurring. In checking the discharge capacity of the drainage layer, the term *transmissibility,* defined as the product of k and A, is useful. For a unit length in the longitudinal direction of the drainage layer, its transmissibility is the product of permeability and thickness. Obviously, smaller permeabilities need thicker layers and vice versa, to meet a given discharge capacity.

In some cases, flow nets and analytical solutions may have to be used to determine the discharge capacities of the drainage layers. Examples are (1) design of chimney and blanket drains in dams, (2) design of sloping embankment drains at the outflow, and (3) design of horizontal drainage layers for pavements. In all these cases, the discharge could be estimated using flow nets. In the case of drains used in dams (Fig. 7.5), the chimney drain should have a permeability

$$k_c = \frac{Q_c}{(h_c/L_c)(t_c \times 1)} \tag{7.5}$$

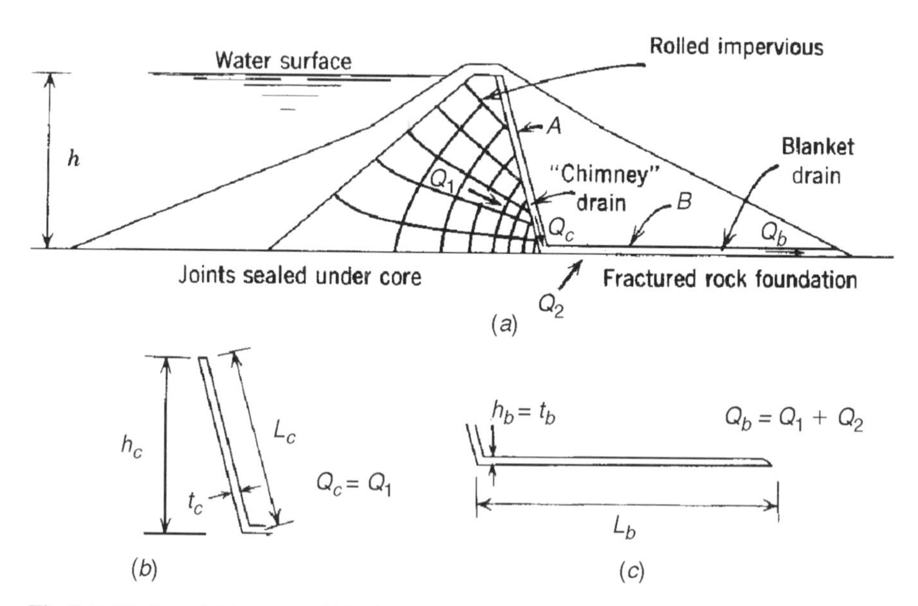

Fig. 7.5. Design of chimney and blanket drains. (Adapted from Cedergren, 1989; copyright © John Wiley & Sons; reprinted with permission.)

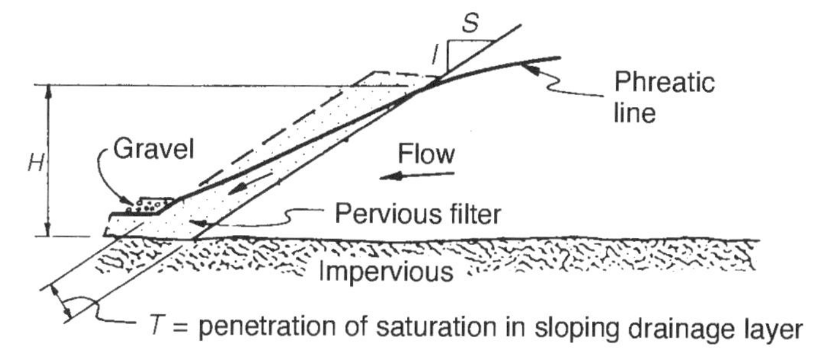

Fig. 7.6. Sloping embankment drains. (Adapted from Cedergren, 1960; copyright © ASCE; reprinted with permission.)

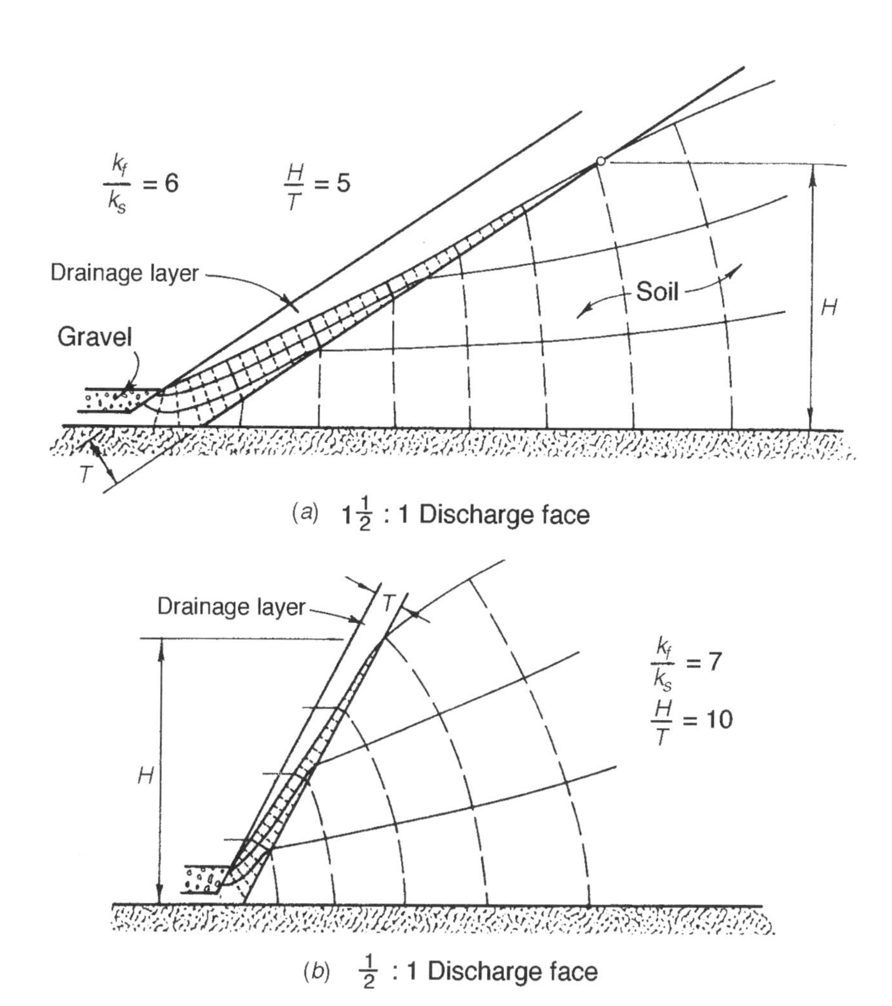

Fig. 7.7. Flow nets in embankments with sloping drains: (*a*) $k_f/k_s = 6$; (*b*) $k_f/k_s = 7$. (Adapted from Cedergren, 1960; copyright © ASCE; reprinted with permission.)

A suitable combination of k_c and t_c may be used, fulfilling Eq. (7.5) to design the drain. The blanket drain may be designed similarly using

$$k_b = \frac{Q_b}{(h_b/L_b)(t_b \times 1)} \tag{7.6}$$

Note that in Eq. (7.6), Q_b is greater than Q_c as a result of seepage from the foundation into the blanket drain. Also, the head difference available for flow in the blanket drain is limited by its thickness. Therefore, Eq. (7.6) may be rewritten as

$$k_b = \frac{Q_b L_b}{t_b^2} \tag{7.7}$$

In the case of sloping embankment drains (Fig. 7.6), the discharge capacity for the material could be determined using composite flow nets. The thickness of the drainage layer, T, should be such that the flow net is totally contained. This is, of course, governed by the permeability contrast of the embankment material and the drainage layer. For two permeability ratios of $k_f/k_s = 6$ and 7, where k_f is the permeability of filter material used as the drainage layer and k_s is the permeability of embankment soil, the flow nets and associated dimensions of the thickness and the length of seepage in the drainage layer are shown in Fig. 7.7*a* and *b*. Using a number of such flow nets, Cedergren (1960) obtained the chart shown in Fig. 7.8, which could be used to determine the proper combination of k_f and T.

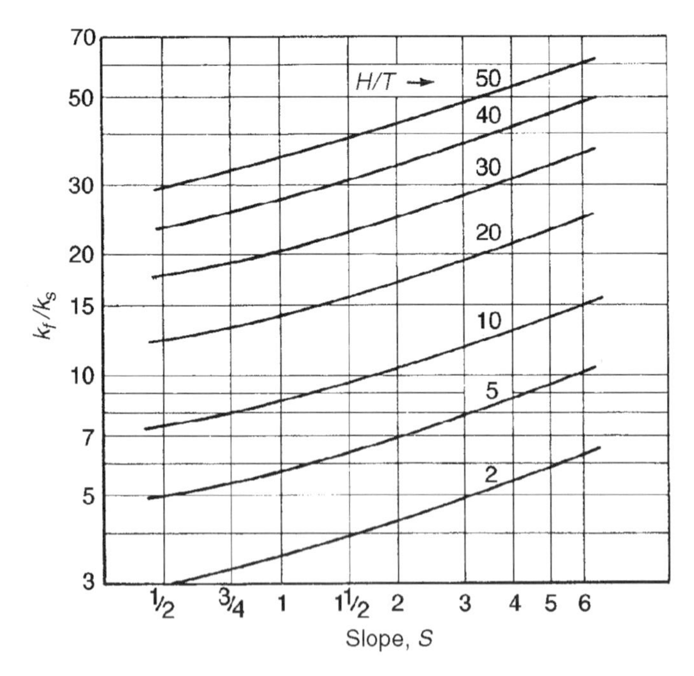

Fig. 7.8. Design chart for sloping embankment drains. (Adapted from Cedergren, 1960; copyright © ASCE; reprinted with permission.)

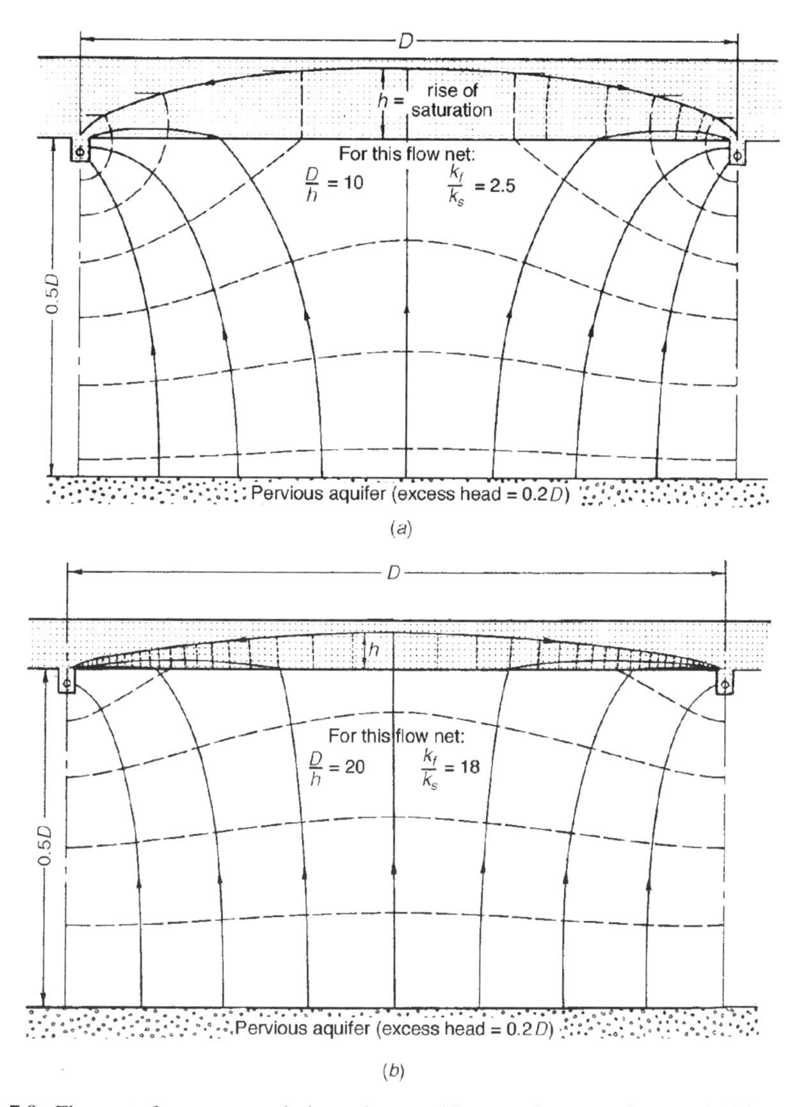

Fig. 7.9. Flow nets for pavement drainage layers with upward seepage from underlying soils: (*a*) $k_f/k_s = 2.5$; (*b*) $k_f/k_s = 18$. (Adapted from Cedergren, 1960; copyright © ASCE; reprinted with permission.)

For a pavement drainage layer with seepage from underlying soils, the thickness and permeability of the layer should be such that the flow net is contained entirely within the drainage layer. Two typical flow nets for vertical seepage into the horizontal drainage blanket are shown in Fig. 7.9*a* and *b*. Using several such flow nets to cover a broad range of permeability ratios, k_f/k_s, Cedergren (1960) developed solutions for the dimensionless quantity D/h, shown in Fig. 7.10. The data in Fig. 7.10 were

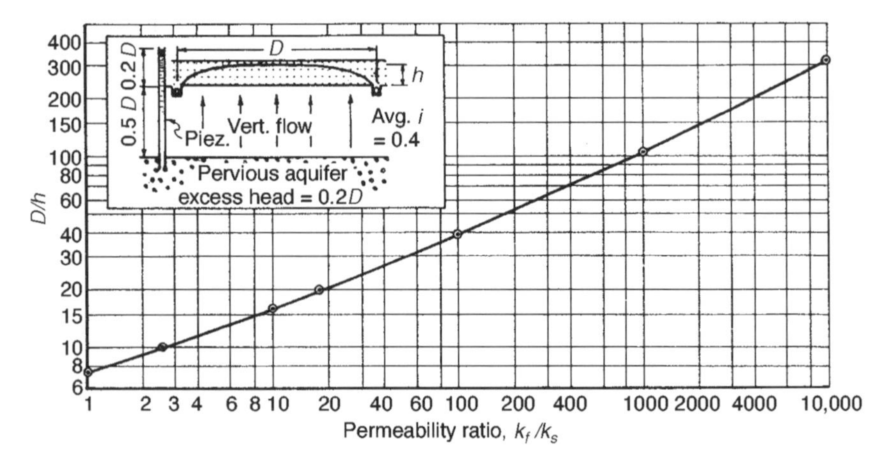

Fig. 7.10. Flow net solutions for pavement drainage layers. (Adapted from Cedergren, 1960; copyright © ASCE; reprinted with permission.)

replotted as shown in Fig. 7.11 to obtain a design chart, which could be used to select the drainage material (in terms of its permeability) and/or D/h. In determining whether a filter material could adequately serve as a drainage layer, the permeability estimates of typical filter gradations shown in Fig. 2.14 will be useful. This figure was developed from data on open graded bases and filter materials.

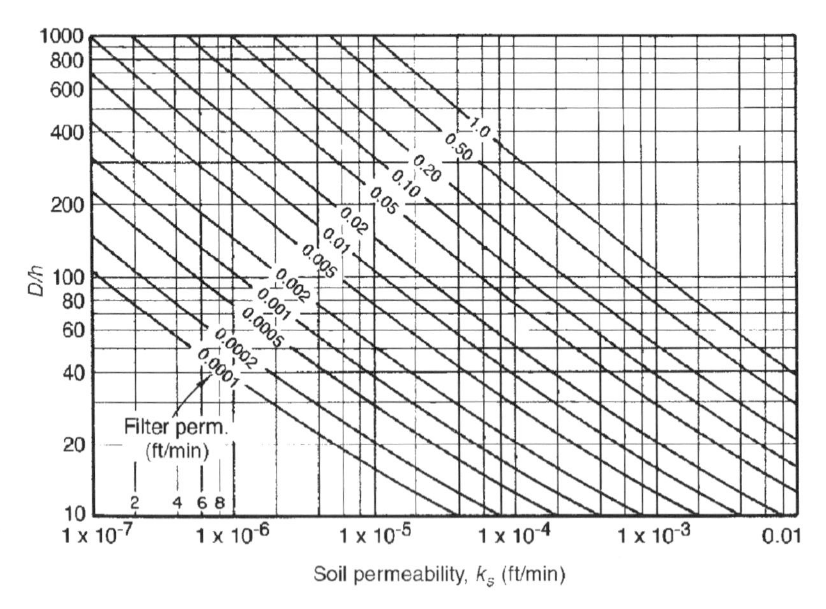

Fig. 7.11. Drain design chart for horizontal pavement drainage layers with shallow collector drains. (Adapted from Cedergren, 1960; copyright © ASCE; reprinted with permission.)

7.5 LABORATORY TESTS FOR FILTER DESIGN

Because of the empiricism in all the existing design criteria, many projects rely on laboratory tests to develop site/soil-specific filter criteria. While some tests are designed for research purposes to develop criteria, others are designed to serve in field projects. We review a few testing methods here to understand the experimental basis for filter design criteria.

7.5.1 U.S. Bureau of Reclamation Test Method

The U.S. Bureau of Reclamation (USBR) uses a test method (U.S. Department of the Interior, 1990) to evaluate the suitability of a soil material as a filter. The purpose of the test is to determine if a filter can prevent unrestrained erosion of a base material with a finer particle-size distribution while maintaining adequate hydraulic conductivity. The test apparatus and a schematic of the flow cell system are shown in Fig. 7.12. The base material tested may be either a compacted specimen of soil or a soil—water slurry. If the base material is cohesive, cylindrical holes are drilled through the compacted base specimen to provide seepage paths for fine particles to move into the filter. The base material is placed over the filter material to be tested, and gravel is placed at the influent and effluent ends of the flow cell to contain the base and filter layers.

Water is percolated downward through the flow cell under varying hydraulic heads. The volume of effluent from the system is measured at several time intervals for each hydraulic head. The turbidity of the effluent is also determined visually and recorded. The total accumulated effluent volume is plotted as a function of time. After conducting the test for a number of different hydraulic heads, the base specimen is removed and the base—filter interface inspected. The filter specimen is excavated in layers, and particle size analyses may be performed on portions of the filter specimen. The results are used to provide a qualitative idea of the filter performance. The same method is used to test the suitability of geotextile materials as filters. The use of geotextiles in filtration and drainage is discussed in Chapter 8.

7.5.2 Soil Conservation Service Tests

More than any other investigation, the tests conducted at the Soil Conservation Service (SCS) of the U.S. Department of Agriculture provided a thorough understanding of the fundamental properties and behavior of filters. Sherard et al. (1984a,b) summarized these tests in their award-winning papers in the American Society of Civil Engineers' *Journal of Geotechnical Engineering*. The test apparatus used to develop filter criteria for cohesionless and cohesive base soils is shown in Fig. 7.13. Uniform sandy soils were used as the base materials in the tests. The filters were mostly uniform sands and gravels with a D_{15} range of 1.0 to 10 mm, consisting of subrounded to subangular particles of alluvial origin. The tests are similar to those of USBR tests except for a "side material" placed between the filter and the cylinder wall to eliminate large voids (pores) at the interface that could be larger than the pore channels

a Aluminum test cylinder system
b Adjustable constant head tank
c Inflow plumbing
d Acrylic test cylinder system
e Outflow plumbing

(*a*)

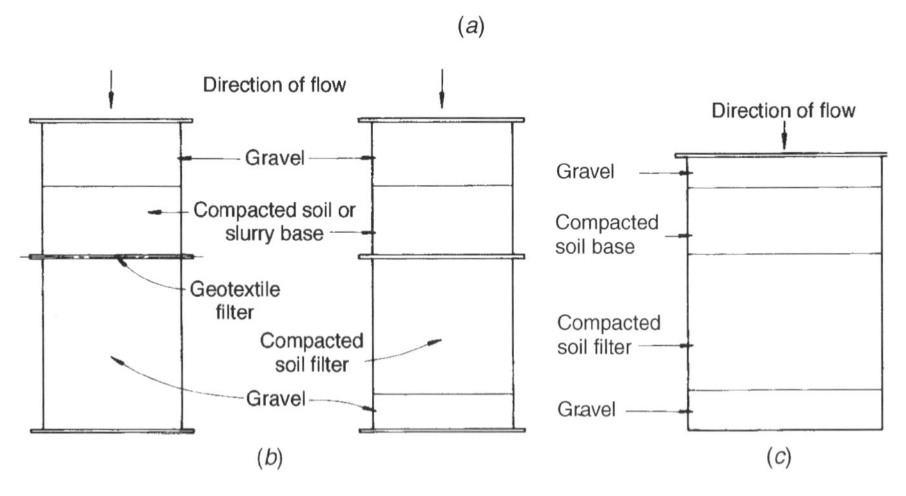

Fig. 7.12. Filter test system and cell configurations: (*a*) test setup; (*b*) acrylic cylinder system; (*c*) aluminum cylinder system. (Adapted from U.S. Department of the Interior, 1990.)

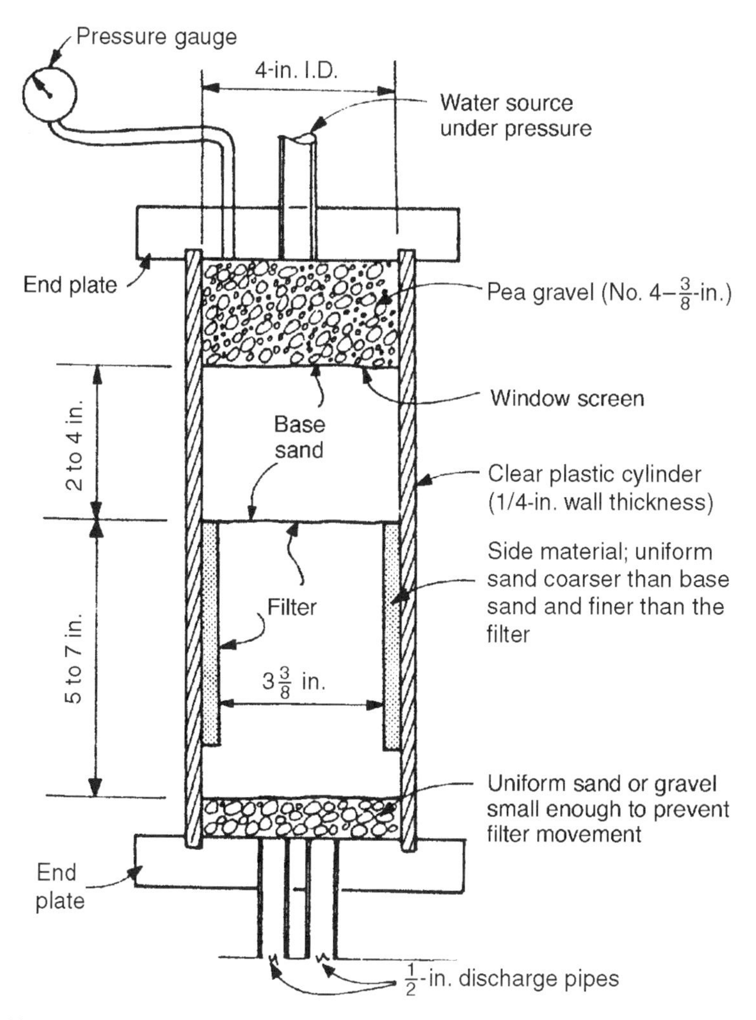

Fig. 7.13. Filter test apparatus details. (Adapted from Sherard et al., 1984a; copyright © ASCE; reprinted with permission.)

inside the filter. In some tests where little or no sand was passing through the filter, the entire apparatus was vibrated on a shaking table with water still flowing through it. The tests were qualitatively judged as follows:

- *Successful.* No significant quantity of base material got through the filter during either the water flow or vibration periods. The thickness and appearance of the base material were unchanged during the test.
- *Failure.* A significant quantity of base material passed through the filter in the first 60 s of flow and continued at about the same rate. If the test was run for a long time, nearly all base material passed through the filter.

- *Borderline.* No significant quantity of base material passed through the filter under the flow of water alone, but a large quantity of base passed through during the vibration.

The test results, plotted in Fig. 7.14, indicate a very narrow boundary between filter failure and success. This boundary is defined by $D_{15}/d_{85} = 9$, which shows that the Terzaghi criterion expressed in Eq. (7.1) is conservative. For tests with silts and clays, Sherard et al. (1984b) used a "slot" test apparatus (Fig. 7.15) designed to evaluate the suitability of a filter when a concentrated leak developed through the base materials, as in the case of a dam core. In the case of filters deemed to be "successful," the flow rate decreased rapidly and the water became progressively clearer, finally sealing the slot completely or stabilizing it at a very small constant flow of clear water. For "unsuccessful" filters, the surge of dirty water continued with no reduction in rate, and the slot expanded in diameter. The tests were conducted for 36 different silts and clays using 25 different filters, with their D_{15} ranging from 0.3 to 9.5 mm. The D_{15B} sizes of the filters corresponding to the failure—success boundaries were plotted as shown in Fig. 7.16. In general, the scatter is much wider than in the case of cohesionless base

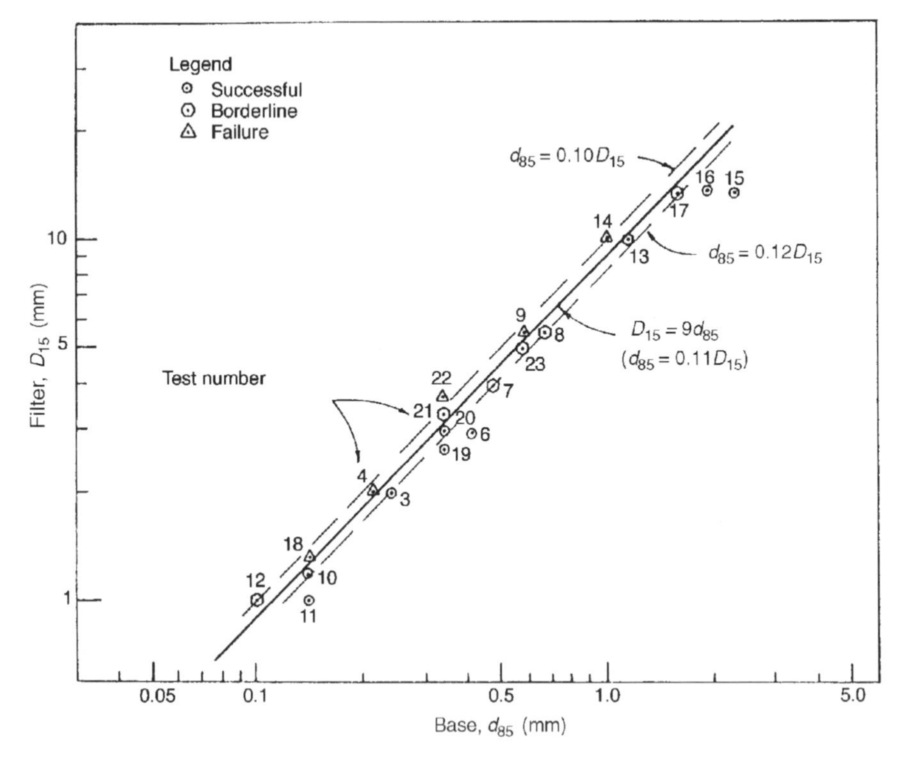

Fig. 7.14. Relationship between D_{15} and d_{85} in SCS filter tests. (Adapted from Sherard et al., 1984a; copyright © ASCE; reprinted with permission.)

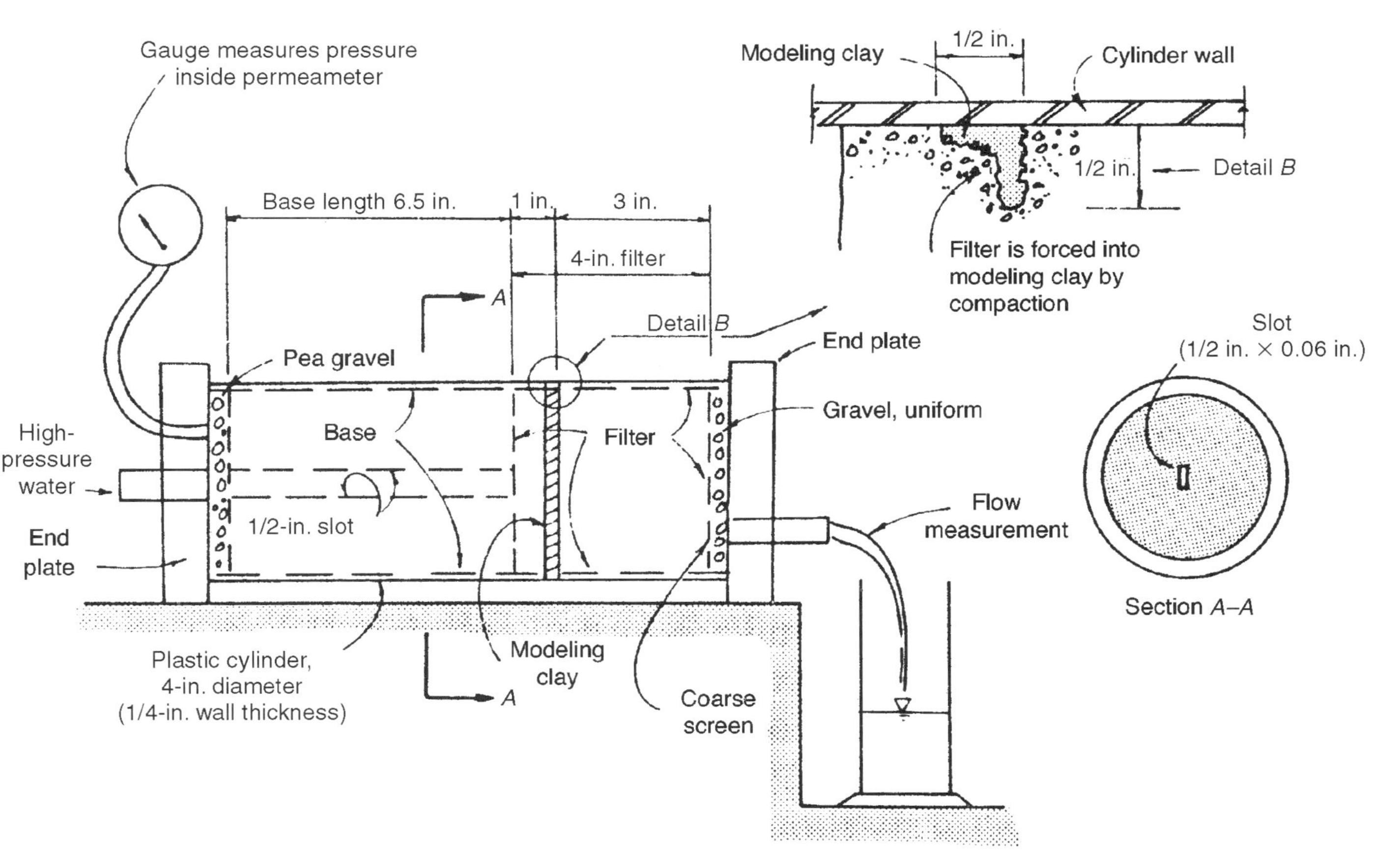

Fig. 7.15. High-pressure slot test apparatus details. (Adapted from Sherard et al., 1984b; copyright © ASCE; reprinted with permission.)

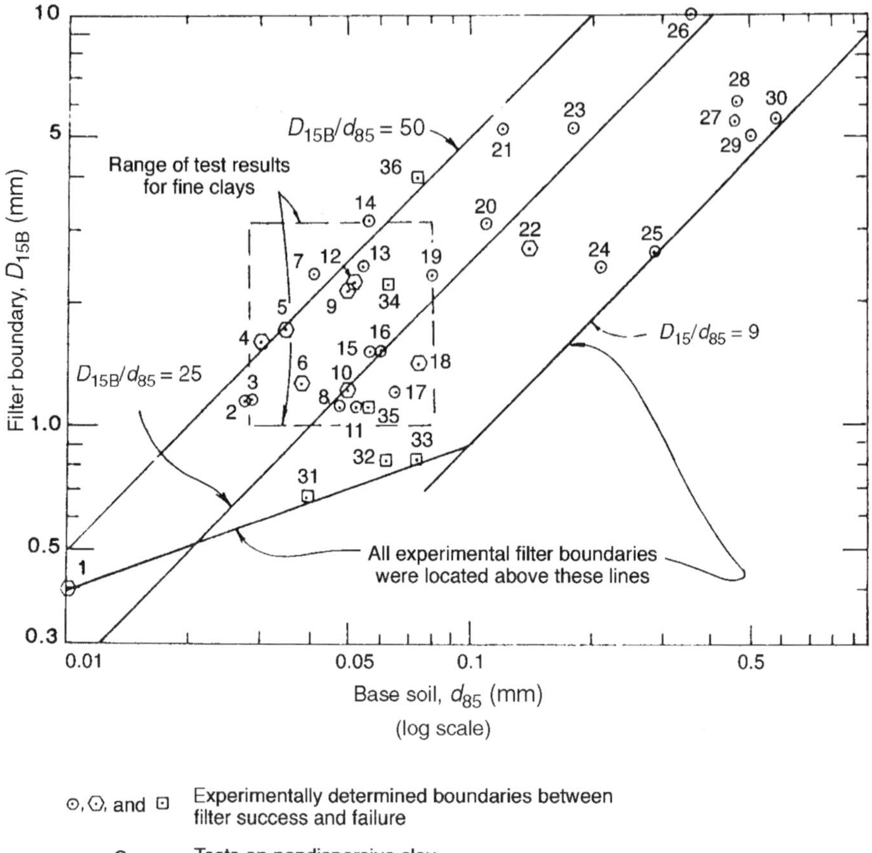

⊙, ⨀, and ⊡	Experimentally determined boundaries between filter success and failure
⊙	Tests on nondispersive clay
⨀	Tests on dispersive clay
⊡	Tests on silts (ML and CL–ML)
⊙ 20	Test on soil No. 20

Fig. 7.16. Summary of results from slot tests. (Adapted from Sherard et al., 1984b; copyright © ASCE; reprinted with permission.)

soils, with D_{15B}/d_{85} being greater than 9 in all cases and exceeding 25 in most cases. Sherard et al. (1984b) concluded that "some property of clay other than the d_{85} size influences the filter needed, but it is not the plasticity as measured by the Atterberg limits."

Subsequent investigations by the same authors (Sherard and Dunnigan, 1989) led to the development of the *no erosion filter test* (NEF test), shown in Fig. 7.17, which is useful for both cohesionless and cohesive base soils. This test is generally recognized to be capable of simulating the most severe conditions that can develop inside a dam when a concentrated erosive leak through the core discharges into a filter. The filter boundary D_{15B} separating successful and unsuccessful tests for a given

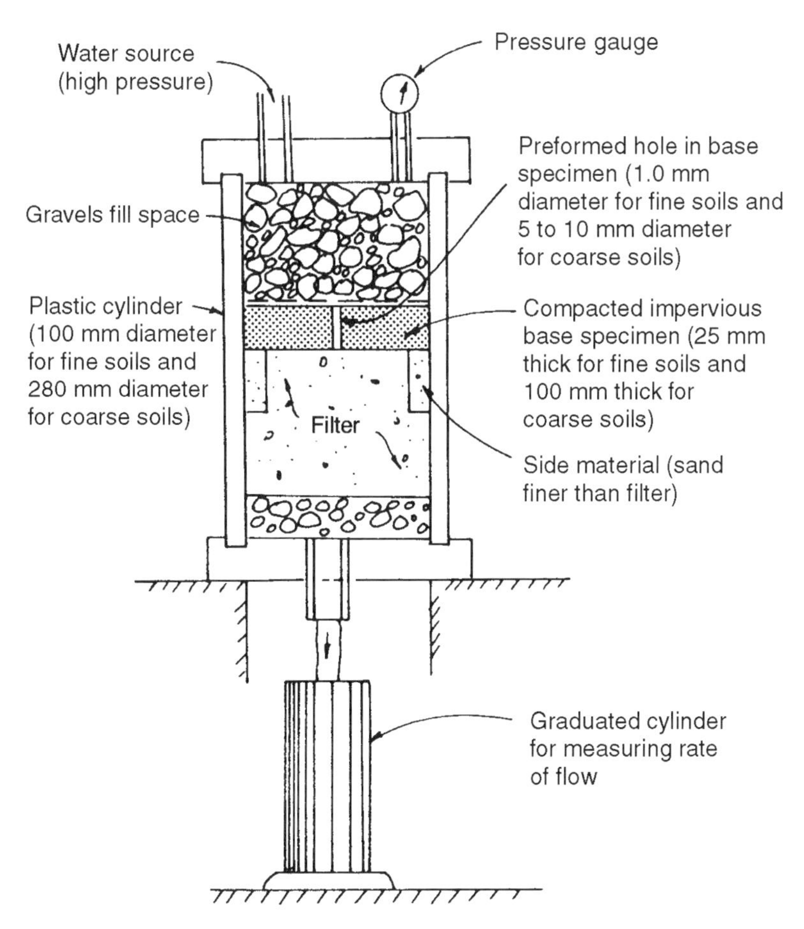

Fig. 7.17. No erosion filter test. (Adapted from Sherard and Dunnigan, 1989; copyright © ASCE; reprinted with permission.)

soil, as determined by the NEF test, is unique and independent of the dimensions of the laboratory apparatus. The results from these tests formed the basis for dividing base soils into four general categories for the sake of generalizing filter criteria (see Table 7.2). There have been several subsequent laboratory investigations, which followed Sherard's basic testing methods. These include the tests conducted by the Australian team of investigators led by Foster and Fell (1999). In general, the research efforts have all been consistent with regard to the grouping of base soils into the four categories.

7.5.3 Flow Pump Tests

Reddi et al. (2000) conducted laboratory tests using concrete sands, typically used as filters for cohesive base soils, and slurries containing kaolinite particles or polystyrene

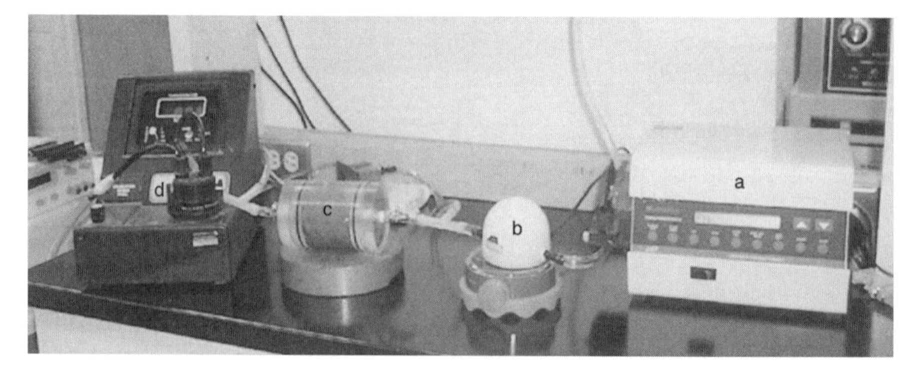

Fig. 7.18. Flow pump test on soil filters: (*a*) flow pump; (*b*) pulse dampener; (*c*) filter soil specimen; (*d*) turbidimeter.

spheres. The objective of these tests was to determine the reductions in hydraulic conductivity of the filter sands due to fine particle clogging of the filter pores. Particles in the influent suspensions used in these tests were colloid sized and were smaller than the majority of pore sizes of the filter. The test apparatus is shown in Fig. 7.18.

Results from these tests indicated that the hydraulic conductivity of the filters could be reduced by one order of magnitude, even when the migrating particles were smaller than the majority of soil filter pores (Fig. 7.19). In the case of broadly graded filters, the particles rearrange themselves during flow, and this process (often referred to as *self-filtration*) alone could cause an order-of-magnitude reduction in hydraulic

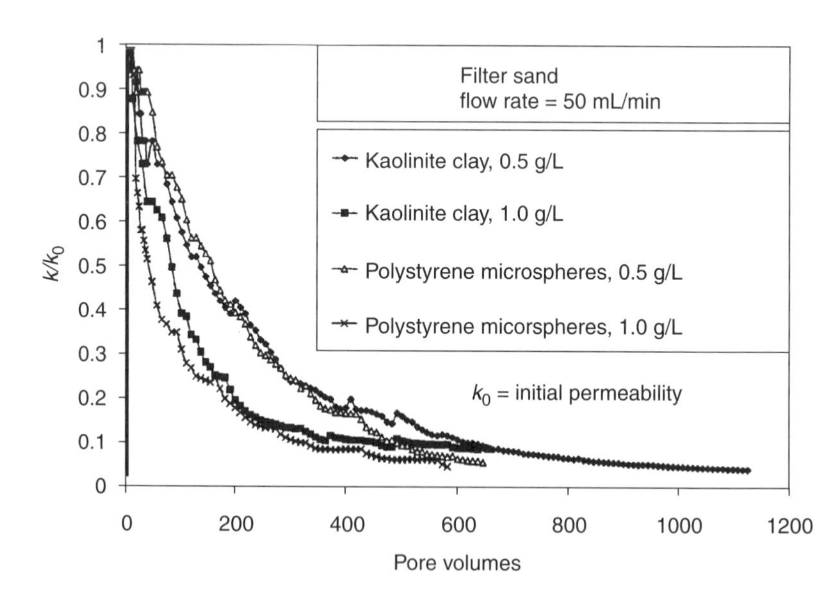

Fig. 7.19. Hydraulic conductivity reduction of filters due to particulate clogging. (Adapted from Reddi et al., 2000.)

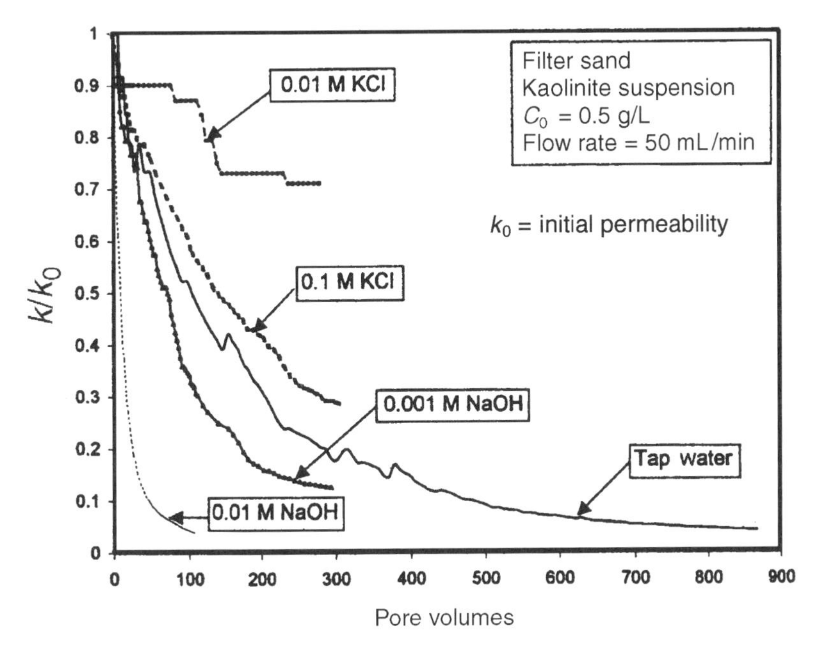

Fig. 7.20. Permeability reduction profiles for different ionic strength conditions of the influent. (Adapted from Hajra et al., 2002.)

conductivity. These tests also attempted to assess the significance of chemical and biological clogging of soil filters. The pore fluid composition as indicated by the ionic strength and pH has a marked effect on filter behavior. The migrating particles flocculate under certain chemical conditions of the pore fluid. The flocculation behavior and intricate balance of electrochemical forces cause differences in fine particle clogging, resulting in different hydraulic conductivity reductions (Fig. 7.20). Biological clogging of soil filters is a much lesser understood mechanism. Growth of microorganisms in filters clogs the pores in much the same way as fine particles from base soils. However, the growth of microorganisms is a function of food sources available in the pore fluid. The rate at which clogging occurs, and the morphology of biomass growth, not only in pores but also at pore constrictions, are subjects of only recent interest. Studies conducted to this date do show, however qualitative they may be, that biological clogging results in a noticeable reduction in drainage capacities of soil filters (Fig. 7.21). The results shown in Fig. 7.21 correspond to flow pump tests conducted with *Pseudomonas aeruginosa* as the bacterial culture and polystyrene spheres, concrete sand, and Ottawa sand as the filter media.

7.5.4 Delft Hydraulics Laboratory Tests

The tests described above involve flow perpendicular to the interface between a filter and base/protected soils. In certain applications, flow takes place parallel or

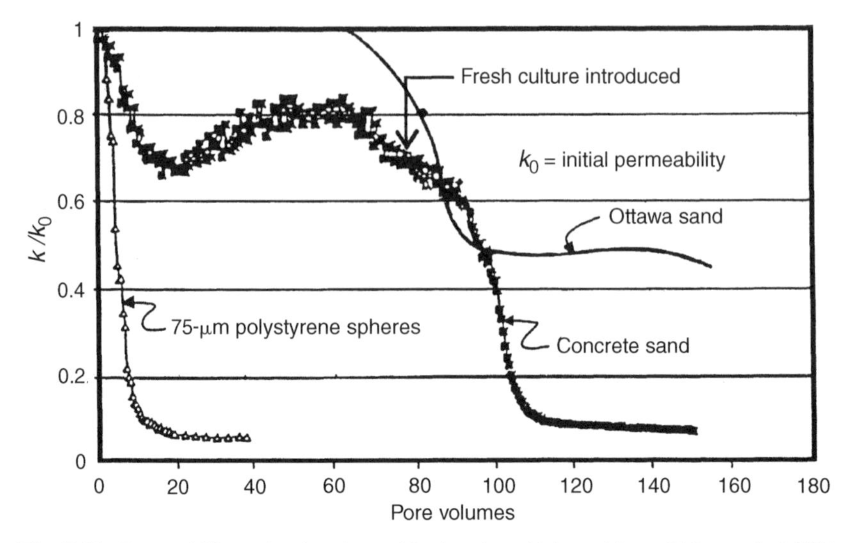

Fig. 7.21. Permeability reduction due to bioclogging. (Adapted from Hajra et al., 2000.)

approximately parallel to the interface. ICOLD (1994) lists these applications as follows:

- At the contacts between bedding filters and base materials and between bedding filter and riprap or revetment on the upstream slopes of embankment dams
- At the contacts between gravel—cobble slope protection and base material on the downstream slopes of embankment dams
- At the contacts between sand—gravel layers and silt—clay layers within alluvial foundations below embankment dams, locations where seepage is parallel or nearly parallel to the slope of the layers
- At the contacts between coarse filters and fine filters within high-flow-capacity filter/drain blankets on downstream foundations

The Delft Hydraulics Laboratory filter box with inclined interface (Bakker et al., 1990) (Fig. 7.22) is an example of filter tests where flow takes place parallel to the interface. These tests in general reveal that the hydraulic gradient plays an important role in the retention criteria of the filters. Accordingly, the test apparatus is designed such that the slope of the device controlling the hydraulic gradient can be varied. Figure 7.23 shows a summary of the results from this test, where the hydraulic gradient is plotted against the ratio D_{15}/d_{50} for two base soils represented by $d_{50} = 0.15$ mm and $d_{50} = 0.82$ mm. Thus, for a base soil with a $d_{50} = 0.15$ mm placed on a slope of 5(H):1(V), D_{15}/d_{50} can be as high as 35. In general, these criteria allow considerably coarser filters than those allowed by the Terzaghi criterion, Eq. (7.1).

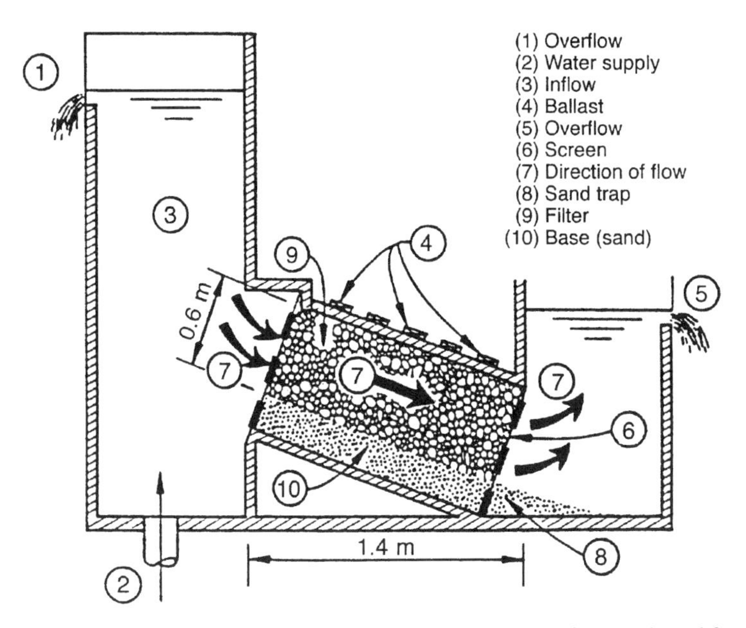

Fig. 7.22. Delft Hydraulics Laboratory filter test with inclined interface. (Adapted from Bakker et al., 1990.)

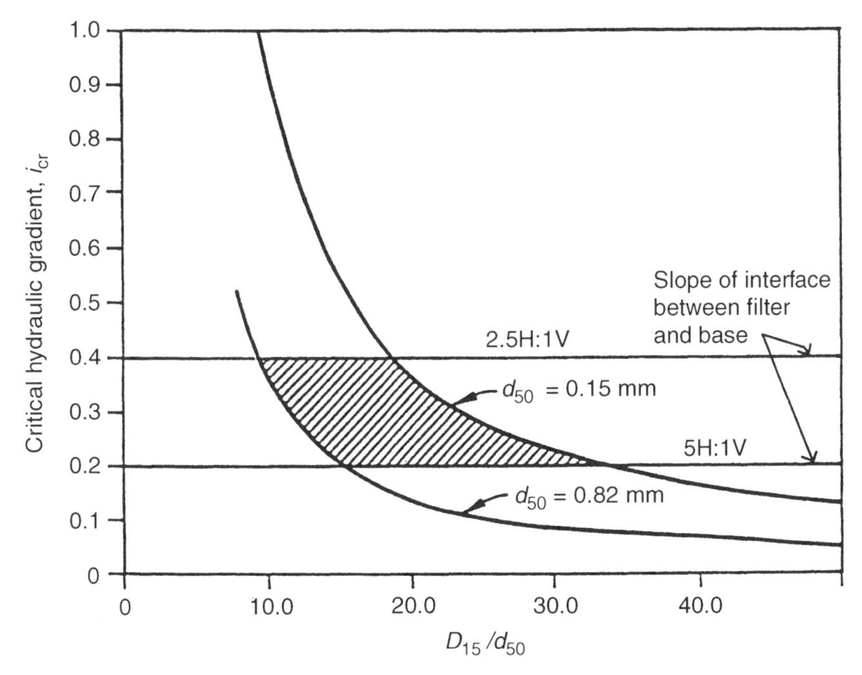

Fig. 7.23. Summary of results from filter tests with flow parallel to interface. (Adapted from Bakker, 1987.)

7.6 CASE STUDIES OF FILTER FAILURES

ICOLD (1994) documented several case histories of failed and successful filters in earth dam projects. In cases where filters failed, the failure was generally attributed to internal erosion from cracks and channels in the earth core, unsuitable filter gradations, and insufficient discharge capacity of chimney and blanket drains. To demonstrate the nature of failures due to internal erosion, we use the case of the failure of Balderhead dam (northern England) reported by Vaughan and Soares (1982). The failure, which occurred in 1967, was due to cracking and internal erosion of the clay core. The filter drain provided downstream of the core (Fig. 7.24*a*) was found to be ineffective in preventing this erosion. At the time of the design of the dam, there was no general agreement on filter design methods for clay cores. The limits of the filter specified for the dam, based on filter drains found to be successful elsewhere (Selset dam), are shown in Fig. 7.24*b*. Investigations made after the damage was discovered showed that the filter placed was somewhat coarser than the specification.

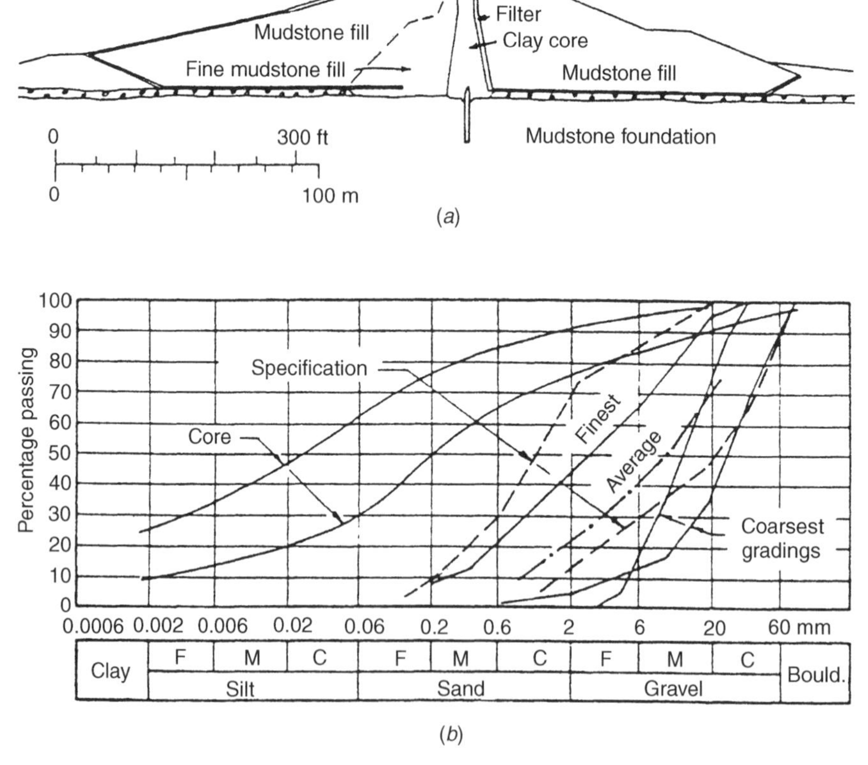

Fig. 7.24. Details of Balderhead dam: (*a*) cross section; (*b*) core and filter gradings. (Adapted from Vaughan and Soares, 1982; copyright © ASCE; reprinted with permission.)

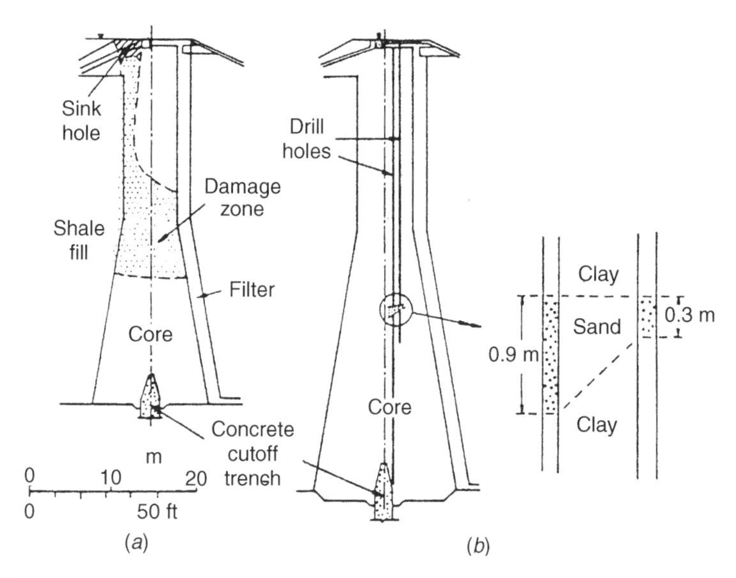

Fig. 7.25. Erosion damage in the Balderhead dam core. (Adapted from Vaughan and Soares, 1982; copyright © ASCE; reprinted with permission.)

The damage, as investigated by rotary drilling using a mud flush, is shown in Fig. 7.25. The crack began near the base of the damage zone and enlarged slowly by erosion until it reached a size at which the cohesive strength of clay was no longer sufficient to prevent a sinkhole formation at the top (Fig. 7.25*a*). The damage migrated upward rapidly. Damage was also noted at a location where no surficial damage was observed. The filter had allowed its invasion by the fines of the core. Segregated fine sands, which were retained by the filter, occupied the damaged portion of the core (Fig. 7.25*b*).

As an exercise, readers should apply the extended criteria for this case to identify the base soil category and relate the design filters to the filters used at the site. Numerous tests performed at the Selset dam had shown that the filter specified was satisfactory for intact clay cores of the type used at the Balderhead dam. The lessons from this case study are that filters differ in their gradation between specified and as-placed conditions in constructions of large magnitude, and that filters may operate successfully where the core remains intact but may not succeed in preventing erosion when the core cracks. The NEF test, proposed by Sherard and Dunnigan (1989), could have provided valuable guidance at the design stage to identify a suitable filter for the clay cores and the operating conditions expected at this dam.

PROBLEMS

7.1. The materials and their gradations used in the Guavio dam are shown in Fig. P7.1. Use the filter criteria of Sections 7.2 and 7.3 and check the adequacy

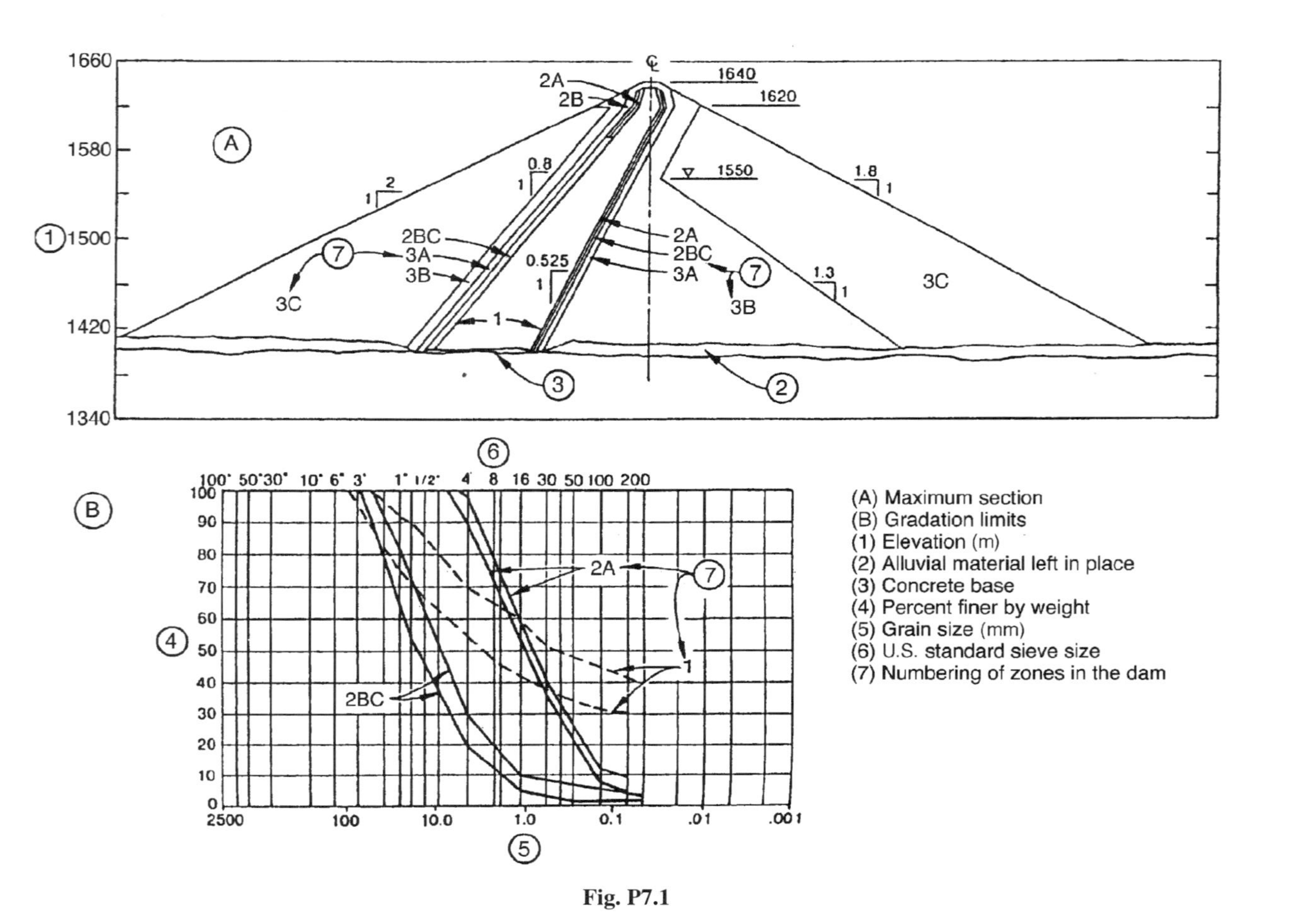
A
1660
1580
1500
1420
1340
1
2A
2B
1640
1620
1550
2
1
0.8
1
1.8
1
0.525
1
1.3
1
2BC
3A
3B
7
3C
2A
2BC
3A
7
3B
3C
1
3
2
B
6
100" 50"30" 10" 6" 3" 1" 1/2" 4 8 16 30 50 100 200
100
90
80
70
60
50
40
30
20
10
0
4
2500
100
10.0
1.0
0.1
.01
.001
5
2A
7
1
2BC
(A) Maximum section
(B) Gradation limits
(1) Elevation (m)
(2) Alluvial material left in place
(3) Concrete base
(4) Percent finer by weight
(5) Grain size (mm)
(6) U.S. standard sieve size
(7) Numbering of zones in the dam

Fig. P7.1

of filters used in the dam. (From the Columbia National Committee Response, ICOLD, 1994.)

7.2. The materials and their gradations used in the Salvajina dam are shown in Fig. P.7.2. Use the filter criteria of Sections 7.2 and 7.3 and check the adequacy of filters used in the dam. (From the Columbia National Committee Response, ICOLD, 1994.)

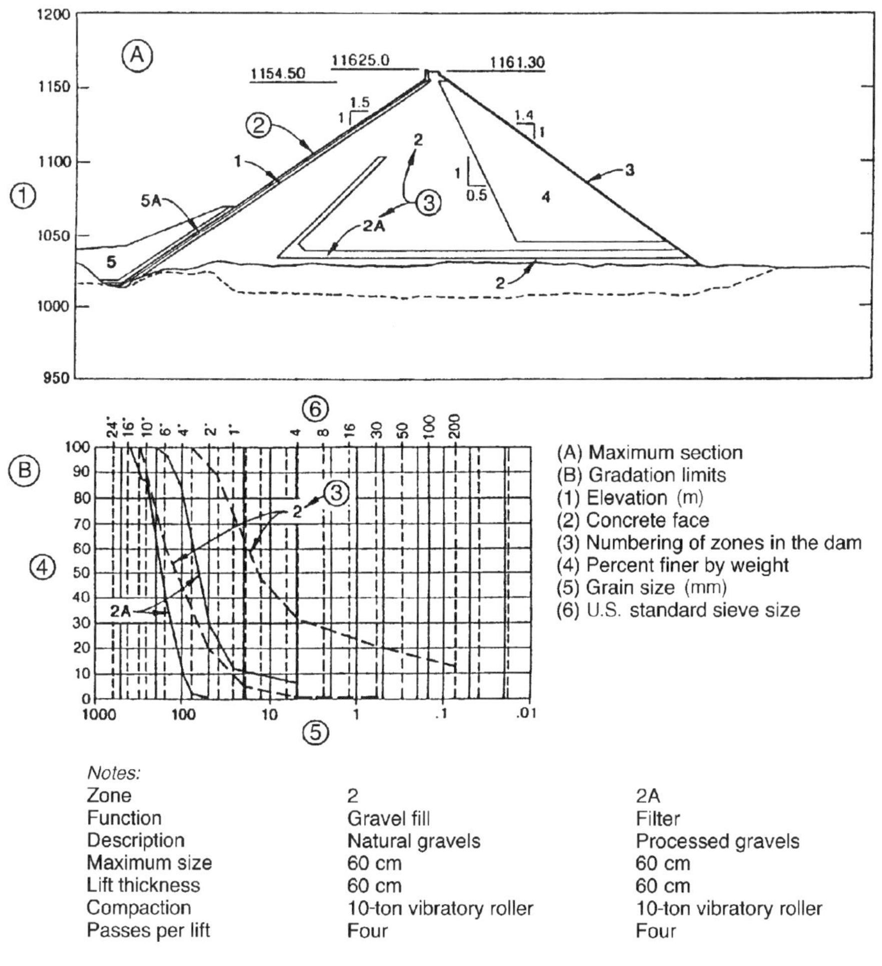

Notes:

Zone	2	2A
Function	Gravel fill	Filter
Description	Natural gravels	Processed gravels
Maximum size	60 cm	60 cm
Lift thickness	60 cm	60 cm
Compaction	10-ton vibratory roller	10-ton vibratory roller
Passes per lift	Four	Four

Fig. P7.2

7.3. Refer to Problem 3.13. For the drainage capacities obtained in the case of Fig. P3.13*a*, design the drainage layer in terms of its geometry, longitudinal slope, and hydraulic conductivity.

CHAPTER 8

GEOSYNTHETICS IN SEEPAGE

Geosynthetics are plastic sheet products, which have found rapid application in civil engineering construction in recent years. Since its skeptic beginnings during the late 1960s and early 1970s, the geosynthetic industry has become a billion dollar industry, with more than 600 different geosynthetic products currently available in the United States alone. Their widespread use is reflected in the very broad definition of a *geosynthetic* as "a planar product manufactured from polymeric material used with soil, rock, earth, or other geotechnical engineering related material as an integral part of a man-made project, structure, or system" (ASTM D 4439). Use of these synthetic products in geotechnical infrastructure represents perhaps the first large-scale use of manufactured products with field soils.

Our interest in this chapter is to explore the use of geosynthetics in engineering applications involving seepage. Specifically, we attempt to focus on the filtration and drainage processes and study how geosynthetics could be used as an alternative to soil filters and drainage layers discussed in Chapter 7. The reader should therefore expect the same outline for this chapter as in Chapter 7. It is also worthwhile to compare the developments in design criteria for geosynthetic filters with those discussed for soil filters in Chapter 7 and draw parallels between the two.

8.1 CLASSIFICATION OF GEOSYNTHETICS

Several subfamilies of geosynthetics have been developed in the relatively short time period of the existence of the geosynthetic industry. The broad categories are listed below with their ASTM definitions:

- *Geocomposite:* a product composed of two or more materials, at least one of which is a geosynthetic
- *Geofoam:* a block or planar rigid cellular foam polymeric material used in geotechnical engineering applications
- *Geogrid:* a geosynthetic formed by a regular network of integrally connected elements with apertures greater than 6.35 mm ($\frac{1}{4}$ in.) to allow interlocking with surrounding soil, rock, earth, and other surrounding materials to function primarily as reinforcement
- *Geomembrane:* an essentially impermeable geosynthetic composed of one or more synthetic sheets
- *Geonet:* a geosynthetic consisting of integrally connected parallel sets of ribs overlying similar sets at various angles for planar drainage of liquids or gases
- *Geosynthetic clay liner:* a manufactured hydraulic barrier consisting of clay bonded to a layer or layers of geosynthetic materials
- *Geotextile:* a permeable polymeric material comprised of synthetic fibers and textile yarns

Out of the subfamilies listed above, *geotextiles* are the most versatile materials, serving several functions in geotechnical engineering applications, including separation, filtration, drainage, reinforcement, and protection. In this chapter we are concerned primarily with geotextiles.

Geosynthetics are usually made from synthetic polymers of polypropylene, polyester, or polyethylene, which are generally known to be resistant to biological and chemical degradation. Although less common, polyamides (nylon) and glass fibers are used in some geosynthetics. Natural fibers such as cotton and jute are finding increasing applications, particularly in Asian countries. Geosynthetics are characterized by several classification schemes and are generally identified by:

- Density of polymer [high density or low density, e.g., high-density polyethylene (HDPE)]
- Type of basic element constituting the product (e.g., filament, yarn, strand, rib, coated rib, etc.)
- Manufacturing process (e.g., woven, needle-punched nonwoven, heat-bonded nonwoven, stich-bonded, extruded, knitted, roughened sheet, smooth sheet)
- Physical properties describing the material (e.g., mass per unit area, thickness, and/or opening size)

As shown in Fig. 8.1, a wide variety of basic elements and manufacturing processes characterize geosynthetics. Geotextiles form perhaps the largest group of geosynthetics. Although some are knitted, the majority of geotextiles are either woven or nonwoven. Woven geotextiles are made using standard weaving machinery with monofilament, multifilament, fibrillated, or slit film yarns. Nonwoven geotextiles are

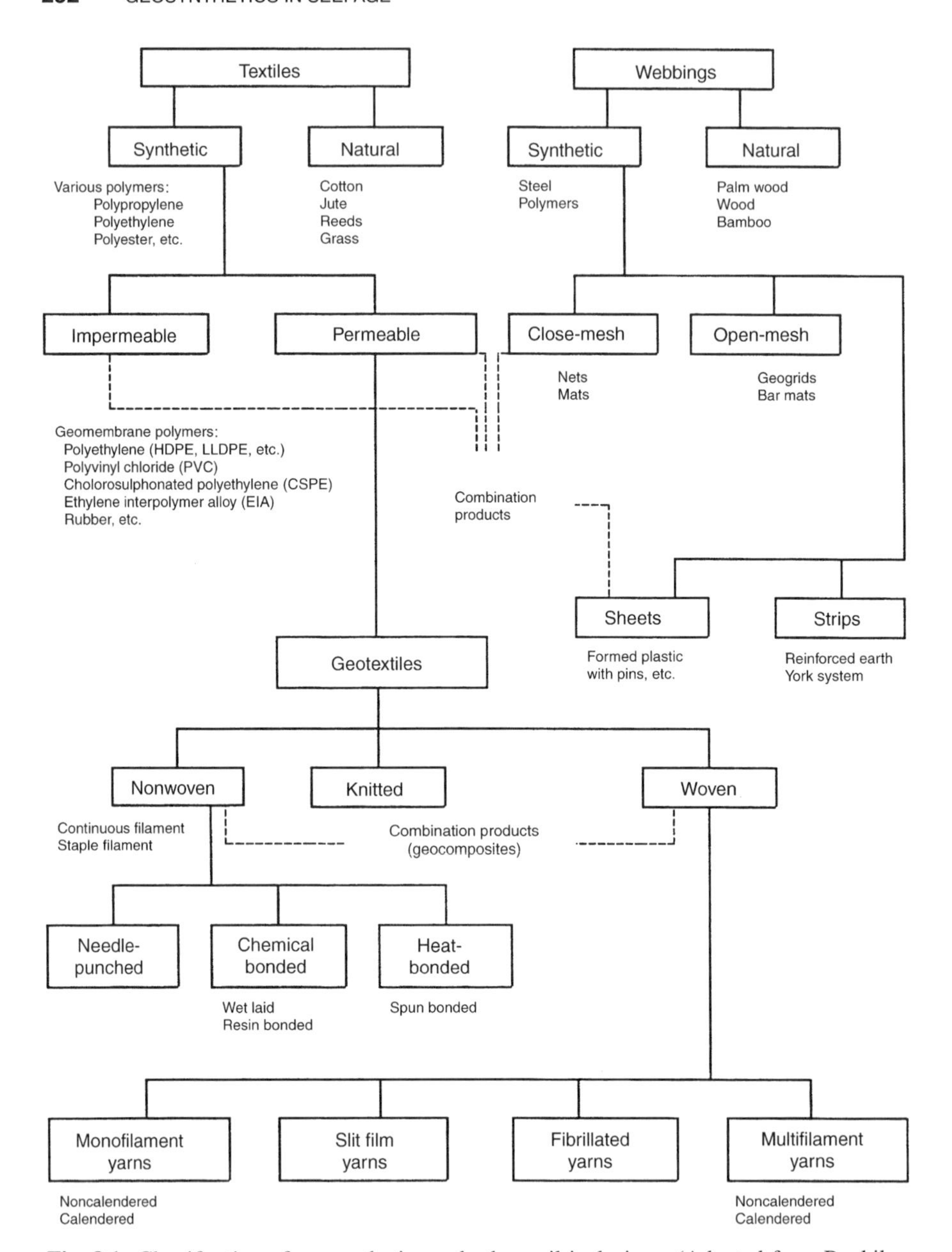

Fig. 8.1. Classification of geosynthetics and other soil inclusions. (Adapted from Rankilor, 1981; copyright © John Wiley & Sons88; reprinted with permission.)

formed by matting the fibers together. In these geotextiles, the filaments or fibers are either needle punched, in which small needles are used to entangle the fibers mechanically, or heat bonded or chemical bonded, in which the fibers are welded together using thermal or chemical processes. Samples of geotextiles are shown in Fig. 8.2 together with representative photomicrographs of woven and nonwoven fabrics.

Because of the wide variety in basic elements and manufacturing processes used, geosynthetics exhibit significant variability in their properties and engineering performance. Several properties are used to characterize geosynthetics. Selection of the properties and tests used to determine these properties are governed by the specific engineering function of the geosynthetics. The broad functions of geosynthetics are filtration, drainage, separation, reinforcement, fluid barrier, and protection. The filtration and drainage functions correspond to particle filtration during water flow and water transmission, respectively. These are the same functions as those served by natural soils, described in Chapter 7. When used to serve the separation function, geosynthetics prevent one material from penetrating another (in the absence of hydraulic flow), such as to prevent road base materials from penetrating the underlying soft subgrade

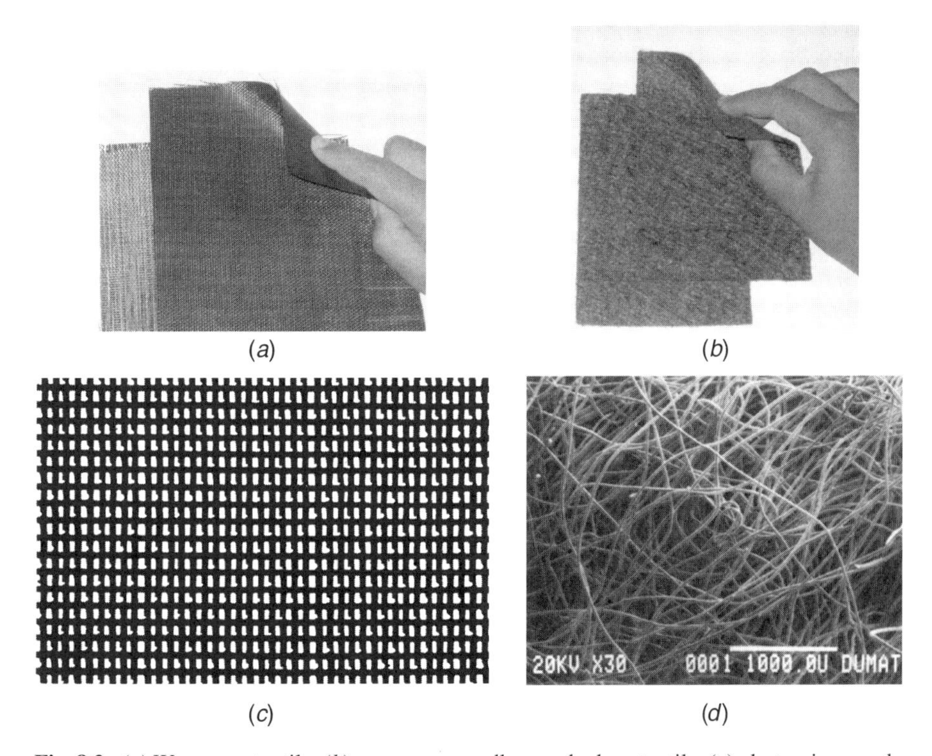

Fig. 8.2. (*a*) Woven geotextile; (*b*) nonwoven needle-punched geotextile; (*c*) photomicrograph of a woven monofilament geotextile (×5); (*d*) photomicrograph of a nonwoven needle-punched geotextile (×30). [(*c*, *d*) Adapted from Koerner, 1998; copyright © Pearson Education; reprinted with permission.]

TABLE 8.1 Type of Geosynthetic versus Possible Functions

	Possible Functions Served[a]				
Geosynthetic Type	Separation	Reinforcement	Filtration	Drainage	Barrier
Geotextiles	P or S	P or S	P or S	P or S	—
Geogrids	—	P	—	—	—
Geonets	—	—	—	P	—
Geomembranes	—	—	—	—	P
Geosynthetic clay liner	—	—	—	—	P
Geocomposites	P or S	P or S	P or S	P or S	P or S
Geopipe	—	—	—	P	—

Source: Adapted from Koerner and Soong, 1995.

[a]P, usual primary function; S, possible secondary function.

soils. Geosynthetics serving the reinforcement function provide tensile strength to a soil matrix, much like what steel does to concrete. When used as fluid barriers, geosynthetics impede the flow of liquids in applications such as waste containment. In the last function of protection, geosynthetics are used as stress relief layers (e.g., as blankets to reduce surface erosion due to rainfall impact). Geosynthetics are used to serve more than one function in a number of applications. For instance, geotextiles used as drainage layers generally serve the filtration function also. Often, geosynthetics are characterized by their primary function, although they serve one or more of the other secondary functions. The various families of geosynthetics are categorized according to their function in Table 8.1. A list of common engineering applications and the controlling functions of geosynthetics is given in Table 8.2. Narrowing down our interest to filtration and drainage, we note that the engineering applications where geotextiles found use are the same applications where we considered soils as filters and drainage layers in Chapter 7. Three examples are shown in Fig. 8.3. The geotextile used in Fig. 8.3*a* serves the filtration function of protecting the core of the dam. A separate soil drainage layer is used on top of the fabric to meet the demands of transmission capacities. In the case of a retaining wall application (Fig. 8.3*b*), the geotextile is expected to serve as a filter and drain to transmit water from behind the wall. When geotextiles are used in highway and railroad underdrains, they not only perform the filtration function but also serve as separators protecting the drainage material/pipe. Figure 8.3*c* shows a number of different configurations of underdrains where geotextiles have been used successfully.

8.2 GENERAL PROPERTIES OF GEOSYNTHETICS

The properties of geosynthetics required in their design will depend on the specific engineering functions they are expected to perform. The design criteria and parameters required for geosynthetic evaluation are listed in Table 8.3. The geosynthetic properties required in filtration and drainage fall predominantly into hydraulic design

TABLE 8.2 Representative Applications and Controlling Functions of Geosynthetics

Primary Function	Application	Secondary Function(s)
Separation	Unpaved roads (temporary and permanent)	Filter, drains, reinforcement
	Paved roads (secondary and primary)	Filter, drains
	Construction access roads	Filter, drains, reinforcement
	Working platforms	Filter, drains, reinforcement
	Railroads (new construction)	Filter, drains, reinforcement
	Railroads (rehabilitation)	Filter, drains, reinforcement
	Landfill covers	Reinforcement, drains, protection
	Preloading (stabilization)	Reinforcement, drains
	Marine causeways	Filter, drains, reinforcement
	General fill areas	Filter, drains, reinforcement
	Paved and unpaved parking facilities	Filter, drains, reinforcement
	Cattle corrals	Filter, drains, reinforcement
	Coastal and river protection	Filter, drains, reinforcement
	Sports fields	Filter, drains, protection
Filter	Trench drains	Separation, drains
	Pipe wrapping	Separation, drains, protection
	Base course drains	Separation, drains
	Frost protection	Separation, drains, reinforcement
	Structural drains	Separation, drains
	Toe drains in dams	Separation, drains
	High embankments	Drains
	Filter below fabric-form	Separation, drains
	Silt fences	Separation, drains
	Silt screens	Separation
	Culvert outlets	Separation
	Reverse filters for erosion control:	Separation
	Seeding and mulching	
	Beneath gabions	
	Ditch armoring	
	Embankment protection, coastal	
	Embankment protection, rivers and streams	
	Embankment protection, lakes	
	Vertical drains (wicks)	
Drainage-transmission	Retaining walls	Separation, filter
	Vertical drains	Separation, filter
	Horizontal drains	Reinforcement
	Below membranes (drainage of gas and water)	Reinforcement, protection
	Earth dams	Filter
	Below concrete (decking and slabs)	Protection

(*continued*)

TABLE 8.2 Representative Applications and Functions of Geosynthetics *(Continued)*

Primary Function	Application	Secondary Function(s)
Reinforcement	Pavement overlays	—
	Subbase reinforcement in roadways and railways	Filter
	Retaining structures	Drains
	Membrane support	Separation, drains, filter, protection
	Embankment reinforcement	Drains
	Fill reinforcement	Drains
	Foundation support	Drains
	Soil encapsulation	Drains, filter, separation
	Net against rockfalls	Drains
	Fabric retention systems	Drains
	Sandbags	—
	Reinforcement of membranes	Protection
	Load redistribution	Separation
	Bridging nonuniformity soft soil areas	Separation
	Encapsulated hydraulic fills	Separation
	Bridge piles for fill placement	—
Fluid barrier	Asphalt pavement overlays	—
	Liners for canals and reservoirs	—
	Liners for landfills and waste repositories	—
	Covers for landfill and waste repositories	—
	Cutoff walls for seepage control	—
	Waterproofing for tunnels	—
	Facing for dams	—
	Membrane-encapsulated soil layers	—
	Expansive soils	—
	Flexible formwork	—
Protection	Geomembrane cushion	Drains
	Temporary erosion control	Fluid barrier
	Permanent erosion control	Reinforcement, fluid barrier

Source: Adapted from Holtz et al. (1998).

and constructability and longevity, requirements. Table 8.4 shows the complete set of properties and parameters and associated test methods used to evaluate these properties. The properties are grouped under:

- General properties, usually supplied by manufacturers
- Index properties, which are themselves grouped under mechanical strength, endurance, and hydraulic properties
- Performance properties

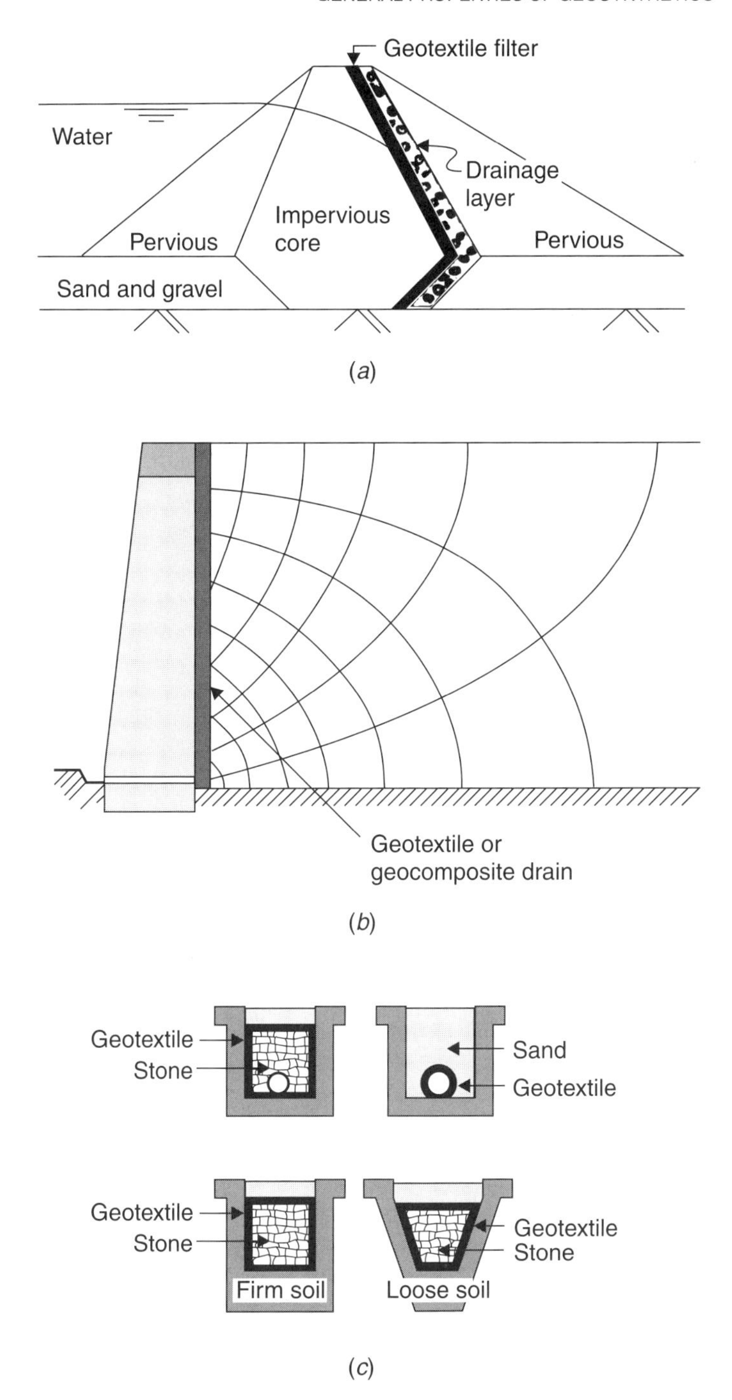

Fig. 8.3. Engineering applications of geotextiles: (*a*) as a filter in zoned dams; (*b*) as a drain behind retaining walls; (*c*) as highway and railroad underdrains. [(*c*) Adapted from Koerner and Soong, 1995; copyright © ASCE; reprinted with permission.]

TABLE 8.3 Important Criteria and Principal Properties Required for Geosynthetic Evaluation

Criteria and Parameter	Property[a]	Function					
		Filtration	Drainage	Separation	Reinforcement	Barrier	Protection
Design requirements:							
Mechanical strength							
Tensile strength	Wide width strength	—	—	—	×	×	—
Tensile modulus	Wide width modulus	—	—	—	×	×	—
Seam strength	Wide width strength	—	—	—	×	×	—
Tension creep	Creep resistance	—	—	—	×	×	—
Compression creep	Creep resistance	—	×[b]	—	—	—	—
Soil–geosynthetic friction	Shear strength	—	—	—	×	×	×
Hydraulic characteristics							
Flow capacity	Permeability	×	×	×	×	×	—
	Transmissivity	—	×	—	—	—	×
Piping resistance	Apparent opening size	×	—	×	×	—	×
Clogging resistance	Porimetry	×	—	—	—	—	×
	Gradient ratio or long-term flow	×	—	—	—	—	×
Constructability requirements:							
Tensile strength	Grab strength	×	×	×	×	×	×
Seam strength	Grab strength	×	×	×	—	×	—
Bursting resistance	Burst strength	×	×	×	×	×	×
Puncture reistance	Rod or pyramid puncture	×	×	×	×	×	×
Tear resistance	Trapezoidal tear	×	×	×	×	×	×

Longevity (durability):							
Abrasion resistance[c]	Reciprocating block abrasion	×	—	—	—	—	—
UV stability[d]	UV resistance	×	—	—	×	×	×
Soil environment[e]	Chemical	×	×	?	×	×	?
	Biological	×	×	?	×	×	?
	Wet–dry	×	×	—	—	—	×
	Freeze–thaw	×	×	—	—	×	—

Source: Adapted from Holtz et al. (1998).

[a]See Table 8.4 for specific procedures.

[b]Compression creep is applicable to some geocomposites.

[c]Erosion control applications where armor stone may move.

[d]Exposed geosynthetics only.

[e]Where required.

TABLE 8.4 Geosynthetic Properties and Parameters

Property	Test Method[a]	Units of Measure
General properties (from manufacturers)		
Type and construction	N/A	—
Polymer	N/A	—
Mass per unit area	ASTM D 5261	g/m^2
Thickness (geotextiles and geomembranes)	ASTM D 5199	mm
Roll length	Measure	m
Roll widths	Measure	m
Roll weight	Measure	kg
Roll diameter	Measure	m
Specific gravity and density	ASTM D 792 and D 1505	g/m^3
Surface characteristics	N/A	—
Index properties		
Mechanical strength: Uniaxial loading		
Tensile strength (quality control)		
Grab strength (geotextiles and CSPE reinforced geomembranes)	ASTM D 4632	N
Single rib strength (geogrids)	GRI:GG1	N
Narrow strip (geomembranes)		
EDPM, CO, IR, CR	ASTM D 412	N
HDPE	ASTM D 638	N
PVC, VLDPE	ASTM D 882	N
Tensile strength (load–strain characteristics)		
Wide strip (geotextiles)	ASTM D 4595	N
Wide strip (geogrid)	No standard	N
Wide strip strength (geomembranes)	ASTM D 4885	N
2% secant modulus (PE geomembranes)	ASTM D 882	N
Junction strength (geogrids)	GRI:GG2	%
Dynamic loading	No standard	

Creep resistance	ASTM D 5262 (*Note:* interpretation required)	Creep strain: % ∈/h Creep rupture: kN/m
Index friction	GRI:GS7	Dimensionless
Seam strength		
Sewn (geotextiles)	ASTM D 4884	% efficiency
Factory peel and shear (geomembranes)	ASTM D 4545	kg/mm
Field peel and shear (geomembranes)	ASTM D 4437	kg/mm
Tear strength		
Trapezoid tearing (geotextile)	ASTM D 4533	N
Tear resistance (geomembranes)	ASTM D 1004	N
Mechanical strength: rupture resistance		
Burst strength		
Mullen burst (geotextiles)	ASTM D 3786	Pa
CBR (geotextiles, geonets, geomembranes)	GRI:GS1	Pa or N
Large-scale hydrostatic (geomembranes and geotextiles)	ASTM D 5514	Pa
Puncture resistance		
Index (geotextiles and geomembranes)	ASTM D 4833	N
Pyramid puncture (geomembranes)	ASTM D 5494	N
CBR (geotextiles, geonets, geomembranes)	GRI:GS1	N
Penetration resistance (dimensional stability)	No standard	
Geosynthetic cutting resistance	No standard	
Flexibility (stiffness)	ASTM D 1388	mg/cm^2
Endurance properties		
Abrasion resistance (geotextile)	ASTM D 4886	%
Ultraviolet (UV) radiation stability		
Xenon-arc apparatus (geotextile)	ASTM D 4355	%
Outdoor exposure	ASTM D 5970	%

(*continued*)

TABLE 8.4 Geosynthetic Properties and Parameters (*Continued*)

Property	Test Method[a]	Units of Measure
Index properties (*continued*)		
Chemical resistance		
Chemical immersion	ASTM D 5322	N/A
Oxidative induction time	ASTM D 5885	minutes
Environmental exposure	EPA 9090	% change
Biological resistance		
Biological clogging (geotextile)	ASTM D 1987	m^3/s
Biological degradation	ASTM G 21 and G 22	
Soil burial	ASTM D 3083	% change
Wet and dry stability	No standard	
Temperature stability		
Temperature stability (geotextile)	ASTM D 4594	% change
Dimensional stability (geomembrane)	ASTM D 1204	% change
Hydraulic properties		
Opening characteristics (geotextile)		
Apparent opening size (AOS)	ASTM D 4751	mm
Porimetry (pore-size distribution)	Use AOS for O_{95}, O_{85}, O_{50}, O_{15}, and O_5	mm
Percent open area (POA)	See Christopher and Holtz (1985)	%
Porosity (n)	(V_{voids}/V_{total}) 100	%
Permeability (k) and permittivity (ψ)	ASTM D 4491	m/s and s^{-1}
Soil retention ability	Empirically related to opening characteristics	
Clogging resistance	ASTM D 5101 and GRI:GT8	
In-plane flow capacity (transmittivity, θ)	ASTM D 4716	m^2/s

Performance Properties		
Stress–strain characteristics		kN/m and % strain
Tension test in soil	see McGown et al. (1982)	
Triaxial test method	see Holtz et al. (1982)	
CBR on soil fabric system	see Christopher and Holtz (1985)	
Tension test in shear box	see Christopher and Holtz (1985)	
Creep tests		kN/m and % strain
Extension test in soil	see McGown et al. (1982)	
Triaxial test method	see Holtz et al. (1982)	
Extension test in shear box	see Christopher and Holtz (1985)	
Pullout method	see Christopher et al. (1990)	
Friction/adhesion		
Direct shear (soil–geosynthetic)	ASTM D 5321	degrees (°)
Direct shear (geosynthetic–geosynthetic)	ASTM D 5321	degrees (°)
Pullout (geogrids)	GRI:GG5	dimensionless
Pullout (geotextiles)	GRI:GT6	dimensionless
Anchorage embedment (geomembranes)	GRI:GM2	kN/m
Dynamic and cyclic loading resistance	No standard procedures	N/A
Puncture		
Gravel, truncated cone or pyramid	ASTM D 5494	kPa
Chemical resistance		
In situ immersion testing	ASTM D 5496	N/A
Soil retention and filtration properties		
Gradient ratio method for noncohesive sand and silt type soils	ASTM D 5101	dimensionless
Hydraulic conductivity ratio (HCR) for fine-grained soils	ASTM D 5567	dimensionless
Slurry method for silt fence applications	ASTM D 5141	%

Source: Adapted from Holtz et al. (1998).

[a]ASTM, American Society for Testing and Materials; GRI, Geosynthetic Research Institute; N/A, not applicable.

The index tests do not produce an actual design property in most cases; however, they are valuable to estimate or infer the design property of interest. Several standardized tests (primarily by ASTM) are available to conduct these tests. Performance tests are designed to test the geosynthetics with the site-specific soils and for the design conditions intended.

Not all the tests are relevant in all engineering applications. The AASHTO standard specifications for highway applications (M 288-96) require a specific classification of geotextiles (class 1, class 2, or class 3) based on strength properties (Table 8.5). Geotextiles selected for a specific engineering application should conform to one of the three classes listed in Table 8.5. For instance, geotextiles used in filtration and drainage applications are required to meet the properties of Class 2 in Table 8.5. The properties required for each class depend on geotextile elongation. The numerical values in Table 8.5 represent minimum average roll values (MARVs) in the weaker principal direction. When sewn seams are required, the seam strength, as measured in accordance with ASTM D 4632, is required to be equal to or greater than 90% of the grab strength specified.

The Corps of Engineers recommended minimum strength values required for survivability of geotextiles in engineering applications. These values, shown in Table 8.6, should be used only as preliminary guidelines, and specific applications may require additional testing.

8.3 HYDRAULIC INDEX PROPERTIES OF GEOSYNTHETICS

The hydraulic properties of geotextiles relevant in seepage applications relate to their ability to allow water flow and to retain soil particles while allowing seepage. As listed in Table 8.4, the corresponding index properties are (1) opening characteristics as determined by the apparent opening size (AOS), porimetry, percent open area (POA), and/or porosity; (2) permeability and permittivity; (3) soil retention ability; (4) clogging resistance; and (5) in-plane flow capacity (transmissivity). We provide a description of these properties in the following sections.

8.3.1 Opening Characteristics

Open pores in geotextiles provide a direct measure of their ability to convey water. The size of the largest pores in geotextiles also reflects their ability to retain soil particles. *Apparent opening size* (AOS), defined as a property which indicates the size of the largest particle that would effectively pass through a geotextile (ASTM D 4751), is one of the key index properties governing filtration and drainage functions of geotextiles. To determine AOS, also denoted as O_{95}, a geotextile specimen is placed in a mechanical sieve shaker and sized glass beads are placed on the geotextile surface. The geotextile is shaken laterally so that the jarring motion will induce the beads to pass through the test specimen. The procedure is repeated on the same specimen with glass beads of various sizes until the apparent opening size has been determined. A detailed description of this test method may be found in ASTM standards (D 4751-99). AOS is sometimes referred to as *equivalent opening size* (EOS).

TABLE 8.5 Geotextile Strength Property Requirements

Property	Test Method	Units of Measure	Geotextile Class[a]					
			Class 1		Class 2		Class 3	
			Elongation $< 50\%$[b]	Elongation $\geq 50\%$[b]	Elongation $< 50\%$[b]	Elongation $\geq 50\%$[b]	Elongation $< 50\%$[b]	Elongation $\geq 50\%$[b]
Grab strength	ASTM D 4632	N	1400	900	1100	700	800	500
Sewn seam strength[c]	ASTM D 4632	N	1260	810	990	630	720	450
Tear strength	ASTM D 4533	N	500	350	400[d]	250	300	180
Puncture strength	ASTM D 4833	N	500	350	400	250	300	180
Burst strength	ASTM D 3786	kPa	3500	1700	2700	1300	2100	950
Permittivity	ASTM D 4991	sec^{-1}	Minimum property values for permittivity, AOS, and UV stability are based on geotextile application. Refer to Table 8.11.					
Apparent opening size	ASTM D 4751	mm						
Ultraviolet stability	ASTM D 4355	%						

Source: Adapted from Holtz et al. (1998).

[a]The severity of installation conditions for the application generally dictate the required geotextile class. Class 1 is specified for more severe or harsh installation conditions where there is greater potential for geotextile damage, and Classes 2 and 3 are specified for less severe conditions.

[b]As measured in accordance with ASTM D 4632.

[c]When sewn seams are required.

[d]The required MARV tear strength for woven monofilament geotextiles is 250 N.

TABLE 8.6 U.S. Army Corps of Engineers Recommended Geotextile Strength Requirements

	Strength Requirements (Minimum Values)[a]				
Geotextile Use	ASTM D 4632 Tensile (N)	ASTM D 4335 Ultraviolet Degradation at 500 h (%)	ASTM D 4833 Puncture (N)	ASTM D 4533 Tear (N)	ASTM D 4886 Abrasion (N)
Riprap slope protection filter with more than 102 mm (4 in.) of bedding	515	50	180	180	110
Riprap slope protection filter without bedding	900	50	360	180	250
Drainage trench	515	50	180	110	Not required
Slab drain	515	50	180	110	Not required
Articulated mattress or interlocking block slope protection filter	515	50	180	180	250

Source: Adapted from Al-Hussaini and Perry (1996).
[a]Strength values are for the weaker principal direction.

Percent open area (POA) is also used to characterize the openings in a geotextile. POA is the ratio of the total open area of the geotextile to the total specimen area. The total open area could be determined projecting light through a geotextile onto a screen or a cardboard and mapping the lighted areas using a planimeter. This property is valid only for woven monofilament geotextiles, since for nonwovens the overlapping yarns prevent the projection of light through the geotextile. For a complete distribution of pore sizes in geotextiles, the following methods are used.

Sieving Sieving methods are a modification of test method ASTM D 4751, designed to determine the AOS. The test could be conducted starting with very fine glass beads in a mechanical sieve shaker and performing the test with ever-coarser glass beads. Related methods are *hydrodynamic sieving,* whereby glass beads of known sizes are sieved by alternating water flow through immersion and emersion of geotextiles in water, and *wet sieving,* whereby water is sprayed on particles and geotextiles during shaking.

Intrusion Methods Intrusion techniques are based on the capillary principle, which relates the pressure required to force a nonwetting fluid (such as mercury) into the pores of a porous medium with the radii of the pores intruded. The *bubble point method* is a related method used originally to determine pore size characteristics of membrane filters. In this method, the flow rate of gas at various pressures is used to

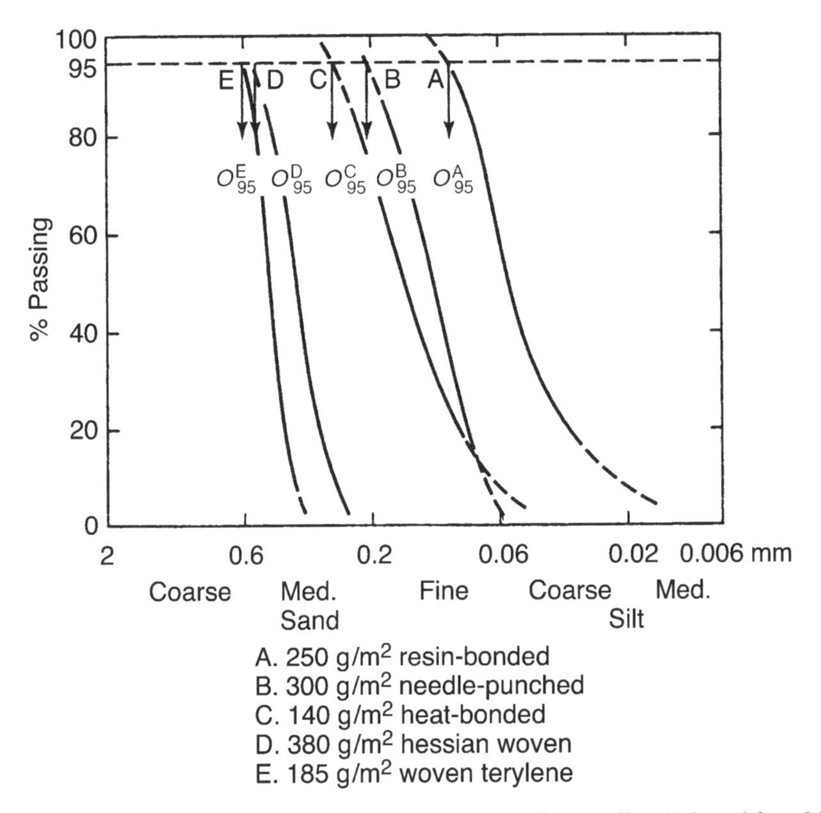

Fig. 8.4. Typical pore-size distributions of different types of geotextiles. (Adapted from McGown, 1976; Koerner, 1998.)

infer the pore sizes. The principle behind this method is similar to that of mercury intrusion methods used for soils. For a fluid to penetrate a soil mass, the pressure applied must exceed the capillary pressure in the largest pore. For cylindrical pores, the pore diameter and gas pressure are related. This method was first applied for nonwoven geotextiles by Bhatia and Smith (1995).

In addition to the methods discussed above, image analyses are sometimes used to make direct observations of the microstructure of geotextiles. The images could be analyzed for a number of measurements, such as pore sizes and number and area of fibers/filaments. For a detailed comparison of the pore size distributions of geotextiles obtained using various methods, the reader is referred to Bhatia and Smith (1995) and Fischer et al. (1996). Typical pore-size distributions of geotextile filters are shown in Fig. 8.4. It is seen that geotextiles represent a broad spectrum of pore sizes, ranging through several orders of magnitude, much like soils. The pore sizes of geotextiles control their ability to retain soil particles and transmit water. Therefore, the accuracy of the test used to determine these sizes is an important consideration in filtration and drainage applications.

Porosity is an aggregate parameter that characterizes the total pore volume of a geotextile. It is a ratio of void volume to the total volume of the geotextile. Bhatia et al. (1993) used digital image analysis to measure porosity. It can also be estimated using the relationship (Koerner, 1998)

$$n = 1 - \frac{m}{\rho t} \tag{8.1}$$

where n is the porosity (dimensionless), m the mass per unit area (M/L^2), ρ the density (M/L^3), and t the thickness of the geotextile (L). The porosity of geotextiles has the same meaning as in the case of soils; however, in the case of geotextiles, n is related to the thickness of the geotextile. As implied in this relationship, overburden stresses may influence porosity directly because of the resulting compression of geotextile and consequent reduction in thickness.

8.3.2 Permeability, Permittivity, and Transmissivity

Permeability of geotextiles carries the same meaning as in the case of soils. It indicates the rate at which the flow of a liquid occurs under a differential pressure. Using Darcy's law for flow, permeability k of geotextiles is the rate of flow across a unit cross-sectional area under unit gradient,

$$k = \frac{q}{iA} \tag{8.2}$$

where q is the flow rate (L^3/T), i the hydraulic gradient (dimensionless), and A the area of cross section for flow (L^2). In filtration applications of geotextiles, it is common for flow to occur normal to the plane of geotextiles. A term known as *permittivity* ψ is used to denote the volumetric flow rate of water in the normal direction per unit cross-sectional area and per unit head difference. Thus,

$$\psi = \frac{q}{(\Delta h)A} \tag{8.3}$$

where Δh is the hydraulic head difference applied across the geotextile. Comparing the definitions in Eqs. (8.2) and (8.3), and recognizing that gradient i is equal to the ratio of the hydraulic head difference and the thickness of the geotextile (when flow is normal to the plane of the geotextile), the relationship between permeability and permittivity may be expressed as

$$\psi = \frac{k}{t} \tag{8.4}$$

where t represents the thickness of the geotextile.

ASTM standard D 4491-99 describes the test methods for determining permittivity of geotextiles using constant head or falling head test procedures. A schematic of the apparatus is shown in Fig. 8.5. The *constant head test* is used when the flow rate of

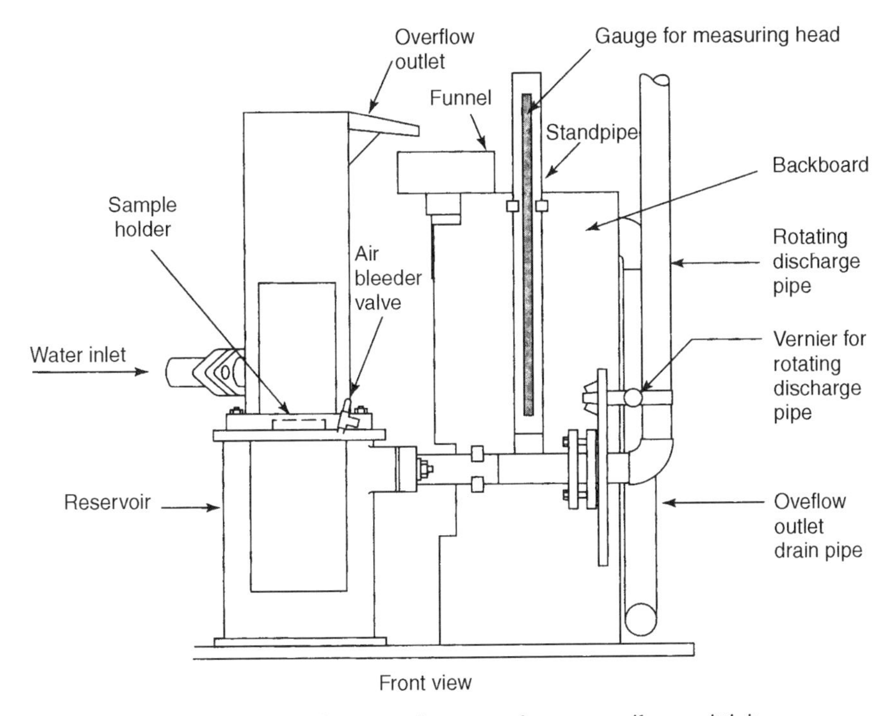

Fig. 8.5. Apparatus for measuring cross-plane geotextile permittivity.

water through the geotextile is so large that it is difficult to obtain readings of head change versus time required in the falling head test. A head of 50 mm (2 in.) of water is maintained on the geotextile throughout the test and the quantity of flow through the geotextile was measured with time. The permittivity is determined using

$$\psi = \frac{QC_t}{\Delta h\ A\ \Delta t} \tag{8.5}$$

where Q is the quantity of flow (L^3), Δh the head of water on the specimen (L), A the cross-sectional area of the test area of the specimen (L^2), Δt the time for flow (T), and C_t the temperature correction factor, determined using

$$C_t = \frac{\eta_t}{\eta_{20°C}} \tag{8.6}$$

where η_t is the water viscosity at test temperature and $\eta_{20°C}$ the water viscosity at 20°C.

In the *falling head test,* a column of water is allowed to flow through the geotextile, and readings of head changes versus time are taken. For this test to be applicable, the flow rate of water through the geotextile must be slow enough so that accurate readings could be taken on head changes. The permittivity is determined using

$$\psi = C_t \frac{a}{A\,\Delta t} \ln \frac{h_1}{h_2} \tag{8.7}$$

where C_t is the temperature correction factor determined using Eq. (8.6), a the cross-sectional area of the standpipe above the specimen, A the cross-sectional test area of the specimen (L^2), Δt the time for the head to drop from h_1 to h_2 (T), h_1 the initial head (L), and h_2 the final head (L). To determine the permittivity of geotextiles under loads representative of in situ overburden stresses, ASTM standard D 5493-93 is used. This test method provides a procedure for measuring the water flow normal to the plane of geotextiles over a range of applied normal compressive stresses. The apparatus used for this purpose accommodates a constant head of water on the geotextile specimen with normal loads applied through a piston and a vertical cylinder on top of the specimen.

In contrast to permittivity, *transmissivity* (θ) indicates the water-carrying capacity of geosynthetics in a direction parallel to the plane. Transmissivity is of importance in applications where geotextiles are used to fulfill the drainage function. It is defined as the volumetric flow rate per unit width of specimen per unit gradient in a direction parallel to the plane of the specimen. Thus,

$$\theta = \frac{q}{iw} \tag{8.8}$$

where w is the width of the geotextile (L). Applying Darcy's law for in-plane flow through geotextiles, one can deduce that

$$\theta = kt \tag{8.9}$$

where the permeability k is now the in-plane permeability of the geotextile. ASTM standard D 4716-99 describes a procedure for determining θ. The test apparatus permits water to flow along the plane of the geotextile under normal stresses and hydraulic gradients representative of field conditions. Three different configurations of in-plane transmissivity tests are shown in Fig. 8.6. In Fig. 8.6*a*, a rectangular sample of geotextile is subjected to hydraulic gradients with a simultaneous application of compressive stresses. Figure 8.6*b* illustrates the simulation of in-plane flow through a cylindrical sample of geotextiles. The fabric is subjected to lateral pressure through an elastic membrane. Radial flow in a circular fabric specimen is illustrated in Fig. 8.6*c*, where the apparatus is similar to a soil consolidation equipment. Water flows radially from the center of the specimen and is collected around the outer periphery of the device. For the simplest case in Fig. 8.6*a*, the transmissivity corresponding to a normal stress may be determined using

$$\theta = \frac{qL}{w\,\Delta h} \tag{8.10}$$

where q is the measured quantity of fluid discharged per unit time (L^3/T), L the length of specimen subjected to the normal compressive stress, w the width of the specimen

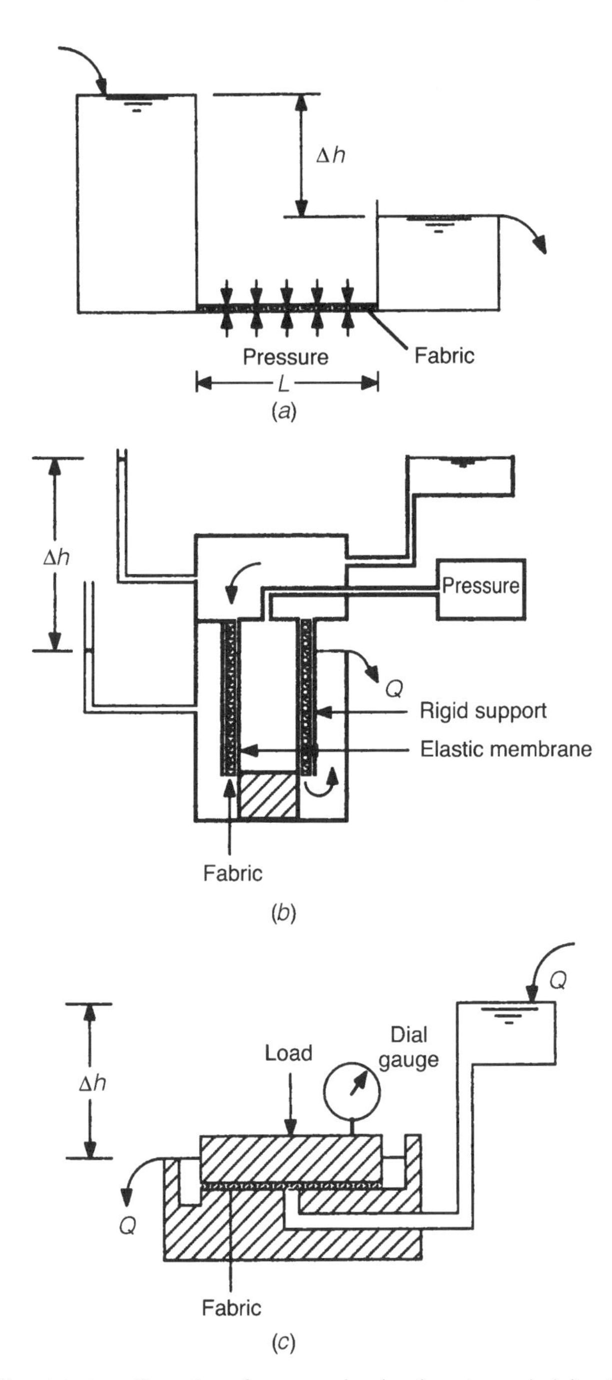

Fig. 8.6. Different test configurations for measuring in-plane transmissivity. (Adapted from Hausmann, 1990.)

(L), and Δh the difference in total head across the specimen. Derivation of equations for in-plane transmittivity in the case of Fig. 8.6*b* and 8.6*c* is left to the student as an exercise.

8.4 HYDRAULIC PERFORMANCE TESTS ON GEOSYNTHETICS

Performance tests are designed to study the hydraulic behavior of geotextiles under site-specific soil and hydrologic conditions. The index properties of permittivity and transmissivity may not adequately reflect the filtration and drainage behavior of geotextiles, particularly in the long term. For instance, filters clog over time, and the potential for a filter to clog in the long term can only be judged by simulating site-specific conditions.

A common test designed to evaluate the filtration performance of geotextiles is the *gradient ratio test* (ASTM D 5101), whose purpose it is to measure the clogging potential of a geotextile when placed next to a site-specific soil. A plastic cylindrical permeameter (Fig. 8.7) is used in this test with a geotextile specimen underlying a site-specific soil sample. Water is passed through this system by applying various differential heads. Measurements of differential heads and flow rates are taken at different time intervals. Head losses in the soil and across the soil–geotextile interface are measured using manometers. The ratio of the head loss across the interface (nominally, 25 mm) to the head loss across 50 mm of soil is termed the *gradient ratio* or *clogging ratio*. Gradient ratios higher than 1 imply that fine soil particles adjacent

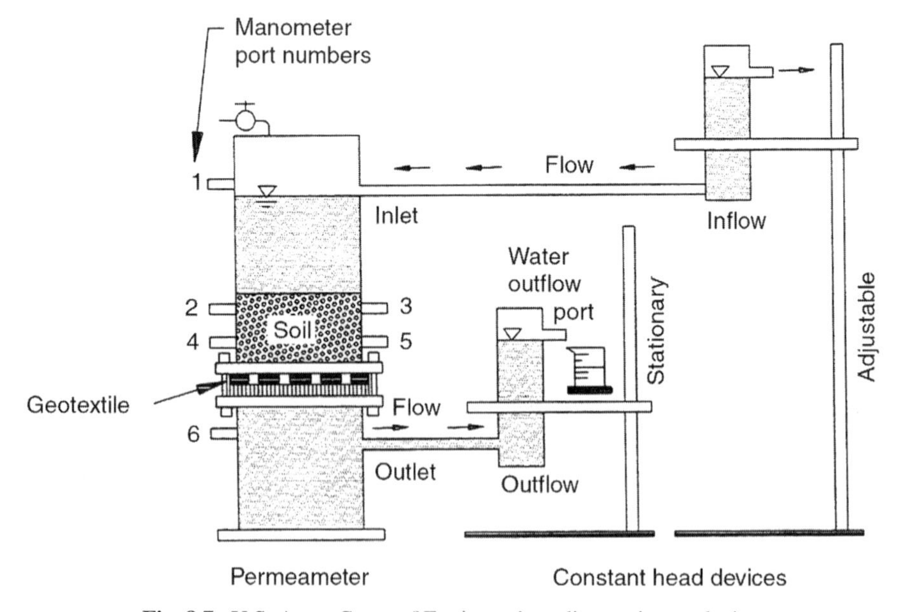

Fig. 8.7. U.S. Army Corps of Engineers' gradient ratio test device.

to the geotextile are trapped inside the geotextile. Gradient ratios above 3 are usually considered to be unacceptable in most filtration applications of geotextiles.

The gradient ratio test uses a rigid-wall permeameter to contain the soil–geotextile system. The associated sidewall leakages are usually negligible for sandy and silty soils with permeabilities greater than 5×10^{-2} cm/s. For fine-grained soils, however, these leakages produce misleading results on gradients and flow. A related test known as the *hydraulic conductivity ratio* (HCR) *test* (ASTM D 5567–94) is used instead. The HCR is defined as the ratio of the hydraulic conductivity of a composite soil–geotextile system to that of the hydraulic conductivity of the soil. The test is recommended for undisturbed or compacted soil specimens whose hydraulic conductivity is less than or equal to 5×10^{-2} cm/s. It requires placement of the soil and geotextile in a flexible-wall permeameter (Fig. 8.8). The soil–geotextile specimen is saturated using

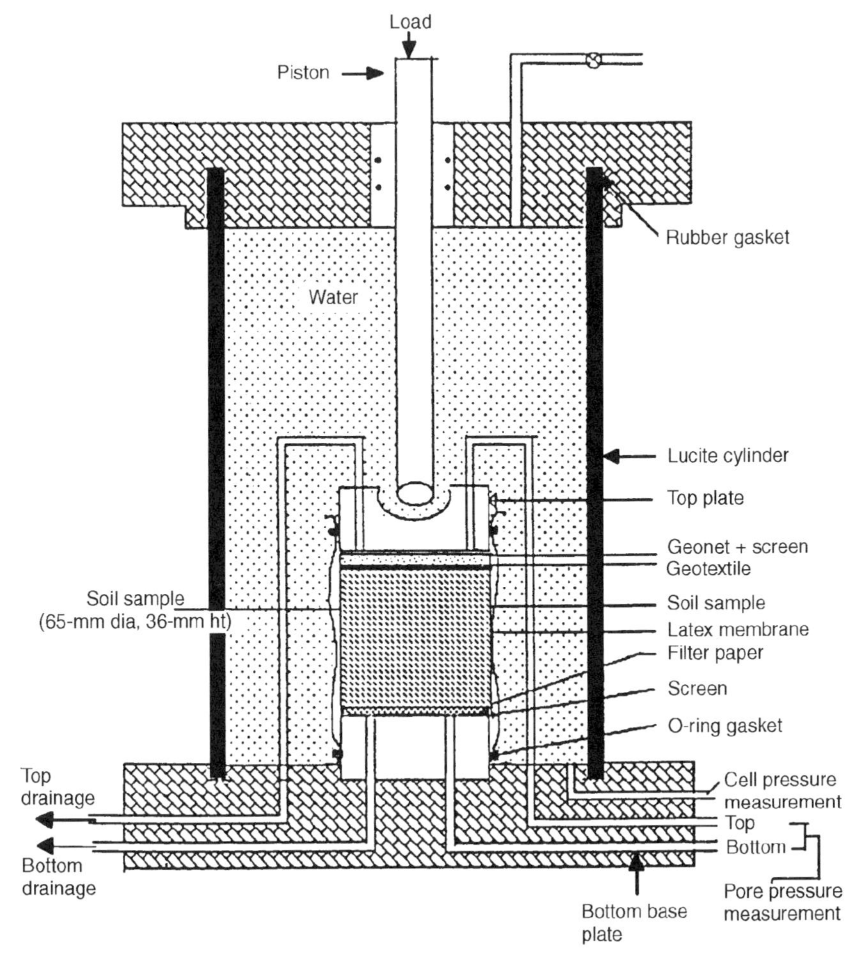

Fig. 8.8. Schematic of hydraulic conductivity ratio test apparatus. (Adapted from Bhatia, 1994.)

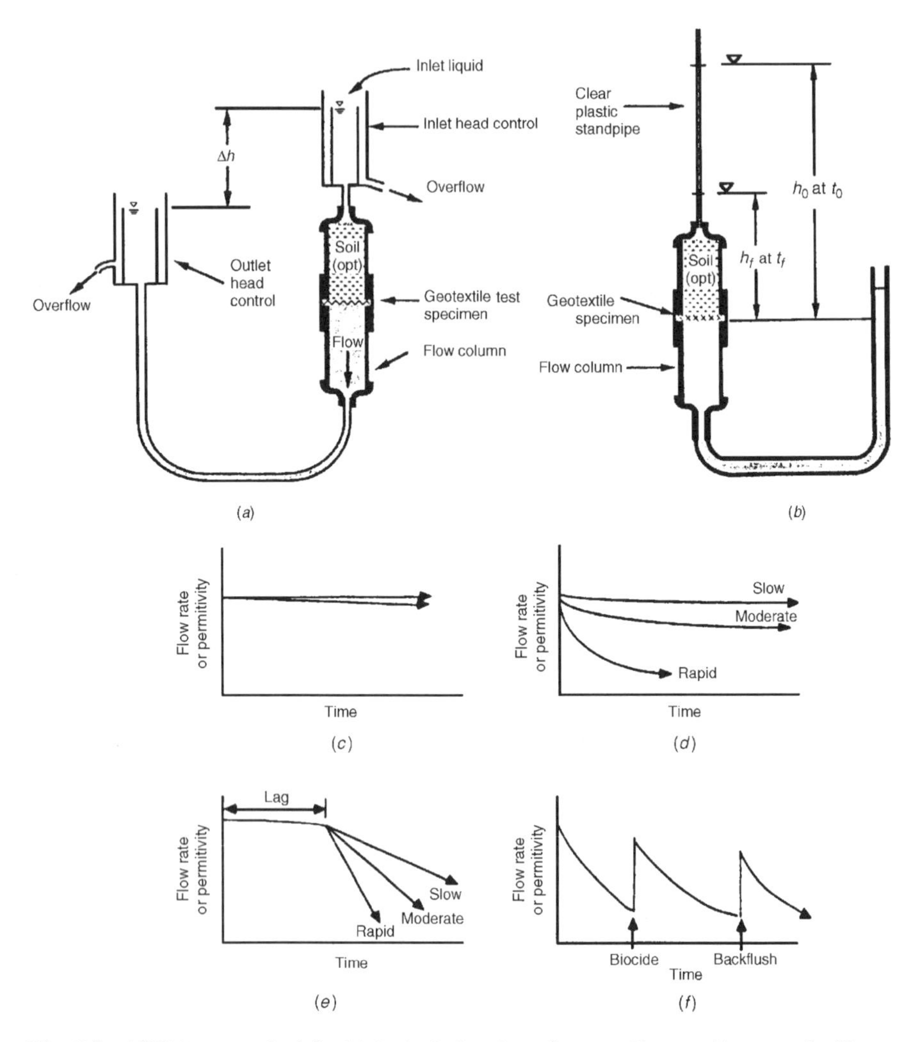

Fig. 8.9. ASTM test method for biological clogging of geotextile or soil–geotextile filters: (*a*) constant head conditions; (*b*) falling head conditions; (*c*) no, or nominal, clogging; (*d*) various degrees of clogging; (*e*) retarded clogging; (*f*) biocide and/or backflushing treatment. (Copyright © ASTM International; reprinted with permission.)

deaired water and backpressure techniques. The specimen is consolidated at the effective stress anticipated in the particular engineering application. Following permeation of the system with water, the hydraulic conductivity of the soil–geotextile system is measured and plotted as a function of elapsed time and volume of water passing through the sample. A reduction in the hydraulic conductivity implies a reduced drainage capacity of the geotextile. The test is continued until a steady hydraulic conductivity is obtained. Other performance tests designed specifically to evaluate the long-term clogging potential of geotextiles include the fine fraction filtration (F^3)

test (Sansone and Koerner, 1992) and the long-term flow (LTF) test (Koerner and Ko, 1982). The merits and demerits of these tests are discussed in Fischer (1994).

In applications where soil–geotextile systems are exposed to fluid streams containing bacteria and other microorganisms, clogging due to biological growth may be of concern. ASTM standard D 1987-95 was designed to determine the potential for, and relative degree of, biological growth that can accumulate on geotextile or geotextile–soil filters. The test method uses the measurement of flow rates over an extended period of time to determine the amount of clogging due to biological growth. A geotextile filter specimen or geotextile/soil filter composite specimen is positioned in a flow column so that a designated liquid (containing microorganisms promoting biological growth) flows through it under either constant or falling head conditions (Fig. 8.9*a* and *b*). The test method can be used for both unsaturated and saturated conditions. Biocides may also be introduced with the backflushing liquid or introduced within the test specimen to evaluate their impact on clogging. Using results from the test, the trends in clogging could be discerned as shown in Fig. 8.9*c* to *f*. Depending on the nature of bioclogging, the permittivity reduction could begin immediately (Fig. 8.9*d*) or could occur after some time lag (Fig. 8.9*e*). Introduction of biocide may kill bacteria and other microorganisms and may increase the permittivity momentarily (Fig. 8.9*f*).

8.5 DESIGN METHODS FOR GEOTEXTILE FILTERS

The filter requirements described for soils in Chapter 7 are also applicable in the case of geotextiles. Like soil filters, geotextile filters must fulfill two functions: particle retention and permeability. Christopher and Fischer (1992) summarize the filtration principles for effective geotextile filters as follows:

1. If the size of the largest pore in the geotextile filter is smaller than the larger particles of soil, the soil will not pass the filter. The larger particles of soil will form a filter bridge over the hole, which in turn, filters smaller particles of soil, thus retaining the soil and preventing piping (Fig. 8.10).

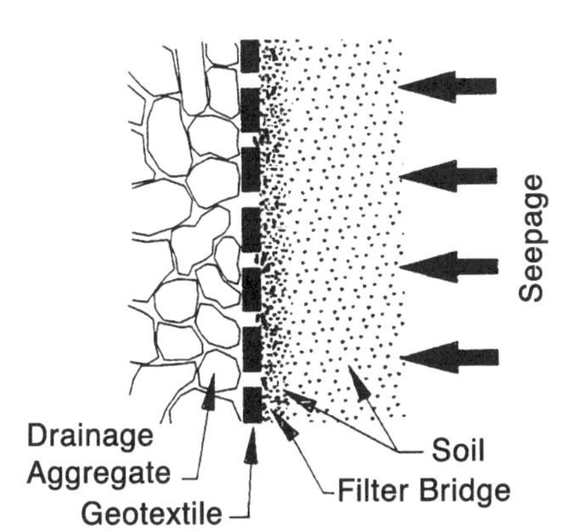

Fig. 8.10. Filter bridge formation. (Adapted from Christopher and Holtz, 1989; copyright © Elsevier Science; reprinted with permission.)

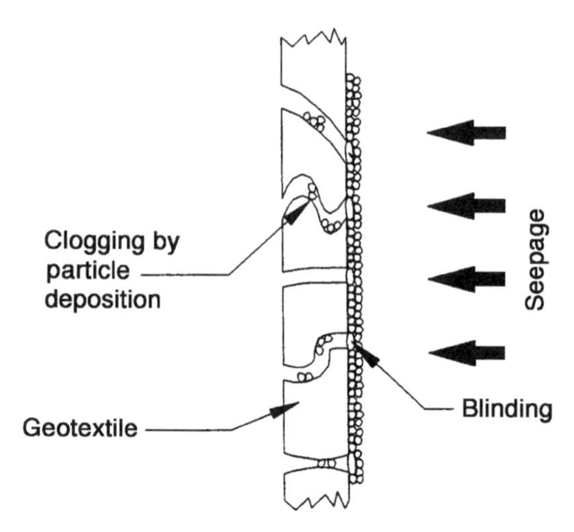

Fig. 8.11. Geotextile clogging and blinding processes. (Adapted from Bell & Hicks, 1980; copyright © Elsevier Science; reprinted with permission.)

2. If the majority of the openings in the geotextile are sufficiently larger than the smaller particles of soil, such that they are able to pass through the filter, the geotextile will not "clog" or "blind" (Fig. 8.11).
3. A large number of openings should be present in the geotextile so that proper flow can be maintained even if a portion of the openings become clogged during the design life of the filter.

These three principles reflect the need for retention, clogging, and permeability criteria, respectively. The criteria are based on those developed earlier for soil media, so the reader is encouraged to draw parallels between the criteria described in this section and those in Section 7.2.

8.5.1 Retention Criteria

Similar to soil filters, the ability of geotextile filters to retain soil particles is judged by comparing the pore size of the geotextile with a representative soil particle size. As shown in Table 8.7, many of the prior investigations used the opening size of the geotextile, O_{95}, O_{50}, O_{90}, or O_{15} (as measured using sieving methods) and the D_{85}, D_{50}, D_{90}, or D_{15} values of the soil. Holtz et al. (1998) summarizes these studies and gives us the following relationship as the retention criterion:

$$\text{AOS or } O_{95(\text{geotextile})} \leq BD_{85(\text{soil})} \tag{8.11}$$

where B is a dimensionless coefficient ranging from 0.5 to 2. B is a function of the type of soil to be filtered, its density, the uniformity coefficient C_u if the soil

TABLE 8.7 Existing Geotextile Retention Criteria

Sources	Criterion	Remarks
Calhoun (1972)	$O_{95}/D_{85} \leq 1$	Wovens soils with $\leq$ 50% passing No. 200 seive
	$O_{95} \leq 0.2$ mm	Wovens, cohesive soils
Zitscher (1975) from Rankilor (1981)	$O_{50}/D_{50} \leq 1.7$–2.7	Wovens, soils with $C_u \leq 2$, $D_{50} = 0.1$ to 0.2 mm
	$O_{50}/D_{50} \leq 25$–37	Nonwoven, cohesive soil
Ogink (1975)	$O_{90}/D_{90} \leq 1$	Wovens
	$O_{90}/D_{90} \leq 1.8$	Nonwovens
Sweetland (1977)	$O_{15}D_{85} \leq 1$	Nonwovens, soils with $C_u = 1.5$
	$O_{15}D_{15} \leq 1$	Nonwovens, soils with $C_u = 4$
Rankilor (1981)	$O_{50}/D_{85} \leq 1$	Nonwovens, soils with $0.02 \leq D_{85} \leq 0.25$ mm
	$O_{15}/D_{15} \leq 1$	Nonwovens, soils with $D_{85} > 0.25$ mm
Schober and Teindl (1979) (with no factor of safety)	$O_{90}/D_{50} \leq 2.5$–4.5	Woven and thin nonwovens, dependent on C_u
	$O_{90}/D_{50} \leq 4.5$–7.5	Thick nonwovens, dependent on C_u
		Silt and sand soils
Millar et al. (1980)	$O_{50}/D_{85} \leq 1$	Wovens and nonwovens
Giroud (1982)	$O_{95}/D_{50} \leq (9$–$18)/C_u$	Dependent on soil C_u and density; assumes fines in soil migrate for large C_u
Carroll (1983)	$O_{95}/D_{85} \leq 2$–3	Wovens and nonwovens
Christopher and Holtz (1985)	$O_{95}/D_{85} \leq 1$–2	Dependent on soil type and C_u
	$O_{95}/D_{15} \leq 1$ or $O_{50}/D_{85} \leq 0.5$	Dynamic, pulsating, and cyclic flow, if soil can move beneath fabric
French Committee on Geotextiles and Geomembranes (1986)	$O_f/D_{85} \leq 0.38$–1.25	Dependent on soil type, compaction, hydraulic, and application conditions
Fischer et al. (1990)	$O_{50}D_{85} \leq 0.8$ $O_{50}D_{15} \leq 1.8$–7.0 $O_{50}D_{50} \leq 0.8$–2.0	Based on geotextile pore size distribution, dependent on C_u of soil

Source: Adapted from Fischer et al., 1990; copyright © Elsevier Science; reprinted with permission.

is granular, the type of geotextile (woven or nonwoven), and the flow conditions. For sands and silts with less than 50% passing the 0.075-mm sieve, B is a function of C_u alone. For silts and clays, with more than 50% passing the 0.075-mm sieve, B is a function of the geotextile parameters. The recommended values for B for the two types of soils are shown in Table 8.8.

In applications where dynamic, cyclic, or pulsating flow conditions prevail, such as in the case of pavement drainage, the soil filter bridge shown in Fig. 8.10 may

TABLE 8.8 Values of Coefficient B for Use in Retention Criteria

Soil Type	C_u	Geotextile Type	B
Sands, gravely sands, silty sands, and clayey sands; $\% - 200 < 50$	$C_u \leq 2$ or ≥ 8 $2 \leq C_u \geq 4$ $4 < C_u < 8$	All	1 $0.5C_u$ $8/C_u$
Silts and clays; $\% - 200 > 50$	Not applicable	Woven Nonwoven Both woven and nonwoven	1 1.8 AOS or $O_{95} \leq 0.3$ mm

Source: Adapted from Holtz et al. (1998).

not develop. In such cases, $B = 1$ may not be conservative, and a value of 0.5 is recommended. Also, in the case of soils that may be internally unstable, the criterion expressed in Eq. (8.11) may not be applicable, and specific performance tests should be conducted to determine the criterion. Generally, broadly graded soils ($C_u > 20$) with concave-upward grain size distribution tend to be internally unstable (Kenney and Lau, 1985, 1986; Lafleur et al., 1989). Luettich et al. (1992) provided a comprehensive set of retention criteria for geotextiles based on the properties of soils retained. The dispersive nature of soil, the Atterberg limits, and the grain size distributions are used to develop the criteria. These are shown in Figs. 8.12 and 8.13 for steady-state flow and dynamic flow conditions, respectively.

8.5.2 Permeability/Permittivity Criteria

Table 8.9 summarizes the permeability/permittivity criteria used or recommended in the literature. In general, the geotextile is chosen to have a higher permeability than that of soil so that excess pore water pressures do not develop. Thus,

$$k_{\text{geotextile}} \geq C\,k_{\text{soil}} \tag{8.12}$$

where C is a dimensionless coefficient ranging from 1 to 10, depending on the importance and severity of filtration in the application. The permittivity requirements suggested are as follows:

$$\psi \geq \begin{cases} 0.5\ \text{s}^{-1} & \text{for} < 15\% \text{ passing } 0.075 \text{ mm} \quad (8.13) \\ 0.2\ \text{s}^{-1} & \text{for } 15 \text{ to } 50\% \text{ passing } 0.075 \text{ mm} \quad (8.14) \\ 0.1\ \text{s}^{-1} & \text{for} > 50\% \text{ passing } 0.075 \text{ mm} \quad (8.15) \end{cases}$$

8.5.3 Clogging Resistance Criteria

Clogging of geotextiles by fine particles of the adjacent soil is a long-term problem that cannot be addressed using either retention or permeability/permittivity criteria.

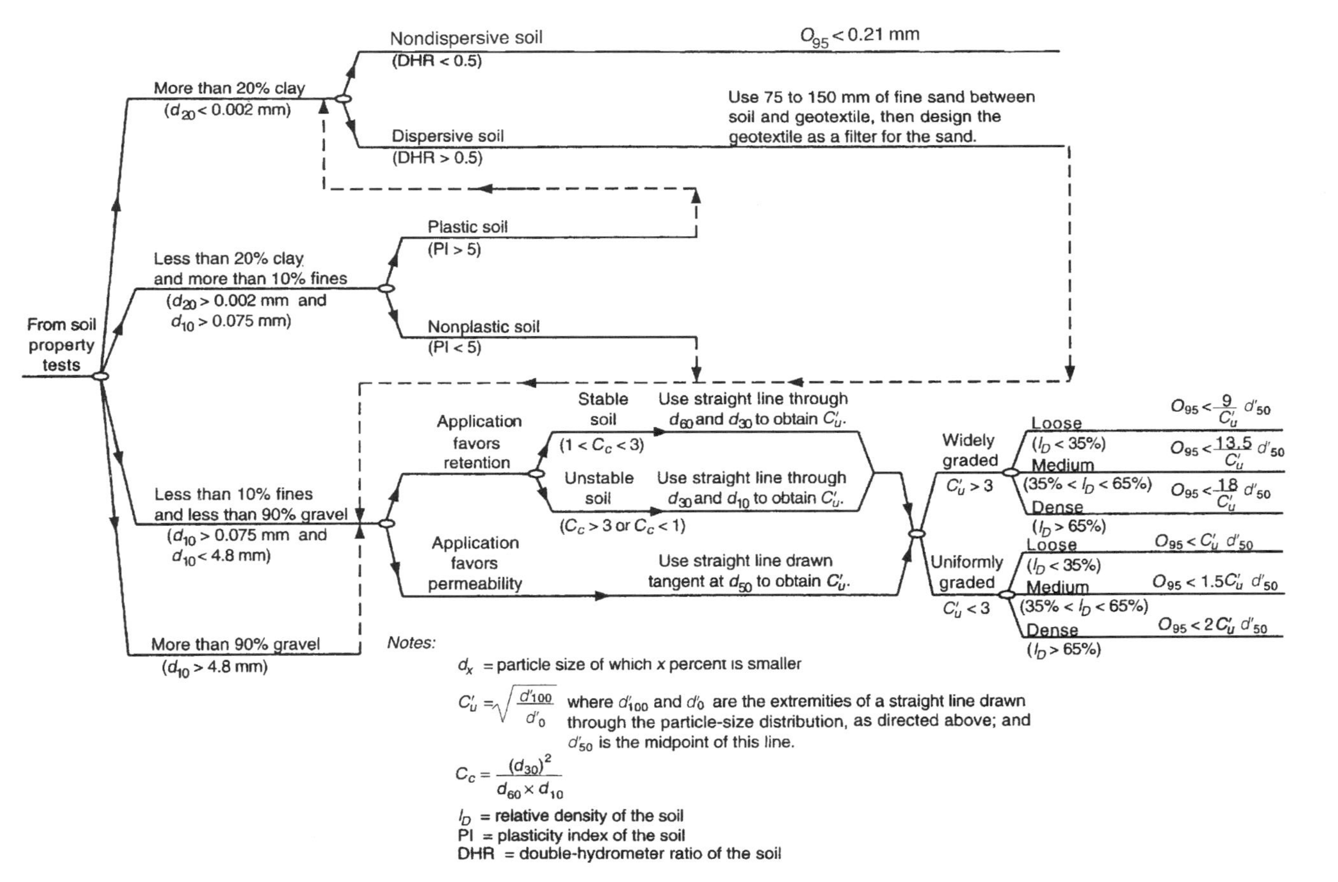

Fig. 8.12. Soil retention criteria for a geotextile filter design using steady-state flow conditions. (Adapted from Luettich et al., 1992; copyright © Elsevier Science; reprinted with permission.).

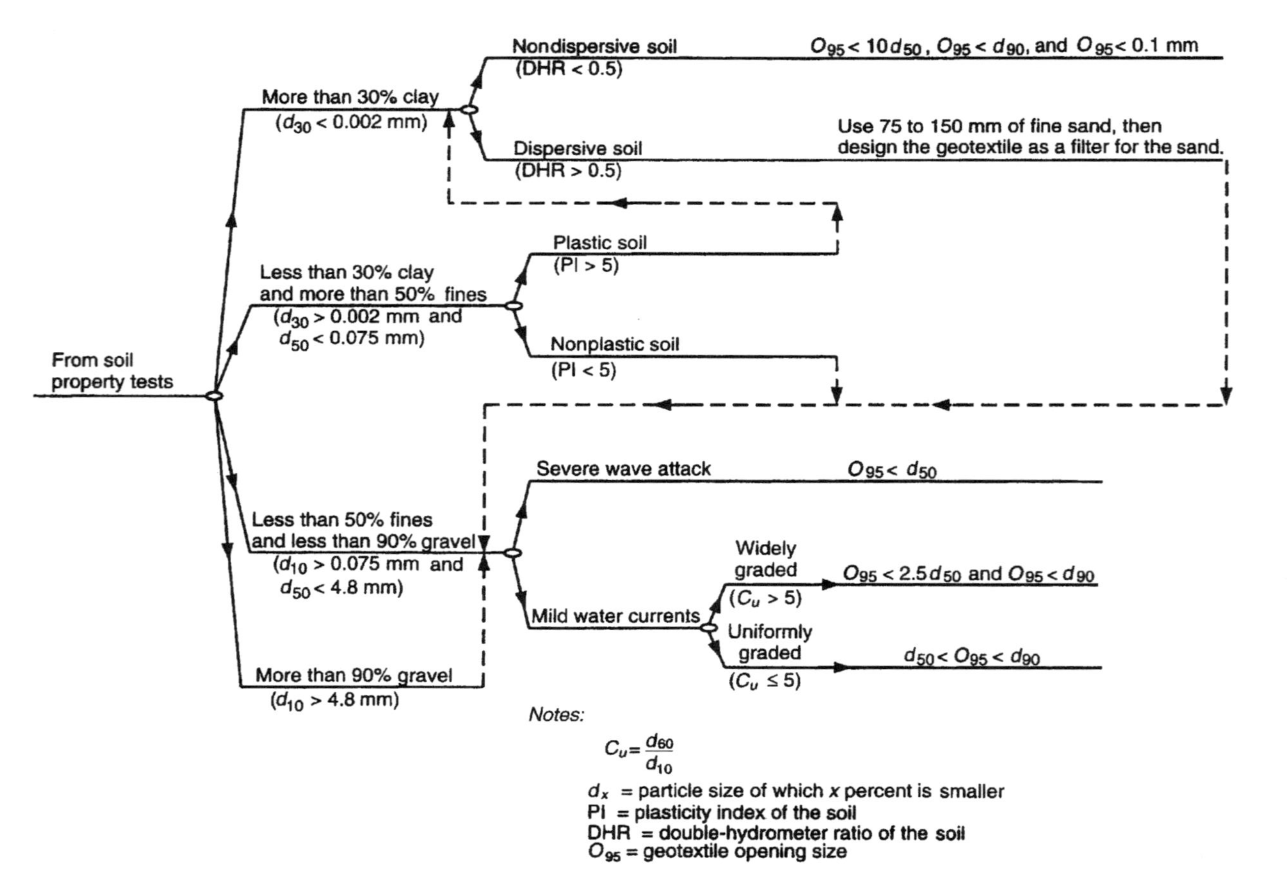

Fig. 8.13. Soil retention criteria for a geotextile filter design under dynamic flow conditions. (Adapted from Luettich et al., 1992; copyright © Elsevier Science; reprinted with permission.)

TABLE 8.9 Existing Geotextile Permeability Criteria

Sources	Criteria	Remarks
Calhoun (1972), Schober and Teindl (1979), Wates (1980); Carroll (1983), Haliburton et al. (1982), Christopher and Holtz (1985), and numerous others	$k_{geotextile} \geq k_{soil}$	Steady-state flow, noncritical application, and nonsevere soil conditions
Carroll (1983), Christopher and Holtz (1985)	$k_{geotextile} \geq 10k_{soil}$	Critical applications and severe soil or hydraulic conditions
Giroud (1982)	$k_{geotextile} \geq 0.1k_{soil}$	No factor of safety
French Committee on Geotextiles and Geomembranes (1986)	Based on permittivity ψ with $\psi \geq 10^{3-5}k_{soil}$	Critical $10^5 k_{soil}$ Less critical $10^4 k_{soil}$ Clean sand $10^3 k_{soil}$
Koerner (1998)	$\psi_{allow} \geq FS\,(\psi_{req'd})$	Factor of safety (FS) based on application and soil conditions

Source: Adapted from Christopher and Fischer, 1992; copyright © Elsevier Science; reprinted with permission.

TABLE 8.10 Clogging Criteria

Critical/severe applications[a]: Perform soil/fabric filtration tests.
(e.g., Calhoun, 1972; Haliburton et al., 1982; Haliburton and Wood, 1982; Giroud, 1982; Carroll, 1983; Christopher and Holtz, 1985, 1989; Koerner, 1998)

Less critical/nonsevere applications

1. Perform soil/fabric filtration tests.
2. Minimum pore-size alternatives for soils containing fines, especially in a noncontinuous matrix:
   (a) $O_{95} \geq 3D_{15}$ for $C_u \geq 3$
   (Christopher and Holtz, 1985; modified, 1989)
   (b) $O_f \geq 4D_{15}$
   (French Committee on Geotextiles and Geomembranes, 1986)
   (c) $O_{15}/D_{15} \geq 0.8$ to 1.2
   $O_{50}/D_{50} \geq 0.2$ to 1
   (Fischer et al., 1990)
3. For $C_u \leq 3$, fabric with maximum opening size from retention criteria should be specified.
4. Apparent open area qualifiers
   (a) Woven fabrics: percent open area $\geq$ 4 to 6%
   (Calhoun, 1972; Koerner, 1998)
   (b) Nonwoven fabrics: porosity $\geq$ 30 to 40%
   (Christopher and Holtz, 1985; Koerner, 1998)

Source: Adapted from Christopher and Fischer, 1992; copyright © Elsevier Science; reprinted with permission.

[a]Filtration tests are performance tests and cannot be performed by the manufacturer, as they depend on specific soil and design conditions. Tests to be performed by specifying agency or its representative.

TABLE 8.11 Geotextile Filtration Requirements

Property	Test Method	Units of Measure	Requirements for Percentage In Situ Soil Passing 0.075 mm[a] < 15	15–50	> 50
Geotextile class	—	—	Class 2 from Table 8.5[b]		
Permittivity[c,d]	ASTM D 4491	s^{-1}	0.5	0.2	0.1
Apparent opening, Size[c,d]	ASTM D 4751	mm	0.43	0.25	0.22[e]
Ultraviolet stability (retained strength)	ASTM D 4355	%	50% after 500 h of exposure		

Source: Adapted from Holtz et al. (1998).

[a]Based on grain size analysis of in-situ soil in accordance with AASHTO T 88.

[b]Default geotextile selection. The engineer may specify a Class 3 geotextile from Table 8.5 for trench drain applications based on one or more of the following: (1) the engineer has found Class 3 geotextiles to have sufficient survivability based on field experience; (2) the engineer has found Class 3 geotextiles to have sufficient survivability based on laboratory testing and visual inspection of a geotextile sample removed from a field test section constructed under anticipated field conditions; (3) subsurface drain depth is less than 2 m, drain aggregate diameter is less than 30 mm, and compaction requirement is less than 95% of AASHTO T 99.

[c]These default filtration property values are based on the predominant particle size of in situ soil. In addition to the default permittivity value, the engineer may require geotextile permeability and/or performance testing based on engineering design for drainage systems in problematic soil environments.

[d]Site-specific geotextile design should be performed, especially if one or more of the following problematic soil environments are encountered: unstable or highly erodible soils such as noncohesive silts; gap-graded soils; alternating sand/silt laminated soils; dispersive clays; and/or rock flour.

[e]For cohesive soils with a plasticity index above 7, geotextile maximum average roll values for apparent opening size is 0.30 mm.

Usually, filtration tests and performance tests such as the gradient ratio test are conducted to evaluate the clogging potential. Some general recommendations are given in Table 8.10 as guidance in the case of less critical/nonsevere applications.

In contrast to the individual criteria described above, the AASHTO M 288 *Standard Specifications for Geotextiles* (1997) provides a comprehensive set of geotextile filtration requirements (Table 8.11). These specifications essentially incorporate all the criteria, including survivability of geotextiles (ultraviolet stability).

Example 8.1 It is proposed to use a geotextile fabric as a filtration layer behind a 5-m-high gabion wall (see Fig. E8.1). The backfill material is a sandy soil with a $\% - 200 < 50$, uniformity coefficient $C_u = 3$, coefficient of permeability $k = 10^{-2}$ cm/s, and $D_{85} = 2$ mm. Check the adequacy of a nonwoven needle-punched geotextile fabric (with a permittivity of 2.0 s^{-1} and an AOS value of 0.30 mm) as the filtration layer. Use the retention and permeability/permittivity criteria discussed in Section 8.5.

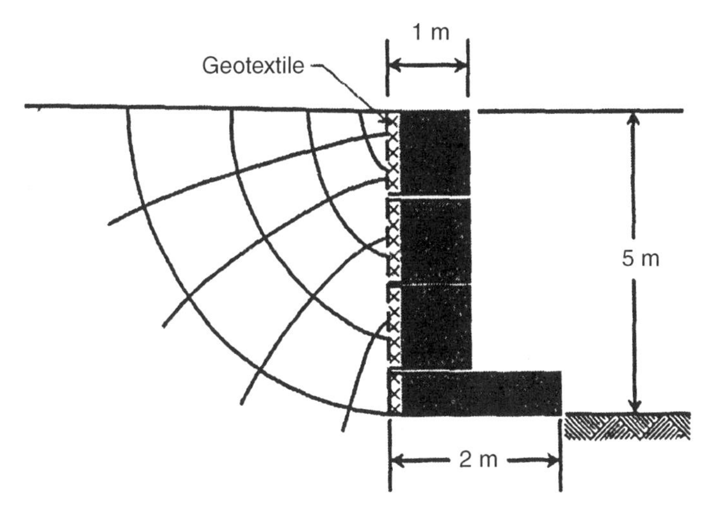

Fig. E8.1 (Adapted from Koerner, 1998; copyright © Pearson Education; reprinted with permission.)

SOLUTION: Check the adequacy using retention criteria. From Table 8.8, $B = 0.5\ C_u = 1.5$. Using Eq. (8.11), AOS or $O_{95} \leq BD_{85}$; $BD_{85} = 3$ mm > AOS, so this geotextile meets the retention criteria.

Check the adequacy using permeability/permittivity criteria. Using the flow net, the flow rate is estimated as

$$q = k\,\Delta h \frac{n_f}{n_e} = 10^{-2}\ \text{cm/s} \times 5\ \text{m} \times \frac{100\ \text{cm}}{1\ \text{m}} \times \frac{4}{5} = 4\ \text{cm}^2/\text{s}$$

(i.e., 4 cm^3/s per unit longitudinal dimension of the wall). The required permittivity is

$$\psi_{\text{req}} = \frac{k}{t} = \frac{q}{\Delta h\ A} = \frac{4\ \text{cm}^2/\text{s}}{500\ \text{cm} \times 500\ \text{cm} \times 1\ \text{cm}} = 1.6 \times 10^{-5}\ \text{s}^{-1}$$

which is smaller than the permittivity of 2.0 s^{-1} of the geotextile. So this geotextile meets the permeability/permittivity criteria.

8.6 GEOSYNTHETICS IN DRAINAGE APPLICATIONS

Because of the small thicknesses of most geotextiles, the in-plane drainage capacities are quite limited compared with those of conventional soil drainage layers. The in-plane flow through geotextiles is several orders less than the seepage capacities of sands and other soil drainage materials. However, the new prefabricated geocomposite drains, which consist of cores of extruded and fluted plastic sheets, three-dimensional meshes and mats, plastic waffles, and nets and channels, are gaining rapid acceptance as replacements for conventional soil drainage layers. Most geocomposite drains generally range in thickness from 5 to 25 mm or greater and have transmission capabilities between 0.0002 and 0.01 m^3/s per linear width of drain. Holtz et al. (1998) list the following as the major application areas for geocomposite drains:

- Edge drains for pavements
- Interceptor trenches on slopes
- Drainage behind abutments and retaining structures
- Relief of water pressures on buried structures
- Substitute for conventional sand drains
- Waste containment systems for leachate collection and gas venting

In addition, some secondary drainage applications of geosynthetics include interceptor drains, transmission of seepage water below pavement base course layers, horizontal and vertical strip drains to accelerate consolidation of soft foundation soils, dissipation of seepage forces in earth and rock slopes, as part of chimney drains in earth dams, dissipaters of pore-water pressures in embankments and fills, and gas venting below containment liner systems.

Koerner (1998) recommends that the allowable flow rates through or within geotextiles, when used in filtration and drainage applications, be reduced because of the several uncertainties in the long-term field performance of geotextiles. The following expression for reduction was suggested:

TABLE 8.12 Recommended Reduction Factor Values to Reduce Flow through Geotextile Filters and Drainage Layers

	Range of Reduction Factors				
Application	Soil Clogging and Blinding[a]	Creep Reduction of Voids	Intrusion into Voids	Chemical Clogging[b]	Biological Clogging
Retaining wall filters	2.0–4.0	1.5–2.0	1.0–1.2	1.0–1.2	1.0–1.3
Underdrain filters	5.0–10	1.0–1.5	1.0–1.2	1.2–1.5	2.0–4.0
Erosion-control filters	2.0–10	1.0–1.5	1.0–1.2	1.0–1.2	2.0–4.0
Landfill filters	5.0–10	1.5–2.0	1.0–1.2	1.2–1.5	5–10[c]
Gravity drainage	2.0–4.0	2.0–3.0	1.0–1.2	1.2–1.5	1.2–1.5
Pressure drainage	2.0–3.0	2.0–3.0	1.0–1.2	1.1–1.3	1.1–1.3

Source: Adapted from Koerner, 1998; copyright © Pearson Education; reprinted with permission.

[a]If stone riprap or concrete blocks cover the surface of the geotextile, use either the upper values or include an additional reduction factor.

[b]Values can be higher particularly for high-alkalinity groundwater.

[c]Values can be higher for turbidity and/or for microorganism contents greater than 5000 mg/L.

$$q_{\text{allow}} = q_{\text{ult}} \frac{1}{\text{RF}_{\text{scb}} \times \text{RF}_{\text{cr}} \times \text{RF}_{\text{in}} \times \text{RF}_{\text{cc}} \times \text{RF}_{\text{bc}}} \tag{8.16}$$

where q_{allow} is the allowable flow rate, q_{ult} the ultimate flow rate, RF_{scb} the reduction factor for soil clogging and blinding, RF_{cr} the reduction factor for creep reduction of void space, RF_{in} the reduction factor for adjacent materials intruding into geotextile's void space, RF_{cc} the reduction factor for chemical clogging, and RF_{bc} the reduction factor for biological clogging. Typical values suggested for these reduction factors are shown in Table 8.12.

Example 8.2 Geotextile fabrics are proposed for the drainage layer behind the 10-m-high cantilever retaining wall shown in Fig. E8.2. The backfill material behind the wall is same as in Example 8.1. Check the adequacy of the following kinds of fabrics for use as the drainage layer: (1) a geotextile fabric with an allowable transmissivity $\theta = 3 \times 10^{-6} m^2/s$, (2) a geonet with an allowable transmissivity $\theta = 8 \times 10^{-4} m^2/s$, and (3) a geocomposite drain with an allowable transmissivity $\theta = 2 \times 10^{-3} \text{m}^2/\text{s}$.

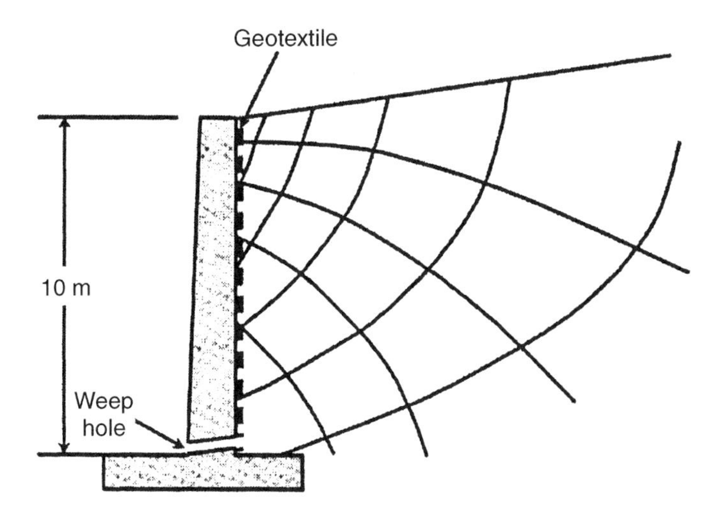

Fig. E8.2 (Adapted from Koerner, 1998; copyright © Pearson Education; reprinted with permission.)

SOLUTION: The flow rate through the geotextile is estimated using the flow net:

$$q = k\,\Delta h \frac{n_f}{n_e} = 10^{-2} \text{ cm/s} \times 10 \text{ m} \times \frac{100 \text{ cm}}{1 \text{ m}} \times \frac{5}{5} = 10 \text{ cm}^2/\text{s}$$

(i.e., 10 cm^3/s per unit longitudinal dimension of the wall).

The flow gradient within the geotextile is $i = \sin 90^\circ = 1$. The required transmissivity (per unit longitudinal dimension of the wall) is

$$\theta_{\text{required}} = kt = \frac{q}{i} = 10 \text{ cm}^2/\text{s} = 10^{-3} \text{ m}^2/\text{s}$$

which is larger than the transmissivity of the geotextile fabric and the geonet and is smaller than that of the geocomposite drain. So the first two do not meet the requirement, while the last one does with a factor of safety of 2.0.

PROBLEMS

8.1. Design geotextiles for highway trench drains (Fig. 8.3*c*), given three representative soil gradations at the site (Fig. P8.1). Estimate the permeability of soils using the grain-size distributions. Use the retention, permittivity, and clogging criteria discussed in Section 8.5 to come up with the required design properties of geotextiles (permeability, permittivity, and AOS). (Adapted from Holtz et al., 1998.)

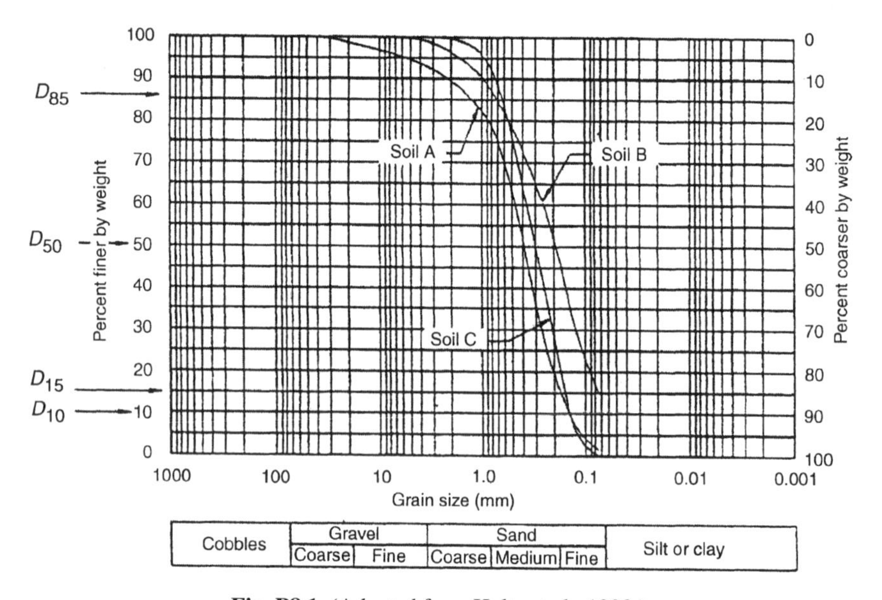

Fig. P8.1 (Adapted from Holtz et al., 1998.)

8.2. Determine the required transmissivity of the fabric (geotextile, geonet, or geocomposite) to be used as the chimney drain in the zoned earth dam shown in Fig. P8.2. The permeability of the core material is 10^{-7} cm/s.

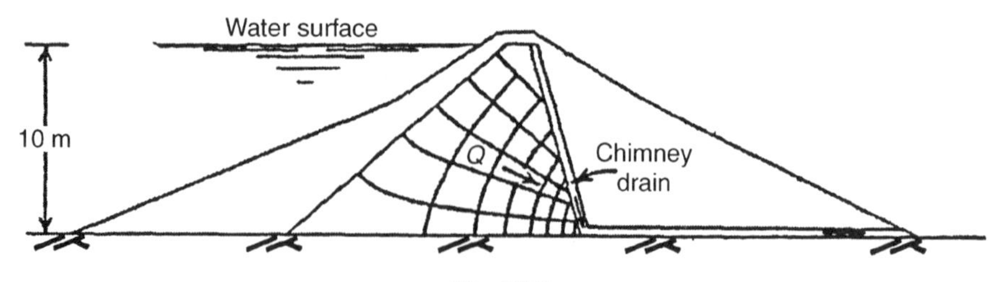

Fig. P8.2

CHAPTER 9

SEEPAGE CONTROL AND DEWATERING

It is necessary in many civil engineering projects to control groundwater flow, in terms of both seepage quantity and direction. At construction sites where foundations need to be placed below the groundwater table, drainage or dewatering is necessary, often for extended periods of time, to keep the excavation dry and to facilitate foundation construction. In pavement drainage systems, interception of groundwater flow is necessary to minimize or eliminate flow into the drainage layer completely. In the case of earth dams, adequate drainage systems should be provided within the structure of the dams to dissipate high energies of upstream water safely. In addition, seepage through foundation soils underneath the dams may have to be reduced using certain cutoff methods.

The purpose of seepage control is to accomplish either of the following two objectives:

1. Drain water from natural soils to facilitate subsurface construction, enhance slope stability (see Chapter 6), or protect pavement structures
2. Cut off water flow either partially or completely to minimize/prevent seepage, such as in the case of dam foundations.

Numerous methods exist to fulfill these objectives. We start with a description of drainage and cutoff methods prior to a study of design of seepage control measures.

9.1 DRAINAGE METHODS

The drainage methods used most often to lower the groundwater table in soils fall under three broad categories: open pumping using ditches and sumps, well systems, and electroosmosis methods. No single method is ideal for all construction sites, so the designer should screen the methods carefully in terms of their limitations and applicability.

9.1.1 Open Pumping

Open pumping is the simplest and most economical method used to dewater soils. As shown in Fig.9.1*a*, an open ditch is used to collect water from excavated slopes. The flow of water occurs under gravity alone; therefore, the method is ideal for coarse-grained soils with high permeabilities (on the order of 10^{-3} cm/s). The water collected in the ditch is pumped out using a sump pump. The upward gradients in the vicinity of the ditch may cause soil instability due to boiling and piping (see Section 6.4). For this reason, filters are used in the ditch (Fig. 9.1*b*), which may provide weight sufficient to counteract the upward seepage forces. The seepage force and associated soil instabilities were discussed in Chapter 6.

9.1.2 Well Systems

Well systems provide drainage by pumping water at depths up to 30 m. Because of the suction energies used, the applicability of these systems is not limited to coarse soils; they could even be employed in silty sands. A simple form of dewatering using this method involves *well points* (Fig. 9.2), which are perforated pipes about $\frac{1}{2}$ to 1 m long and 5 to 8 cm in diameter. A conical steel drive point with jetting holes allows jetting of water for easy driving of the well point into the subsurface. The cavity formed around the pipe is filled with a sand and gravel filter. A series of well points, which are generally spaced between 1 to 2 m, are connected to a common suction header and pump (Fig. 9.3). The well points could be used to bring water to the surface from a maximum depth of about 6 m. To lower the water table more than 6 m, a multiple-stage well point system (Fig. 9.4) is used in which new well points are installed in successive stages. Beyond a dewatering depth of 16 m, it is no longer economical to use well points.

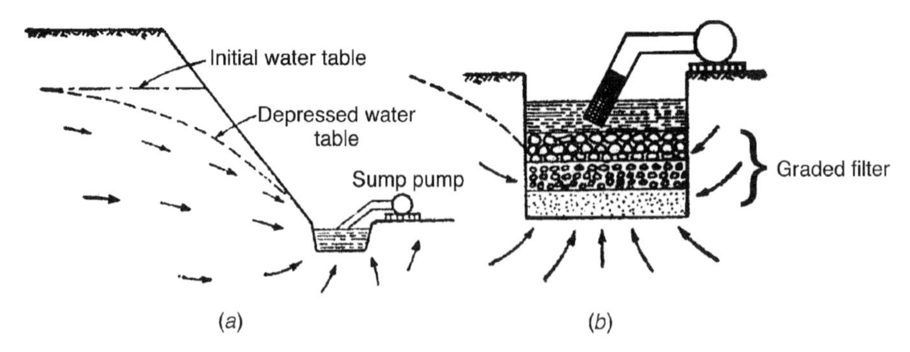

Fig. 9.1. Open pumping: (*a*) perimeter trench and sump pump; (*b*) weighted filter in the trench.

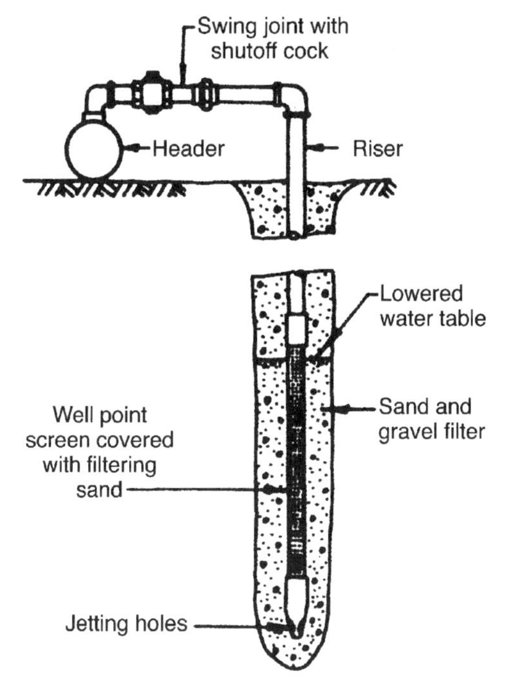

Fig. 9.2. Schematic of well points.

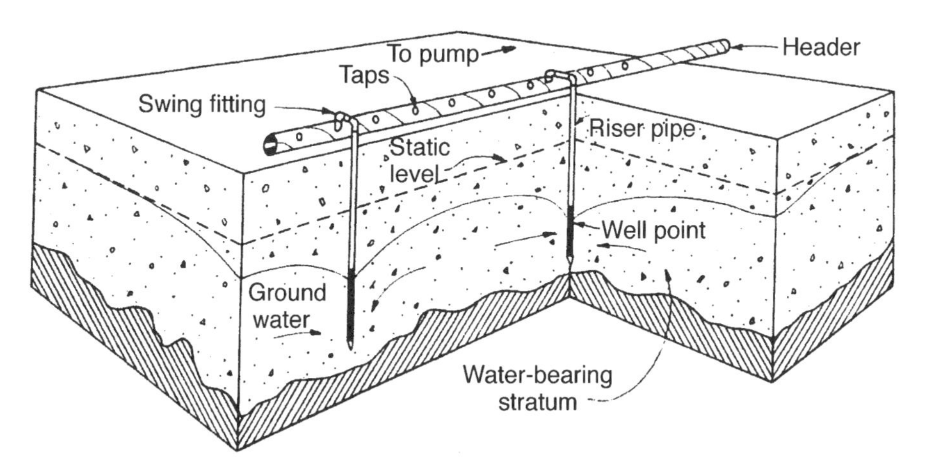

Fig. 9.3. Schematic of a section of a well point dewatering system. (Adapted from U.S. Department of the Interior, 1977.)

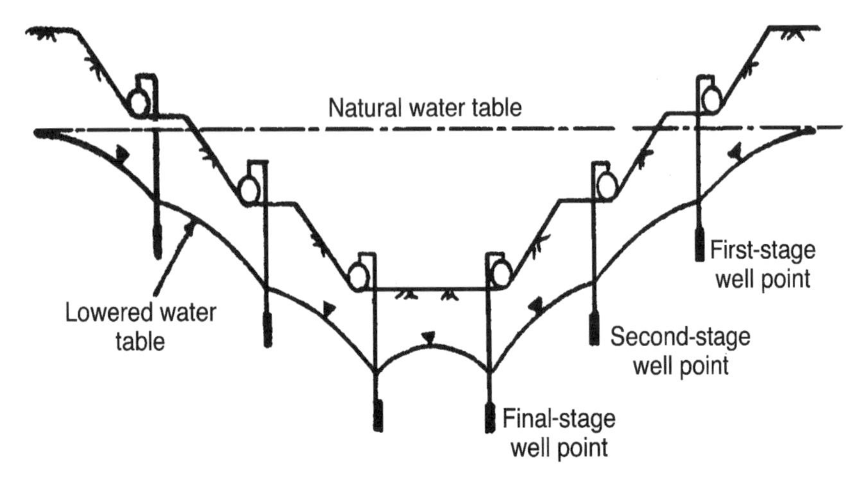

Fig. 9.4. Multiple stage well point system.

For lowering the water table to depths greater than 16 m, deep wells are used, often in combination with well points (Fig. 9.5). To install a deep well, a borehole (with a diameter ranging from 15 to 60 cm) is created and a casing with a screen of length 5 to 25 cm is provided. These wells could bring up water from greater depths than well points because each well has its own pump. The pump used may be of a turbine type, with a motor at the surface, or a submersible pump placed within the well casing at the bottom of the well. Deep wells are ideal when large quantities of water are to be pumped at the site for extended periods of time.

To dewater in fine sands or silty sands with relatively low permeability, it may be necessary to apply suction heads using vacuum pumps at the surface. In such cases, well points and/or deep wells may be used and vacuum maintained in the well with the use of airtight seals.

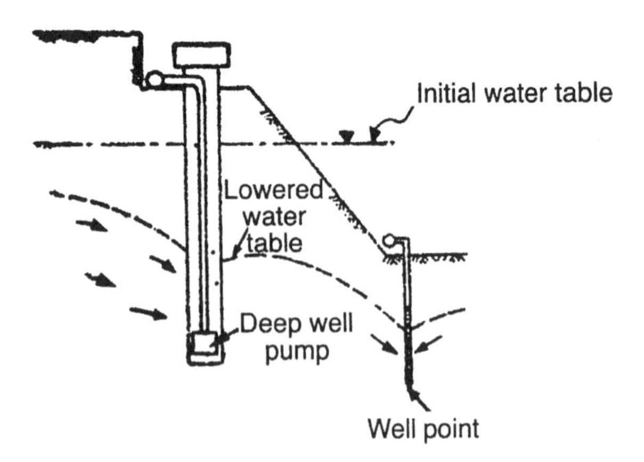

Fig. 9.5. Dewatering using deep wells combined with well points.

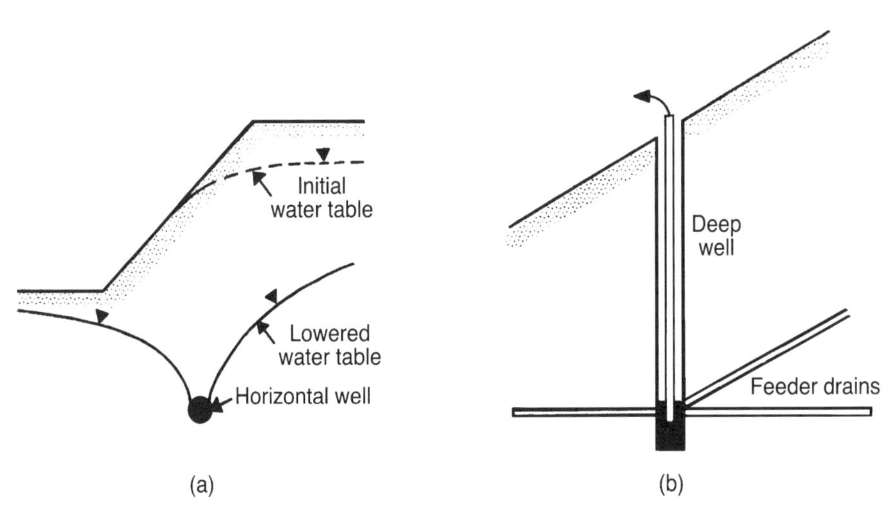

Fig. 9.6. Dewatering using horizontal wells: (*a*) single-well system; (*b*) network of lateral feeder drains connected to a deep well.

Horizontal wells (Fig. 9.6*a*) are also being used increasingly to dewater soils. Drilling of horizontal wells has become an established and common technology. Horizontal wells may influence larger areas than vertical wells and may be ideally suited for certain configurations of excavations. An alternative method to dewater large areas is to use a network of lateral feeder drains connected to a deep well (Fig. 9.6*b*). The pumping capacities of single deep wells govern how extensive the feeder drains could be.

9.1.3 Electroosmosis

Electroosmosis is used to dewater fine-grained soils, which by virtue of their low permeability, render the previous methods inapplicable. The principle behind this method is that when a saturated soil mass is subjected to electric gradients, water will travel from positive electrode (anode) to negative electrode (cathode) (Fig. 9.7*a*). The electrodes could be placed such that the flow due to electroosmosis is in a direction opposite to that of natural flow. For the purpose of collecting and pumping out water, the cathode is made in the form of a well point.

Electroosmosis is a result of electrochemical interactions between negatively charged soil particle surfaces (particularly clays) and pore water. In the presence of water, a diffuse layer of cations (positive ions) such as Na^+, K^+, and Ca^{2+} exists near the negatively charged surface of the soil particle (Fig. 9.7*b*). These cations, hydrated by the dipolar water molecules, are strongly attracted to the negatively charged soil particles. Away from the particle surfaces, the hydrated cations are loosely held in the pore water, although they are fewer in number than those near the particle surface. When an electric potential is applied to the saturated soil mass, the hydrated

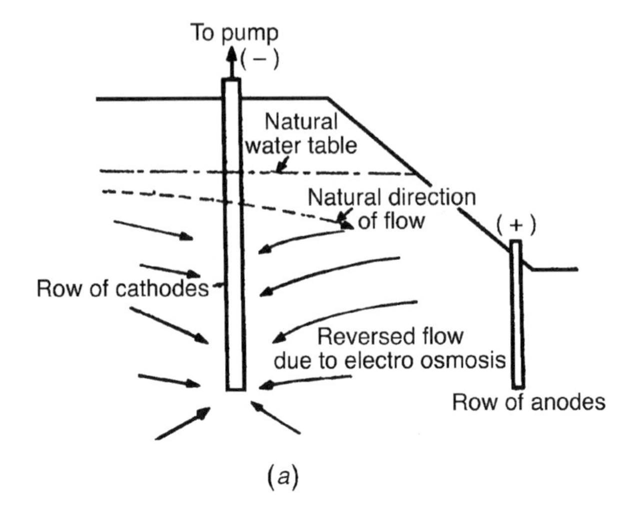

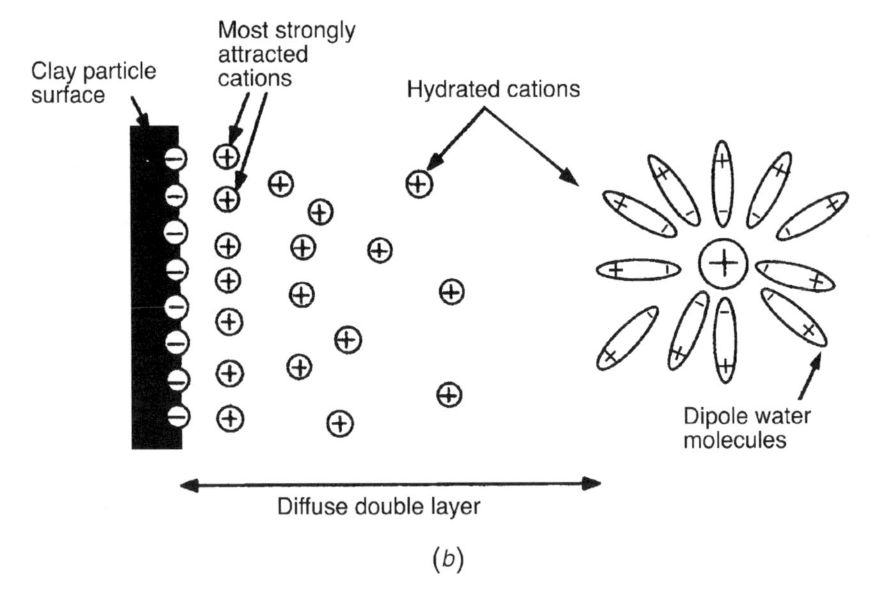

Fig. 9.7. Dewatering using electroosmosis: (*a*) schematic of electrodes used to reverse flow; (*b*) hydrated cations in the diffuse double layer.

positive ions move toward the negative electrode (cathode), dragging the dipolar water molecules with them. The movement of cations occurs in the *diffuse double layer*, where they exist in greater numbers. Thus, the amount of water that could be transported using electroosmosis is governed by the electric potential applied, the surface area and the negative charge of soil particles, and the type of cations present in the soils.

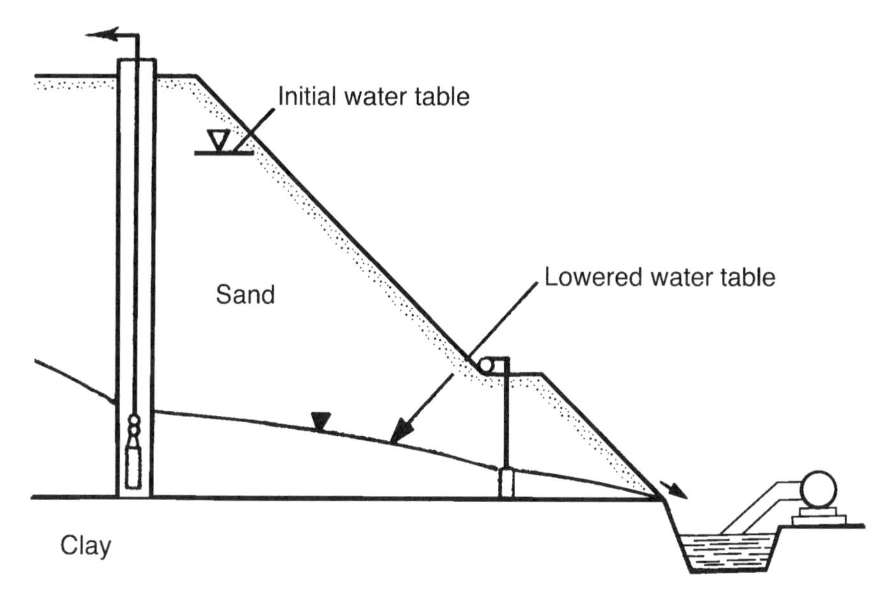

Fig. 9.8. Combined dewatering system using deep wells, well points, and open pumping.

The three types of methods described above are often used in combination to speed up the dewatering process. Figure 9.8 shows an example where deep wells, well points, and open drainage are used to dewater an excavated slope. The deep well lowers the water table to an extent where a single-stage well point system could pick up with the help of open pumping from the ditch.

The choice of dewatering method to be employed at a specific site should be based on a number of factors, including the nature of the soil, the size and depth of excavation, the depth to which the groundwater table should be lowered, the project schedule, and economics. Permeability of soils is by far the most important factor controlling the choice of dewatering method. Since particle size is a good indicator of permeability, the applicability of dewatering systems may be based on the grain size distribution of soils (Fig. 9.9).

9.2 SEEPAGE CUTOFF METHODS

The purpose of seepage cutoff methods is to decrease the seepage quantities or completely prevent seepage in soils. This is accomplished by installing barriers in the subsurface. The barriers range from simple configurations of sheetpile walls to complicated and energy-intensive methods of ground freezing using ammonium brine or liquid nitrogen. The applicability of the various methods, and the characteristics and requirements of the methods, are summarized in Table 9.1.

Installation of sheetpiles is one of the simplest methods of seepage cutoff. This could be a temporary measure to facilitate construction of foundations in deep excavations, or in other cases, the sheetpile could become part of the completed structure.

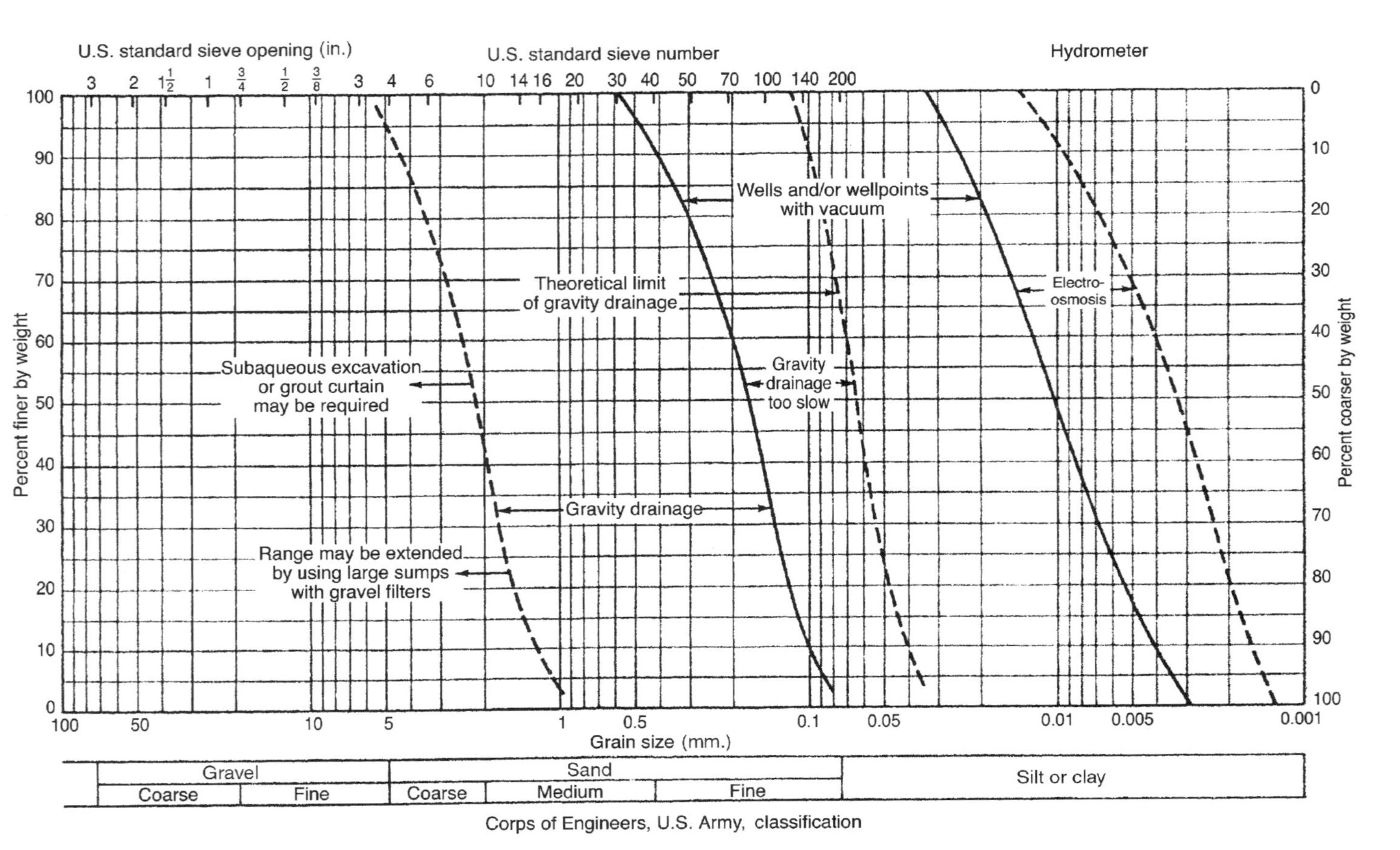

Fig. 9.9. Dewatering systems applicable to different soils. (Adapted from Moretrench Corp.)

TABLE 9.1 Cutoff Methods for Seepage Control

Method	Applicability	Characteristics and Requirements
Sheet pile cutoff wall	Suited especially for stratified soils with high horizontal and low vertical permeability or pervious hydraulic fill materials. May be easily damaged by boulders or buried obstructions. Tongue-and-groove wood sheeting utilized for shallow excavation in soft to medium soils. Interlocking steel sheet piling is used for deeper cutoff.	Steel sheeting must be driven carefully to maintain interlocks tight. Steel H-pile soldier beams may be used to minimize deviation of sheeting in driving. Some deviation of sheeting from plumb toward the side with least horizontal pressure should be expected. Seepage through interlocks is minimized where tensile force acts across interlocks. For straight wall sheeting an appreciable flow may pass through interlocks. Decrease interlock leakage by filling interlocks with sawdust, bentonite, cement grout, or similar material.
Compacted barrier of impervious soil	Formed by compacted backfill in a cutoff trench carried down to impervious material or as a core section in earth dams.	Layers or streaks of pervious material in the impervious zone must be avoided by careful selection and mixing of borrow materials, scarifying lifts, aided by sheepsfoot rolling. A drainage zone downstream of an impervious section of the embankment is necessary in most instances.
Grouted or injected cutoff	Applicable where depth or character of foundation materials make sheet pile wall or cutoff trench impractical. Utilized extensively in major hydraulic structures. May be used as a supplement below cutoff sheeting or trenches.	A complete positive grouted cutoff is often difficult and costly to attain, requiring a pattern of holes staggered in rows with carefully planned injection sequence and pressure control.

(continued)

TABLE 9.1 Cutoff Methods for Seepage Control *(Continued)*

Method	Applicability	Characteristics and Requirements
Slurry trench method	Suited for construction of impervious cutoff trench below groundwater or for stabilizing trench excavation. Applicable whenever cutoff walls in earth are required. Is replacing sheet pile cutoff walls.	Vertical-sided trench is excavated below groundwater as slurry with specific gravity generally between 1.2 and 1.8 is pumped back into the trench. Slurry may be formed by mixture of powdered bentonite with fine-grained material removed from the excavation. For a permanent cutoff trench, such as a foundation wall or other diaphragm wall, concrete is tremied to bottom of trench, displacing slurry upward. Alternatively, well-graded backfill material is dropped through the slurry in the trench to form a dense mixture that is essentially an incompressible mixture; in working with coarser gravels (which may settle out), to obtain a more reliable key into rock and a narrower trench, use a cement–bentonite mix.
Impervious wall of mixed-in-place piles	Method may be suitable to form cofferdam wall where sheet pile cofferdam is expensive or cannot be driven to suitable depths, or has insufficient rigidity, or requires excessive bracing.	For a cofferdam surrounding an excavation, a line of overlapping mixed-in-place piles are formed by a hollow-shaft auger or mixing head rotated into the soil while cement grout is pumped through the shaft. Where piles cannot be advanced because of obstructions or boulders, supplementary grouting or injection may be necessary.
Freezing-ammonium brine or liquid nitrogen.	All types of saturated soils and rock. Forms ice in voids to stop water. Ammonium brine is better for large applications of long duration. Liquid nitrogen is better for small applications of short duration where quick freezing is needed.	Gives temporary mechanical strength to soil. Installation costs are high and refrigeration plant is expensive. Some ground heave occurs.

Source: Adapted from U.S. Department of the Navy, Naval Facilities Engineering Command (1982).

An important consideration is the exit gradients and possible piping or heaving at the downstream side of sheetpiles. The depth of penetration of sheetpiles should be chosen to minimize the possibility of soil heaving or piping. This was discussed in the context of seepage forces in Chapter 6.

Compacted clay could be used in cutoff trenches as an impervious material, as a core section in earth dams, or as a liner in waste impoundments to minimize seepage. Although this is one of the most economical methods of seepage cutoff, it is also one of the most difficult methods to assure quality of construction. The type of clay material, molding water contents, and type and energy of compaction control the effectiveness of compacted clay barriers.

Grouted or injected cutoff is one of the fastest-growing methods of seepage cutoff. The basic categories of soil grouting are shown in Fig. 9.10. The hydrofracture grouting involves injecting cement-based grouts at pressures as high as 4 MPa. The grout fills up voids and creates impermeable barriers, predominantly horizontal. In compaction grouting, stiff low-mobility soil–cement mixes are injected at high pressures to densify soft or loose soils. In the case of permeation grouting, the grout is introduced into soil pores such that the original soil structure and volume are not altered. Finally, jet grouting is a method used to mix or replace soil completely with cement. A special drill bit with horizontal and vertical high-speed water jets is used to excavate soils. Hard, impervious columns are produced by pumping grout through the horizontal nozzles, which mixes with foundation material as the drill bit is withdrawn.

At waste containment structures such as landfills, slurry walls are used to minimize seepage and prevent groundwater pollution (Fig. 9.11). As discussed in Chapter 11, slurry walls play an important role in isolating contaminants and accidental spills. These walls are constructed by first excavating a narrow, straight-sided trench. To stabilize the trench and prevent walls from caving in, it is filled with a slurry of bentonite. The bentonite slurry is eventually replaced with a soil–bentonite mix

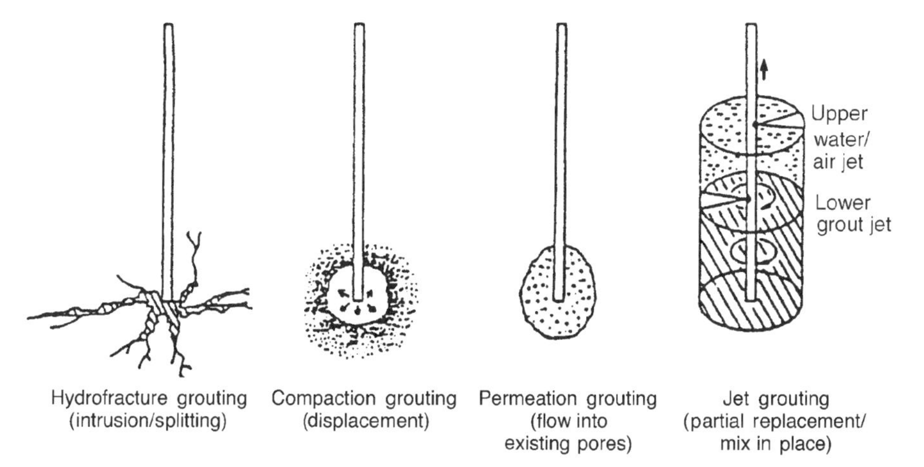

Fig. 9.10. Basic categories of soil grouting. (Adapted from Xanthakos et al., 1994; copyright © John Wiley & Sons; reprinted with permission.)

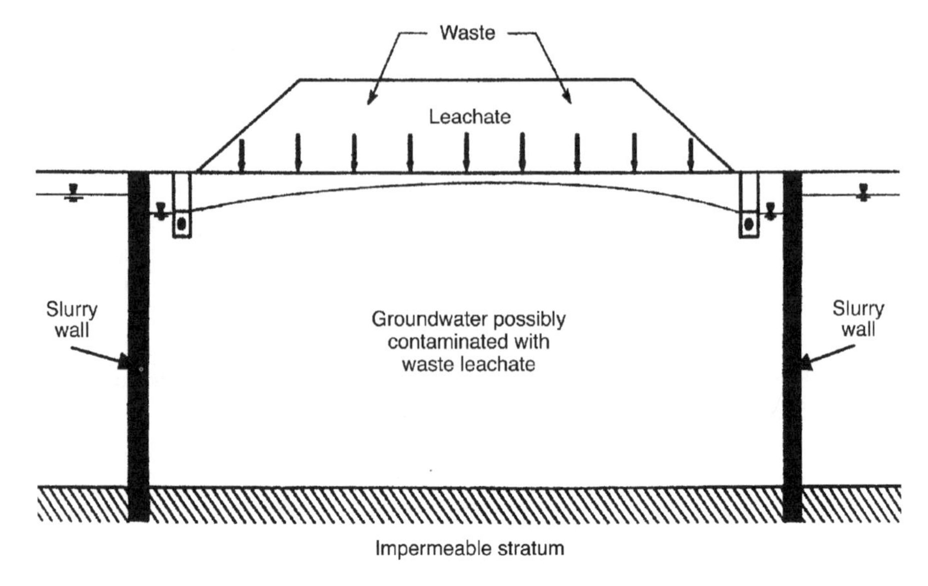

Fig. 9.11. Seepage cutoff at waste containment systems using slurry walls.

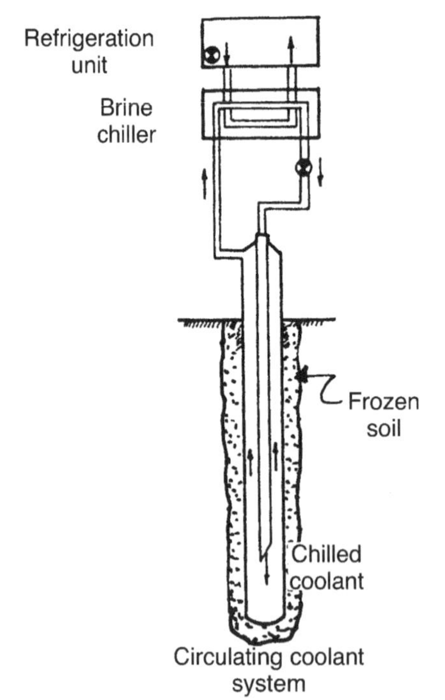

Fig. 9.12. Basic refrigeration system for ground freezing. (Adapted from Xanthakos et al., 1994; copyright © John Wiley & Sons; reprinted with permission.)

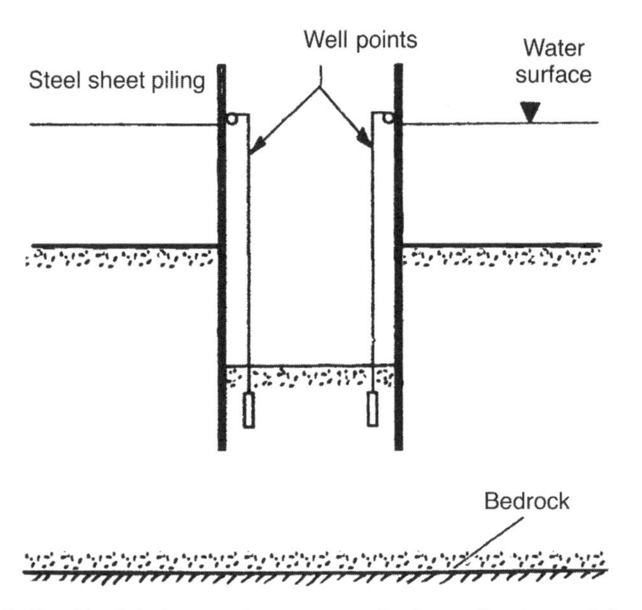

Fig. 9.13. Combined drainage and seepage cutoff using well points and sheet piles.

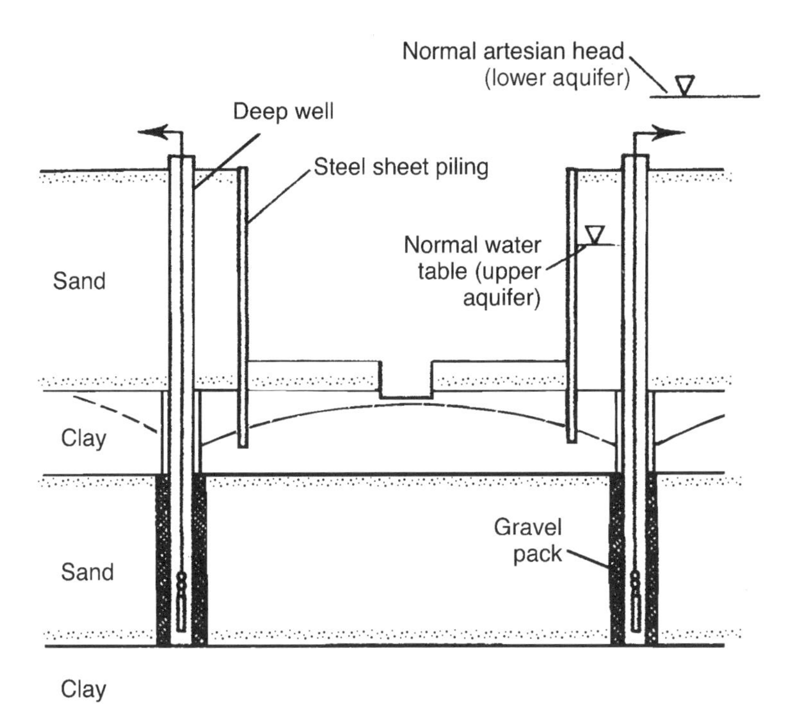

Fig. 9.14. Combined drainage and seepage cutoff using deep wells and sheet piles.

yielding a relatively impermeable barrier. Slurry walls are discussed further in Chapter 11.

Ground freezing involves freezing the moisture in soil pores, which then acts as a cementing agent binding the soil particles together. Freezing pipes or "probes" are placed in the zone to be frozen, and a coolant is circulated, cooled, and recirculated continuously through the heat removal system (Fig. 9.12). Freezing occurs radially outward from the probe, forming a frozen cylinder along the length of the probe. Several such adjoining cylinders form a continuous wall enclosing the excavation area to be protected from seepage.

Again, a combination of the various drainage and cutoff methods may be the most cost-effective solution at several sites. Examples of such combinations are shown in Figs. 9.13 to 9.15. Well points offer a convenient way to stabilize soils at toes of sheetpiles and perhaps lessen the required depth of penetration of sheetpiles (Fig. 9.13). In Fig. 9.14, deep wells are used to relieve the artesian pressure beneath the clay layer and minimize the risk of soil heaving. Figure 9.15 depicts an innovative system where deep wells embedded in slurry walls are used in combination with soil grouting to cutoff seepage into the excavation.

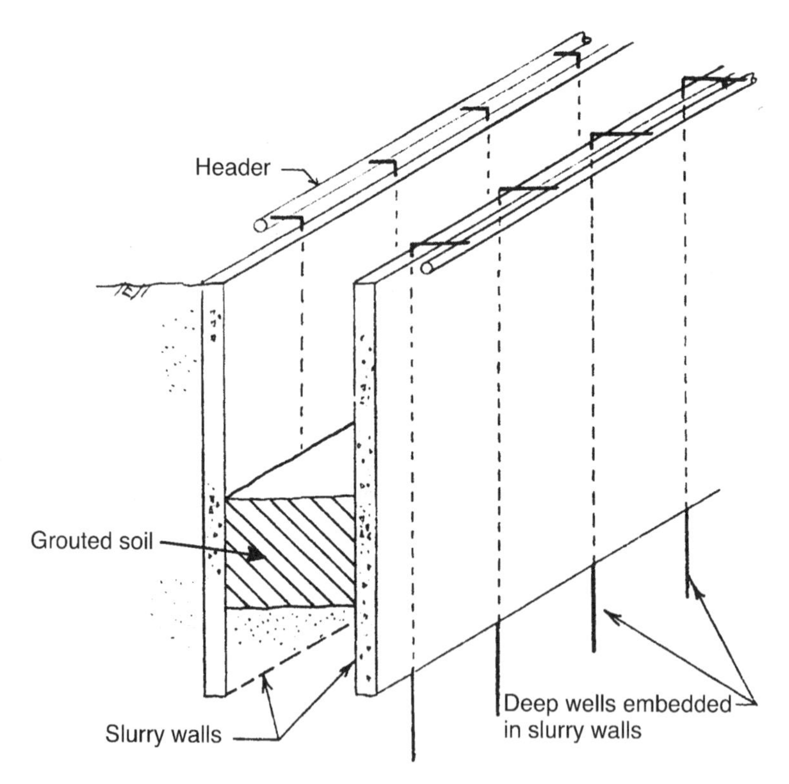

Fig. 9.15. Combined drainage and seepage cutoff using deep wells embedded in slurry walls and grouting. (Adapted from Xanthakos et al., 1994; copyright © John Wiley & Sons; reprinted with permission.)

9.3 SEEPAGE CONTROL IN EARTH DAMS

Seepage control is the most important consideration in the design of earth dams. The high potential energy of upstream water has to be dissipated in a safe manner. Considering that it is impossible to cutoff seepage of water toward the downstream completely, it is important to provide adequate internal drainage. The methods available to control seepage in earth dams may be categorized under two broad categories:

1. Methods that prevent or reduce seepage through the earth dam and its foundation:
   a. Sheet piles at the upstream and/or downstream
   b. Cutoff trenches and grout curtains
   c. Impermeable upstream blankets
2. Drainage methods to control seepage in a way that uplift pressures and internal erosion are minimized:
   a. Embankment zoning (with filters and drains)
   b. Longitudinal drains and blankets
   c. Chimney drains
   d. Partially penetrating toe drains
   e. Relief wells

The methods under the first category are similar to those discussed under seepage cutoff methods in Section 9.2. These methods are illustrated in Fig. 9.16 for earth dams. Again, the reader will find the methods used in combination. The primary cutoff mechanism is through the use of clay-filled trench at the center of the dam (Fig. 9.16*a*), which may be augmented by sheetpiles (Fig. 9.16*b*) at the upstream and/or downstream or concrete walls or grout curtains (Fig. 9.16*b* and *c*) in the pervious foundation underneath the core of the dam. An impervious concrete blanket may be used on the upstream slope in all these cases.

Cedergren (1989) provided interesting analyses of seepage reductions using flownets for dams provided with trenches and grout curtains. Figure 9.17 shows the percentage reduction in seepage for various percentages of cutoff trench penetration. These results are useful in the design phase to determine the trench penetration depth for a targeted seepage reduction. A similar analysis for seepage reduction using grout curtains is shown in Fig. 9.18. It was assumed in this analysis that the grout curtain penetrated the jointed rock formation completely. These results highlighted the importance of the relative permeability of grout curtain with respect to the adjoining soil (Fig. 9.18*b*, *c*, and *d*). To achieve 37% of the original seepage, the grout must be 95% effective (Fig. 9.18*d*); in other words, its permeability must be 5% that of the ungrouted rock. The cutoff efficiency of grout curtains and other impermeable structures is reduced drastically if small openings or gaps exist in these structures. Consider, for example, an impervious steel membrane containing openings or slits with a total open space ratio of 5 and 10% (Fig. 9.19*a*). Cedergren (1989)

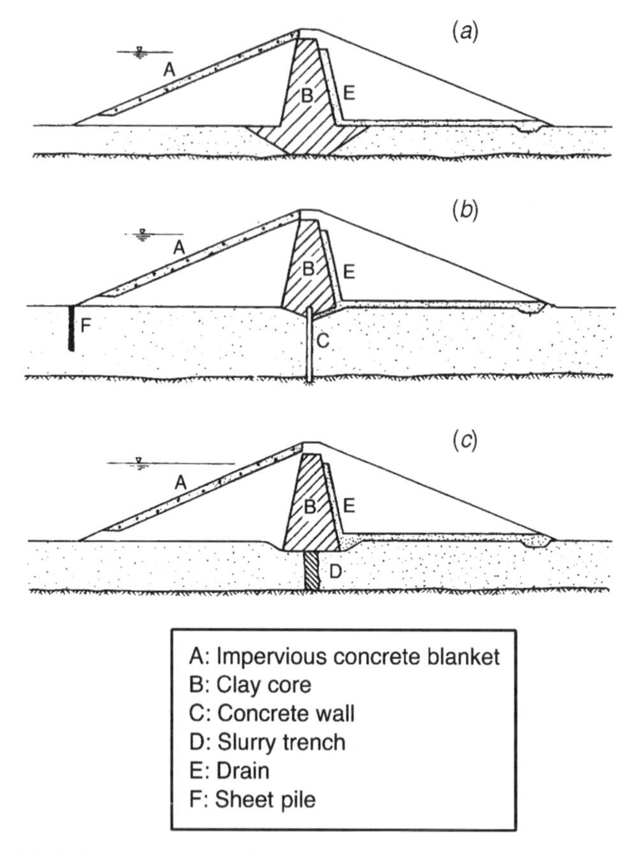

Fig. 9.16. Methods to prevent or reduce seepage through earth dam and its foundation.

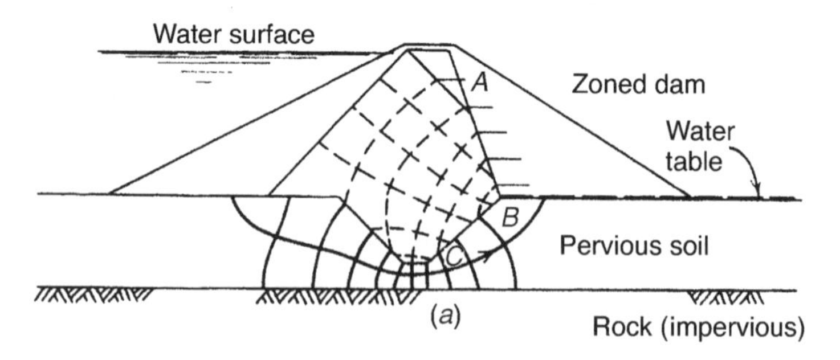

Fig. 9.17. Reduction in seepage underneath dams for different penetrations of cutoff trench. (Adapted from Cedergren, 1989; copyright © John Wiley & Sons; reprinted with permission.)

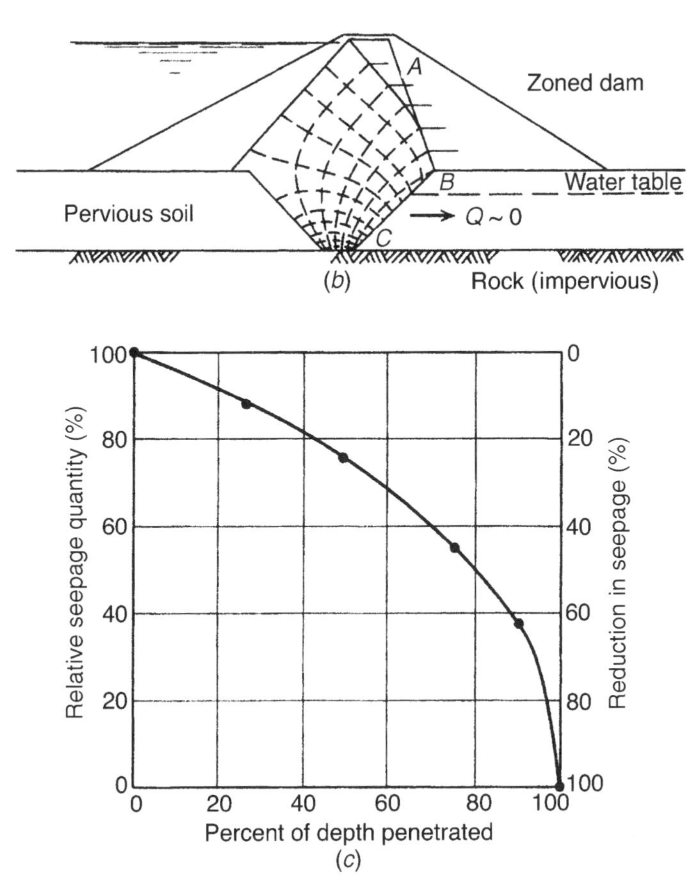

Fig. 9.17. (*Continued*)

presented an analysis using flow nets for such a scenario. The results, shown in Fig. 9.19*b*, indicate that if a 5% open space is the result of one opening, the seepage reduction is slightly over 60%. If eight openings contribute to the open space, the seepage reduction drops to less than 20%.

In addition to seepage cutoff methods, earth dams are protected using drainage provisions. Earth dams are "zoned" using an impervious core (for seepage cutoff) and outer "shells" consisting of stable granular materials (Fig. 9.20). The filter and drainage materials provided on the downstream side of the earth dams provide a safe outlet for water. As shown in Fig. 9.21, the phreatic line on the downstream side could be drawn down significantly by choosing materials relatively more permeable than the clay used in the central core. The various types of earth dam cross sections utilizing combinations of cutoff and drainage methods are shown in Fig. 9.22. Note that drainage provisions are sometimes used at the upstream end as well, mainly to facilitate rapid drawdown of reservoir.

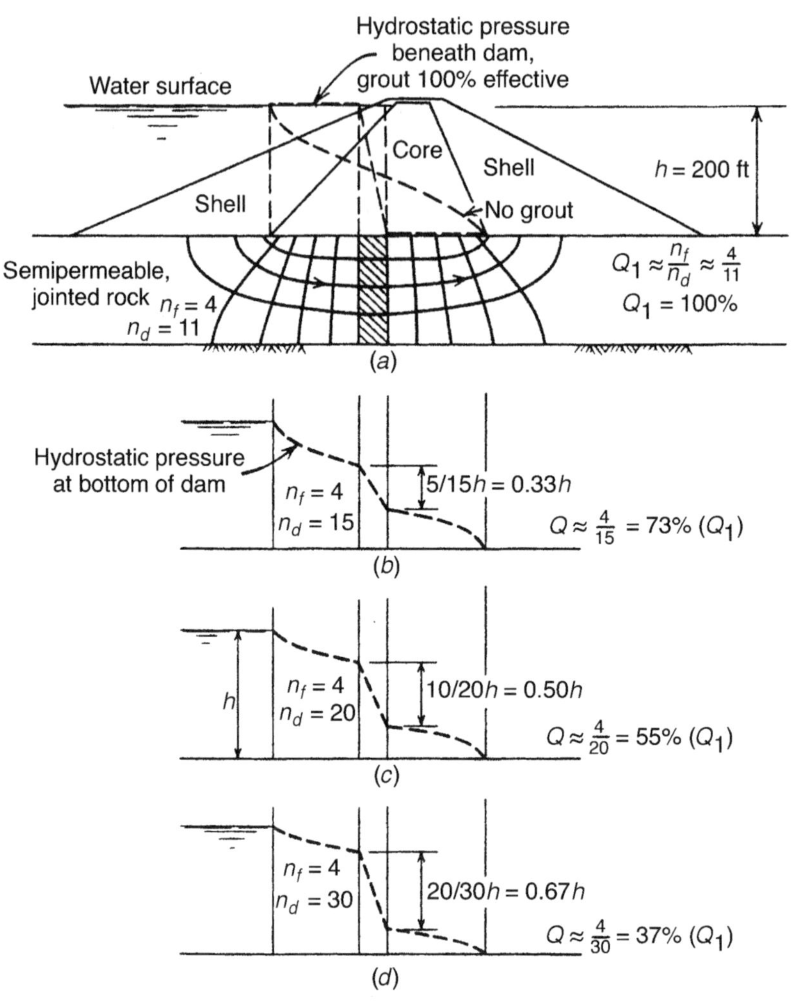

Fig. 9.18. Seepage reduction under earth dams using grout curtains with different relative permeabilities: (*a*) cross section and typical flow net; (*b*) grout 80% effective ($k_g = 0.2k$); (*c*) grout 90% effective ($k_g = 0.1k$); (*d*) grout 95% effective ($k_g = 0.05k$). k_g = permeability of grout; k = permeability of ungrouted rock. (Adapted from Cedergren, 1989; copyright © John Wiley & Sons; reprinted with permission.)

9.4 INTERCEPTING DRAINS

Shallow trenches with collector pipes could be used to lower the water table in a pervious stratum. Examples are using a trench to collect and control surface runoff over a saturated surface layer and using interceptor drains to draw down the water table sufficiently below pavements (Fig. 9.23). In many situations, flow nets could be used to determine the seepage rate and the drawdown of water table. In some cases, the

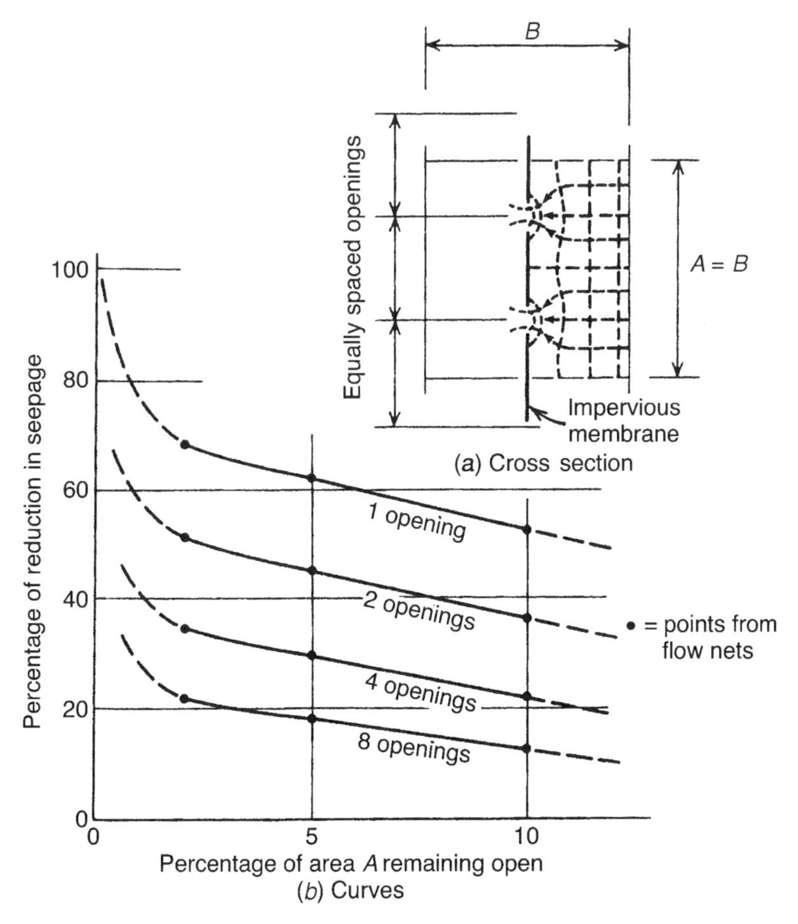

Fig. 9.19. Effect of openings and gaps in cutoff structures on seepage reduction. (Adapted from Cedergren, 1989; copyright © John Wiley & Sons; reprinted with permission.)

results from flow net analyses could be synthesized into nomograms for convenient use. In others, analytical solutions discussed in Chapters 4 and 5 could be used.

9.4.1 Shallow Drains for Ponded Areas

The purpose of the drains (shown schematically in Fig. 9.24) is to prevent ponding of water on the ground surface. The drainage trenches should be designed so that flow into a single trench is equal to or greater than the intensity of rainfall on the area bound by two successive trenches. Kirkham (1960) presented solutions for the two cases when the trenches are keyed into a pervious soil underlying the surface layer (Fig. 9.25) and when the trenches are keyed into an impervious layer underlying the surface layer (Fig. 9.26). The following assumptions were made:

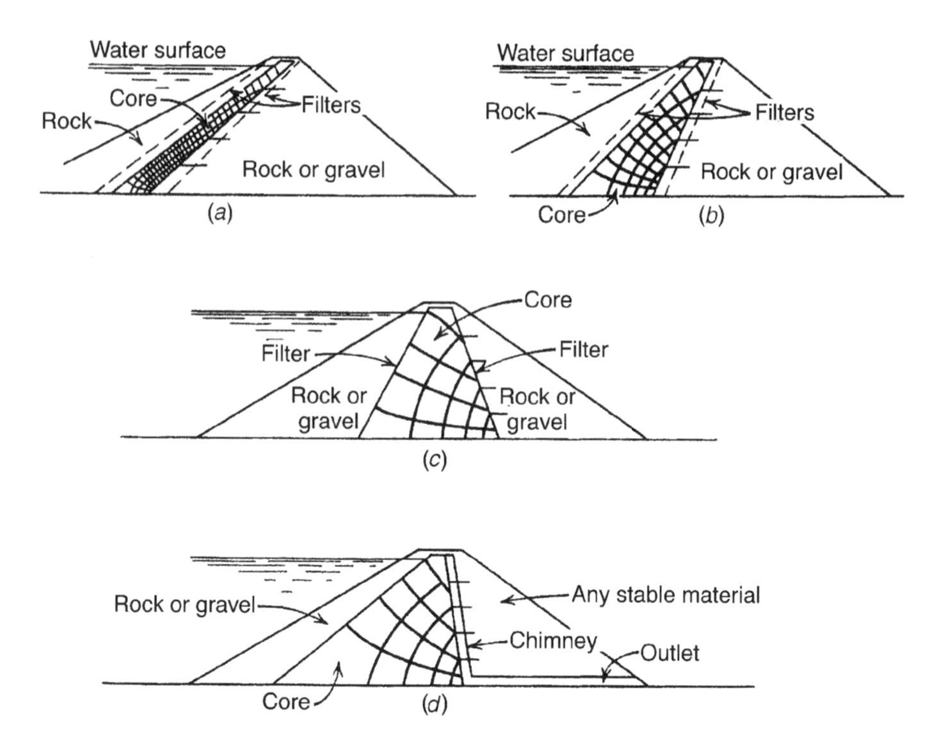

Fig. 9.20. Several kinds of zoned dam: (*a*) thin, sloping core dam; (*b*) thick, sloping core dam; (*c*) central core dam; (*d*) dam with chimney drain and outlet blanket drain. (Adapted from Cedergren, 1989; copyright © John Wiley & Sons; reprinted with permission.)

1. The surface layer is saturated by continuous rainfall.
2. There is no head loss in the trench backfill or underlying pervious layer.
3. Ponding of water on the ground surface is not permitted.

Under these assumptions, the maximum discharge into trench per unit length of trench, q in the case when the surface layer is underlain by a pervious layer, is given by

$$q = kSF \tag{9.1}$$

where k is the permeability of surface layer, $2S$ the spacing of trenches, and F the flow coefficient, which is obtained from Fig. 9.27. When the surface layer is underlain by an impervious stratum into which the trench is keyed, the discharge q is given by

$$q = \frac{16Dk}{\pi^2}F \tag{9.2}$$

where D is the thickness of the surface layer, and flow coefficient F is obtained using Fig. 9.27.

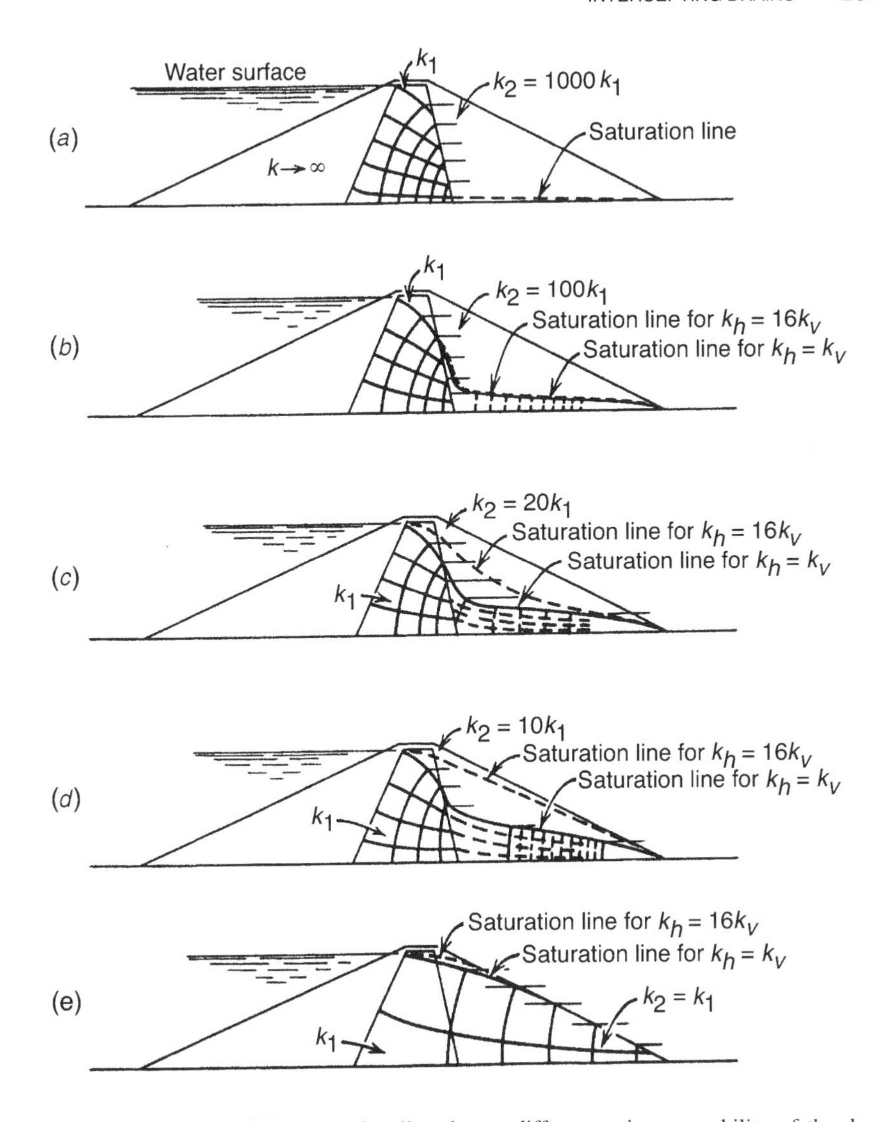

Fig. 9.21. Drawdown of the saturation line due to differences in permeability of the dam core and downstream soils. (Adapted from Cedergren, 1989; copyright © John Wiley & Sons; reprinted with permission.)

9.4.2 Longitudinal Drains for Roadways

We will consider the case of flow toward a single interceptor drain (Fig. 9.28). Donnon (1959) presented the following solution for the shape of the drawdown curve based on Dupuit theory.

$$x = \frac{H \log[(H - H_0)/(H - y)] - (y - H_0)}{S} \tag{9.3}$$

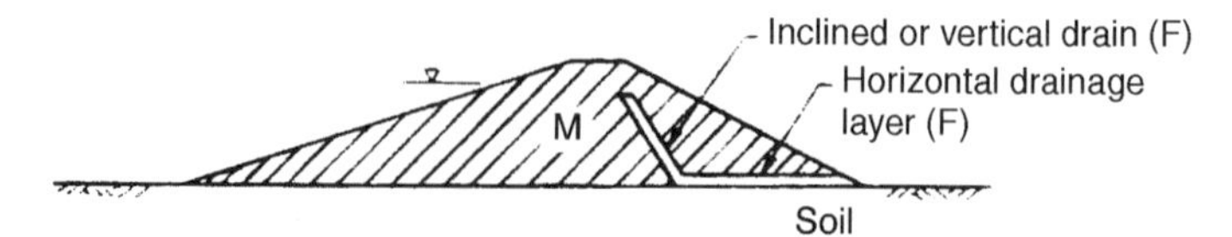

(*a*) Homogeneous dam with integral drainage on impervious foundation

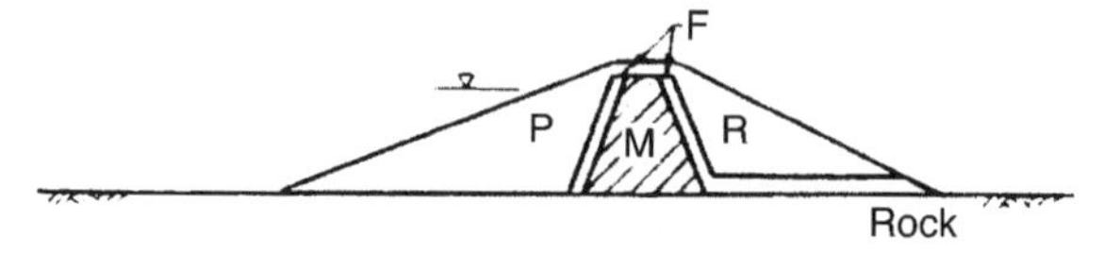

(*b*) Central core dam on impervious foundation

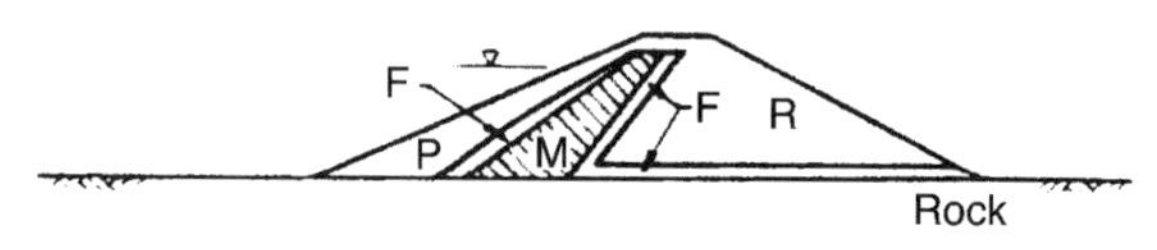

(*c*) Inclined core dam on impervious foundation

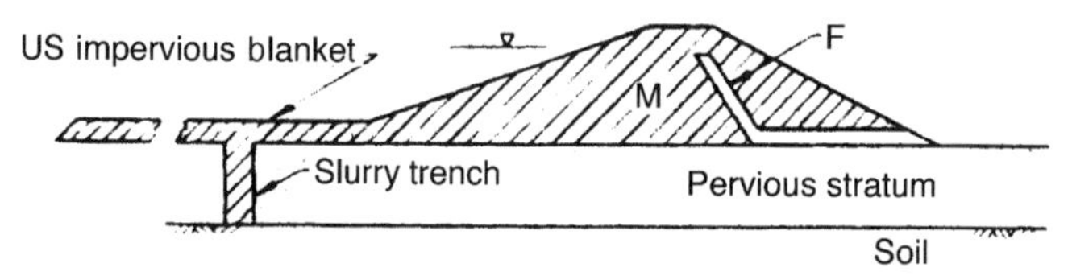

(*d*) Homogeneous dam with internal drainage on pervious foundation

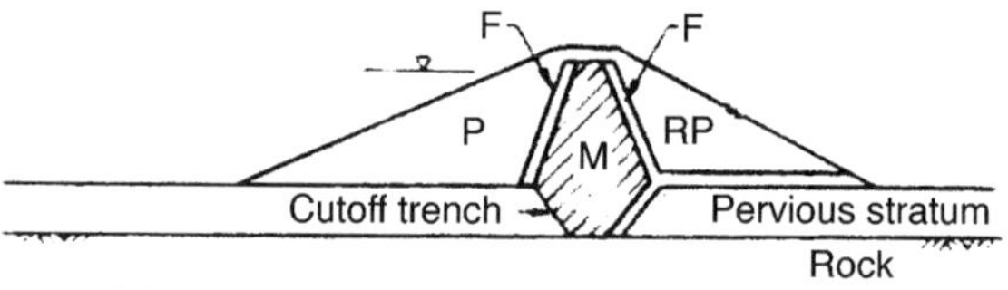

(*e*) Central core dam on pervious foundation

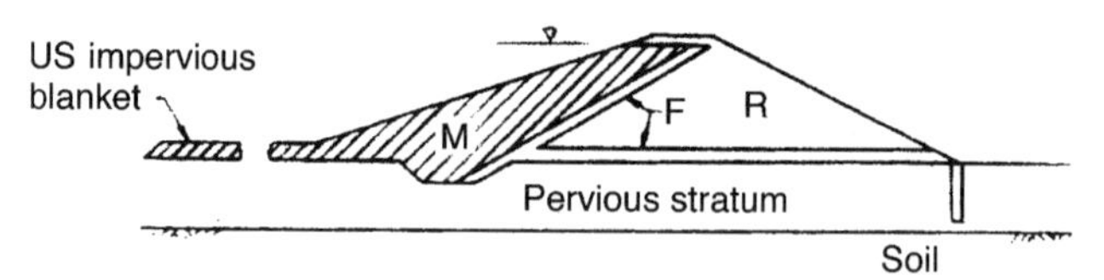

(*f*) Dam with upstream impervious zone on pervious foundation

M: Impervious
P: Pervious
R: Random
F: Select pervious material
US: Upstream

Fig. 9.22. Combinations of cutoff and drainage methods used for earth dams. (Adapted from Wilson and Marsal, 1979; copyright © ASCE; reprinted with permission.)

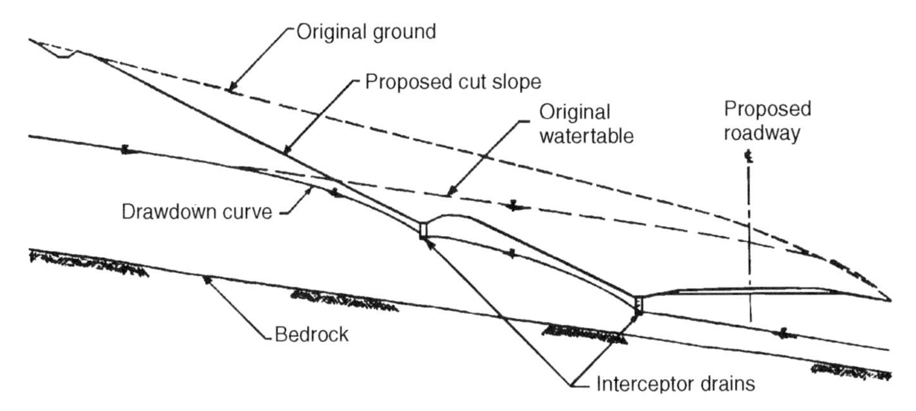

Fig. 9.23. Schematic of a roadway section with interceptor drains used to cutoff seepage. (Adapted from Moulton, 1980.)

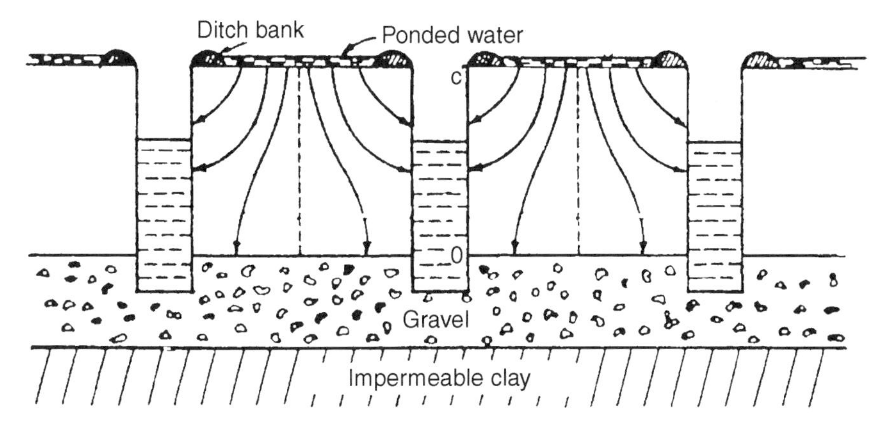

Fig. 9.24. Schematic of shallow drains used at ponded areas. (Adapted from Kirkham, 1960.)

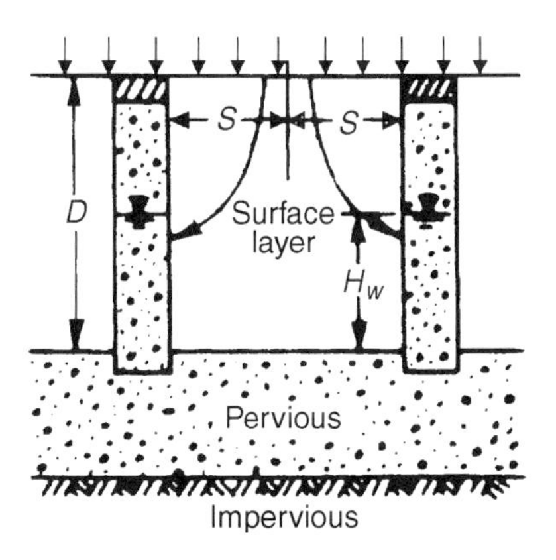

Fig. 9.25. Shallow drains keyed into a pervious soil underlying the surface layer.

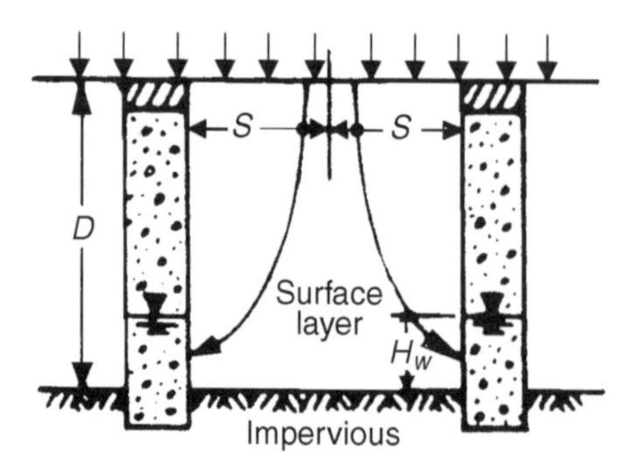

Fig. 9.26. Shallow drains keyed into an impervious layer underlying the surface layer.

where x and y are coordinates of a point on the drawdown curve (see Fig. 9.28), H is the height of the original groundwater table above an impervious boundary of slope S, and H_0 is the height of the drain above the impervious boundary. Equation (9.3) yields a drawdown curve asymptotic to the groundwater table. To make the solution practical, it is assumed that the drawdown is insignificant (i.e., $y = H$) in a finite distance L_i from the drain (see Fig. 9.29). Keller and Robinson (1959) modified this solution using laboratory studies involving a finite source of seepage located at a distance L_i from the drain. Using a point on a fictitious extension of the drawdown curve with height H' above the impervious layer, they obtained

$$SL_i = H' \log \frac{H' - H_0}{H' - H} - (H - H_0) \tag{9.4}$$

and

$$q = q_0 \frac{H' - H_0}{H} \tag{9.5}$$

where q is the quantity of flow into the drain and q_0 is the magnitude of the approach flow given by

$$q_0 = kHS \tag{9.6}$$

where k is the coefficient of permeability of the porous medium. Equation (9.4) gives an estimate of H' which when used in Eq. (9.5) gives us the quantity of flow into the drain. Equations (9.4) and (9.5) are presented in the form of nomograms in Figs. 9.30 and 9.31. q and H' could be obtained from Fig. 9.30 given the values of S, L_i, H, and k. Figure 9.31 could be used to trace the drawdown curve in terms of x and y. The length of influence L_i to be used in these figures is given by

$$L_i = 3.8(H - H_0) \tag{9.7}$$

In the case of multiple interceptor drains, it is recommended that the drains be decoupled and an impervious boundary established for each upper drain roughly parallel to

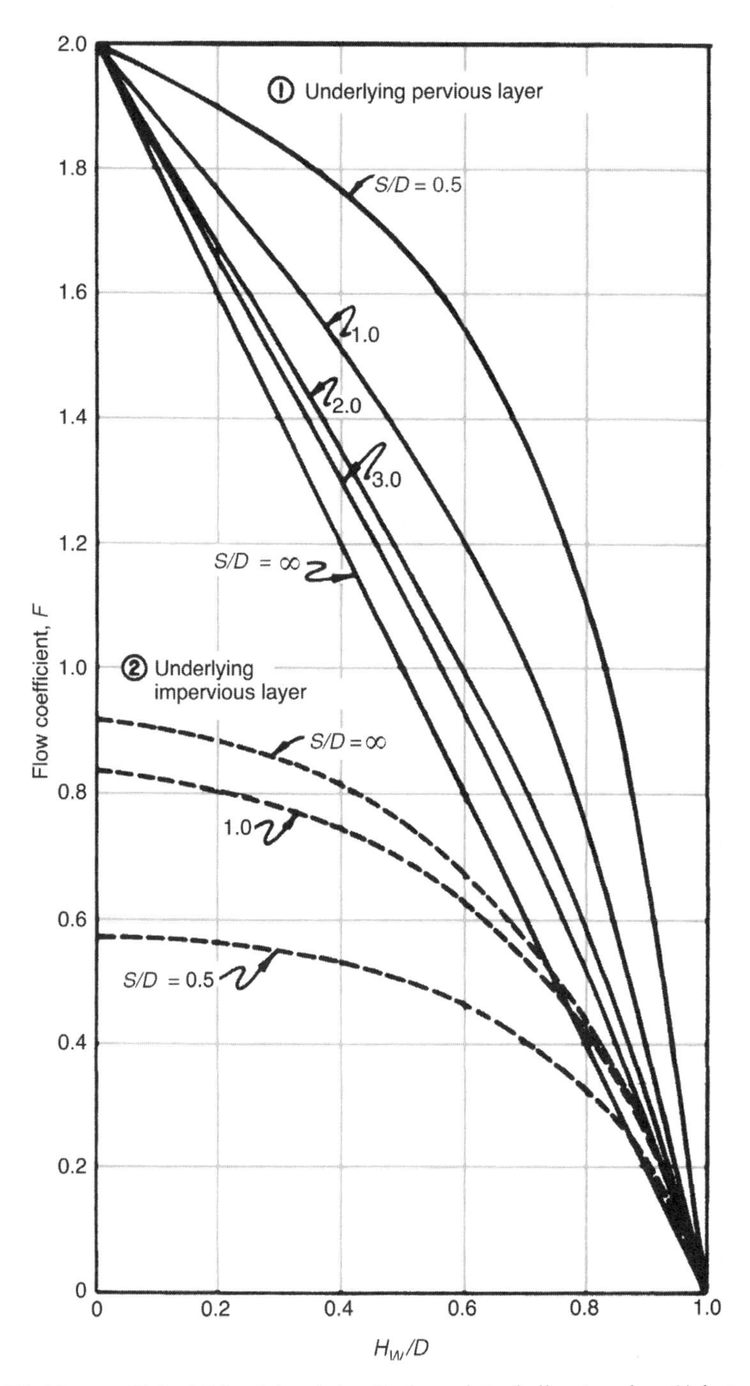

Fig. 9.27. Flow coefficient F for determining discharge into shallow trenches. (Adapted from the U.S. Department of the Navy, 1982.)

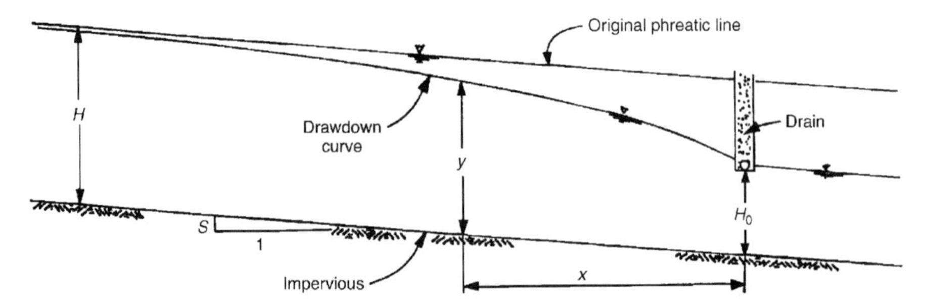

Fig. 9.28. Flow toward a single interceptor drain. (Adapted from Moulton, 1980.)

the lower sloping boundary. The depth of the fictitious impervious boundary below each drain is set equal to $\frac{1}{10}$ to $\frac{1}{12}$ of the drain spacing. This is illustrated in Fig. 9.32 for a two-drain system. Note that H_{02}, the height of the upper drain above the fictitious impervious boundary, is kept equal to $\frac{1}{12}$ of the drain spacing. With the data shown in Fig. 9.32, Figs. 9.30 and 9.31 can now be used separately for the two drains to calculate the flow into the drains and to trace the drawdown curves.

In the case of symmetrical drawdown drains such as those provided at either end of a pavement section, the flow problem could be solved as a combination of fragments. Consider the problem shown in Fig. 9.33. The flow coming into the drain is the sum of the flow components q_1 and q_2. q_1 is from the flow domain above the level of the drain, whereas q_2 is from the domain below the bottom of the drain. Thus, it is assumed that there exists an impervious boundary (with a horizontal streamline) at the level of the drain. Based on the method of fragments (see Chapter 5), the flow from fragment 1, q_1, and the drawdown curve in fragment 1 could be obtained using the following equations.

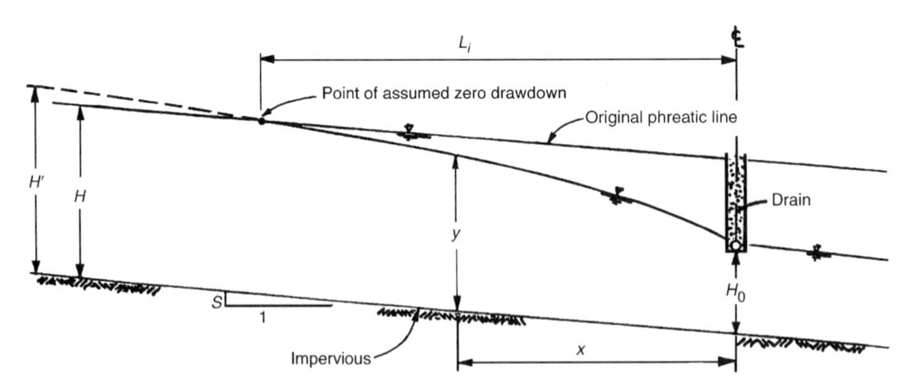

Fig. 9.29. Flow toward a single interceptor drain with drawdown neglected beyond a finite distance L_i from the drain. (Adapted from Moulton, 1980.)

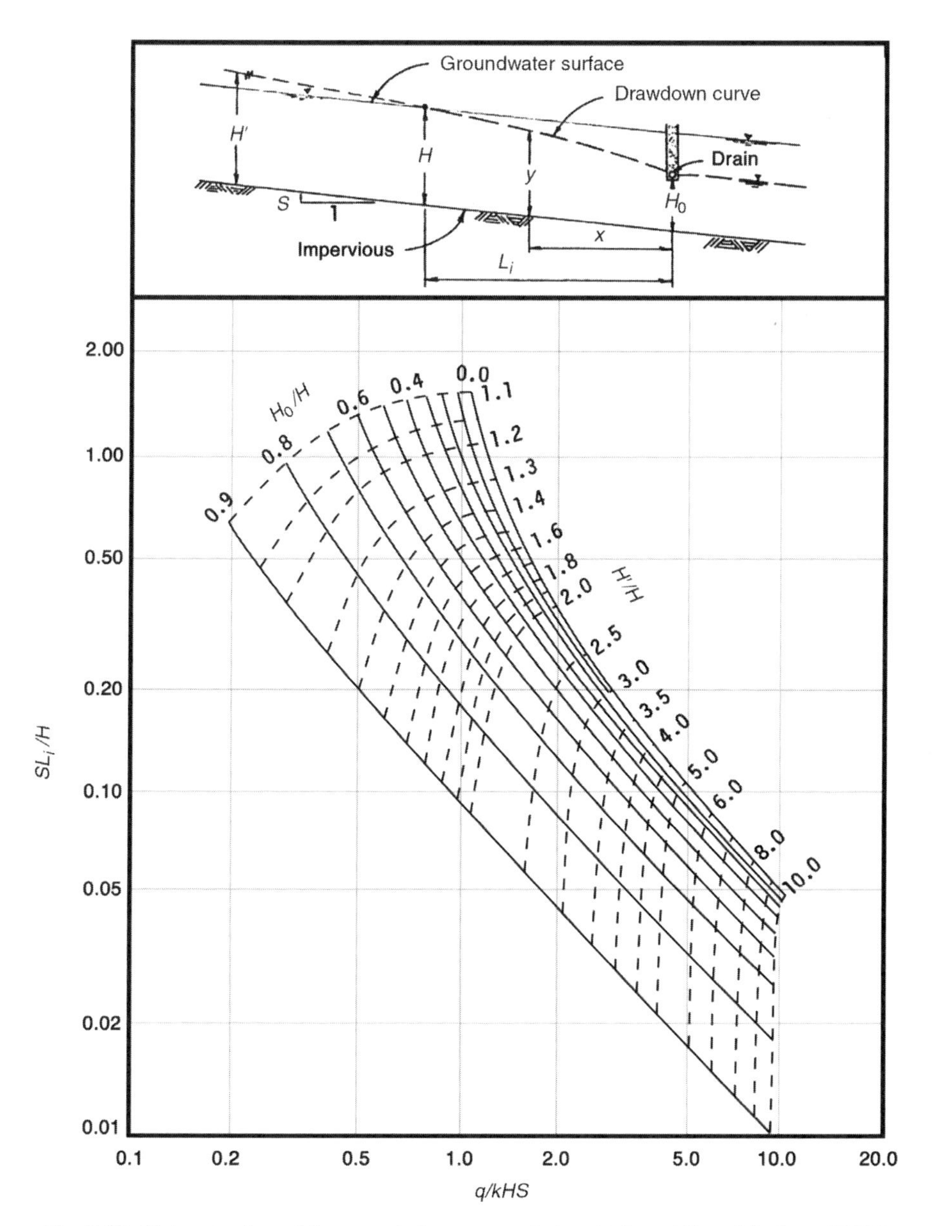

Fig. 9.30. Determination of flow rate in interceptor drains. (Adapted from Moulton, 1979.)

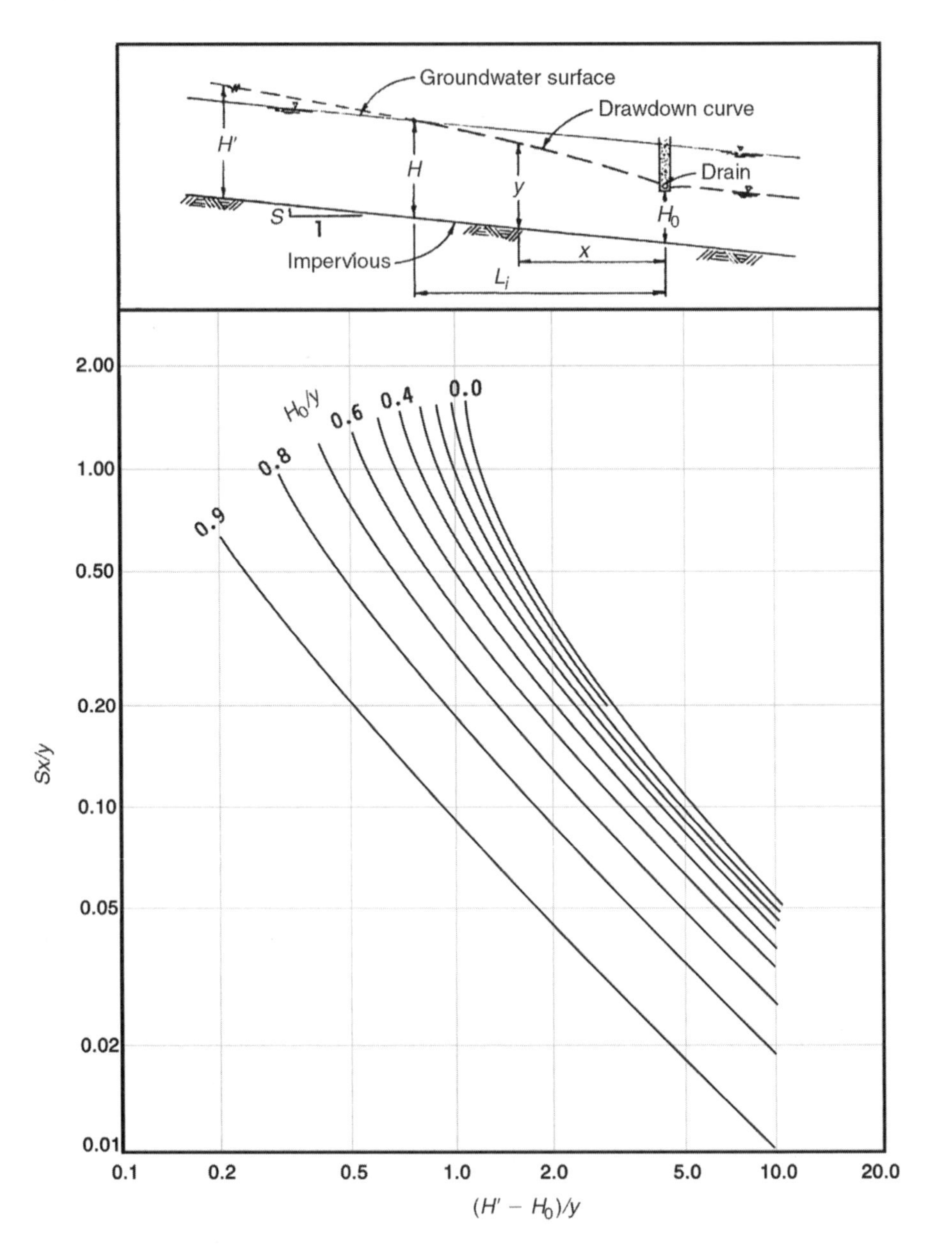

Fig. 9.31. Determination of drawdown curves for interceptor drains. (Adapted from Moulton, 1979.)

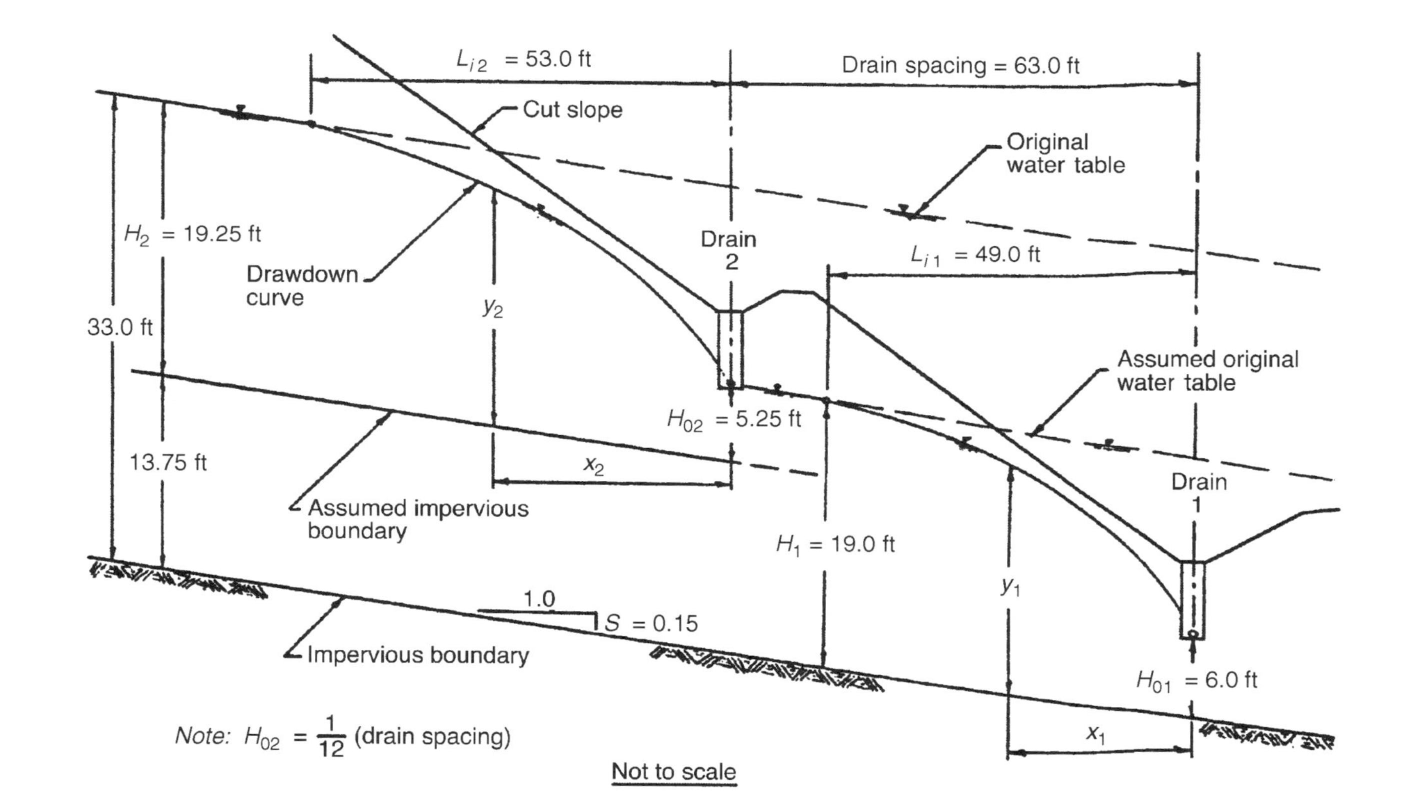

Fig. 9.32. Schematic of a decoupled two-drain system. (Adapted from Moulton, 1980.)

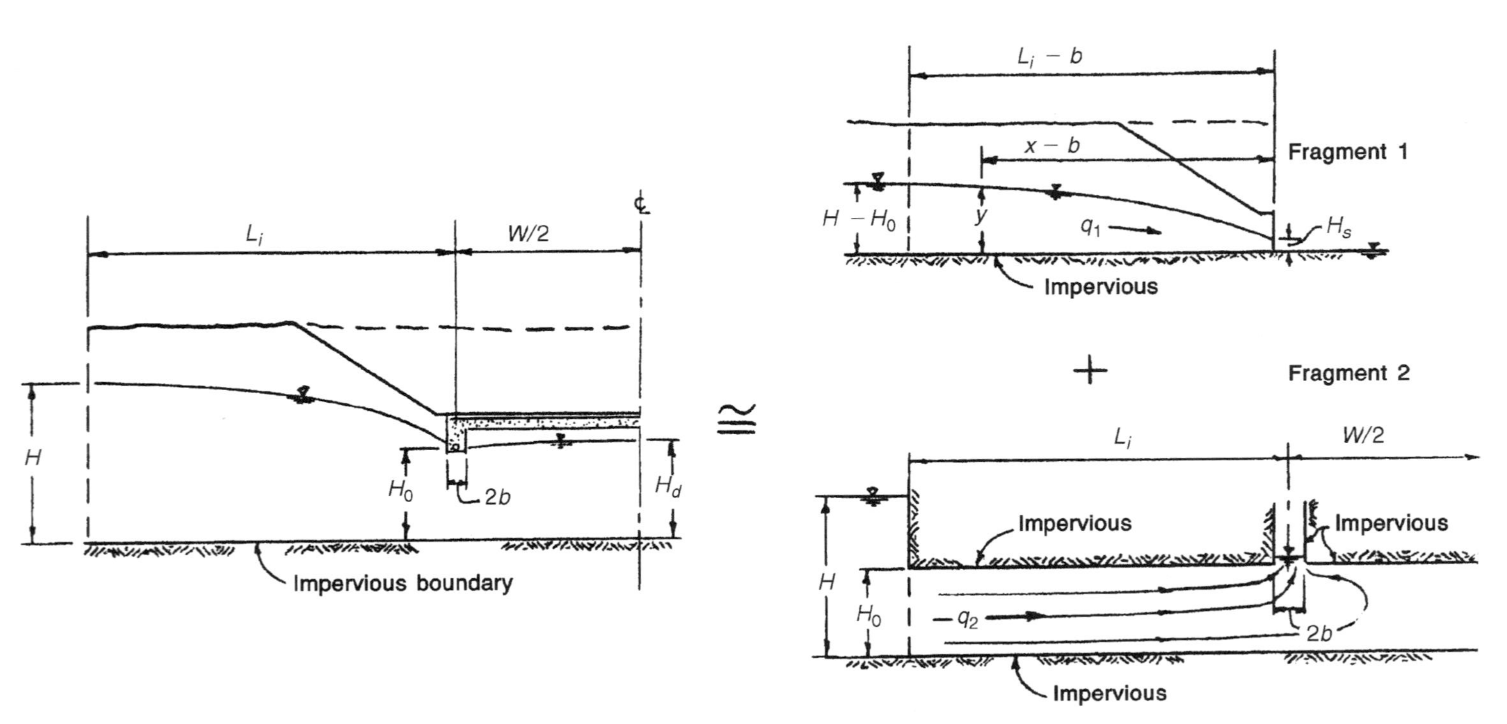

Fig. 9.33. Solution for a symmetrical drawdown drain problem using fragments. (Adapted from Moulton, 1980.)

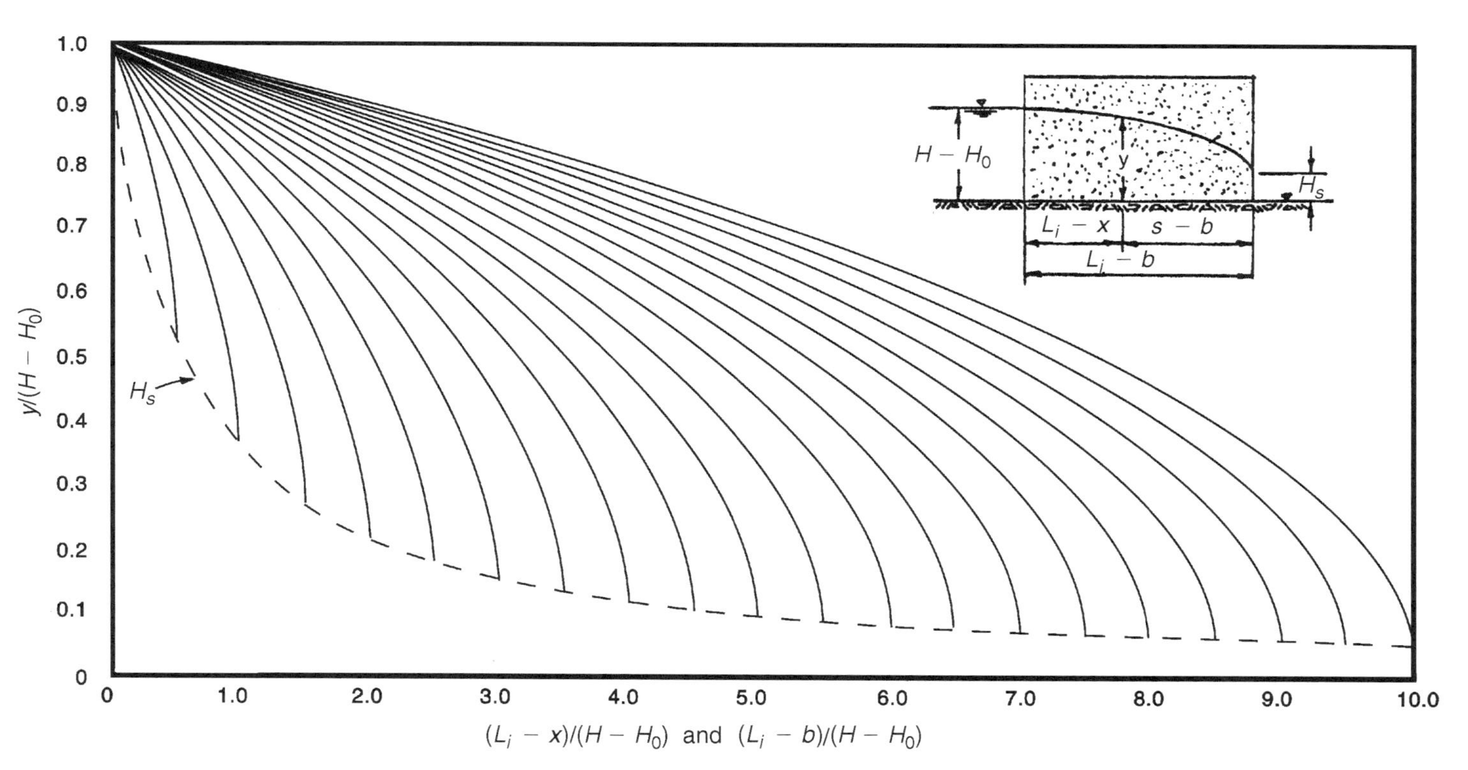

Fig. 9.34. Free water surface in the first fragment of Fig. 9.33. (Adapted from Moulton, 1980.)

$$q_1 = \frac{k(H - H_0)^2}{2(L_i - b)} \tag{9.8}$$

and

$$x = (L_i - b) + \frac{1}{2H_s m}\left[y\sqrt{y^2 - H_s^2 m^2} - (H - H_0)\sqrt{(H - H_0)^2 - H_s^2 m^2} - H_s^2 m^2 \log \frac{y + \sqrt{y^2 - H_s^2 m^2}}{(H - H_0) + \sqrt{(H - H_0)^2 - H_s^2 m^2}}\right] \tag{9.9}$$

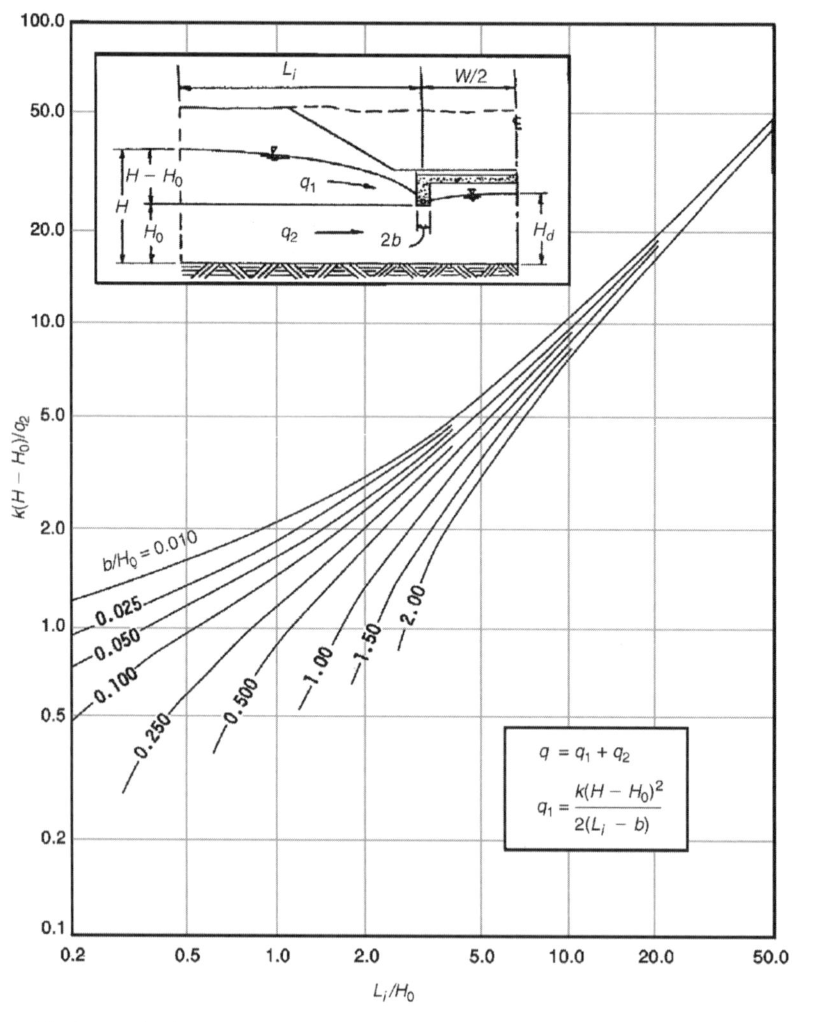

Fig. 9.35. Determination of flow rate in symmetrical underdrains. (Adapted from Moulton, 1980.)

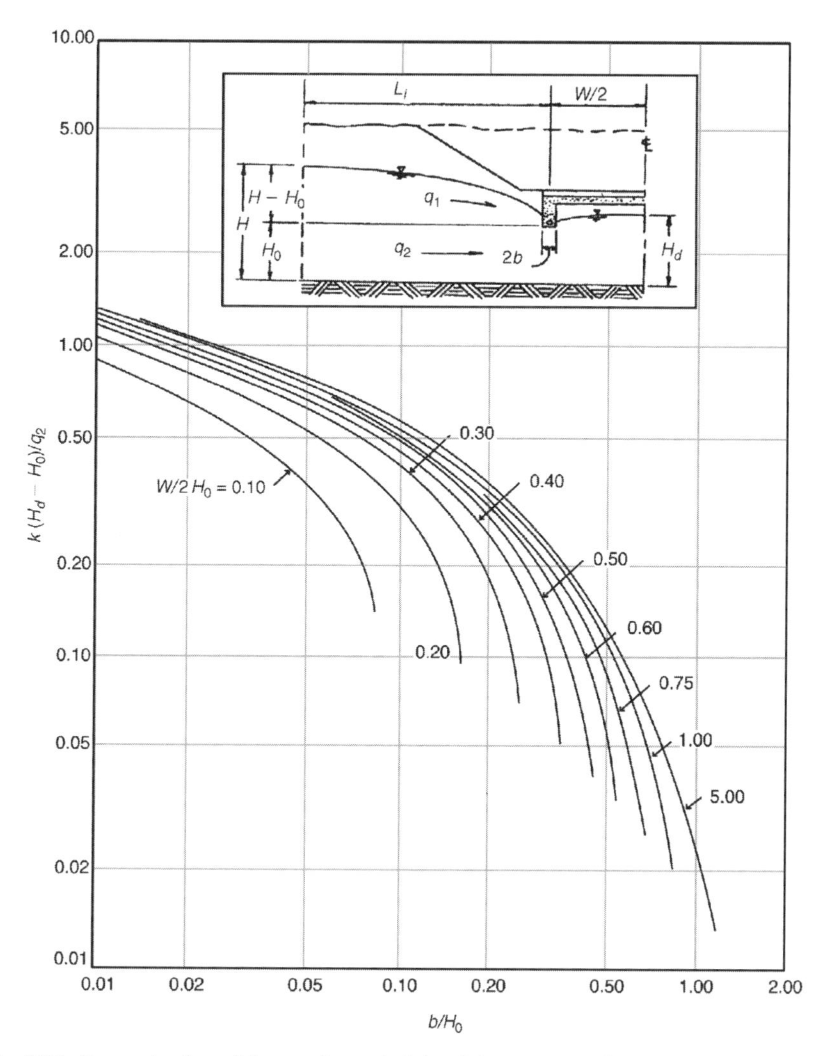

Fig. 9.36. Determination of the maximum height of free water surface between symmetrical underdrains. (Adapted from Moulton, 1980.)

where $m = 0.43\pi$ and H_s is as shown in Fig. 9.33. Figure 9.34 is a nomogram based on Eq. (9.9). In the case of fragment 2, q_2 is given by

$$q_2 = \frac{k(H - H_0)}{(L_i/H_0) - (1/\pi)\log\left[\frac{1}{2}\sinh(\pi b/H_0)\right]} \tag{9.10}$$

and the head at the roadway centerline, $H_d - H_0$, is expressed as

$$(H_d - H_0) = \frac{q_2}{\pi k}\log\left(\coth\frac{\pi b}{2H_0}\right) \tag{9.11}$$

Equations (9.10) and (9.11) are presented in the form of nomograms in Figs. 9.35 and 9.36, respectively. Figure 9.35 could be used to determine flow into the drain, whereas Fig. 9.36 could be used to estimate the maximum height of free water surface between symmetrical underdrains, H_d.

Example 9.1 Refer to Fig. 9.28. Given $H = 20$ ft, $H_0 = 10$ ft, $k = 1$ ft/day, and $S = 0.20$, (a) determine the flow rate into the drain, and (b) plot the drawdown curve.

SOLUTION: (a) To determine the flow rate into the drain using Fig. 9.30, we need to determine SL_i/H and H_0/H first. The influence length L_i is estimated using Eq. (9.7):

$$L_i = 3.8(H - H_0) = 3.8 \times (20\text{ ft} - 10\text{ ft}) = 38\text{ ft}$$

Therefore,

$$\frac{SL_i}{H} = \frac{0.2 \times 38\text{ ft}}{20\text{ ft}} = 0.38$$

and $H_0/H = 0.5$. From Fig. 9.30, $q/kHS = 1.3$, or

$$q = 1.3kHS = 1.3 \times 1\text{ ft/day} \times 20\text{ ft} \times 0.2 = 5.2\text{ ft}^2\text{/day}$$

(b) To plot the drawdown curve, we use either Eq. (9.4) or Fig. 9.31. In either case, H' is needed. From Fig. 9.30, $H' = 36$ ft. The drawdown curve obtained using this H' value is shown in Fig. E9.1.

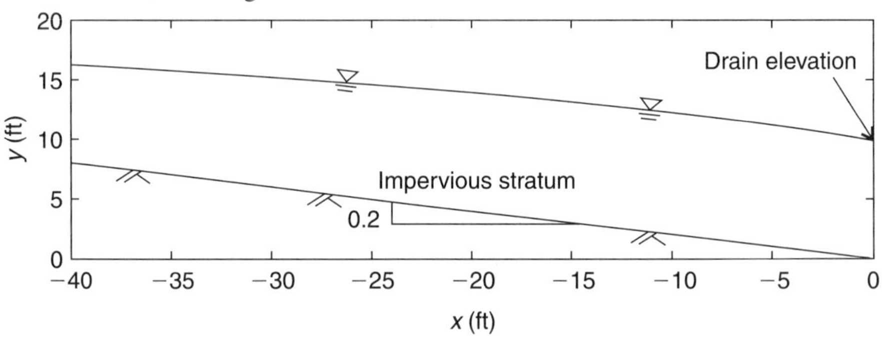

Fig. E9.1

9.5 DESIGN OF DEWATERING SYSTEMS

Design of dewatering systems using deep wells and well points involves determination of the number and configuration of wells/well points and the rate at which water must be pumped to achieve the required drawdown. The costs involved in a dewatering project vary greatly. A free-draining aquifer may be dewatered quickly with a few large wells, whereas a site consisting of stratified soils may require a large number of wells, often in conjunction with other dewatering methods. In small dewatering projects, the analytical solutions described in Chapter 4 could be used for determination of the discharge required to achieve a desired drawdown. Many of these solutions were derived using mathematical idealizations of complex aquifer geometry and invoking the Dupuit–Forchheimer assumptions. For situations that cannot be represented using these idealizations, numerical groundwater models may be necessary.

Many dewatering projects require installation of a number of wells and/or well points. As an approximate guide, nomograms were developed to determine well point spacing for uniform clean sands and gravels (Fig. 9.37) and for stratified clean sands

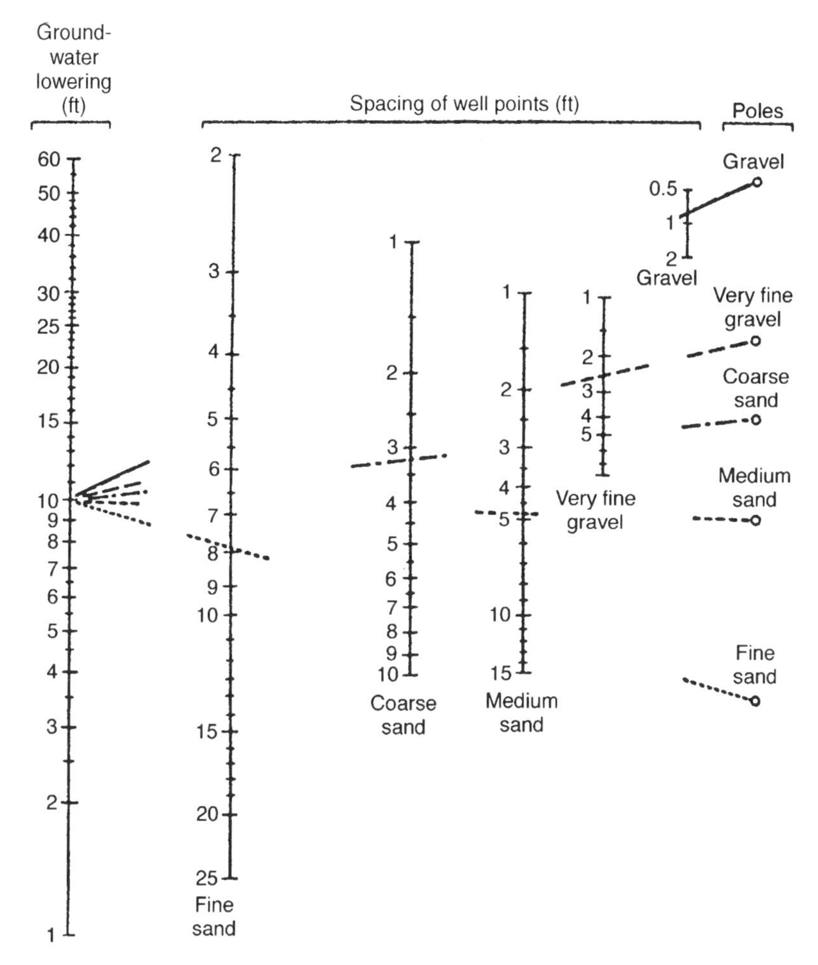

Fig. 9.37. Well point spacing for uniform clean sands and gravels. (Adapted from Moretrench Corp.)

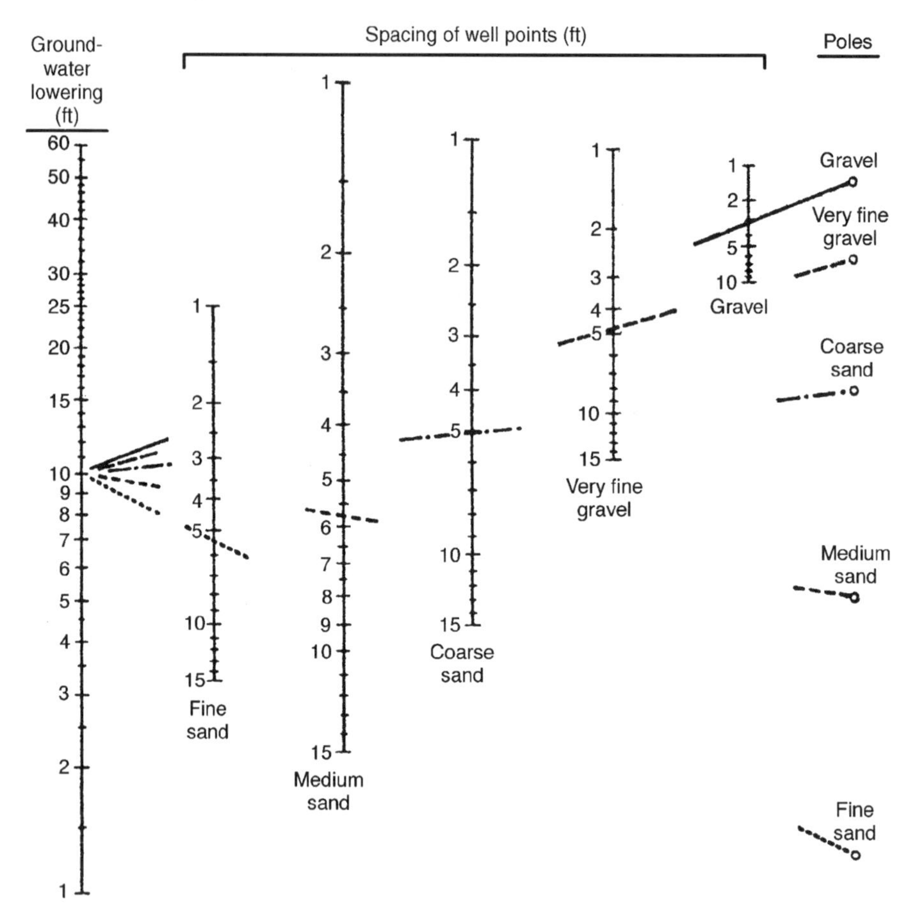

Fig. 9.38. Well point spacing for stratified clean sands and gravels. (Adapted from Moretrench Corp.)

and gravels (Fig. 9.38). These nomograms were based on empirical data, and more detailed analyses are necessary for accurate determination of well spacing. Formulas for multiple well systems, outlined below, should be used for large-scale projects and projects where accuracy is important.

Consider a system of fully penetrating wells with seepage from a circular source. The wells are located on a circular island surrounded by an unlimited source of water. For a system of three wells, the drawdown profiles for the two cases of artesian and gravity wells are shown in Fig. 9.39. The general equation for drawdown $(H - h)$ at any point P due to a system of artesian wells was first given by Forchheimer:

$$H - h = \frac{1}{2\pi kD}\left(q_{w1}\ln\frac{R_1}{r_1} + q_{w2}\ln\frac{R_2}{r_2} + \cdots + q_{wn}\ln\frac{R_n}{r_n}\right) \tag{9.12}$$

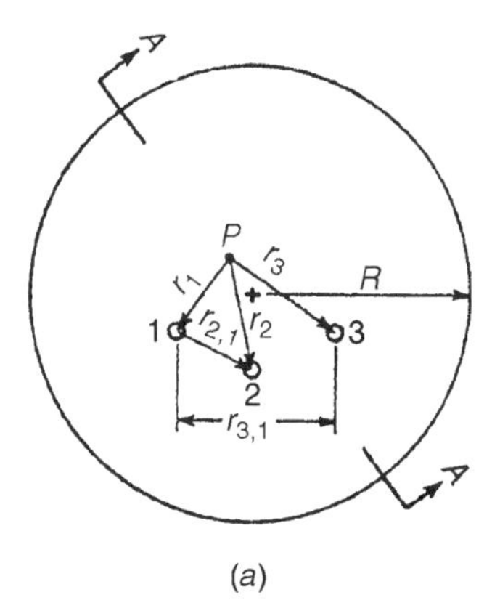

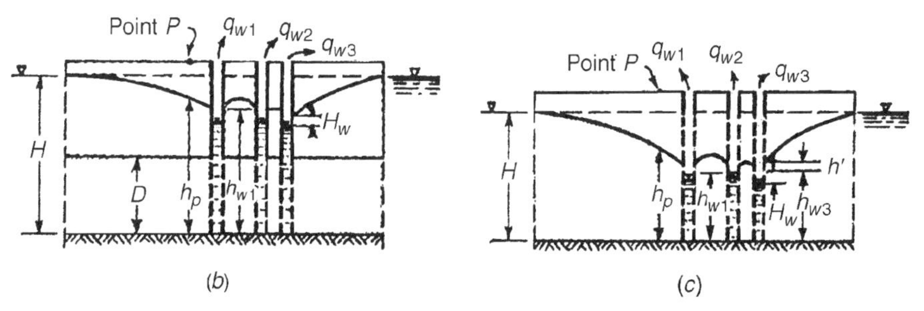

Fig. 9.39. Drawdown for a group of fully penetrating wells with seepage from a circular source: (*a*) plan; (*b*) section *A–A,* artesian case; (*c*) section *A–A,* gravity case. (Adapted from Mansur and Kaufman, 1962; copyright © The McGraw-Hill Companies; reprinted with permission.)

or

$$H - h = \frac{1}{2\pi kD} \sum_{i=1}^{n} q_{wi} \ln \frac{R_i}{r_i} \tag{9.13}$$

where q_{wi} is the discharge from ith well, R_i the radius of influence for ith well, r_i the distance from ith well to point P, and n the total number of wells in the group. The corresponding equation for a group of gravity wells was expressed as

$$H^2 - h^2 = \frac{1}{\pi k} \sum_{i=1}^{n} q_{wi} \ln \frac{R_i}{r_i} \tag{9.14}$$

The summation terms in Eqs. (9.13) and (9.14) denote the discharges and positions of wells in the system relative to the point where drawdown is needed. Using F to denote these summation factors, Eqs. (9.13) and (9.14) may be expressed, respectively, as

$$H - h = \frac{F}{2\pi kD} \tag{9.15}$$

and

$$H^2 - h^2 = \frac{F}{\pi k} \tag{9.16}$$

To determine the head h_{wj} at a specific well j in a system of fully penetrating wells, the following equations are used:

Artesian wells:

$$H - h_{wj} = \frac{1}{2\pi kD}\left(q_{wj} \ln \frac{R_j}{r_{wj}} + \sum_{i=1}^{n-1} q_{wi} \ln \frac{R_i}{r_{i,j}}\right) \tag{9.17}$$

Gravity wells:

$$H^2 - h_{wj}^2 = \frac{1}{\pi k}\left(q_{wj} \ln \frac{R_j}{r_{wj}} + \sum_{i=1}^{n-1} q_{wi} \ln \frac{R_i}{r_{i,j}}\right) \tag{9.18}$$

where q_{wj} is the flow from the well j, R_j the radius of influence of the well j, r_{wj} the effective well radius of well j, and $r_{i,j}$ the distance from each well to well j. Using F_w to represent summation terms, Eqs. (9.17) and (9.18) may be expressed, respectively, as

$$H - h_{wj} = \frac{F_w}{2\pi kD} \tag{9.19}$$

and

$$H^2 - h_{wj}^2 = \frac{F_w}{\pi k} \tag{9.20}$$

It is seen in Eqs. (9.17) and (9.18) that the summation on the right-hand side consists of two components. The first component expresses the drawdown due to pumping from well j, and the second component expresses drawdown due to pumping from the rest of the wells. This indicates that the drawdown at a given point is a linear combination of drawdowns caused at that point by all wells in the system computed as though the wells do not interfere with each other.

The general equations for drawdown in the case where seepage originates from a line source (Fig. 9.40) are:

Artesian wells:

$$H - h = \frac{F'}{2\pi kD} \tag{9.21}$$

Gravity wells:

$$H^2 - h^2 = \frac{F'}{\pi k} \tag{9.22}$$

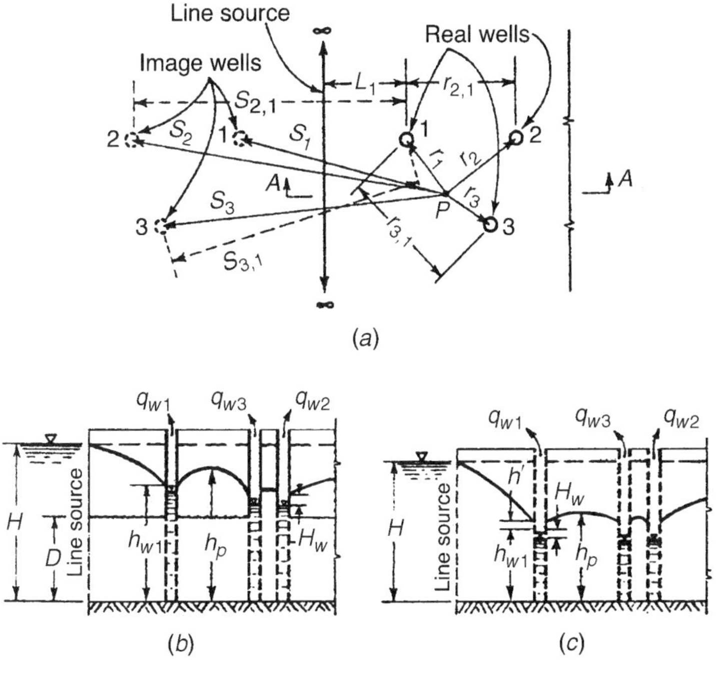

Fig. 9.40. Drawdown for a group of fully penetrating wells with seepage from a line source: (*a*) plan; (*b*) section *A—A*, artesian case; (*c*) section *A—A*, gravity case. (Adapted from Mansur and Kaufman, 1962; copyright © The McGraw-Hill Companies; reprinted with permission.)

where

$$F' = \sum_{i=1}^{n} q_{wi} \ln \frac{S_i}{r_i} \tag{9.23}$$

S_i is the distance from point P to image well i and r_i is the distance from point P to actual well i. The corresponding equations for determination of head h_{wj} at well j are:

Artesian wells:

$$H - h_{wj} = \frac{F'_w}{2\pi kD} \tag{9.24}$$

Gravity wells:

$$H^2 - h_{wj}^2 = \frac{F'_w}{\pi k} \tag{9.25}$$

where

$$F'_w = q_{wj} \ln \frac{2L_j}{r_{wj}} + \sum_{i=1}^{n-1} q_{wi} \ln \frac{S_{i,j}}{r_{i,j}} \tag{9.26}$$

where $2L_j$ is the distance from real well j to image well j, r_{wj} the effective radius of well j, $S_{i,j}$ the distance from real well j to image well i, and $r_{i,j}$ the distance from real well j to real well i. Note that $2L_j$ is taken as the radius of influence of well j in Eq. (9.26).

In cases where dewatering wells are placed in an orderly array, which is usually the case in practice, the equations above could be simplified further. Figures 9.41 and 9.42 present the specific solutions for wells placed in orderly arrays in the cases of circular and line sources, respectively.

In the case of wells where the source of seepage is neither a circular nor a line source of free water, the radius of influence of wells, R in the equations above, must be defined. When a free body source is far away from the dewatering wells, the radius of influence of a well may be defined as the radius of the circle beyond which the well has no significant influence on the original groundwater table. It would depend on the amount of drawdown at the well, duration of dewatering, and the properties of soil (permeability and porosity) at the site. The following empirical equation proposed by Sichardt may be used to estimate R in such cases:

$$R = CS\sqrt{k} \tag{9.27}$$

where R is the radius of influence (ft), S the drawdown at the well (ft), k the coefficient of permeability (cm/s), and C is a dimensionless constant, equal to 3 for gravity and artesian wells. The U.S Army Corps of Engineers found Eq. (9.27) to adequately represent the radius of influence for a single line of well points when a range of 1.5 to 2.0 was used for C. Since a coefficient of permeability depends on the grain size of soils, the radius of influence can also be expressed in terms of the effective grain size, D_{10}. Using a relationship between k and D_{10} proposed by the U.S. Army Corps of Engineers, Mansur and Kaufman (1962) presented a nomogram (Fig. 9.44) for R versus D_{10} with a drawdown of 10 ft.

Another key parameter in the design of dewatering systems is the discharge from a single well, q_w. In determining the capacity of a single well, the possibility of turbulence at the well and the associated filter instability must be considered. To minimize this possibility, Sichardt proposes that the entry gradients in the case of gravity wells should not exceed

$$i_{\max} = \frac{1}{15\sqrt{k}} \tag{9.28}$$

where k is the coefficient of permeability (m/s). Using Darcy's law, the allowable discharge from a gravity well may be expressed in terms of maximum entry gradient $i_{\max}$ as

$$q_w = 2\pi r_w h_w k i_{\max} \tag{9.29}$$

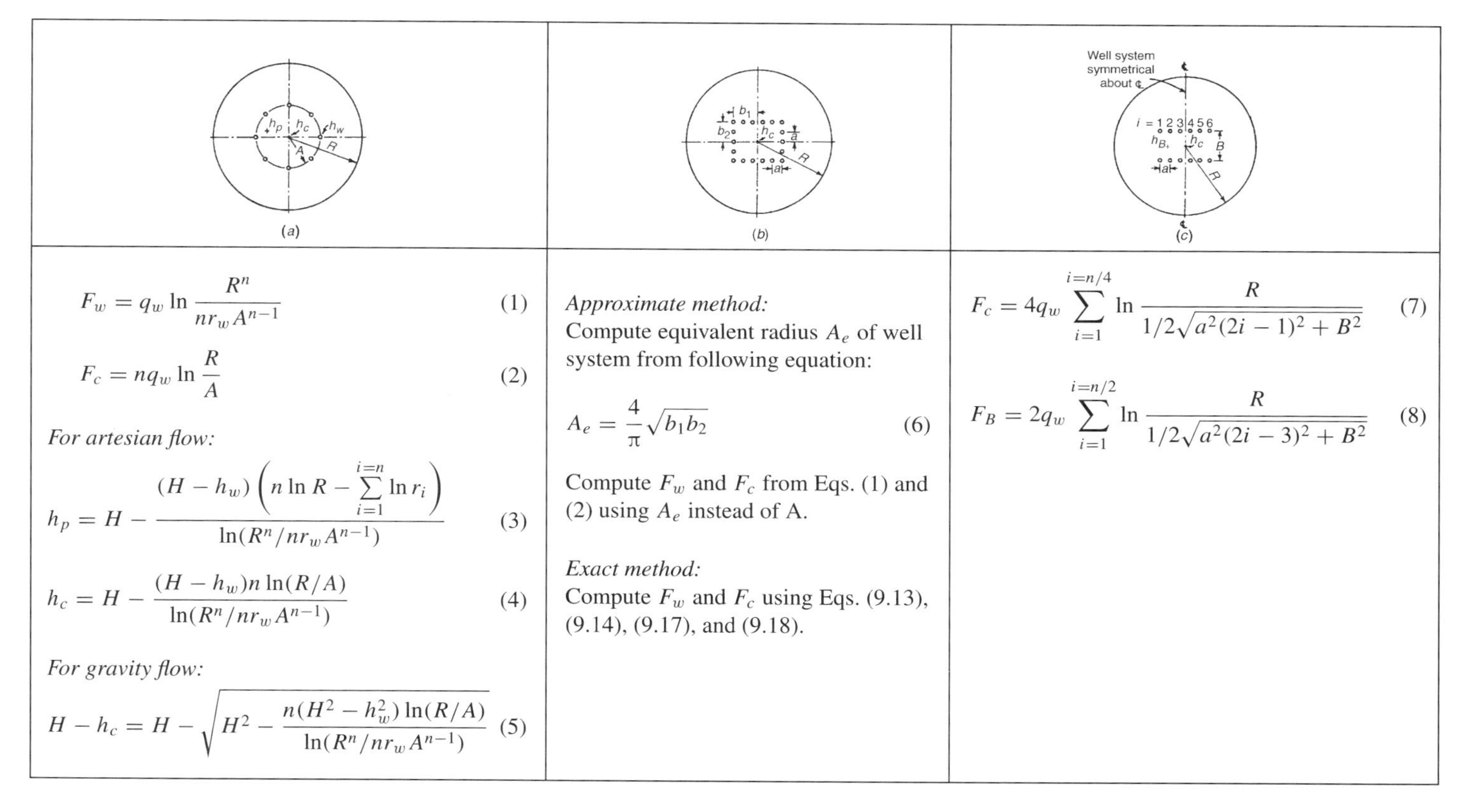

$$F_w = q_w \ln \frac{R^n}{n r_w A^{n-1}} \quad (1)$$

$$F_c = n q_w \ln \frac{R}{A} \quad (2)$$

For artesian flow:

$$h_p = H - \frac{(H - h_w)\left(n \ln R - \sum_{i=1}^{i=n} \ln r_i\right)}{\ln(R^n / n r_w A^{n-1})} \quad (3)$$

$$h_c = H - \frac{(H - h_w) n \ln(R/A)}{\ln(R^n / n r_w A^{n-1})} \quad (4)$$

For gravity flow:

$$H - h_c = H - \sqrt{H^2 - \frac{n(H^2 - h_w^2)\ln(R/A)}{\ln(R^n / n r_w A^{n-1})}} \quad (5)$$

Approximate method:
Compute equivalent radius A_e of well system from following equation:

$$A_e = \frac{4}{\pi}\sqrt{b_1 b_2} \quad (6)$$

Compute F_w and F_c from Eqs. (1) and (2) using A_e instead of A.

Exact method:
Compute F_w and F_c using Eqs. (9.13), (9.14), (9.17), and (9.18).

$$F_c = 4 q_w \sum_{i=1}^{i=n/4} \ln \frac{R}{1/2\sqrt{a^2(2i-1)^2 + B^2}} \quad (7)$$

$$F_B = 2 q_w \sum_{i=1}^{i=n/2} \ln \frac{R}{1/2\sqrt{a^2(2i-3)^2 + B^2}} \quad (8)$$

Fig. 9.41. Analytical solutions for drawdown at groups of wells with a circular source. (Adapted from Mansur and Kaufman, 1962; copyright © The McGraw-Hill Companies; reprinted with permission.)

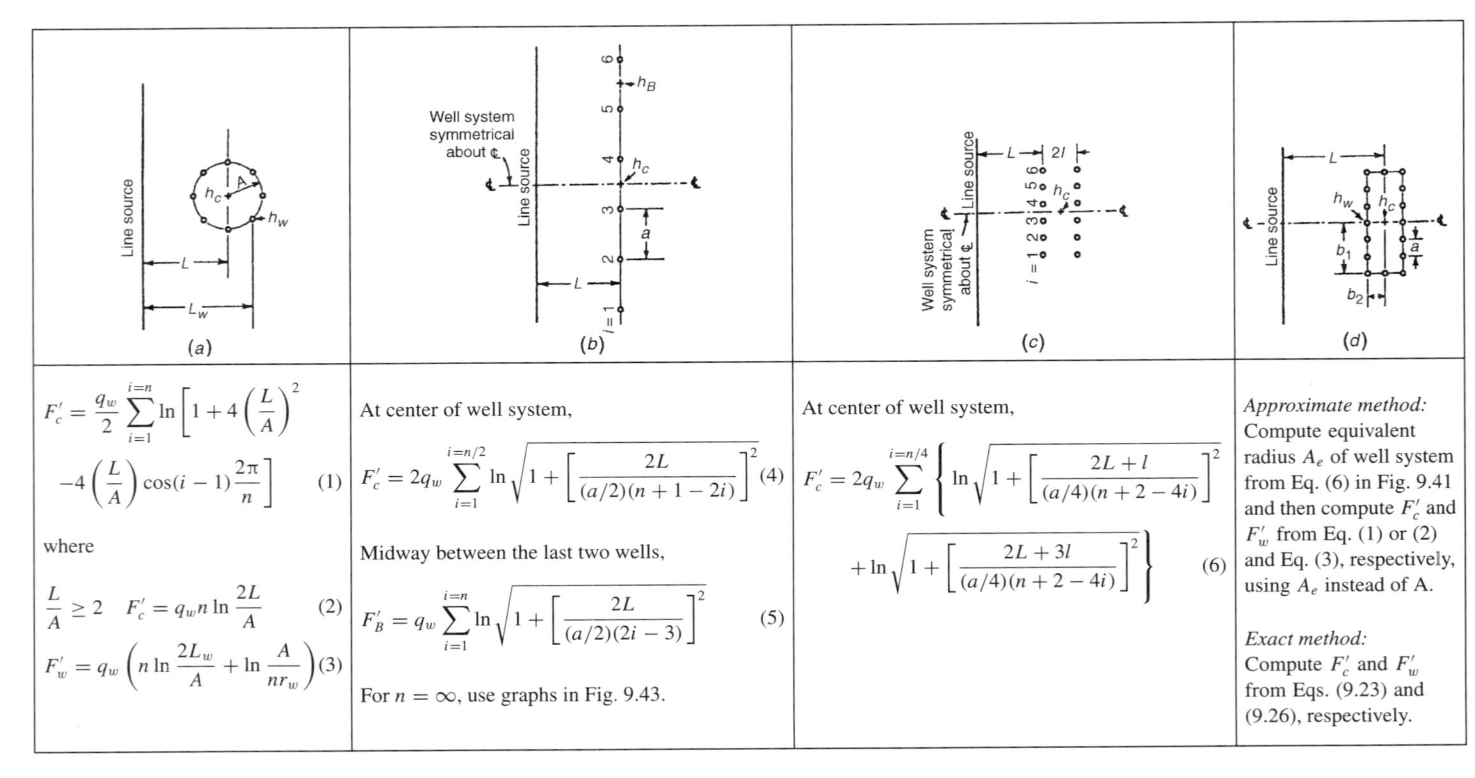

Fig. 9.42. Analytical solutions for drawdown at groups of wells with a line source. (Adapted from Mansur and Kaufman, 1962; copyright © The McGraw-Hill Companies; reprinted with permission.)

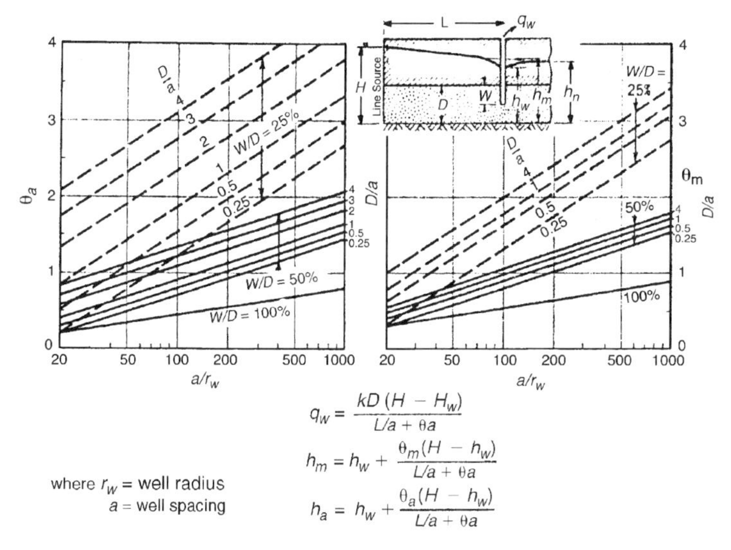

$$q_w = \frac{kD(H - H_w)}{L/a + \theta a}$$

$$h_m = h_w + \frac{\theta_m(H - h_w)}{L/a + \theta a}$$

$$h_a = h_w + \frac{\theta_a(H - h_w)}{L/a + \theta a}$$

Fig. 9.43. Factors for computing flow to and head reduction from an infinite line of artesian wells with a line source of seepage (refer to Fig. 9.42). (Adapted from Mansur and Kaufman, 1962; copyright © The McGraw-Hill Companies; reprinted with permission.)

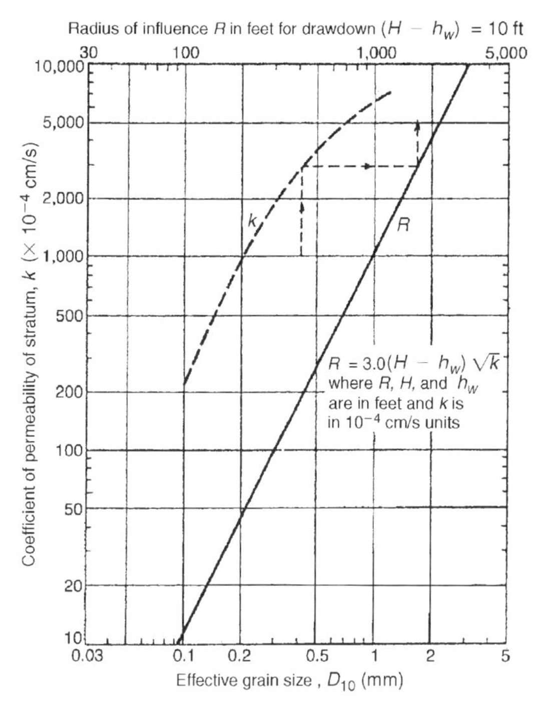

Fig. 9.44. Coefficient of permeability and radius of influence versus effective grain size. (Adapted from Mansur and Kaufman, 1962; copyright © The McGraw-Hill Companies; reprinted with permission.)

where r_w is the radius of the well and h_w the height of water in the well. Substituting Eq. (9.28) for the entry gradient in Eq. (9.29), q_w should therefore be limited to

$$q_w < 2\pi r_w h_w \frac{\sqrt{k}}{15} \tag{9.30}$$

In arriving at the drawdown for single wells using the Dupuit–Forchheimer well discharge formula, the discharge should be limited using the Eq. (9.30). In other words,

$$q_w = \frac{\pi k \left(H^2 - h_w^2\right)}{\ln(R/r_w)} < 2\pi r_w h_w \frac{\sqrt{k}}{15} \tag{9.31}$$

Example 9.2 Refer Fig. E9.2*a*. Design a system of gravity wells along the circular border of an excavation to achieve a drawdown of 20 ft at the center of the excavation. Limit the individual well discharge to 1000 gpm, and assume all wells are of the same radius, $r_w = 1$ ft. Use $k = 200$ ft/day.

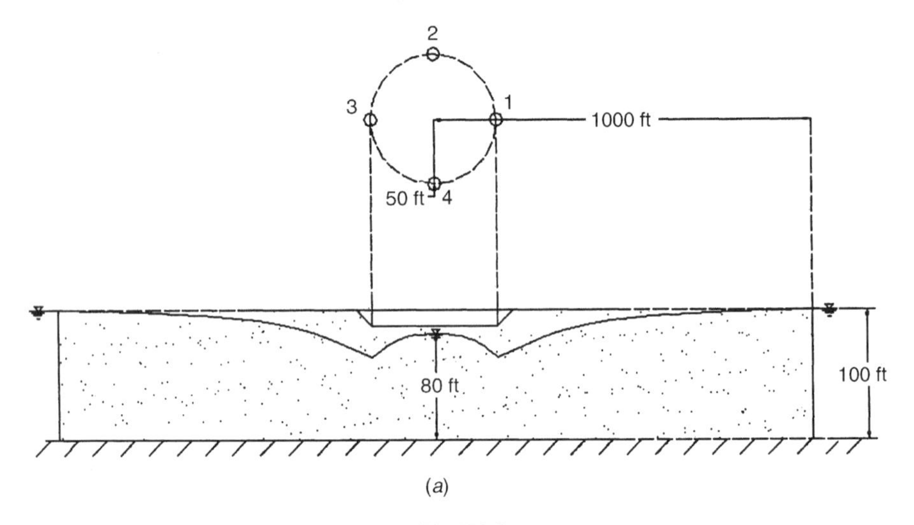

Fig. E9.2

SOLUTION: Estimate the total flow required for an equivalent well of radius 50 ft. The total flow required is estimated as

$$\begin{aligned} q_T &= \frac{\pi k \left(H^2 - h_w^2\right)}{\ln(R/r_w)} \\ &= \frac{\pi \times 200 \text{ ft/day} \times \left[(100 \text{ ft})^2 - (80 \text{ ft})^2\right]}{\ln(1000 \text{ ft}/50 \text{ ft})} \\ &\approx 7.5 \times 10^5 \text{ ft}^3\text{/day} \\ &= 3.9 \times 10^3 \text{ gpm} \end{aligned}$$

Estimate the number of wells needed as

$$n = \frac{q_T}{q_w} = \frac{3.9 \times 10^3 \text{ gpm}}{1000 \text{ gpm}} > 3$$

Use four wells. To check the adequacy of the system, the head is estimated at the center h_c and at each well. The head at the center is

$$h_c = \sqrt{H^2 - \frac{F_c}{\pi k}}$$

where

$$\begin{aligned} F_c &= n q_w \ln \frac{R_i}{r_i} \\ &= 4 \times 1000 \text{ gpm} \times \ln \frac{1000 \text{ ft}}{50 \text{ ft}} \\ &= 1.2 \times 10^4 \text{ gpm} \\ &\approx 2.3 \times 10^6 \text{ ft}^3/\text{day} \end{aligned}$$

so

$$h_c = \sqrt{(100 \text{ ft})^2 - \frac{2.3 \times 10^6 \text{ ft}^3/\text{day}}{\pi \times 200 \text{ ft/day}}} \approx 79.6 \text{ ft} < 80 \text{ ft} \qquad \text{OK}$$

The head at well 1 is

$$h_{w1} = \sqrt{H^2 - \frac{F_w}{\pi k}}$$

where

$$\begin{aligned} F_w &= q_w \ln \frac{R}{r_{w1}} + \sum_{i=2}^{4} q_w \ln \frac{R}{r_{i,1}} \\ &= q_w \left(\ln \frac{R}{r_{w1}} + \ln \frac{R}{r_{2,1}} + \ln \frac{R}{r_{3,1}} + \ln \frac{R}{r_{4,1}} \right) \\ &= q_w \ln \frac{R^4}{r_{w1} r_{2,1} r_{3,1} r_{4,1}} \end{aligned}$$

In Fig. E9.2*b*, $r_{2,1} = r_{4,1} = \sqrt{2}\, r_i = \sqrt{2} \times 50 \text{ ft} \approx 70.7 \text{ ft}$, and $r_{3,1} = 2r_i = 2 \times 50 \text{ ft} = 100 \text{ ft}$. Therefore,

$$\begin{aligned} F_w &= 1000 \text{ gpm} \times \frac{192.5 \text{ ft}^3/\text{day}}{1 \text{ gpm}} \times \ln \frac{(1000 \text{ ft})^4}{1 \text{ ft} \times 70.7 \text{ ft} \times 100 \text{ ft} \times 70.7 \text{ ft}} \\ &\approx 2.8 \times 10^6 \text{ ft}^3/\text{day} \end{aligned}$$

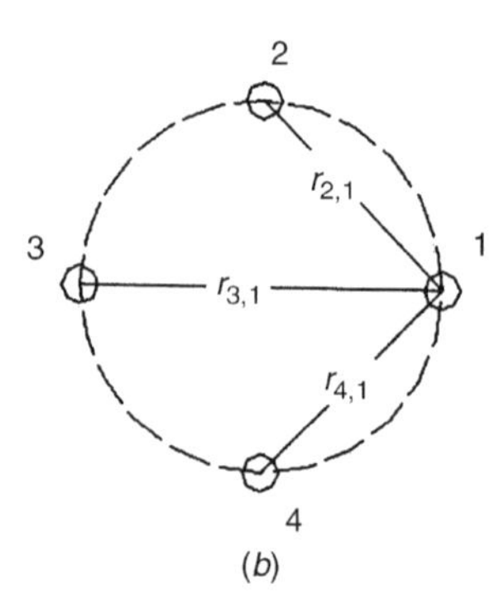

Fig. E9.2

and

$$h_{w1} = \sqrt{(100\text{ ft})^2 - \frac{2.8 \times 10^6\text{ ft}^3/\text{day}}{\pi \times 200\text{ ft/day}}} \approx 74.5\text{ ft} < 80\text{ ft} \qquad \text{OK}$$

In Fig. E9.2*b*, $h_{w,1} = h_{w,2} = h_{w,3} = h_{w,4}$, due to symmetry. So the head at each well is lower than 80 ft. Hence the design meets the dewatering requirement.

Example 9.3 Repeat Example 9.2 with the line source shown in Fig. E9.3*a*.

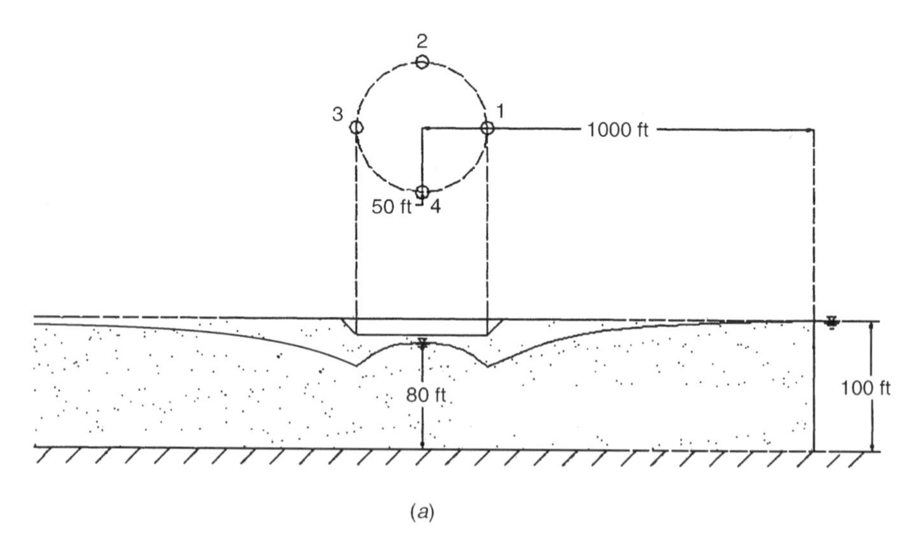

Fig. E9.3

SOLUTION: The total flow required is estimated as

$$
\begin{aligned}
q_T &= \frac{\pi k\left(H^2 - h_c^2\right)}{\ln(2L/r_w)} \\
&= \frac{\pi \times 200 \text{ ft/day} \times \left[(100 \text{ ft})^2 - (80 \text{ ft})^2\right]}{\ln[(2 \times 1000 \text{ ft})/50 \text{ ft}]} \\
&\approx 6.1 \times 10^5 \text{ ft}^3\text{/day} \\
&= 3.2 \times 10^3 \text{ gpm}
\end{aligned}
$$

Estimate the number of wells needed as

$$
n = \frac{q_T}{q_w} = \frac{3.2 \times 10^3 \text{ gpm}}{1000 \text{ gpm}} > 3
$$

Use four wells.

To check the adequacy of the system, the head is estimated at the center h_c and at each well. The head at the center is

$$
h_c = \sqrt{H^2 - \frac{F_c'}{\pi k}}
$$

where

$$
\begin{aligned}
F_c' &= n q_w \ln \frac{S_i}{r_i} \\
&= 4 \times 10^3 \text{ gpm} \times \ln \frac{2 \times 1000 \text{ ft}}{50 \text{ ft}} \\
&\approx 1.5 \times 10^4 \text{ gpm} \\
&\approx 2.9 \times 10^6 \text{ ft}^3\text{/day}
\end{aligned}
$$

so

$$
h_c = \sqrt{(100 \text{ ft})^2 - \frac{2.9 \times 10^6 \text{ ft}^3\text{/day}}{\pi \times 200 \text{ ft/day}}} \approx 73.4 \text{ ft} < 80 \text{ ft} \qquad \text{OK}
$$

The head at well 1 is

$$
h_{w1} = \sqrt{H^2 - \frac{F_w'}{\pi k}}
$$

where

$$
F_w' = q_w \ln \frac{2L}{r_{w1}} + \sum_{i=2}^{4} q_{wi} \ln \frac{S_{i,1}}{r_{i,1}}
$$

$$= q_w \ln \frac{2LS_{2,1}S_{3,1}S_{4,1}}{r_{w1}r_{2,1}r_{3,1}r_{4,1}}$$

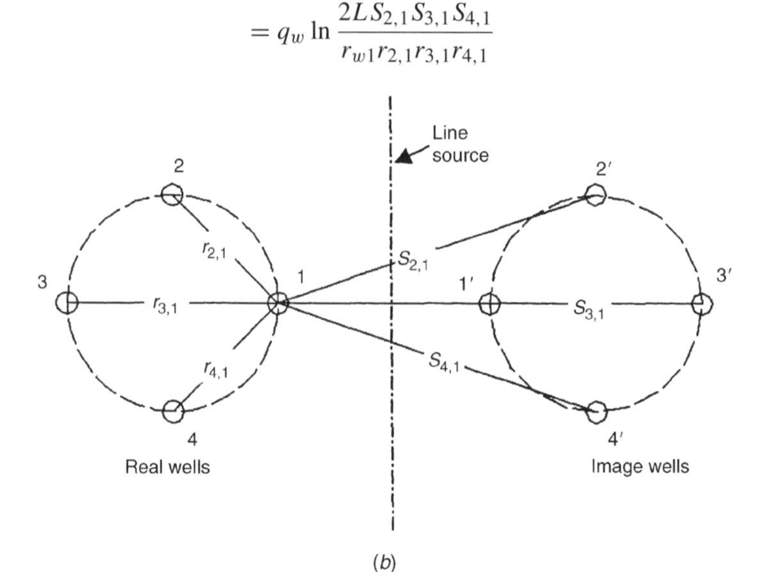

Fig. E9.3

From Fig. E9.3*b*, $r_{2,1} = r_{4,1} = \sqrt{2}\, r_i = \sqrt{2} \times 50$ ft ≈ 70.7 ft, and $r_{3,1} = 2r_i = 2 \times 50$ ft $= 100$ ft, $S_{2,1} = S_{4,1} \approx \sqrt{(50\text{ ft})^2 + (1950\text{ ft})^2} = 1951$ ft, $S_{3,1} = 2000$ ft, and $2L_1 = 2000$ ft $- 100$ ft $= 1900$ ft. Therefore,

$$F'_w = 1000\text{ gpm} \times \frac{192.5\text{ ft}^3/\text{day}}{1\text{ gpm}} \times \ln \frac{1900\text{ ft} \times 195\text{ ft} \times 2000\text{ ft} \times 1951\text{ ft}}{1\text{ ft} \times 70.7\text{ ft} \times 100\text{ ft} \times 70.7\text{ ft}}$$

$$\approx 3.3 \times 10^6\text{ ft}^3/\text{day}$$

so

$$h_{w1} = \sqrt{(100\text{ ft})^2 - \frac{3.3 \times 10^6\text{ ft}^3/\text{day}}{\pi \times 200\text{ ft/day}}} \approx 68.9\text{ ft} < 80\text{ ft} \qquad \text{OK}$$

The head at well 1 is higher than at the other wells. Hence the design meets the requirement of dewatering. It is left as an exercise to the reader to try three wells instead of four.

PROBLEMS

9.1. Determine the flow rate into shallow drains keyed into a pervious stratum if $H_w = 5$ ft, $S = 10$ ft, $D = 10$ ft, and $k = 1$ ft/day.

9.2. Determine the change in flow rate if the shallow drains in Problem 9.1 are keyed into an impervious stratum.

9.3. Refer to Example 9.1. Use flow nets to determine the flow rate into the drain and to plot the location of the drawdown curve. Compare the solutions.

9.4. Refer to Example 9.1. How would the flow rate change as the drain is lowered from $H_0 = 10$ ft to 5 ft?

9.5. Using dimensions shown in Fig. 9.32, determine flow rates into the drains. Use $k = 1$ ft/day.

9.6. Redo Problem 9.5 using flow nets.

9.7. Refer to Fig. 9.33. For $W = 30$ ft, $H = 30$ ft, $H_0 = 20$ ft, and $b = 0.5$ ft, determine the design flow rate to the drain and plot the drawdown curve. Use $k = 2$ ft/day. Also determine maximum height of the free water surface between the symmetrical underdrains, H_d. (*Note:* Problems 10.1 and 10.2 are related to this problem.)

9.8. Redo Problem 9.7 using flow nets.

9.9. Given a scale of 1 in. = 1000 ft for Fig. 9.40*a*, compute F' corresponding to point P and F'_w, corresponding to well 1. Assume that the discharge from all the wells is constant at 50 cfm, and all wells are of the same radius, 1 ft.

9.10. Assuming that the wells are gravity type in Problem 9.9, determine drawdown at well 1 and at point P if $H = 75$ ft. The coefficient of permeability $k = 3$ ft/day.

9.11. Redo the design in Example 9.2 for $h_c = 70$ ft.

9.12. Redo the design in Example 9.3 for $h_c = 70$ ft.

CHAPTER 10

SUBSURFACE DRAINAGE IN PAVEMENT SYSTEMS

Subsurface drainage in pavement systems is an essential element of highway design and maintenance. Groundwater is encountered at many highway construction sites in wet cuts and in hilly terrains. At such sites, seepage should be controlled, not only to facilitate pavement construction but also to keep the pavements dry during their operation. There are several sources of seepage at pavements. Rainfall and runoff water may infiltrate through pavement cracks, moisture from the underlying water table may flow upward into the pavements as a result of capillary action and evaporation; and seepage may occur from higher ground and from ditches or creeks in the vicinity.

The importance of keeping pavements dry seemed to be well recognized by the Romans, Greeks, and Egyptians in their road construction thousands of years ago. Nearly 200 years ago, in his report to the London Board of Agriculture, John McAdam, a popular British pavement designer, cautioned against overemphasis on the structural strength of pavements (in lieu of drainage provision): "The erroneous opinion . . . that a road may be made sufficiently strong, artificially, to carry carriages though the subsoil be in a wet state . . . has produced most of the defects of the roads of Great Britain."

Despite such early recognitions, the importance of drainage in pavements was not always well accepted, and it has been a subject of intense debate in the past 50 years. Harry Cedergren led a crusade in the last half of the twentieth century to highlight the need for adequate drainage provision in pavement design. Using FHWA repair cost data, he estimated that poor drainage practices represent losses of $15 billion a year in the United States (Cedergren, 1994). Cedergren likened poorly drained U.S. pavements to the world's longest bathtubs, and pointed out that damage from traffic and the environment can reduce the life of undrained pavements to one-third or less of the life of well-drained pavements.

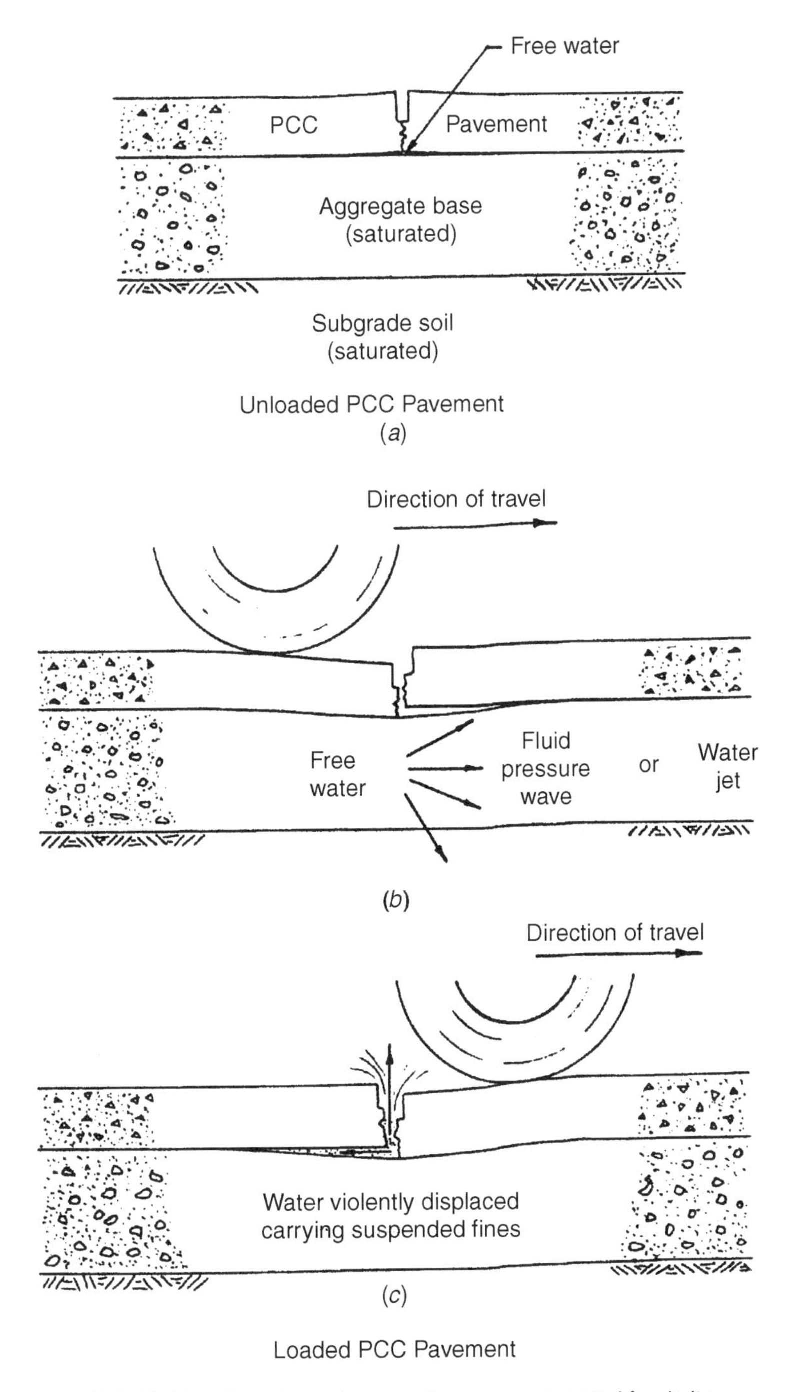

Fig. 10.1. Pumping phenomena under portland cement concrete pavements. (Adapted from Cedergren et al., 1973.)

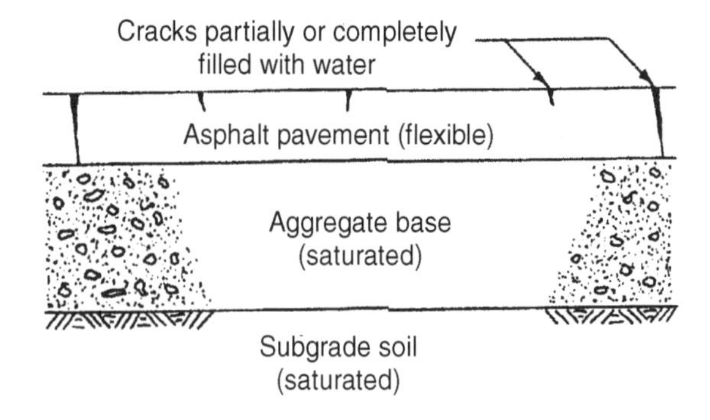

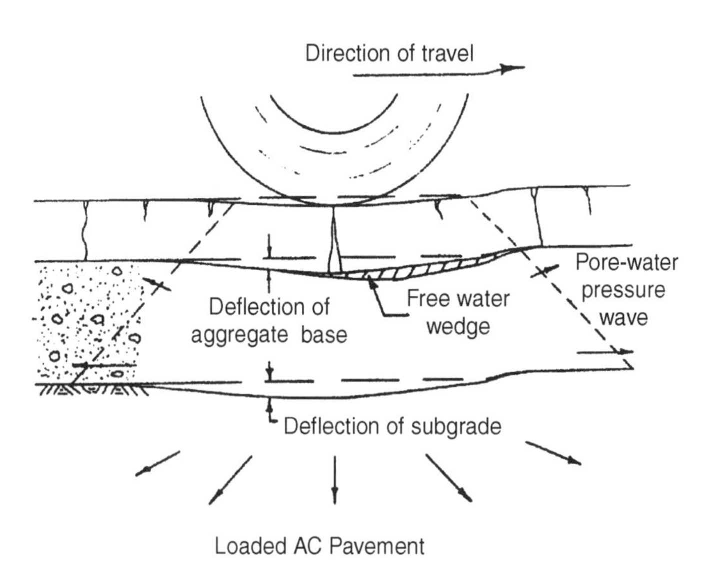

Fig. 10.2. Action of free water in AC pavement structural sections under dynamic loading. (Adapted from Cedergren et al., 1973.)

To understand how water influences the life of pavement systems, consider Figs. 10.1 and 10.2. The dynamic loads on the pavement due to traffic cause momentarily high pore pressures in the base layer. Figure 10.1 shows the commonly accepted mechanism of portland cement concrete (PCC) pavement distress due to the presence of undrained water. Due to approaching wheel load, the trailing edge of the slab at a crack deflects downward, sending a fluid pressure wave or water jet in a forward direction. As the wheel passes over the joint, the trailing slab rebounds upward and the leading edge of the next slab is deflected downward. This results in ejection of water and consequent pumping and erosion of fines from underneath the slab. Repeated pumping of water and fines eventually causes the pavement slabs to crack due to lack of adequate subgrade support. In asphalt concrete (AC) pavement systems, the pumping effect is generally not as dramatic (Fig. 10.2). Although the ejection of water and pumping of fines may still occur in these systems, the flexible slabs may fail by cracking before localized ejection of water occurs.

Perhaps because of work in this area by Cedergren and others, it is now generally accepted that accumulation of free water in pavements is detrimental to their performance. Given that we cannot keep water away from the pavements, we must take steps to prevent water from accumulating under the pavements and drain it. A layer of high-permeability material becomes essential under the pavement sections to drain water out of the structure. The other components of pavement drainage design are a filter or separator underneath the drainage layer to prevent soil particles from penetrating the drainage layer, and collector pipes and outlet pipes to provide outlet for water drained from the drainage layer. In this chapter we look into the various pathways of water flow into and out of pavements and outline the design methods for pavement drainage.

10.1 CLASSIFICATION OF DRAINAGE SYSTEMS

Pavement drainage systems may be classified in three different ways based on the source of subsurface water they are designed to control, the function they perform, and their location and geometry.

10.1.1 Classification Based on Source of Drained Water

There are two primary sources of water that drainage systems are designed to control: water infiltrating from pavement sections and cracks, and groundwater seeping from the vicinity toward the pavement systems. Often, drainage systems are designed to drain water from both of these sources. An *infiltration control system* designed to drain the first source contains a drainage layer under the pavement section; however, a *groundwater control system* designed to drain the groundwater seepage may consist of one or more interceptor drains (Fig. 10.3) in addition to the drainage layer underneath the pavement section.

10.1.2 Classification Based on Function

Pavement drainage systems generally perform one or more of three functions: interception or cutoff of seepage, drawdown or lowering of the water table, and collection of flow from drainage layers. Often, drainage systems serve more than one function. The examples shown in Fig. 10.3*b* and *c* serve multiple functions of lowering the water table and collecting flow from drainage layers.

10.1.3 Classification Based on Location and Geometry

This is the most common classification basis used to designate pavement drainage systems. The most familiar types of drains under this classification are (1) *longitudinal drains* located parallel to the roadway in both horizontal and vertical alignment

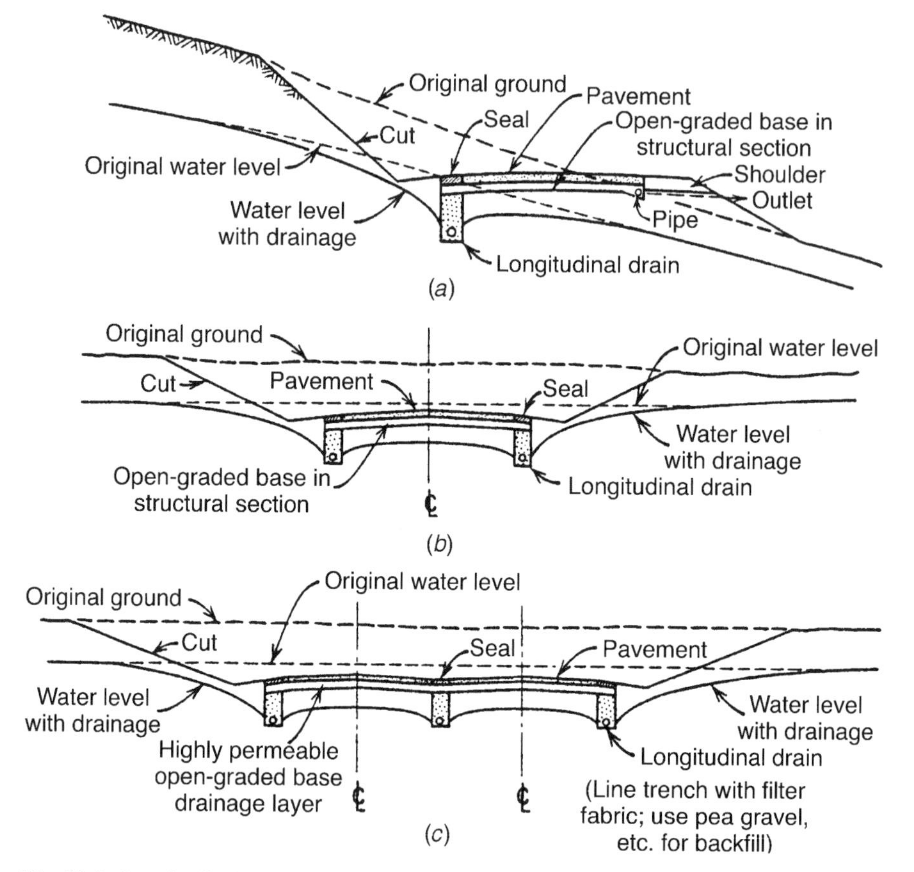

Fig. 10.3. Longitudinal highway drains: (*a*) sidehill construction; (*b*) narrow road in flat terrain; (*c*) wide road in flat terrain. (Adapted from Cedergren, 1989; copyright © John Wiley & Sons; reprinted with permission.)

(examples shown in Fig. 10.3); (2) *transverse and horizontal drains*, which run laterally beneath the roadway designed to drain infiltration and groundwater in bases and subbases when flow takes place more in the longitudinal direction than in the lateral direction (Fig. 10.4); and (3) *drainage blankets*, which are permeable layers provided beneath pavement structures to drain groundwater from artesian sources (Fig. 10.5*a*) or infiltered water alone (Fig. 10.5*b*).

10.2 DESIGN METHODOLOGY

The design elements in a pavement drainage system are drainage layers underneath pavement sections, filters and separators to protect the drainage layers, and collection systems to provide outlet for water in the drainage layers. Moulton (1980) recommends a systematic procedure for analysis and design of pavement drainage systems. The procedure, modified here to include design of filters, is as follows.

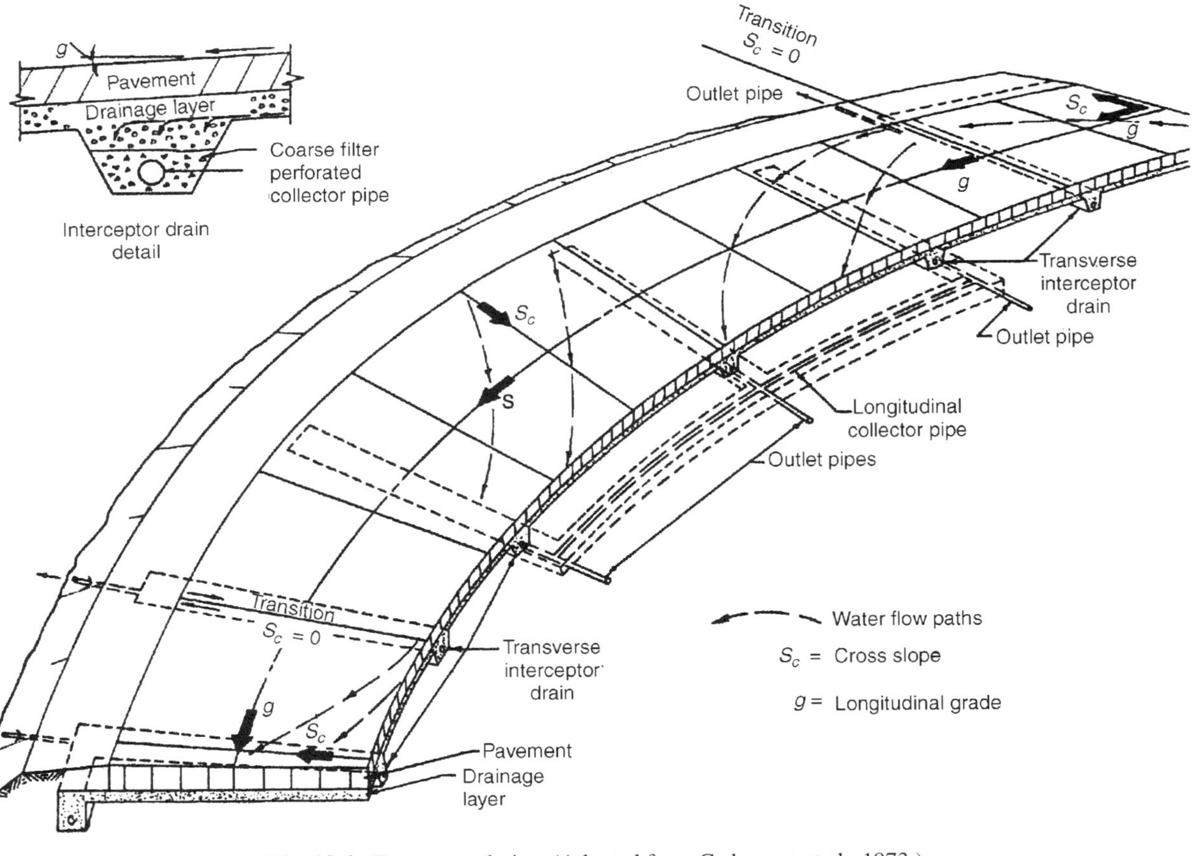

Fig. 10.4. Transverse drains. (Adapted from Cedergren et al., 1973.)

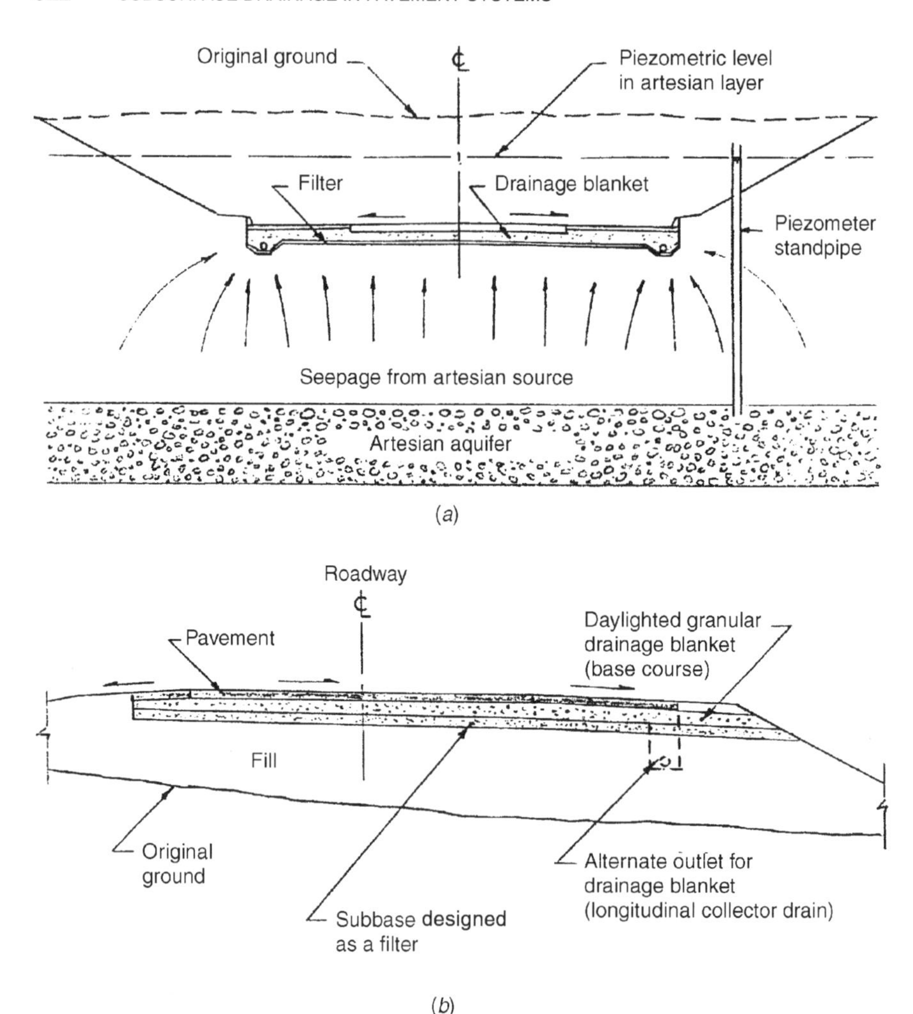

Fig. 10.5. Applications of horizontal drainage blankets: (*a*) drainage blanket for seepage from an artesian source; (*b*) drainage blanket for infiltered water alone. (Adapted from Moulton, 1980.)

Step 1: Assemble all available data on highway geometry and surface topography, soil characteristics, groundwater table elevation, precipitation, frost penetration, and any other miscellaneous considerations.

Step 2: Determine the quantity of water inflow into and outflow from pavement sections, considering all possible sources of water.

Step 3: Analyze and/or design pavement drainage layers, in accordance with one of the two accepted design criteria: the time for a certain percentage of drainage of the pavement base, beginning with the completely saturated condition, should be less than a certain value; or an inflow–outflow criterion where the base or subbase should be capable of draining the water at a rate equal to or more than the inflow rate without becoming completely saturated.

Step 4: Analyze and/or design collection systems to provide for the disposal of water removed by the drainage layers designed above.

Step 5: Design soil and/or geotextile filters and/or separators to protect the drainage layers and collection systems selected in steps 3 and 4.

Step 6: Conduct a critical evaluation of the results of steps 3 to 5, with respect to long-term performance, construction, maintenance, and economics. Long-term clogging of filters and drainage layers, which hinders their performance, should be a key consideration.

We follow this approach and elaborate on the methods necessary to carry out steps 2 to 5 in the following sections.

10.3 ESTIMATION OF INFLOW INTO PAVEMENT SYSTEMS

The sources of water contributing to inflow into pavement systems are several. The two primary sources are the water migrating from the reservoirs or ditches in the vicinity, and surface infiltration through joints and cracks in the pavement section (Fig. 10.6). Other sources of water are due to capillary action, which transports water

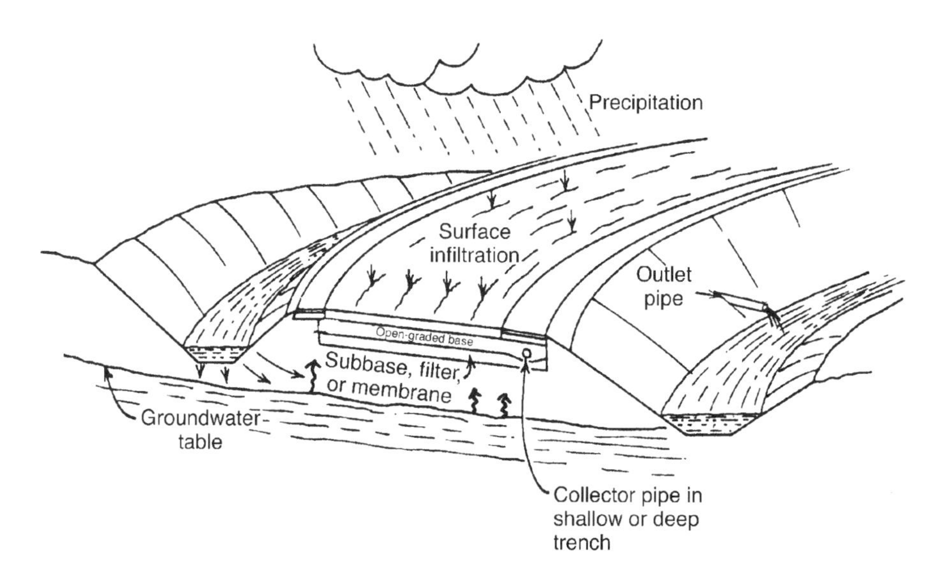

Fig. 10.6. Sources of water at pavement systems. (Adapted from Cedergren, 1989.)

from shallow groundwater tables into the pavement section, and melting of ice layers during the thawing cycle in frost-susceptible areas. Flow nets and simple analytical solutions discussed in Chapters 3 and 4 are generally applicable to determine the inflow rates. Moulton (1980) outlines a variety of solutions available to determine flow rates into pavement sections. The following sections summarize these solutions.

10.3.1 Infiltration

To estimate the inflow due to infiltration, Cedergren et al. (1973) recommended that the design infiltration rate be found by multiplying the 1-hour-duration 1-year-frequency precipitation rate (shown in Fig. 10.7) by a coefficient varying from 0.50 to 0.67 for portland cement concrete pavements and 0.33 to 0.50 for bituminous concrete pavements. However, this approach does not consider rainfall duration and does not relate infiltration to cracking. An alternative approach relating infiltration to cracking was given by Ridgeway (1976). The suggested equations for infiltration i are

$$i = \begin{cases} q\left(N + 1 + \dfrac{W}{S_r}\right) & \text{for rigid pavements} \quad (10.1) \\[2ex] q\left(N + 1 + \dfrac{W}{S_f}\right) & \text{for flexible pavements} \quad (10.2) \end{cases}$$

where i is the infiltration through pavement (ft^3/hr per linear foot of pavement), q the inflow rate through a crack (a value of 0.1 ft^3/hr per foot of crack is recommended), N the number of lanes, W the lane width (ft), S_r the transverse joint spacing in rigid pavements (ft), and S_f the mean spacing of transverse cracks in flexible pavements (an estimate of 40 ft may be used). Equations (10.1) and (10.2) are only approximate; the actual infiltration depends on the permeability of the pavement structure, the slope of the pavement surface, and the intensity and duration of rainfall. The permeability of portland cement concrete is low (on the order of 10^{-9} cm/s), and it is generally believed that most infiltrated water enters through cracks and joints or other discontinuities in the pavement surface.

10.3.2 Groundwater Flow

In many situations, groundwater flow into the pavement section could be estimated using flow nets. Two types of flow are identified in general: gravity drainage and artesian flow. For the simple case of a symmetrical configuration shown in Fig. 10.8, the design inflow rate from gravity drainage q_g (ft^3/day per square foot of drainage layer) can be estimated as

$$q_g = \frac{q_2}{0.5W} \tag{10.3}$$

where q_2 is the total upward flow into one half of the drainage blanket (ft^3/day per linear foot of roadway) estimated using Fig. 10.8, and W is the width of the drainage layer (ft). The length of influence, L_i in Fig. 10.8, may be estimated using

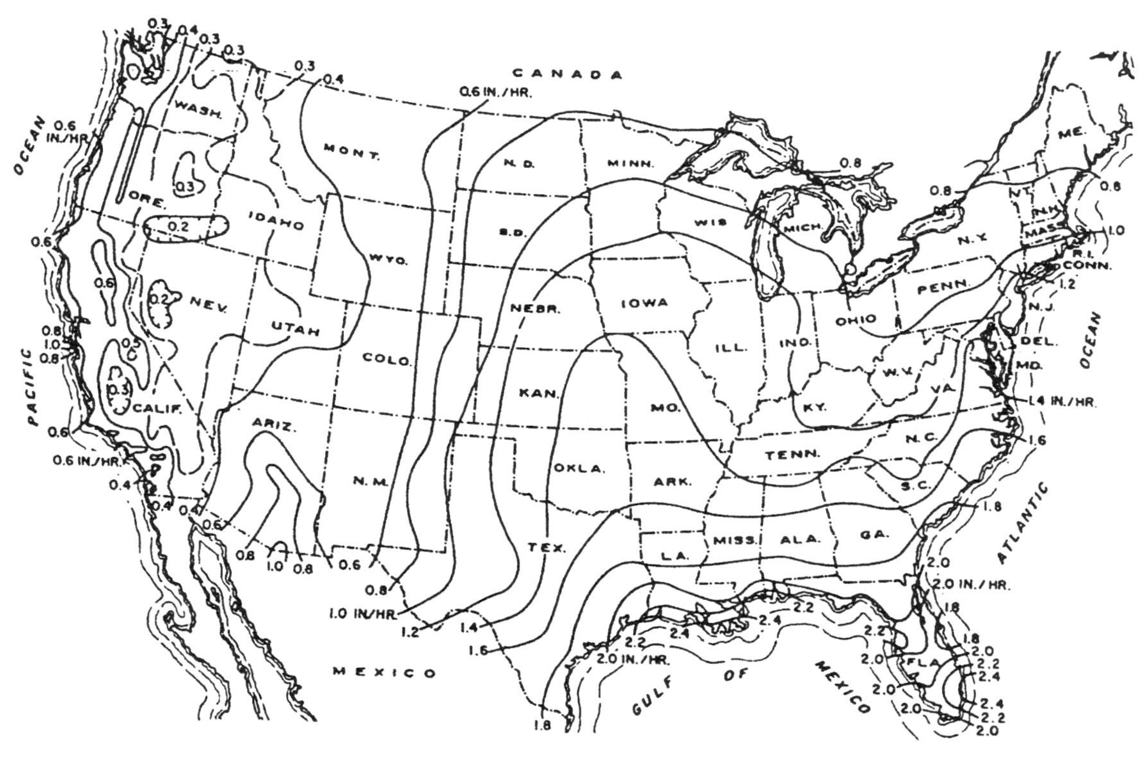

Fig. 10.7. Precipitation rate, 1-hour duration and 1-year frequency. (Adapted from Cedergren et al., 1972.)

$$L_i = 3.8(H - H_0) \tag{10.4}$$

where $H - H_0$ is the amount of total drawdown as shown in Fig. 10.8.

The artesian flow is shown schematically in Fig. 10.9. The inflow rate due to artesian flow can be estimated using Darcy's law:

$$q_a = k\frac{\Delta h}{H_0} \tag{10.5}$$

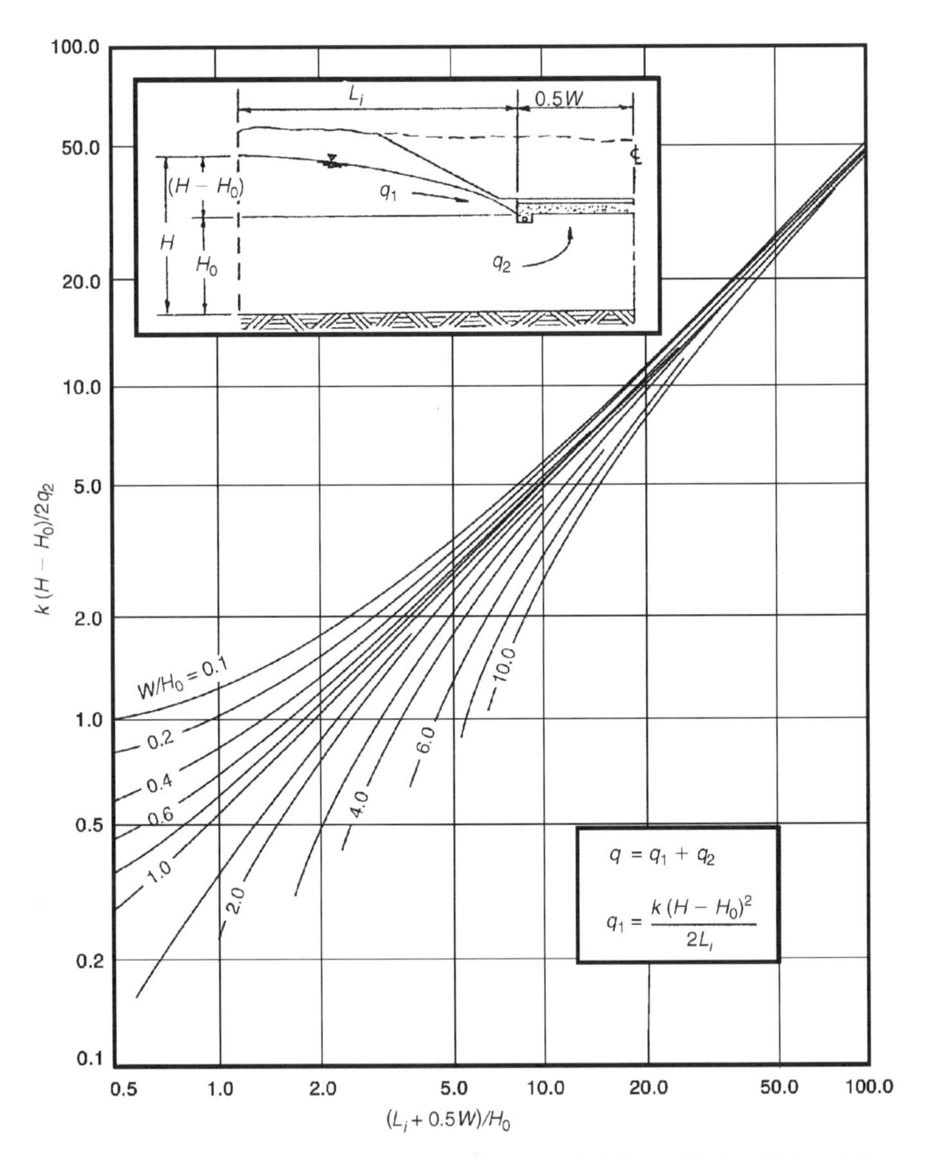

Fig. 10.8. Chart for determining flow rate in a horizontal drainage blanket. (Adapted from Moulton, 1980.)

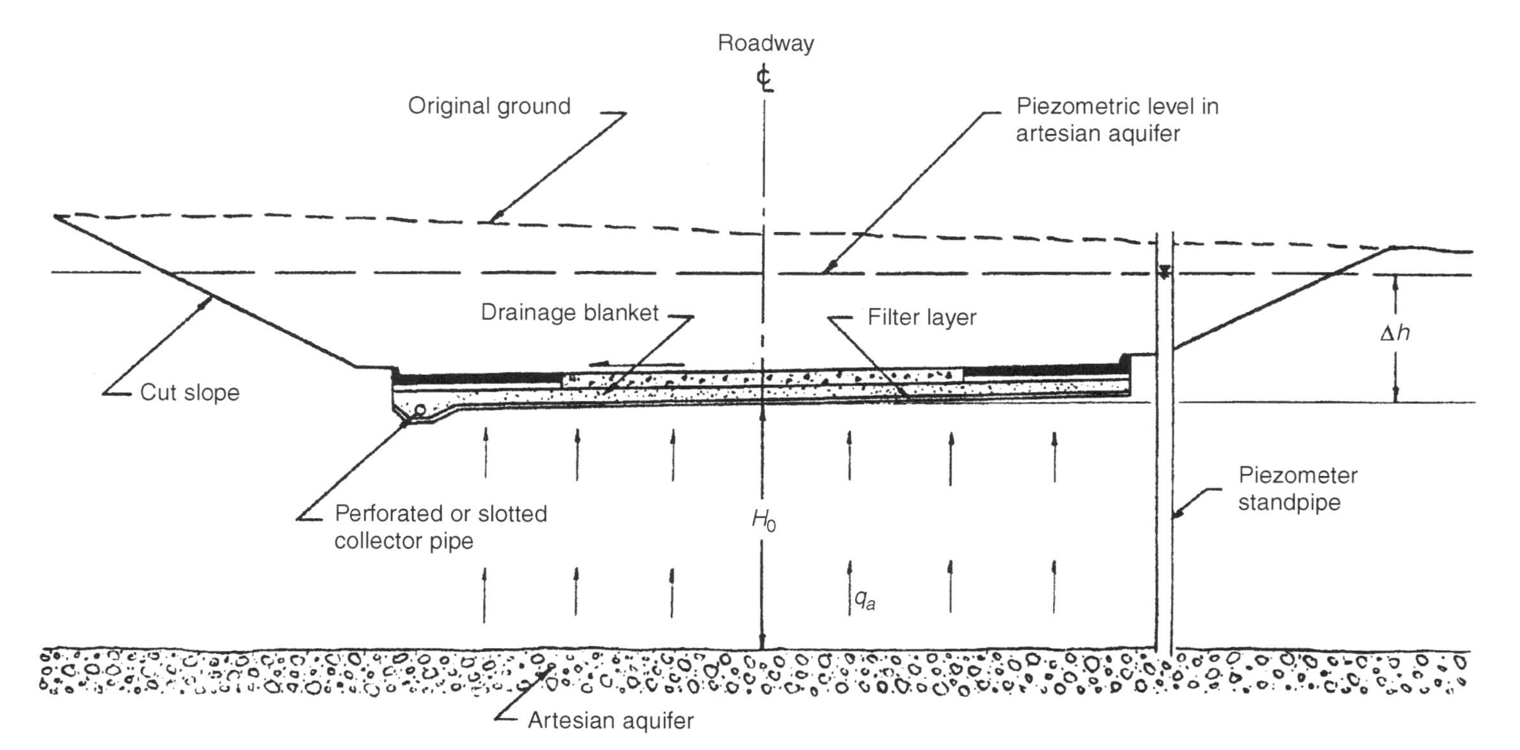

Fig. 10.9. Artesian flow of groundwater into a pavement drainage layer. (Adapted from Moulton, 1980.)

where q_a is the design inflow rate from artesian flow (ft^3/day per square foot of drainage layer), Δh the excess artesian head (ft), and H_0 the thickness of subgrade soil between the artesian aquifer and the drainage layer (ft).

10.3.3 Frost Melt

The inflow due to frost melt is largely a function of the frost susceptibility of the subgrade soils and is dependent on the rate of thawing, the permeability of the thawed soil, and the stresses imposed by the overlying pavement structure and vehicular traffic. Frost susceptibility of soils may be determined using laboratory freezing tests. In the absence of such tests, Table 10.1 may be used as a guidance to estimate the rate at which soils heave. This table was based on results from standard laboratory freezing

TABLE 10.1 Guidelines for Selection of Heave Rate or Frost Susceptibility Classification for Use in Fig. 10.10

Unified Classification		Percent	Heave Rate	Frost Susceptibility
Soil Type	Symbol	< 0.02 mm	(mm/day)	Classification
Gravels and sandy gravels	GP	0.4	3.0	Medium
	GW	0.7–1.0	0.3–1.0	Negligible to low
		1.0–1.5	1.0–3.5	Low to medium
		1.5–4.0	3.5–2.0	Medium
Silty and sandy gravels	GP–GM	2.0–3.0	1.0–3.0	Low to medium
	GW–GM	3.0–7.0	3.0–4.5	Medium to high
	GM			
Clayey and silty gravels	GW–GC	4.2	2.5	Medium
	GM–GC	15.0	5.0	High
	GC	15.0–30.0	2.5–5.0	Medium to high
Sands and gravely sands	SP	1.0–2.0	0.8	Very low
	SW	2.0	3.0	Medium
Silty and gravely sands	SP–SM	1.5–2.0	0.2–1.5	Negligible to low
	SW–SM	2.0–5.0	1.5–6.0	Low to high
	SM	5.0–9.0	6.0–9.0	High to very high
		9.0–22.0	9.0–5.5	
Clayey and silty sands	SM–SC	9.5–35.0	5.0–7.0	High
	SC			
Silts and organic silts	ML–OL,	23.0–33.0	1.1–14.0	Low to very high
	ML	33.0–45.0	14.0–25.0	Very high
		45.0–65.0	25.0	Very high
Clayey silts	ML–CL	60.0–75.0	13.0	Very high
Gravely and sandy clays	CL	38.0–65.0	7.0–10.0	High to very high
Lean clays	CL	65.0	5.0	High
	CL–OL	30.0–70.0	4.0	High
Fat clays	CH	60.0	0.8	Very low

Source: Adapted from Moulton (1980).

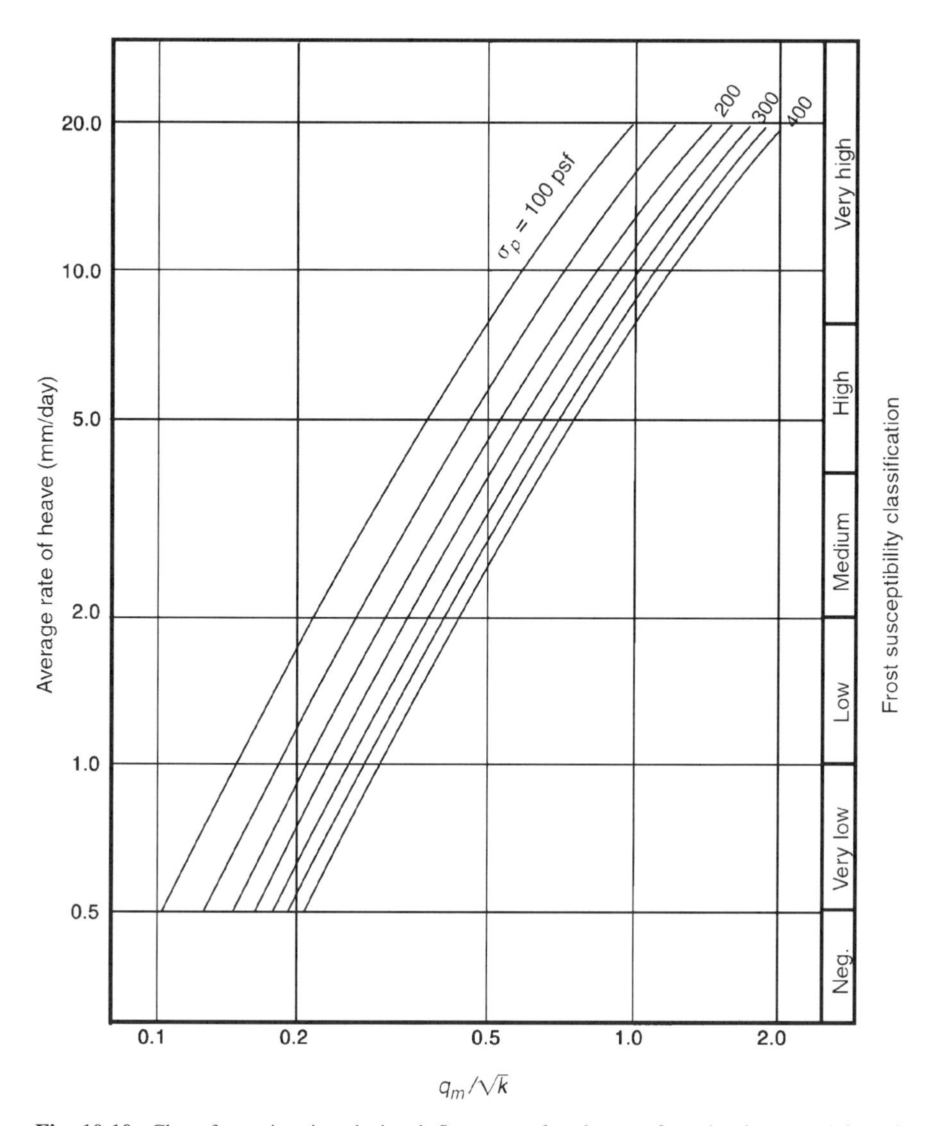

Fig. 10.10. Chart for estimating design inflow rate of meltwater from ice lenses. (Adapted from Moulton, 1980.)

tests performed by the Corps of Engineers between 1950 and 1970. With the estimated heave rate, Fig. 10.10 may be used to determine the design inflow rate, q_m, due to frost melt. In developing Fig. 10.10, Moulton (1980) assumed that the drainage of a thawed subgrade soil could be represented by the Terzaghi's one-dimensional consolidation model. σ_p in Fig. 10.10 represents the stress imposed on the subgrade soil by the pavement structure; therefore, it is proportional to the unit weight of the pavement material. The seepage rate from the consolidating subgrade soils is high immediately following thawing and decreases rapidly with time. Moulton (1980) recommends that q_m estimated from Fig. 10.10 be used as the average occurring during the first day following thawing.

10.4 ESTIMATION OF OUTFLOW FROM PAVEMENT SYSTEMS

Some of the water that has entered the pavement section may flow out of the system as a result of the hydrogeological conditions and the soil strata around the pavement section. Although the outflow can be estimated drawing flow nets for several configurations, no single method could be prescribed to generalize the flow process.

Three broad scenarios are generally identified to classify the cases under which seepage out of the pavement sections can take place. These are:

1. The flow is directed toward a water table, either horizontal or sloping, existing at some depth below the pavement section.
2. The subgrade soil or embankment is underlain at some depth by a stratum whose permeability is very high relative to that of the subgrade material, thus promoting nearly vertical flow.
3. The flow is directed vertically and laterally through the underlying embankment and its foundation to exit through a surface of seepage on the embankment slope and/or through the foundation.

In general, the flow scenarios are unsteady with flow domains changing and saturation lines moving with respect to time. Also, changes in infiltration into the pavement section cause unsteady conditions below the pavement section. However, when infiltration is sustained for long periods of time, outflow rates could be estimated using simple flow nets. Figures 10.11 to 10.13 show simple nomograms developed from flow nets. The q_v value estimated using these figures could serve as a reliable first approximation in a number of situations.

Using the estimates of inflow and outflow from Sections 10.3 and 10.4, the net inflow rate, q_n, that the pavement drainage layer has to be designed for may now be estimated. It is important to identify all of the sources contributing to inflow in the computations of net inflow rate.

10.5 DESIGN CRITERIA FOR PAVEMENT DRAINAGE SYSTEMS

Design of pavement drainage systems is carried out using one of the following two criteria:

1. The time for a certain percentage of drainage of the base or subbase beginning with the completely saturated condition should be less than a certain value.
2. An inflow–outflow criterion where the base or subbase should be capable of draining the water at a rate equal to or more than the inflow rate without becoming completely saturated.

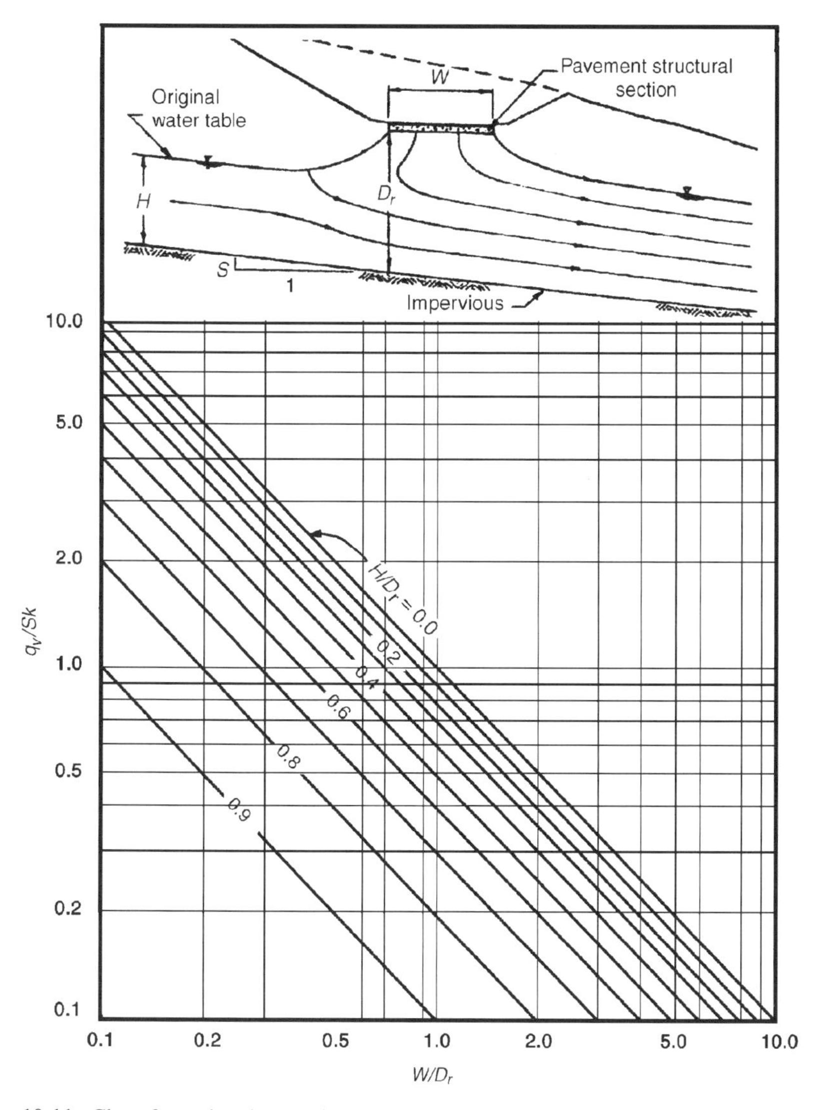

Fig. 10.11. Chart for estimating vertical outflow from a pavement structural section through subgrade soil to a sloping underlying water table. (Adapted from Moulton, 1980.)

The first criterion dictates the time required for a desired extent of drainage, whereas the second criterion relates the rate of drainage to the inflow rate. There is no general consensus among transportation industries/agencies on the time periods to be used in the first criterion. Cedergren, for instance, suggested that for airport pavement drainage design, the time required for 50% drainage of free water from the base course be not more than 10 days. This criterion may not be sufficient, however, for highway pavements, which are subjected to frequent repetition of loads. Other

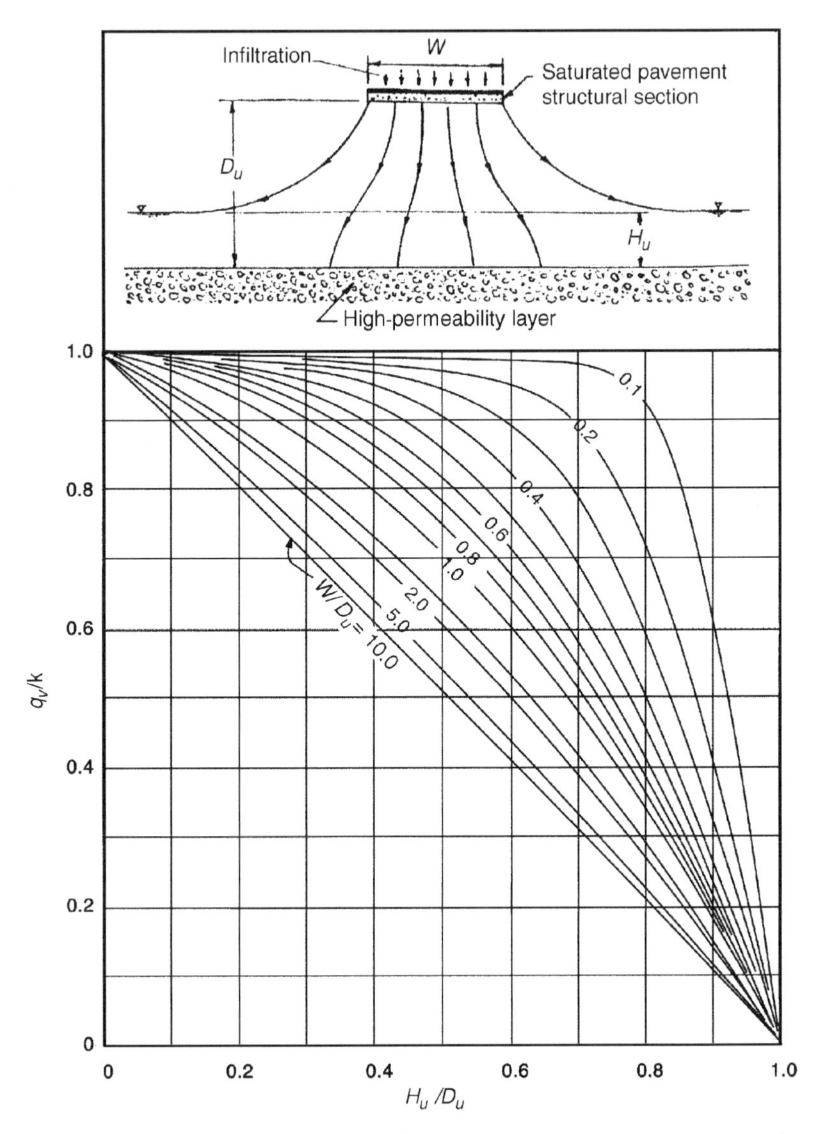

Fig. 10.12. Chart for estimating vertical outflow from a pavement structural section through the subgrade to an underlying high-permeability layer. (Adapted from Moulton, 1980.)

recommendations are to limit the time to 1 hour for 50% drainage or to 5 hours for 85% drainage. To estimate the time for a specific percentage (or degree) of drainage, Fig. 10.14 is commonly used. As shown in the figure, the degree of drainage is a function of slope factor S_1, which represents the geometry of the drainage layer, and an m factor, which depends on the permeability k and yield capacity n_e of the drainage material among other variables. For a specified criterion, Fig. 10.14 may be used to

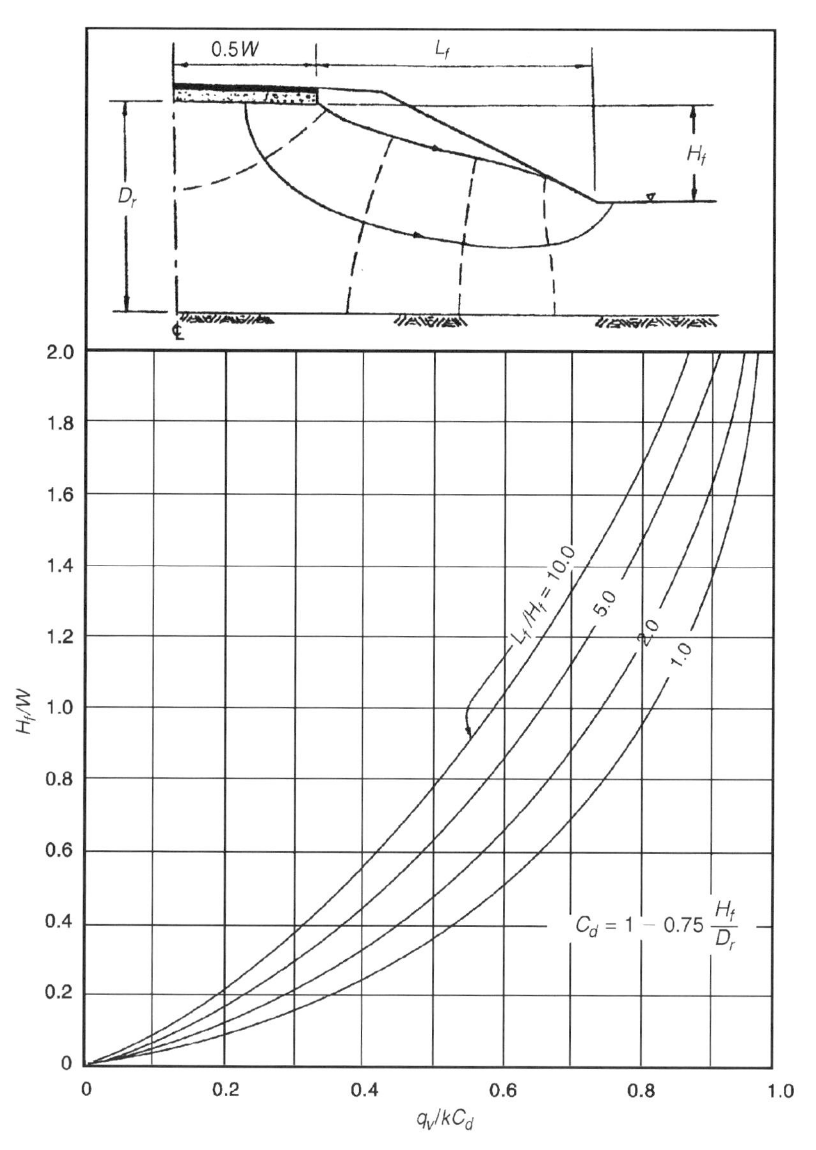

Fig. 10.13. Chart for estimating vertical outflow from a pavement structural section through embankment and foundation soil. (Adapted from Moulton, 1980.)

obtain the design geometry (H, L, or S) given the properties of the design material (k, n_e), or vice versa.

In general, most highway agencies recommend designs based on time to drain 50% of the drainable water. For instance, FHWA (1992) recommends the criterion of 1 hour to drain 50% of drainable water for the highest-class roads with the greatest amount of traffic. For most other interstate highways and freeways, a time limit of

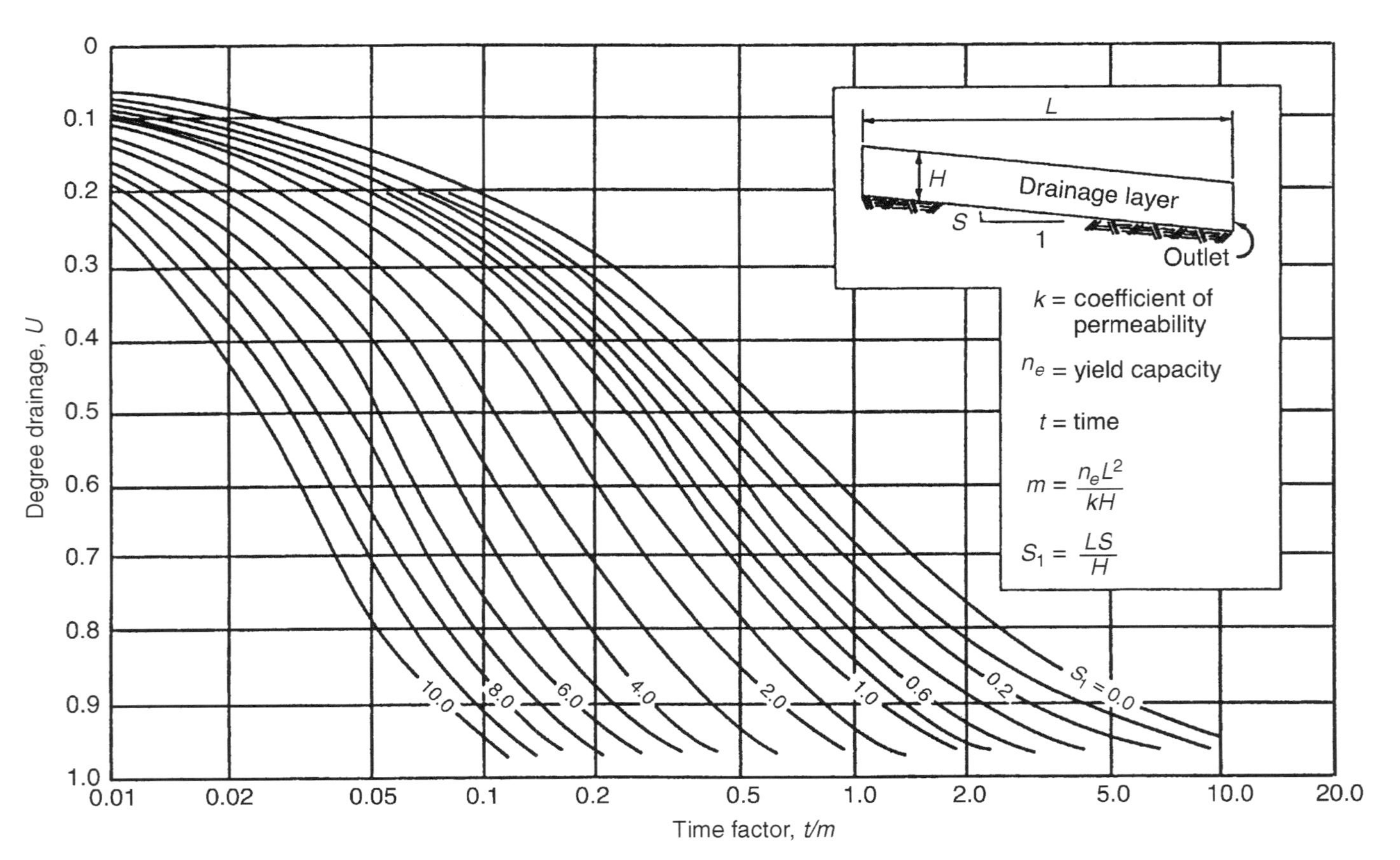

Fig. 10.14. Time-dependent drainage of the saturated layer. (Adapted from FHWA, 1992.)

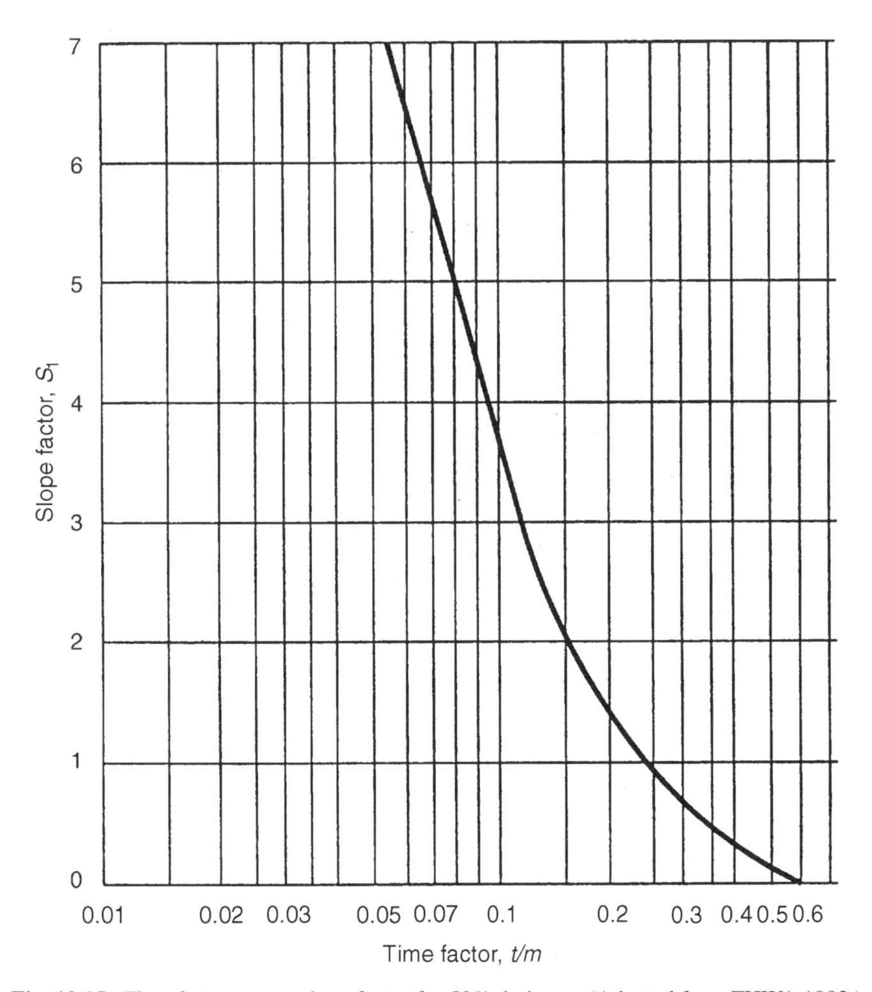

Fig. 10.15. Time factor versus slope factor for 50% drainage. (Adapted from FHWA,1992.)

2 hours was recommended to drain 50% of the drainable water. To use these criteria, the solutions plotted in Fig. 10.14 are replotted in Fig. 10.15 for 50% drainage. The two properties of the drainage material required to use these figures are permeability k and yield capacity n_e. The permeability of the material may be determined using laboratory experiments or estimated using the empirical expressions discussed in Chapter 2. In particular, Figures 2.13 and 2.14 are useful for the types of materials used for the drainage layer. The *yield capacity,* or *effective porosity,* is an indicator of the pore volume occupied by mobile water. This is used to exclude the immobile pore water such as the water in dead-end pores or the water that is strongly bound to fine-textured porous matrix. Based on published measurements on soils with varying gradations and densities, Moulton (1980) correlated effective porosity with coefficient of permeability. This correlation, shown in Fig. 10.16, may be used to obtain n_e.

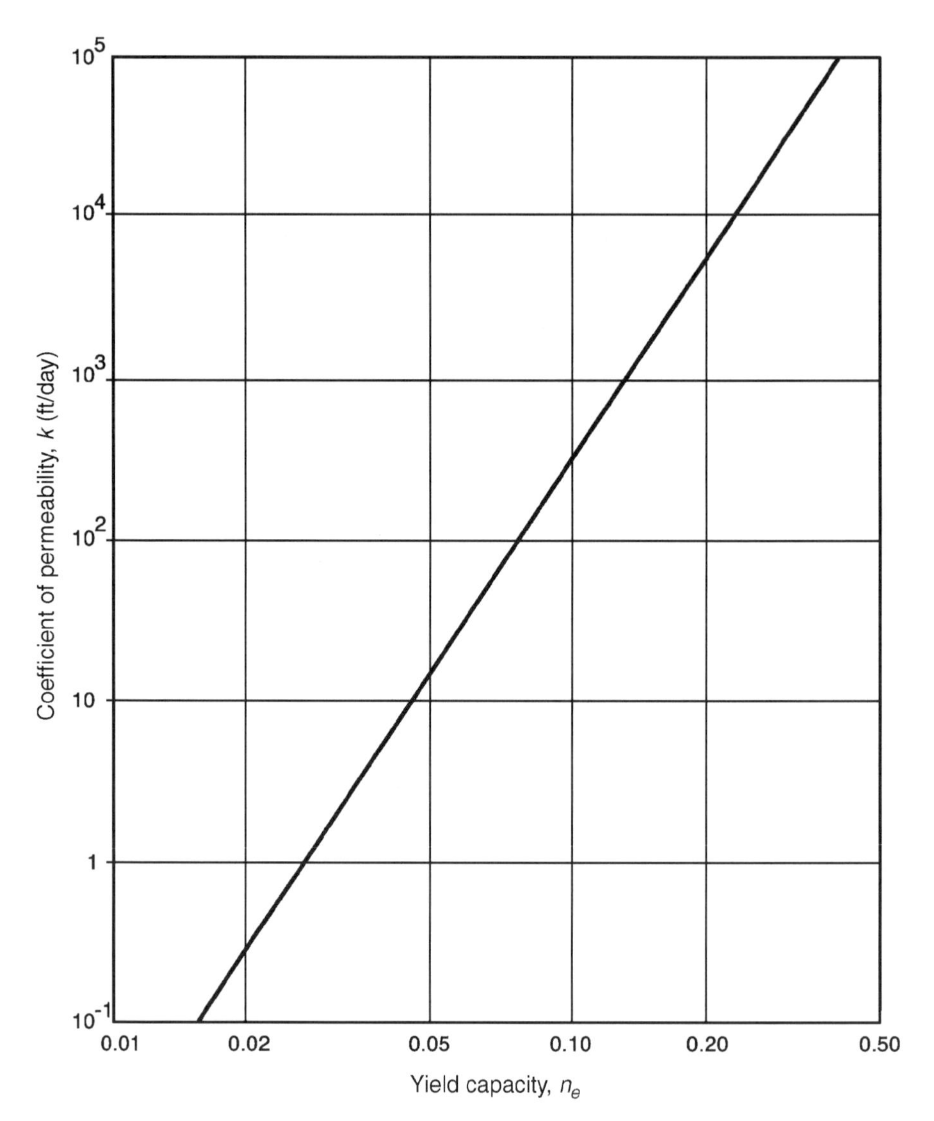

Fig. 10.16. Yield capacity (effective porosity) versus coefficient of permeability. (Adapted from Moulton, 1980.)

According to the second criterion, the drainage layer characteristics must be chosen so that the layer is capable of draining all the inflow to a suitable collection system. To design according to this criterion, Fig. 10.17 is commonly used. In developing Fig. 10.17, it was assumed that the inflow was steady and was uniformly distributed across the surface of the pavement section. The maximum mound height in the drainage layer H_m controls the thickness of the drainage layer. It is possible to determine the permeability required of the drainage layer once the maximum mound height and the geometrical characteristics (length of flow path L and slope of drainage layer S) are

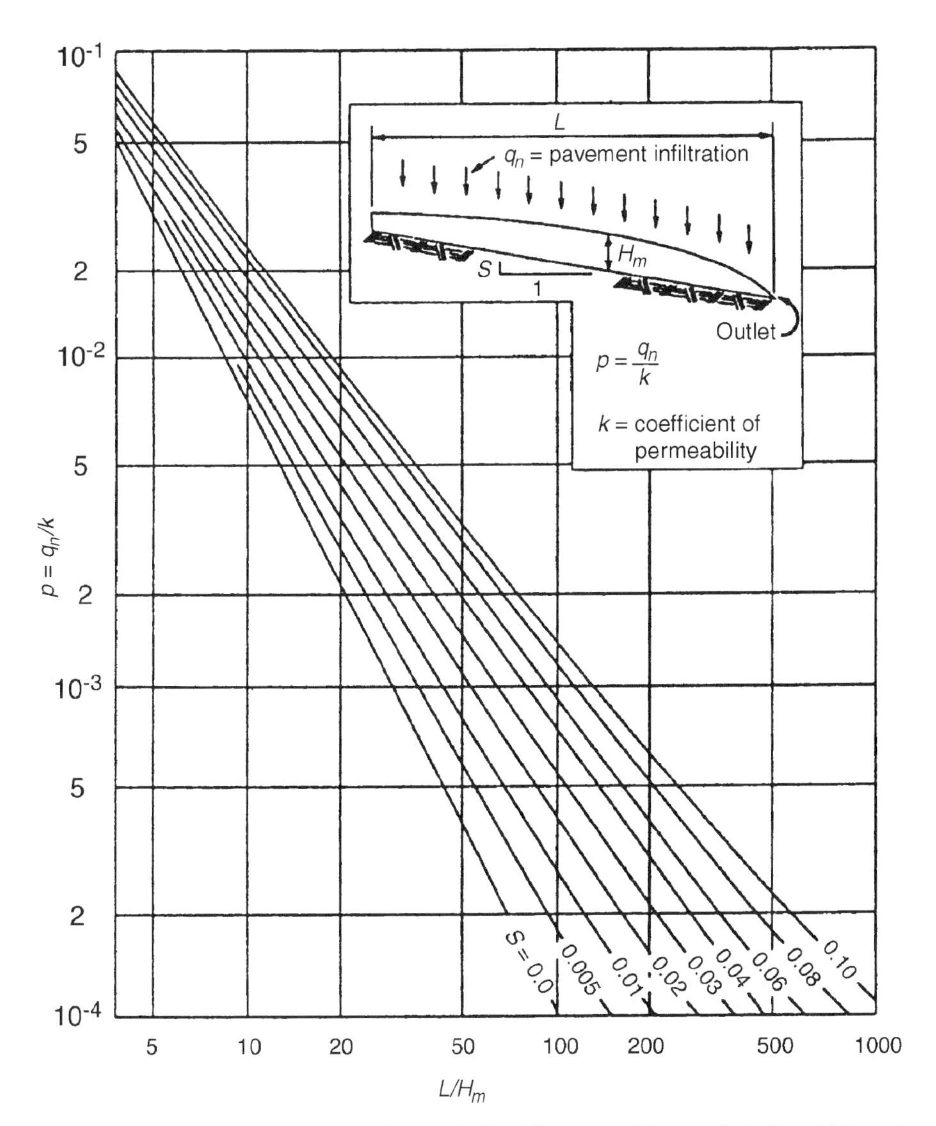

Fig. 10.17. Chart for estimating maximum depth of flow caused by steady inflow. (Adapted from FHWA, 1992.)

known. Conversely, the required thickness of the drainage layer could be determined from H_m once the permeability and geometrical characteristics of the drainage layer are specified.

Example 10.1 The length and thickness of a pavement drainage layer are 25 ft and 0.5 ft, respectively. Its slope $S = 0.025$. The coefficient of permeability of the material, $k = 2000$ ft/day. Determine the time to drain 50% of the permeable base.

SOLUTION: Figure 10.14 is used to determine the time to drain 50% of the permeable base.

$$S_1 = \frac{LS}{H} = \frac{25 \text{ ft} \times 0.025}{0.5 \text{ ft}} = 1.25$$

and $U = 0.5$, so $t/m = 0.22$. From Figure 10.16, $n_e = 0.165$; therefore,

$$m = \frac{n_e L^2}{kH} = \frac{0.165 \times (25 \text{ ft})^2}{2000 \text{ ft/day} \times 0.5 \text{ ft}} = 0.1 \text{ day}$$

and $t = 0.22 \times 0.1$ day $= 0.022$ day $= 0.53$ hr. The time to drain 50% of the permeable base is about half an hour.

Example 10.2 Refer to Example 10.1. Use the second criterion for pavement drainage design and determine the maximum inflow rate that the permeable base is capable of draining. Assume the maximum mound height to be equal to the thickness of the layer (i.e., $H_m = 0.5$ ft).

SOLUTION: Figure 10.17 is used to determine the maximum inflow rate. With $L/H_m = 25$ ft/0.5 ft $= 50$ and $S = 0.025$, $q_n/k = 1.5 \times 10^{-3}$. So the maximum inflow rate is

$$q_n = 1.5 \times 10^{-3} k = 1.5 \times 10^{-3} \times 2000 \text{ ft/day} = 3 \text{ ft/day}$$

10.6 DESIGN OF COLLECTION SYSTEMS

To avoid saturation of the drainage layers, the water drained must be collected using drains and outlet pipes. The design methodologies used in Section 10.5 are based on the assumption that the drains and outlet pipes have sufficient capacities to transport the drainable water from the drainage layers. Usually, the location of collector systems is governed by the longitudinal roadway grade. It is generally recommended that the location of the collection systems be established before designing the drainage layers. In designing the collection systems, the following considerations are kept in mind:

1. The type of pipe to be utilized
2. The location and depth of longitudinal and transverse collectors and their outlets
3. The slope of collector pipes
4. The size of the pipes
5. The provisions for adequate filter protection to ensure long-term and sufficient drainage capacity

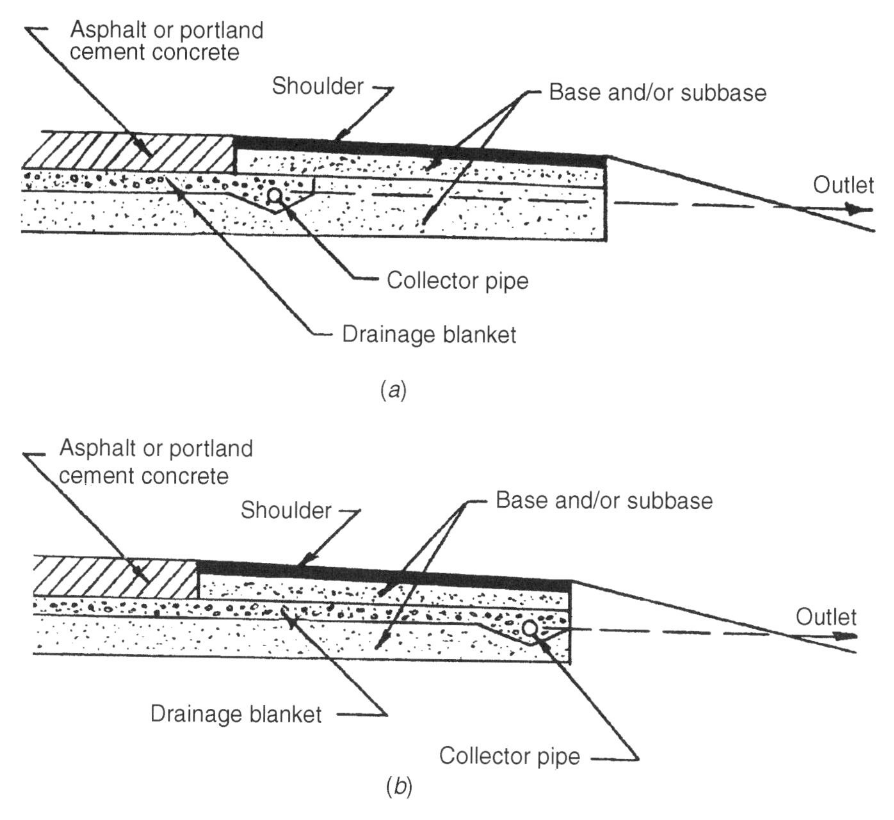

Fig. 10.18. Location of shallow longitudinal collector pipes: (*a*) drain placed just outside the joint; (*b*) drain placed at the outer edge of the shoulder. (Adapted from Moulton, 1980.)

Figures 10.18 and 10.19 show typical configurations of collector pipes. In Fig. 10.18 the collector pipes are placed in shallow trenches. This is recommended as well as economical when frost penetration depth is not significant and it is not necessary to draw down a high water table. In Fig. 10.19 the collector pipes are placed in the deeper trenches because of shallow frost penetration depths and/or the necessity to draw down the water table. Usually, the collector drains are placed just outside the joint between pavement and shoulder to protect the joint against pumping (Figs. 10.18*a* and 10.19*a*). However, it is also common to locate the collectors at the outer edge of the shoulders (Figs. 10.18*b* and 10.19*b*) in situations where shoulders are also required to be protected and drained. It is important to note that this alters the flow length L, which in turn affects the design of the drainage layer.

The collector pipes should be sized based on the quantity of water entering the pipe per running foot, the pipe gradient, the distance between outlets, and the hydraulic characteristics of the pipe. Figure 10.20 shows a nomogram relating these parameters.

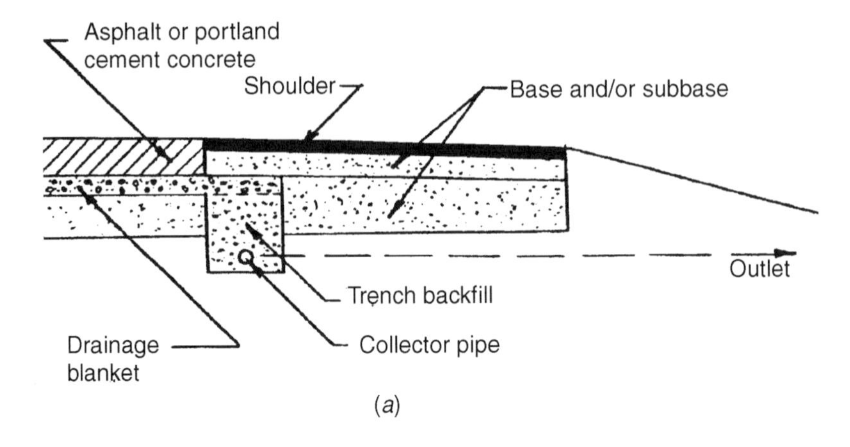

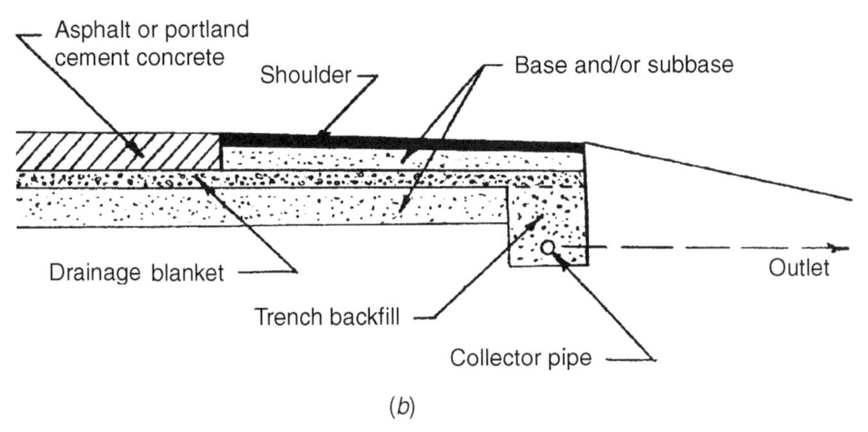

Fig. 10.19. Location of deep longitudinal collector pipes: (*a*) drain placed just outside the joint; (*b*) drain placed at the outer edge of the shoulder. (Adapted from Moulton, 1980.)

The flow rate in the drain, q_d, is estimated by multiplying the net design inflow rate and the length of flow path:

$$q_d = q_n L \tag{10.6}$$

Using the nomogram, the required pipe diameter could be determined once the pipe gradient and outlet spacing are specified. Conversely, the outlet spacing could be determined if the other parameters (pipe size and gradient) are known.

10.7 FILTER DESIGN FOR PAVEMENT DRAINAGE SYSTEMS

The drainage layer and the collection systems designed in Sections 10.5 and 10.6 must be protected using proper filters to ensure their long-term performance. Filters play a

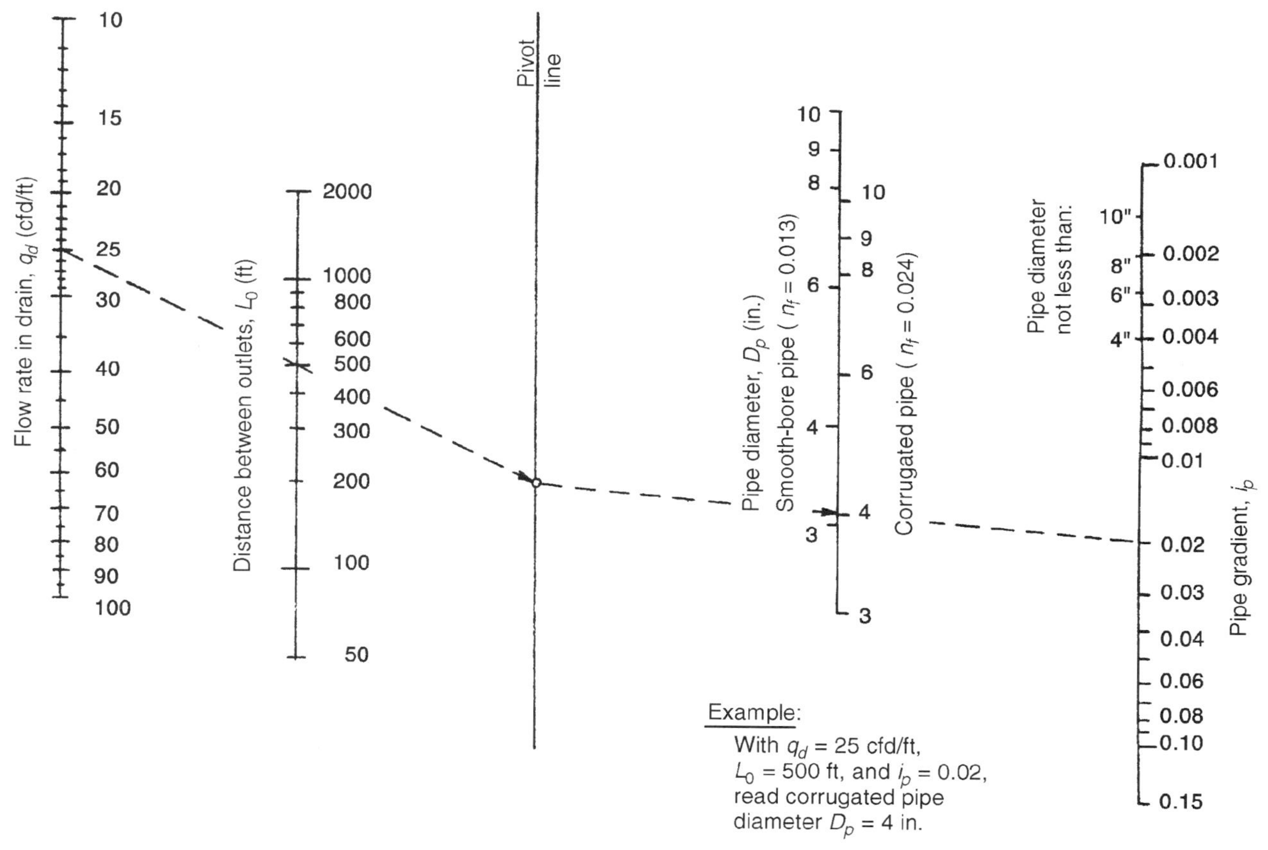

Fig. 10.20. Nomogram relating collector pipe size with flow rate, outlet spacing, and pipe gradient. (Adapted from Cedergren et al., 1973.)

crucial role in pavement drainage systems. In several cases of failed drainage systems, they were proved to be the weakest link in design. A common failure of drainage systems is due to their clogging by subgrade soil particles. As discussed in Chapter 7, fine colloid particles trapped in drainage layers may reduce their permeability by as much as one order of magnitude. A safe practice would be to use a reduced value of permeability (reduced from the value obtained in the laboratory, or estimated empirically) in the design of drainage layers. Filters and separators used underneath the drainage layers, or around the drains, must be designed using criteria outlined in Chapters 7 and 8. In addition to using the retention and permeability criteria, the filters and separators must also be designed to withstand dynamic loads due to traffic applied on the pavement structure. The strength of the filter layer, as assessed in its saturated state, must be sufficient to retain its structural integrity under the dynamic loads.

PROBLEMS

10.1. In Fig. 10.8, use $H = 25$ ft, $H_0 = 15$ ft, $W = 36$ ft, and $k = 0.5$ ft/day. Assuming that q_1 is drained to the longitudinal drain, determine the average inflow rate from gravity drainage q into the drainage blanklet.

10.2. If the longitudinal drain is provided on only one side of the permeable base in Problem 10.1, q_1 from the other side must also be drained through the drainage blanket. Determine q_1 and the total inflow rate into the drainage blanklet.

10.3. Refer to Fig. 10.11. Determine the vertical outflow q_v if $D_r = 30$ ft, $H = 10$ ft, $W = 40$ ft, and $S = 0.1$. The coefficient of permeability of the subgrade soil $k = 3.0$ ft/day.

10.4. Refer to Problem 10.3. How would the vertical outflow q_v vary as the water table rises from $H = 10$ ft to 25 ft?

10.5. Refer to Fig. 10.12. Determine the vertical outflow q_v if $D_u = 40$ ft, $H_u = 10$ ft, $W = 34$ ft, and $k = 3.0$ ft/day.

10.6. Refer to Problem 10.5. How would the vertical outflow q_v vary as the water table rises from $H_u = 10$ ft to 30 ft?

10.7. Refer to Fig. 10.13. Determine the vertical outflow q_v if $L_f = 40$ ft, $W = 40$ ft, $H_f = 15$ ft, $D_r = 40$ ft, and $k = 3.0$ ft/day.

10.8. Refer to Problem 10.7. How would q_v change if $D_r = 15$ ft?

10.9. Draw flow nets to solve Problems 10.7 and 10.8 and compare the solutions.

10.10. For a permeable base with $L = 30$ ft, $H = 0.6$ ft, and $S = 0.025$, the time to drain 50% of the drainable water should be no more than 1 hour. Determine the permeability required of the drainage material.

10.11. A permeable base with $L = 20$ ft, $H = 0.4$ ft, and $S = 0.02$ is required to drain an inflow rate of 0.5 ft/day. Determine the permeability required of the drainage layer.

10.12. Draw coefficient of permeability versus time to drain for 50% drainage of permeable bases given $S = 0.025$, $H = 0.5$ ft, and $L = 25$ ft. Vary k from 500 to 4000 ft/day.

10.13. Draw slope S versus time to drain for 50% drainage of permeable bases given $k = 2000$ ft/day, $H = 0.5$ ft, and $L = 25$ ft. Vary S from 0.005 to 0.10.

10.14. Draw the length of the drainage layer L versus time to drain for 50% drainage of permeable bases given $k = 2000$ ft/day, $S = 0.025$, and $H = 0.5$ ft. Vary L from 20 to 50 ft.

10.15. Draw the thickness of the drainage layer H versus time to drain for 50% drainage of permeable bases given $k = 2000$ ft/day, $S = 0.025$, and $L = 25$ ft. Vary H from 5 to 15 in.

10.16. For the parameters given in Example 10.1, determine the time required to drain 85% of the drainage base. Also, determine the times required to drain 25%, 75%, 90%, and 100%, and draw a curve showing the degree of drainage versus time to drain.

10.17. For each of Problems 10.12 to 10.15, use the second criterion for pavement drainage design and plot the maximum inflow rate that the permeable base is capable of draining. Assume the maximum mound height to be equal to the thickness of the layer.

CHAPTER 11

SEEPAGE AT WASTE CONTAINMENT FACILITIES

Waste containment is one of the most challenging issues facing the engineering community today. Rapid industrialization during recent decades created a waste disposal problem of immense proportions. With an estimated 200 million tons or more of municipal solid waste being generated every year in the United States alone, space for landfilling the waste is becoming limited. The key objective in the design of waste containment facilities is to prevent or minimize transport of contaminants to the surrounding soil and groundwater. The periphery of the containment system should be such that it allows minimal outward seepage. In the case of solid waste facilities (Fig. 11.1), this would mean that the seepage through the bottom liner and top cover systems would be accurately quantified and proper drainage provided to discharge leachate generated within the facility. To isolate buried waste or accidental spills, the containment typically would involve construction of cutoff slurry walls and top cover systems with or without extraction wells (Fig. 11.2.). The walls are keyed into a bedrock or a low-permeability layer. In Fig. 11.2*a*, the slurry wall isolates the buried waste, allowing only inward flow to the extent allowed by the permeability of the wall material. In Fig. 11.2*b*, extraction wells control the groundwater table elevation so that the groundwater does not come into contact with the buried waste.

The design of waste containment facilities has to be carried out in accordance with Public Law 94-580, the Resource Conservation and Recovery Act (RCRA), passed by the U.S. Congress in 1976. The intent of this act is to give a legal basis for implementation of guidelines and standards for waste storage, treatment, and disposal. Regulations concerning the design of municipal solid waste landfills (MSWLFs) are established in *Subtitle D* program (40 CFR 258); regulations concerning design of hazardous waste landfills are established in *Subtitle C* program (40 CFR 264 and 265).

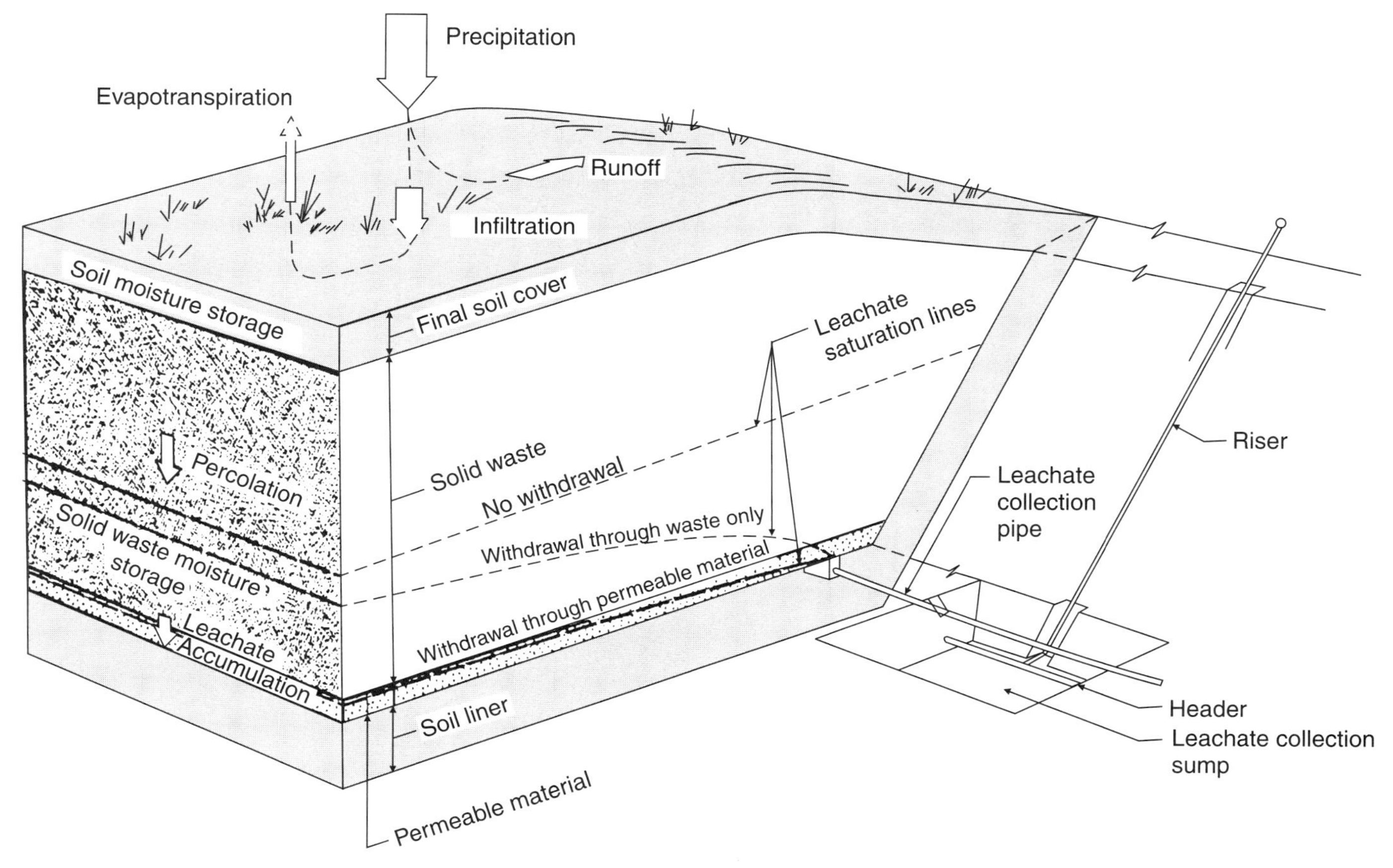

Fig. 11.1. Accumulation, containment, collection, and withdrawal of landfill leachate showing saturation levels for different conditions. (Adapted from U.S. EPA, 1980.)

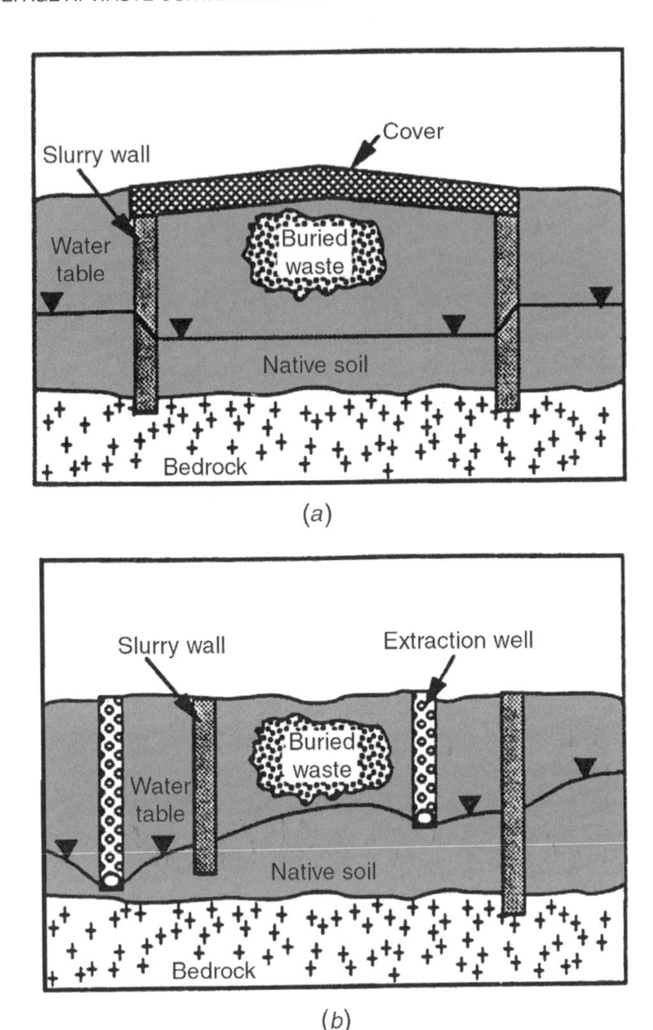

Fig. 11.2. Configurations of slurry walls: (*a*) used with surface cover; (*b*) used with extraction wells.

11.1 DESIGN CRITERIA

Subtitle C and D programs specify the design criteria for various elements of waste containment facilities. Criteria involving location, operation, and maintenance of waste storage facilities are also specified by these programs. The important elements constituting the waste containment system are the (1) liner system at the bottom of the facility, (2) leachate collection and removal system, and (3) top cover system. The design criteria for a MSWLF facility according to Subtitle D regulations are shown in Fig. 11.3. The purpose of the leachate collection system is to maintain

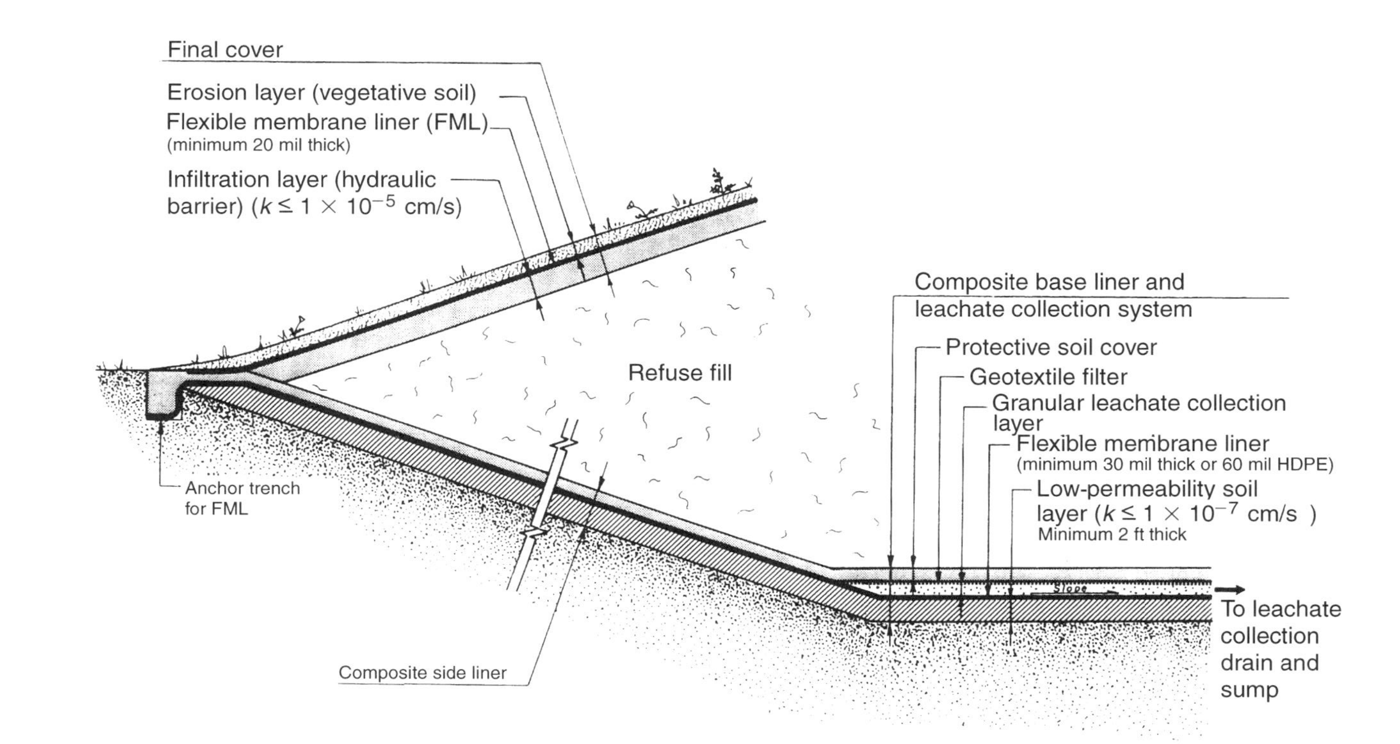

Fig. 11.3. Subtitle D liner and final cover for MSWLF. (Adapted from Sharma and Lewis, 1994; copyright © John Wiley & Sons; reprinted with permission.)

less than a 12-in. (30-cm) depth of leachate over the liner. The composite liner must consist of a minimum 30-mil-thick flexible membrane liner (FML) [or at least 60-mil high-density polyethylene (HDPE)] overlying a minimum 2-ft-thick compacted soil with hydraulic conductivity no greater than 1×10^{-7} cm/s. The purpose of the *final cover* system is to minimize infiltration and erosion. The erosion layer must consist of a minimum of 6 in. of earthen material capable of sustaining the growth of native plants. The infiltration layer must consist of a minimum of 18 in. of earthen material that has hydraulic conductivity less than 1×10^{-5} cm/s. The regulations also contain provisions for approval of alternative cover systems that are capable of reducing the infiltration to the same extent as the layer mentioned above.

The Subtitle C program regulates the design of hazardous waste impoundments and landfills. A solid waste is termed *hazardous* if (1) the organic and inorganic constituents present in the waste are listed in Appendix VIII of 40 CFR 261, and (2) the waste meets specific criteria on ignitability, corrosivity, reactivity, and toxicity. Figure 11.4 shows the design criteria for a hazardous waste surface impoundment. The unit must have two or more liners and a leachate collection and removal system between these liners. The liner system must be a composite system with an FML overlying a compacted soil liner. The requirements for a hazardous waste landfill unit are similar to those for the impoundment (Fig. 11.5). The system must consist of a leachate collection and removal system (LCRS); a top FML; a leachate detection, collection, and removal system (LDCRS); and a composite liner system. A schematic of the cover system at the top of the hazardous waste landfill is shown in Fig. 11.6. Again, the purpose of the cover system is to minimize infiltration into the landfill, promote drainage, and minimize erosion of the cover. The important components

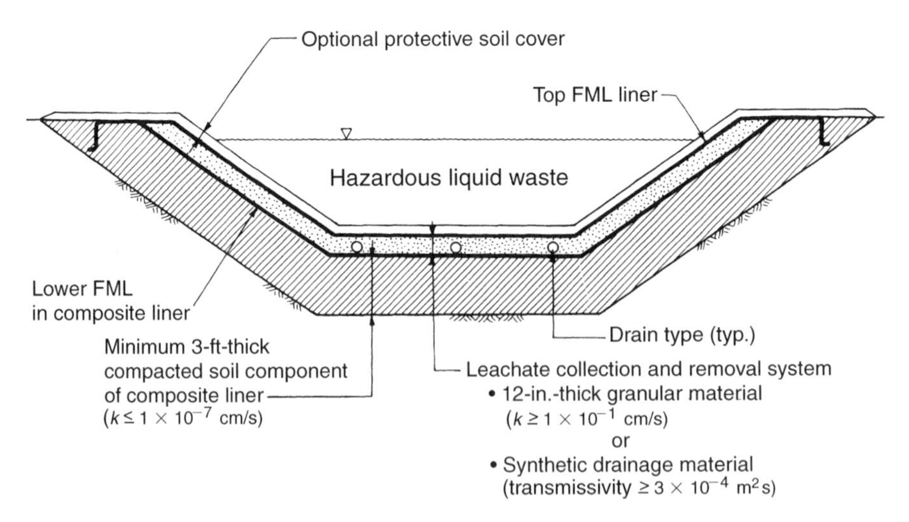

Fig. 11.4. Schematic of composite liner system for surface impoundment for disposal of hazardous waste. (Adapted from Sharma and Lewis, 1994; copyright © John Wiley & Sons; reprinted with permission.)

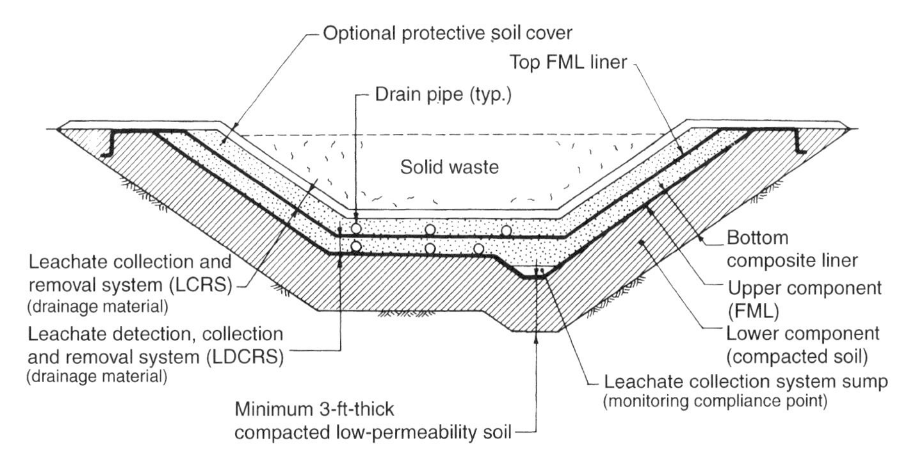

Fig. 11.5. Schematic of double liner and leachate collection system for a hazardous waste landfill. (Adapted from U.S. EPA, 1989.)

are the vegetative cover, drainage layer, and a low-permeability layer. The design standards for each of these components are shown in Table 11.1.

Controlling seepage through the various elements of waste containment facilities requires a thorough assessment of water balance at the facilities. Thus we will necessarily have to take up this problem from a system scale to a component scale. We look first at the water balance of the system prior to a discussion of seepage through the liners, lechate collection and removal systems (LCRs), and slurry walls. We restrict our attention to water flow only. Mass transport considerations are equally important for a waste containment system design. The reader is referred to Reddi and Inyang (2000) and other sources for discussion of contaminant mass transport through these elements.

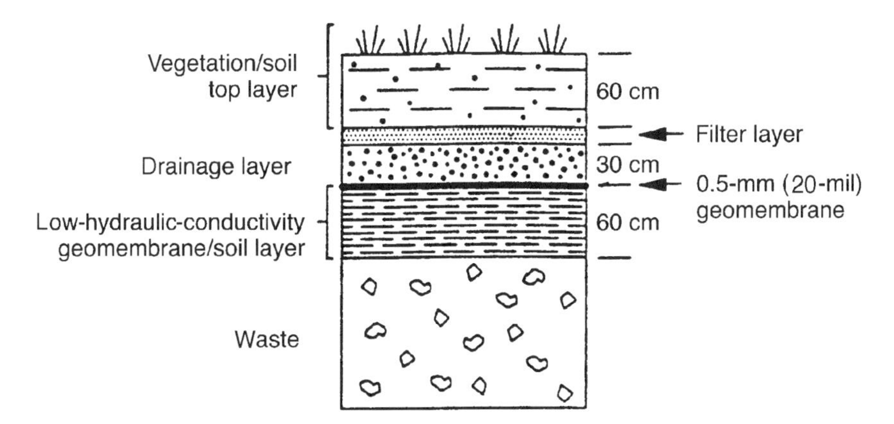

Fig. 11.6. Recommended cover system for landfills. (Adapted from U.S. EPA, 1991.)

TABLE 11.1 Cover System Design Standards for Hazardous Waste Landfills

Vegetative cover
- Thickness $\geq$ 60 cm (24 in.)
- Minimal erosion and maintenance (e.g., fertilization, irrigation)
- Vegetative root growth not to extend below 2 ft
- Final top slope between 3 and 5% after settlement or subsidence. Slopes greater than 5% not to exceed 2.0 tons/acre erosion (USDA universal soil loss equation)
- Surface drainage system capable of conducting run-off across cap without rills and gullies

Drainage layer design
- Thickness $\geq$ 30 cm (12 in.)
- Saturated hydraulic conductivity $\geq 10^{-3}$ cm/s
- Bottom slope $\geq$ 2% (after settlement/subsidence)
- Overlain by graded granular or synthetic filter to prevent clogging
- Allow lateral flow and discharge of liquids

Low-permeability layer design

FML component:
- Thickness $\geq$ 20 mils
- Final upper slope $\geq$ 2% (after settlement)
- Located wholly below the average depth of frost penetration in the area

Soil component:
- Thickness $\geq$ 60 cm (24 in.)
- Saturated hydraulic conductivity $\leq 1 \times 10^{-7}$ cm/s
- Installed in 15-cm (6-in.) lifts

Source: Adapted from U.S. EPA (1989).

11.2 LEACHATE GENERATION AT LANDFILLS

As water from rainfall or runoff infiltrates a waste containment system, it reacts with the solid and liquid constituents of the waste. During the percolation process, the infiltrating water acquires some characteristics of the waste. Depending on the type of waste and the reactions between the percolating water and the waste, this percolated water, commonly termed *leachate*, may become highly contaminated. The various reaction processes occurring within the waste are (1) dissolution of certain solid forms of the waste and subsequent precipitation, (2) decomposition and disintegration of the solids, and (3) reactions between the original liquids contained in the waste and the percolating water. The concentrations of various chemicals in the leachate are governed by a number of mass transfer processes operating simultaneously. For instance, an acidic pH condition of the liquid may trigger a number of processes, such as dissolution and precipitation, ion exchange, and sorption, simultaneously.

Apart from the reaction processes occurring within the waste, the concentrations of the leachate produced also depend on the waste disposal technique and time elapsed after the waste disposal. In general, the concentrations of chemicals in leachate are believed to reach a peak value sometime after the placement of waste, and then decrease with time. This decrease in concentrations with time is due to dilution, biochemical

TABLE 11.2 Ranges of Constituent Concentrations in Leachate from Municipal Waste Landfills

Constituent	Concentration Range[a]	Constituent	Concentration Range[a]
COD	50–90,000	Total P	0.1–150
BOD	5–75,000	Organic P	0.4–100
Total organic carbon	50–45,000	Nitrate nitrogen	0.1–45
Total solids	1–75,000	Phosphate (inorganic)	0.4–150
TDS	725–55,000	Ammonia nitrogen (NH_3-N)	0.1–2000
Total suspended solids	10–45,000	Organic N	0.1–1000
Volatile suspended solids	20–750	Total Kjeldahl nitrogen	7–1970
Total volatile solids	90–50,000	Acidity	2700–6000
Fixed solids	800–50,000	Turbidity (Jackson units)	30–450
Alkalinity (as $CaCo_3$)	0.1–20,350	Cl	30–5000
Total coliform (CFU/100 mL)	$0–10^5$	pH (dimensionless)	3.5–8.5
Fe	200–5500	Na	20–7600
Zn	0.6–220	Cu	0.1–9
Sulfate	25–500	Pb	0.001–1.44
Ni	0.2–79	Mg	3–15,600
Total volatile acids (TVA)	70–27,700	K	35–2300
Mn	0.6–41	Cd	0–0.375
Fecal coliform (CFU/1000 mL)	$0–10^5$	Hg	0–0.16
Specific conductance (mhg/cm)	960–16,300	Se	0–2.7
Ammonium nitrogen (NH_4-N)	0–1106	Cr	0.02–18
Hardness (as $CaCo_3$)	0.1–36,000		

Source: Adapted from U.S. EPA (1986).
[a]In mg/L unless noted.

processes leading to breakdown of chemicals, and to continuous removal of leachate from a collection point (Lu et al., 1985; Rowe, 1991; Ehrig and Scheelhaase, 1993).

Because of these various factors, leachate quality in a containment system usually exhibits tremendous spatial and temporal variability. A knowledge of the predominant constituents in the leachate is essential in order to design waste containment barriers. Table 11.2 shows the typical ranges of constituent concentrations found in leachate from municipal waste landfills. In contrast to municipal waste landfills, hazardous waste landfills typically contain elevated amounts of heavy metals, organic compounds, and other toxic substances. Table 11.3 shows the representative hazardous constituents that may be expected in industrial leachate.

11.3 WATER BALANCE AT WASTE CONTAINMENT SYSTEMS

The total quantity of leachate generated at a given waste containment system is primarily a function of the quantity of water infiltrated into the system and quantity of fluids generated inherently within the waste. The former is in turn dependent on a

TABLE 11.3 Representative Hazardous Substances within Industrial Waste Streams

	Hazardous Substances										
Industry	Arsenic	Cadmium	Chlorinated Hydro-carbons[a]	Chromium	Copper	Cyanides	Lead	Mercury	Miscellaneous Organics[b]	Selenium	Zinc
Battery		×		×	×						×
Chemical manufacturing			×	×	×			×	×		
Electric and electronic			×		×	×	×	×		×	
Electroplating and metal finishing		×		×	×	×					×
Explosives	×				×		×	×	×		
Leather				×					×		
Mining and Metallurgy	×	×		×	×	×	×	×		×	×
Paint and dye		×		×	×	×	×	×	×	×	
Pesticide	×		×			×	×	×	×		×
Petroleum and coal	×		×				×				
Pharmaceutical	×							×	×		
Printing and duplicating	×			×	×		×		×	×	
Pulp and paper								×	×		
Textile				×	×				×		

Source: Adapted from Matrecon, Inc. (1980).

[a]Including polychlorinated biphenyls.

[b]For example: acrolein, chloropicrin, dimethyl sulfate, dinitrobenzene, dinitrophenol, nitroaniline, and pentachlorophenol.

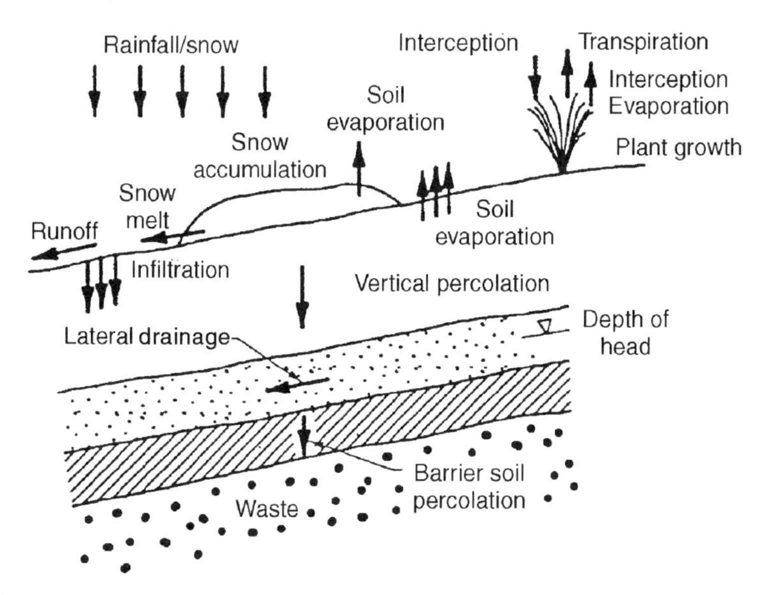

Fig. 11.7. Water pathways for a waste containment system. (Adapted from U.S. EPA, 1991.)

number of climatological and hydrological processes, primarily rainfall, runoff, and evaporation. Thus, to estimate the quantity of leachate, one needs to conduct a water balance for the entire system. This will essentially involve a bookkeeping procedure to account for the final disposition of total precipitation at a given site, through a number of pathways, such as evaporation, infiltration, and runoff.

Figure 11.7 shows the various pathways through which precipitation is disposed of in a waste containment system. Precipitation in the form of snow or rainfall is partitioned into interception by vegetation (for subsequent evapotranspiration), temporary storage followed by subsequent runoff from the surface of the system, and infiltration into the cover. The proportion of the precipitation that infiltrates the system will alter the water storage in the topsoil, which will undergo possible evapotranspiration from the vegetation and the soils. In addition to these, a proportion of the infiltrated water may be carried through a lateral drainage system if one is provided in the surface cover system. Eventually, a portion of infiltrated water may travel down into the waste, past the topsoil cover and barrier, and contribute to leachate percolation.

It is possible to estimate the water consumed in each of the pathways above and obtain the quantity of percolation through the waste using water-routing methods of hydrology. Fenn et al. (1975), Lutton et al. (1979), Lu et al. (1985), and Oweis and Khera (1990) demonstrated a water balance method using monthly hydrologic and climatological data to track changes in infiltration, evapotranspiration, and soil-water storage. The water balance for the entire system is analyzed using the known retention and transmission characteristics of the soil cover and refuse in a method proposed by Thornthwaite and Mather (1957). Based on water conservation at the site, the quantity of the infiltrated water I is expressed as

$$\mathrm{I} = P + \mathrm{SR} - R \tag{11.1}$$

where P is the precipitation, SR the water carried by surface runoff into the system, and R the surface runoff out of the system. The pathway of interception by vegetation is ignored in Eq. (11.1). R may be estimated by any of the empirical methods available in the hydrology literature. Rational formula offers a simple way of estimating R as a proportion of precipitation,

$$R = CP \tag{11.2}$$

where C is the runoff coefficient, which is a function of soil type, vegetation, and surface topography. Table 11.4 shows typical values of C. Once the quantity of infiltrated water I is known via Eq. (11.1), the quantity percolated out of the soil cover into the waste can be estimated applying water conservation principle to the soil cover. Thus,

$$\mathrm{PER} = \mathrm{I} - \mathrm{AET} - \Delta S_c \tag{11.3}$$

where PER is the quantity of water percolated out of the soil cover, AET the actual evapotranspiration from the cover, and ΔS_c the change in storage of the cover as a result of infiltration. We make a distinction here between the potential evapotranspiration (PET) and actual evapotranspiration (AET). PET occurs when more than adequate moisture is available to meet the evaporation demand of the atmosphere. AET, on the other hand, is the actual amount of evapotranspiration that takes place when the soil is dry (I < PET) and the evaporation demand cannot be met. Thus, AET is always less than or equal to PET.

The monthly potential evapotranspiration may be estimated using the *Thornthwaite equation,*

$$\mathrm{PET\ (mm)} = 16\left(\frac{10\,T}{\mathrm{TE}}\right)^{a} \tag{11.4}$$

TABLE 11.4 Runoff Coefficient as Affected by Cover Material and Slope

	Runoff Coefficient C		
Type of area	Flat: Slope < 2%	Rolling: Slope 2–10%	Hilly: Slope > 10%
Grassed areas	0.25	0.3	0.3
Earth areas	0.6	0.65	0.7
Meadows and pasturelands	0.25	0.3	0.35
Cultivated land			
Impermeable (clay)	0.5	0.55	0.6
Permeable (loam)	0.25	0.3	0.35

Source: Adapted from Perry (1976).

where T is the temperature of the month under consideration and TE is the temperature efficiency index, given as the summation of the heat indices of the 12 months in the year. Considering that heat index for a given month is expressed as $(T/5)^{1.514}$, TE may be written as

$$\mathrm{TE} = \sum_{i=1}^{12} \left(\frac{T_i}{5} \right)^{1.514} \tag{11.5}$$

a in (11.4) is an empirical coefficient given by

$$a = 6.75 \times 10^{-7}\,(\mathrm{TE})^3 - 7.7 \times 10^{-5}\,(\mathrm{TE})^2 + 1.79 \times 10^{-2}\,(\mathrm{TE}) + 0.49239 \tag{11.6}$$

PET, as given by (11.4), also depends on the hours of sunlight, in addition to the temperature and the heat index. To account for the unequal durations of sunlight (daylight times) during the year, PET is usually multiplied by an adjustment factor. Values of this factor, listed in Table 11.5, depend on the month of the year and latitude of the location under consideration.

Equation (11.3) is implemented as follows to determine the percolation quantities. The quantity ($I-$ PET) is first calculated. If this quantity is positive, the evaporation demand by the atmosphere is met, and the surplus goes down into the soil cover and makes it wetter. In this case, AET = PET. If the water content of the soil cover is already high, say at its field capacity, the soil can store no additional water and the surplus will percolate into the waste (note that the field capacity is the maximum water content that a soil can retain under gravitational draining). On the other hand, if ($I-$ PET) is negative, the evaporation demand by the atmosphere is not met, and AET $<$ PET. The soil gives up its water content to the atmosphere if it is wet, and gets dryer. The amount of drying depends not only on the magnitude of ($I-$ PET), but also on the water content of the soil cover. Under these conditions, no percolation

TABLE 11.5 Adjustment Factors for Potential Evapotranspiration Computed by the Thornthwaite Equation

Latitude	Jan.	Feb.	Mar.	Apr.	May	June	July	Aug.	Sept.	Oct.	Nov.	Dec.
0	1.04	0.94	1.04	1.01	1.04	1.01	1.04	1.04	1.01	1.04	1.01	1.04
10	1.00	0.91	1.03	1.03	1.08	1.06	1.08	1.07	1.02	1.02	0.98	0.99
20	0.95	0.90	1.03	1.05	1.13	1.11	1.14	1.11	1.02	1.00	0.93	0.94
30	0.90	0.87	1.03	1.08	1.18	1.17	1.20	1.14	1.03	0.98	0.89	0.88
35	0.87	0.85	1.03	1.09	1.21	1.21	1.23	1.16	1.03	0.97	0.86	0.85
40	0.84	0.83	1.03	1.11	1.24	1.25	1.27	1.18	1.04	0.96	0.83	0.81
45	0.80	0.81	1.02	1.13	1.28	1.29	1.31	1.21	1.04	0.94	0.79	0.75
50	0.74	0.78	1.02	1.15	1.33	1.36	1.37	1.25	1.06	0.92	0.76	0.70

Source: Adapted from Criddle (1958).

TABLE 11.6 Soil Moisture Retention after Potential Evapotranspiration Has Occurred

	S_T (mm)[b]								
$\sum$ neg(I − PET)[a]	25	50	75	100	125	150	200	250	300
0	25	50	75	100	125	150	200	250	300
10	16	41	65	90	115	140	190	240	290
20	10	33	57	81	106	131	181	231	280
30	7	27	50	74	98	122	172	222	271
40	4	21	43	66	90	114	163	213	262
50	3	17	38	60	83	107	155	204	254
60	2	14	33	54	76	100	148	196	245
70	1	11	28	49	70	93	140	188	237
80	1	9	25	44	65	87	133	181	229
90	1	7	22	40	60	82	127	174	222
100		6	19	36	55	76	120	167	214
150		2	10	22	37	54	94	136	181
200		1	5	13	24	39	73	111	153
250			2	8	16	28	56	91	130
300			1	5	11	20	44	74	109
350			1	3	7	14	34	61	92
400				2	5	10	26	50	78
450				1	3	7	20	41	66
500				1	2	5	16	33	56
600					1	3	10	22	40
700						1	6	15	28
800						1	4	10	20
1000							1	4	10

Source: Adapted from Oweis and Khera, 1990; copyright © Elsevier Science; reprinted with permission.

[a]neg(I − PET) is lack of infiltration water needed for vegetation.

[b]S_T, soil moisture storage at field capacity.

occurs until there arises a situation when (I − PET) is again positive and the topsoil cover is brought to its field capacity.

The amount of drying from the soil cover that takes place during the dry spell (when there is no percolation) depends on the type of soil and the cumulative water deficit. This is essentially a problem of flow through unsaturated soil. Thornthwaite and Mather (1957) provide tables using which one can estimate the soil moisture retention after evapotranspiration occurred from a dry soil. Table 11.6 provides an abbreviated form as documented in Oweis and Khera (1990). The soil moisture storage required in this table may be obtained from Fig. 11.8, which summarizes the water-holding characteristics of various soils.

The final estimate of leachate q, coming out of the refuse, may now be obtained by applying water conservation principle to the waste body. Thus,

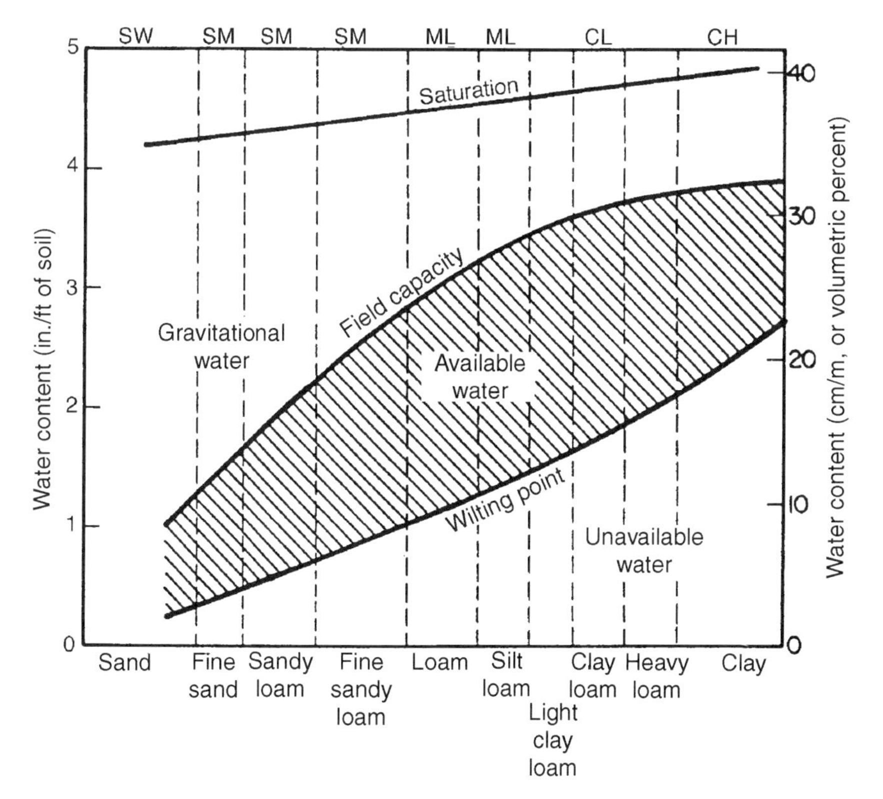

Fig. 11.8. Water storage capacities of USDA soils. (Adapted from Lutton et al., 1979.)

$$q = \text{PER} + W_d - \Delta S_w \tag{11.7}$$

where W_d is the water generated from waste decomposition and ΔS_w is the change in moisture storage of the waste. Until the waste attains its field capacity, no leachate may be expected at its bottom. The percolation will first satisfy the moisture-holding capacity of the waste and the surplus will then leach out of the waste. The water-holding capacity of waste is highly variable, as shown in Table 11.7 for some common types of refuse. Depending on the permeability of the barrier at the bottom of the refuse, a leachate mound may form, which provides a hydraulic head for leakage through the barrier.

The monthly water balance method thus requires accounting for how the precipitation is apportioned, one month at a time. The simple expressions involved in the calculations make it suitable for spreadsheet programming. Detailed case studies using this method are given in Fenn et al. (1975), Lutton et al. (1979), Lu et al. (1985), and Oweis and Khera (1990). Example 11.1 illustrates the method.

TABLE 11.7 Water Absorption Ranges for Solid Waste Components

	Moisture Content (% dry weight)					
	Water Absorption Capability			Total Moisture-Holding Capability[a]		
Component	Maximum	Average	Minimum	Maximum	Average	Minimum
Newsprint[b]	—	290	—	—	290[c]	—
Cardboard (solid and corrugated)[b]	—	170	—	—	170[c]	—
Other miscellaneous paper	400	—	100	400[c]	—	100[c]
Lawn clippings (grass and leaves)	200	—	60	370	—	140
Shrubbery, tree prunings	100	—	10	250	—	10
Food waste (kitchen garbage)	100	—	0	300	—	0
Textiles (cloth of all types, rope)	300	—	100	300[c]	—	100[c]
Wood, plastic, glass, metal (all inorganic)	—	0	—	—	0	—

Source: Adapted from Stone (1974).

[a]Calculated from water absorption plus initial moisture content in as-received samples.

[b]Sample variation was negligible.

[c]Initial moisture contents as received were less than 6% in the laboratory tests; therefore, they were considered negligible compared to the variation in moisture absorbed.

Example 11.1 A municipal solid waste landfill, located at 40° latitude, is covered by 2 ft (0.6 m) of silt loam cover. Using the following weather information given in Table E11.1A and assuming a runoff coefficient $C = 0.3$, determine the monthly percolation through the cover system.

TABLE E11.1A

	July	Aug.	Sept.	Oct.	Nov.	Dec.	Jan.	Feb.	Mar.	Apr.	May	June
Mean monthly temperature (°C)	24.4	23.5	19.8	13.7	7.4	1.7	0.8	1.0	5.2	11.2	17.1	21.9
Mean monthly precipitation (mm)	108	142	100	74	90	77	86	75	102	85	90	103

SOLUTION: Thickness of soil cover = 2 ft = 0.6 m. From Fig. 11.8, the field capacity of silt loam = 27.8 cm/m and the wilting point = 11.7 cm/m. The total storage capacity of the top cover is

$$S_T = (27.8 - 11.7) \times 0.6 = 9.66 \text{ cm}$$

The remaining calculations are shown in Tables E11.1B and E11.1C. From row 2, the temperature efficiency index (TE) = $\sum (T_i/5)^{1.514}$ = 56.47. The empirical coefficient a for use in Eq. (11.4) is obtained using Eq. (11.6):

$$a = 6.75 \times 10^{-7}\,\text{TE}^3 - 7.7 \times 10^{-5}\,\text{TE}^2 + 1.79 \times 10^{-2}\,\text{TE} + 0.49239 \approx 1.38$$

Monthly percolation from cover is shown in row 15. The total annual percolation is 204.6 mm.

TABLE E11.1B

Row	Parameter	Estimation
1	Temperature (°C)	Given
2	Temp. efficiency index	$(T_i/5)^{1.514}$
3	Potential evapotranspiration	Eqs. (11.4) and (11.5)
4	Adjustment factor for PET	Read from Table 11.5 for a latitude of 40°
5	Adjusted PET	Row 4 × row 3
6	Precipitation P (mm)	Given
7	Runoff coefficient C	Given
8	Runoff R	Row 7 × row 6
9	Infiltration I	Row 6 − row 8
10	I-adjusted PET	Row 9 − row 5
11	$\sum$ neg (I-adjusted PET)	Negative values of row 10
12	Cover storage S_c	Read from Table 11.6 for total storage S_T = 96.6 mm
13	Change in cover storage ΔS_c	Row 12 for current month − row 12 for preceding month
14	Actual evapotranspiration (mm)	Row 5 if row 10 is greater than or equal to 0; row 9 + abs(row 13) if row 10 is less than 0
15	Percolation	(Row 9 − row 13 − row 14) if row 13 is positive; zero if row 13 is negative

Alternatively, the computer program "Hydrological Evaluation of Landfill Performance" (HELP) can be used to estimate q. Details of the program implementation, applications, and limitations are provided by Schroeder et al. (1984a,b) and U.S. EPA (1991). The program has become popular during recent years primarily because of its ability to incorporate a number of layers in a waste containment system and also to account for the lateral drainage in individual layers. Besides providing accurate estimates of water budget components, HELP is useful in evaluating and comparing

TABLE E11.1C. Calculation of Monthly Percolation

Row	Parameter	July	Aug.	Sept.	Oct.	Nov.	Dec.	Jan.	Feb.	Mar.	Apr.	May	June	Sum
1	Temperature (°C)	24.4	23.5	19.8	13.7	7.4	1.7	0.8	1.0	5.2	11.2	17.1	21.9	
2	Temp. efficiency index	11.0	10.4	8.0	4.6	1.8	0.2	0.1	0.1	1.1	3.4	6.4	9.4	56.5
3	Potential evapotranspiration	120.6	114.5	90.4	54.4	23.2	3.1	1.1	1.5	14.3	41.2	73.8	103.9	
4	Adjustment factor for PET	1.3	1.2	1.0	1.0	0.8	0.8	0.8	0.8	1.0	1.1	1.2	1.3	
5	Adjusted PET	153.2	135.1	94.0	52.2	19.3	2.5	0.9	1.2	14.7	45.7	91.6	129.9	
6	Precipitation P (mm)	108.0	142.0	100.0	74.0	90.0	77.0	86.0	75.0	102.0	85.0	90.0	103.0	
7	Runoff coefficient C	0.3	0.3	0.3	0.3	0.3	0.3	0.3	0.3	0.3	0.3	0.3	0.3	
8	Runoff R	32.4	42.6	30.0	22.2	27.0	23.1	25.8	22.5	30.6	25.5	27.0	30.9	
9	Infiltration I	75.6	99.4	70.0	51.8	63.0	53.9	60.2	52.5	71.4	59.5	63.0	72.1	
10	I-adjusted PET	−77.6	−35.7	−24.0	−0.4	43.7	51.4	59.3	51.3	56.7	13.8	−28.6	−57.8	
11	$\sum$ neg (I-adjusted PET)	−77.6	−113.3	−137.4	−137.8							−28.6	−86.3	
12	Cover storage S_c	45.0	33.0	25.0	25.0	68.7	96.6	96.6	96.6	96.6	96.6	75.0	40.0	
13	Change in cover storage ΔS_c	0.0	−12.0	−8.0	0.0	43.7	27.9	0.0	0.0	0.0	0.0	−21.6	−35.0	
14	Actual evapotranspiration (mm)	75.6	111.4	78.0	51.8	19.3	2.5	0.9	1.2	14.7	45.7	84.6	107.1	
15	Percolation	0.0	0.0	0.0	0.0	0.0	23.5	59.3	51.3	56.7	13.8	0.0	0.0	204.6

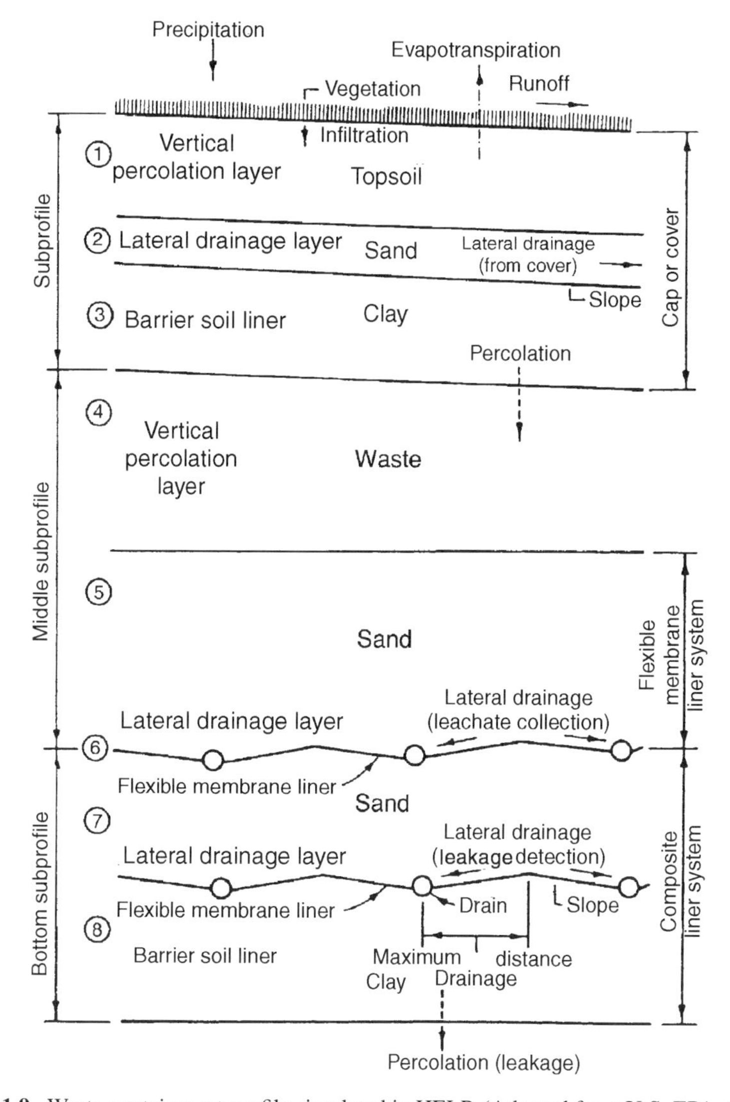

Fig. 11.9. Waste containment profile simulated in HELP. (Adapted from U.S. EPA, 1991.)

alternative containment systems. A typical hazardous waste containment profile that HELP is capable of simulating is shown in Fig. 11.9. The model uses methods similar to those described earlier. It apportions the precipitation into several water budget components as water percolates through each layer. These components include not only surface runoff, evapotranspiration, and water content changes in the soils, but also lateral drainage collected in each drain system and percolation through each layer of the system. It can generate daily, monthly, annual, and long-term average water budgets. A number of physical and hydrologic methods are used in HELP to simulate individual water budget components.

The following input data are required by HELP:

1. *Climatological data:* daily precipitation, daily mean temperature, daily solar evaporation, maximum leaf area index, growing season, and evaporative zone depth
2. *Soil and design data:* porosity, field capacity, wilting point, and saturated hydraulic conductivity of each layer; SCS runoff curve number, surface area, number of layers in the profile, and thicknesses of the layers

The program includes default daily weather data for 102 U.S. cities and is capable of synthetically generating weather data for 183 U.S. cities. In addition, default data are included for 15 soil types as well as solid waste. The output from the HELP consists primarily of percolation or leakage through each layer and depth of saturation on the surface of liners. Incremental and cumulative quantities of water budget for the various components are also provided. The HELP model allows one to look at the sensitivity of the water balance to numerous design variables. Case studies were reported in U.S. EPA (1991) to demonstrate the sensitivity analyses. The climatological regimes at three different locations (Santa Maria, California; Schenectady, New York; and Shreveport, Louisiana) were used to study the water balance for two typical landfill covers (Fig. 11.10). Two types of topsoils with different thicknesses were also

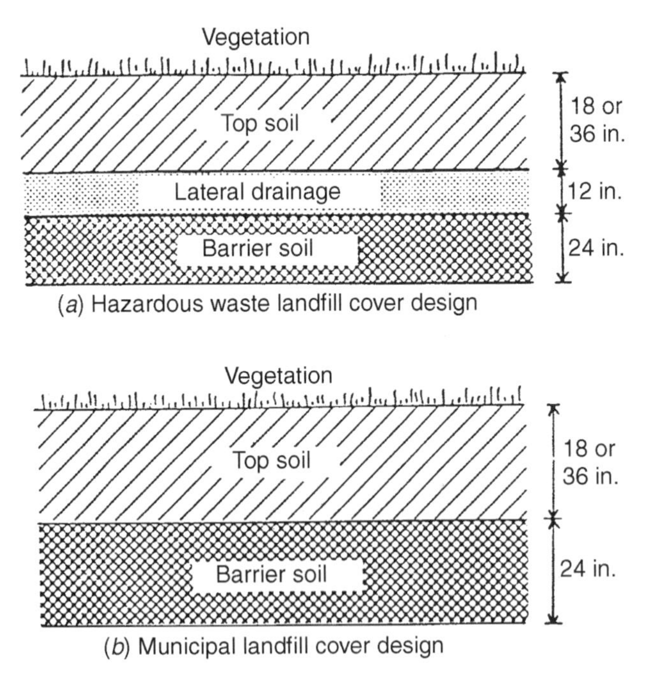

Fig. 11.10. Two types of landfill covers tested for water balance using HELP. (Adapted from U.S. EPA, 1991.)

TABLE 11.8 Effects of Climate and Vegetation on Water Balance

	% Precipitation for:					
	Two-Layer Cover Design[a]			Three-Layer Cover Design[b]		
	CA	LA	NY	CA	LA	NY
Poor grass						
Runoff	5.6	4.6	5.5	3.0	4.4	2.2
Evapotranspiration	51.8	53.0	52.1	51.6	51.9	50.3
Lateral drainage	—	—	—	41.2	40.6	44.0
Percolation	42.6	42.4	42.4	4.2	3.1	2.5
Good grass						
Runoff	3.1	0.2	3.5	0.0	0.2	0.0
Evapotranspiration	55.0	57.2	55.3	52.6	53.0	51.0
Lateral drainage	—	—	—	43.2	43.7	45.5
Percolation	42.9	42.6	41.2	4.2	3.1	2.5

Source: Adapted from U.S. EPA (1991).

[a]900 mm of sandy loam topsoil and 0.6 m of 10^{-6}-cm/s clay liner.

[b]450 mm of sandy loam topsoil, 300 mm of 0.03-cm/s sand with 60 m drain length at 3% slope, and 0.6 m of 10^{-7}-cm/s clay liner.

studied. The results from the simulations are shown in Tables 11.8 to 11.10. Table 11.8 shows the effects of climate and vegetation with and without lateral drainage. It is seen that vegetation decreases runoff and increases evapotranspiration but tends to have little effect on the rest of the water balance. A three-layer cover design (which allows for lateral drainage) is definitely superior to a two-layer design, indicating that the design of the cover is far more important than the climatological and vegetation factors. It should be noted that although vegetation is shown to have little effect on percolation in these simulations, its importance lies elsewhere in the context of erosion prevention.

Table 11.9 shows the effects of topsoil thickness on the water balance for the cover design where lateral drainage is absent. The effects of topsoil thickness are similar for all three locations. Runoff and evapotranspiration were greater for the thinner topsoil, indicating that the head above the barrier maintained higher moisture contents

TABLE 11.9 Effects of Topsoil Thickness on Water Balance[a]

	% Precipitation for:					
	45.7 cm of Topsoil			91.4 cm of Topsoil		
	CA	LA	NY	CA	LA	NY
Runoff	11.2	7.5	13.4	5.6	4.6	5.5
Evapotranspiration	51.9	56.9	54.5	51.8	53.0	52.1
Percolation	36.9	35.6	32.1	42.6	42.4	42.4

Source: Adapted from U.S. EPA (1991).

[a]Sandy loam topsoil with a poor stand of grass underlain by 0.6 m of 10^{-6}-cm/s clay liner.

TABLE 11.10 Effects of Topsoil Type on Water Budget Components

	% Precipitation for Three-Layer Cover Design[a]					
	Sandy Loam			Silty Clayey Loam		
	CA	LA	NY	CA	LA	NY
Runoff	3.0	4.4	2.2	21.6	22.3	19.2
Evapotranspiration	51.6	51.9	50.3	61.2	64.4	58.6
Lateral drainage	41.2	40.6	44.0	15.0	11.3	20.3
Percolation	4.2	3.1	2.5	2.2	2.0	1.9

Source: Adapted from U.S. EPA (1991).

[a]450 mm of topsoil with poor stand of grass, 300 mm of 0.03-cm/s sand with 60-m drain length at 3% slope, and 0.6 m of 10^{-7} cm/s clay liner.

in the evaporative zone. The percolation was consequently less than in the cases with greater topsoil thickness. Although thin topsoil is in general favorable for reduced percolation, it is important to provide adequate thickness for the top cover to support vegetation, maintain soil stability, and control erosion.

Table 11.10 shows the effects of topsoil type on water budget components. Clayey topsoil increased both runoff and evapotranspiration, which in turn greatly decreased lateral drainage and percolation. Considering the California site, runoff increased from 3% to 22% of the precipitation, and evapotranspiration increased from 52% to 61% of precipitation, when the topsoil was changed from sandy loam to a silty clayey loam. The lower hydraulic conductivity of clayey topsoil slowed down the lateral drainage, which was reduced from about 41% to 15% of precipitation. Because of these effects, percolation differed only slightly, from 4% to 2%, between the two soil types. Figure 11.11 shows another sensitivity analysis to illustrate how the percolation through the top liner is dependent on thickness and permeability of the liner and drainage provision in the protective cover. In general, reducing the liner's permeability by an order of magnitude lowers the percolation considerably. In contrast, using a thicker liner does not result in comparable reductions in percolation. As outlined above, thicker protective covers are associated with greater quantities of percolation. Drains could be used in the protective cover to reduce percolation in the case of thicker covers and/or higher liner permeabilities.

11.4 LEACHATE COLLECTION AND REMOVAL SYSTEMS

The flow of leachate through barriers at the bottom of the containment structure can be minimized in two ways: by minimizing the leachate accumulated at the top of the barrier via a collection system and/or by designing the barrier such that its permeability is very low. Design of the leachate collection system is an important element in the overall design of the waste containment system. It includes sizing and spacing of the pipes used to collect and remove the leachate. The total leachate percolated out of the waste is apportioned between the leachate collection system and the liner leakage. Thus the problem of leachate collection above the liner and that

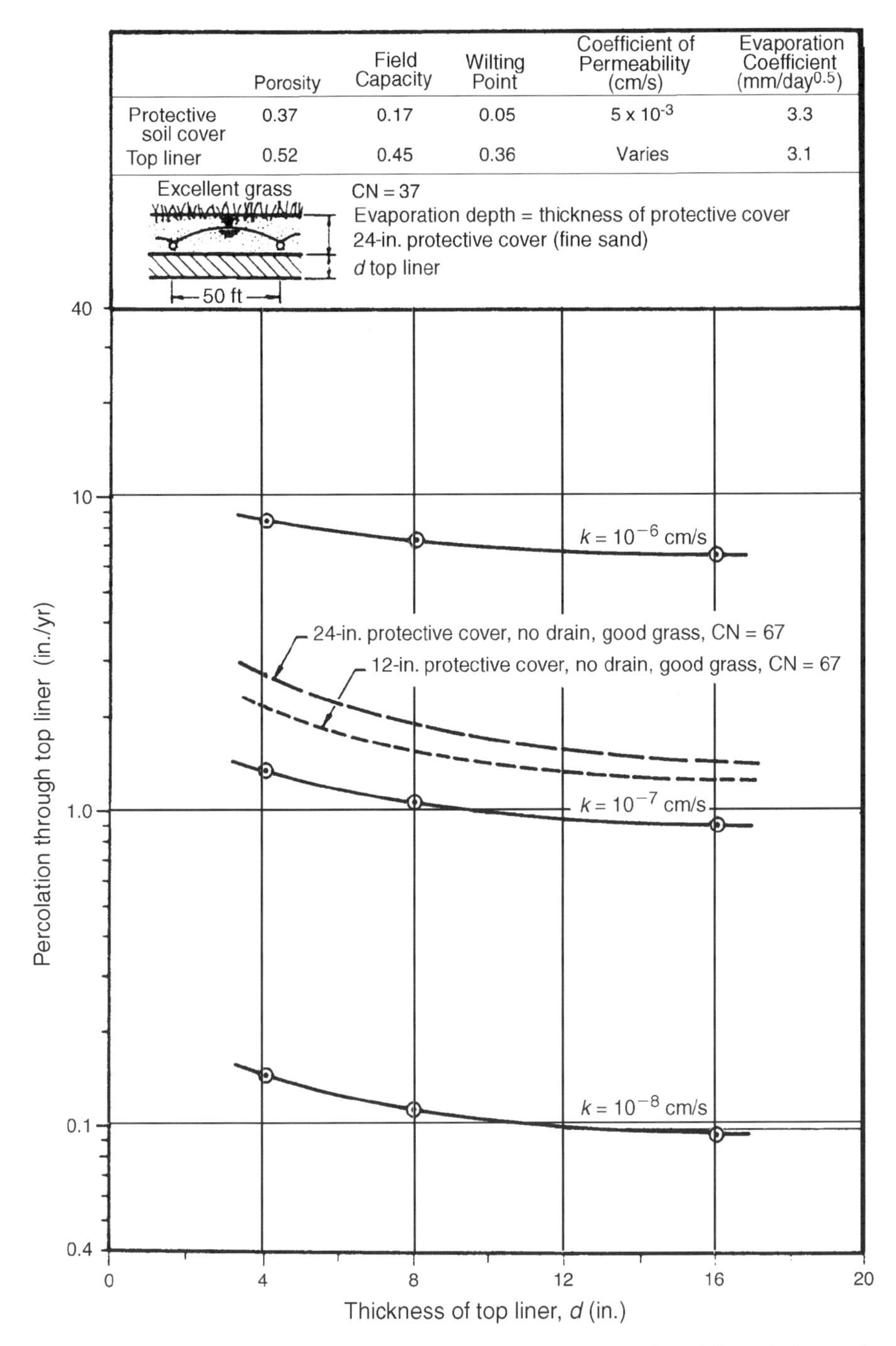

Fig. 11.11. Sensitivity analysis on a cover system with drains. (Adapted from Ardaman & Associates, Inc., 1989.)

of flow through the barrier is an integrated one and cannot be decoupled. However, in practice, simpler methods are used to predict the height of the leachate mound assuming either a constant leakage rate through the liner or no leakage at all. Leakage through the liner is often ignored for a conservative design of the leachate collection and treatment systems.

The maximum height of leachate mound above the barrier depends on how the collection system is laid out. Figure 11.12 shows the typical configurations of a collection system. The configuration shown in Fig. 11.12*a* offers a simple expression for maximum height $h_{\max}$ of the leachate mound. Applying Darcy's law to estimate flow in the drainage layer, the height of the mound $h(x)$ can be estimated as

$$h(x) = \left[\frac{q}{k_d}(L-x)x\right]^{0.5} \tag{11.8}$$

where q is the percolation rate of leachate coming onto the drainage layer, k_d the saturated hydraulic conductivity of the drainage layer, and L the drain spacing. The maximum height of the mound occurs at the midpoint between the two drains; therefore,

$$h_{\max} = \frac{L}{2}\left(\frac{q}{k_d}\right)^{0.5} \tag{11.9}$$

From a design standpoint, the drains must be capable of discharging (qL) per unit length, since percolation over one-half of the drain spacing flows from either side of each drain. When one puts constraints on $h_{\max}$ to ensure that the mound lies totally within the drainage layer, Eq. (11.9) allows estimation of the required spacing L or hydraulic conductivity k_d.

Figure 11.12*b* and *c* illustrate more common configurations of the collection system, where the barriers are typically graded to enhance flow toward leachate collection drains. Moore (1980) developed a solution for the case shown in Fig. 11.12*b*,

$$h_{\max} = \frac{L\sqrt{c}}{2}\left(\frac{\tan^2\alpha}{c} + 1 - \frac{\tan\alpha}{c}\sqrt{\tan^2\alpha + c}\right) \tag{11.10}$$

where $c = q/k_d$ and α is the slope angle of the liner. As stated earlier, Eq. (11.10) may also be used to obtain the required design parameters given $h_{\max}$.

For the configuration shown in Fig. 11.12*c*, where drains are provided at the top of the slope also, McBean et al. (1982) developed the following solution for h:

$$x = x_{h\max}\left\{1 - \frac{\left(h_0^2/x_{h\max}^2 - mh_0/x_{h\max} + c\right)^{1/2}}{\left(h^2/(x_{h\max}-x)^2 - [mh/(x_{h\max}-x)] + c\right)^{1/2}} \times \exp\left[\frac{m}{\sqrt{4c+m^2}}\left\{\tan^{-1}\frac{(2h_0/x_{h\max}) - m}{\sqrt{4c+m^2}} - \tan^{-1}\frac{[2h/(x_{h\max}-x)] - m}{\sqrt{4c+m^2}}\right\}\right]\right\} \tag{11.11}$$

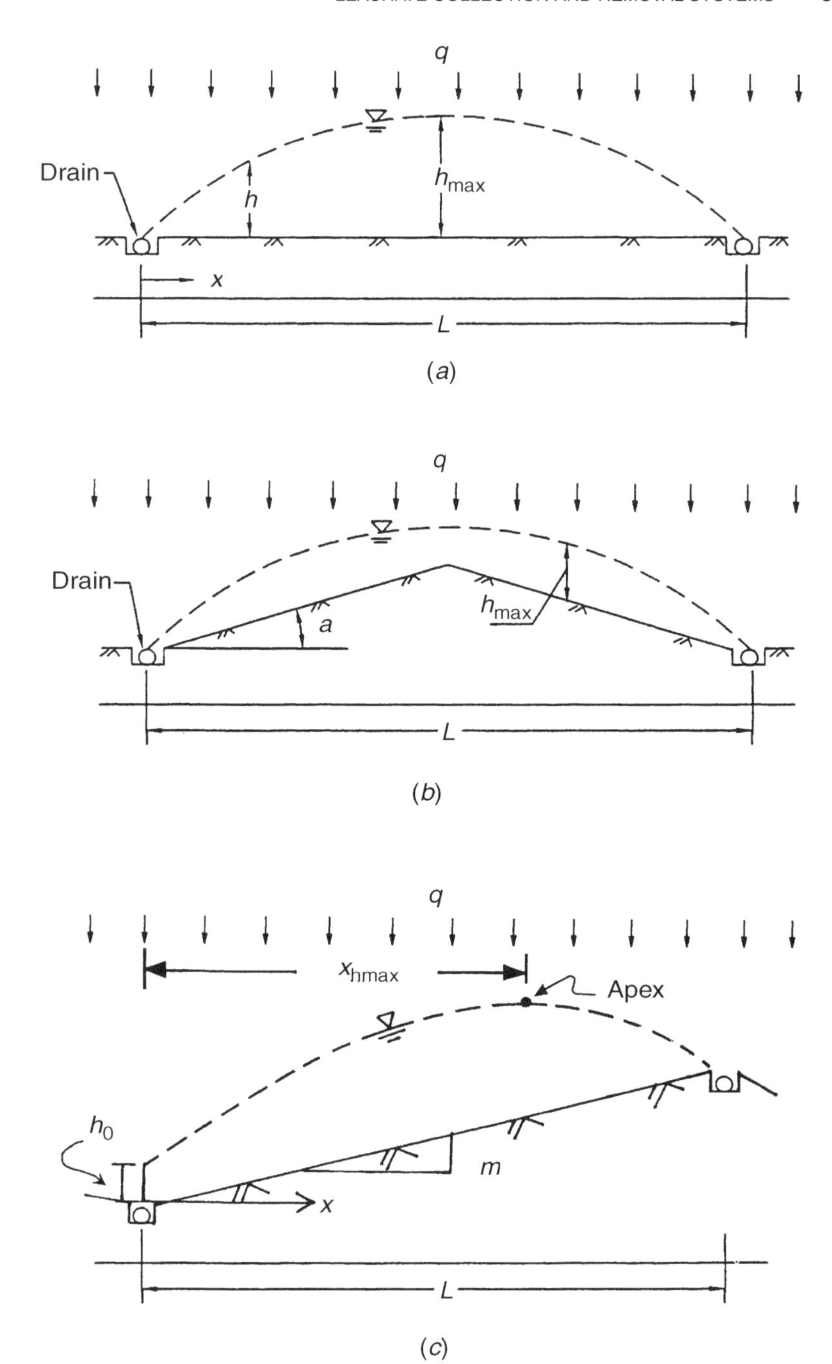

Fig. 11.12. Leachate collection system configurations.

where $h = h(x)$. For slope $m = 0$, the solution simplifies to

$$h(x) = \sqrt{h_0^2 + 2\left(x_{h\max}x - \frac{x^2}{2}\right)} \tag{11.12}$$

McBean et al. (1982) note that when $c(= q/k_d)$ is large, the computed water table profile in the drainage layer is insensitive to h_0, which is the boundary condition at $x = 0$. The derivation of Eqs. (11.10) and (11.11) is left to the reader as an exercise.

It should be noted that the solutions above are approximate since the leakage through the barrier was not considered in their development. When the barrier leaks a portion of q, say q_i, the actual amount of leachate collected in the drains is equal to $(q - q_i)L$. Methods coupling leakage through barrier and flow in the drainage layer are available in the literature (Wong, 1977; Dematrocopoulos et al., 1984; Korfiatis and Demetracopoulos, 1986; McEnroe and Schroeder, 1988). Wong (1977) developed some of the earlier equations enabling apportionment of leachate between drainage collection and leakage through the barrier. For the landfill system shown schematically in Fig. 11.13, Wong made the following assumptions to simplify the mathematical treatment.

1. The drainage layer and the refuse–cover mixture above the liner have the same conductivity.

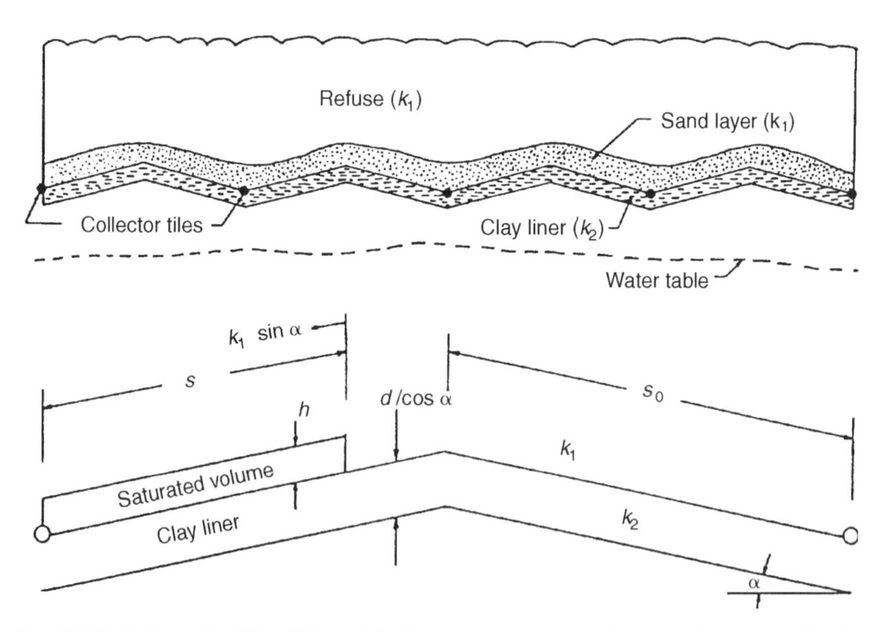

Fig. 11.13. Schematic of landfill model showing parameters relevant to leachate collection system. (Adapted from Wong, 1977.)

2. All layers are at field capacity, so that any infiltration of precipitation into the refuse results in gravity drainage to the bottom of the landfill.
3. The percolation of leachate results in an instantaneous input onto the clay liner, so that a saturated volume of rectilinear shape is formed.
4. The saturated volume above the liner retains its rectilinear shape as drainage toward the drains and leakage through the barrier take place.

Under these assumptions, Wong arrived at the following equations to describe the movement of leachate quantitatively:

$$\frac{s}{s_0} = 1 - \frac{t}{t_1} \tag{11.13}$$

$$\frac{h}{h_0} = \left(1 + \frac{d}{h_0 \cos\alpha}\right) e^{-\kappa t/t_1} - \frac{d}{h_0 \cos\alpha} \qquad 0 \le t \le t_1 \tag{11.14}$$

where

$$t_1 = \frac{s_0}{k_1 \sin\alpha} \tag{11.15}$$

$$\kappa = \frac{s_0}{d}\frac{k_2}{k_1} \cot\alpha \tag{11.16}$$

s is the length of the saturated volume of soil at time t (see Fig. 11.13), h the thickness of the saturated volume at time t, s_0 and h_0 the initial dimensions of the saturated volume, k_1 the hydraulic conductivity of the material above the liner, k_2 the hydraulic conductivity of the clay liner, α the slope angle of the liner, and d the thickness of the liner. Note that the equations are applicable as long as s/s_0 and h/h_0 are positive values and the time is measured from the instant when the saturated volume appears on the liner.

The solution expressed in Eqs. (11.13) and (11.14) is shown in Fig. 11.14. The information shown in Fig. 11.14 may be used to apportion the leachate volume as follows. At any time t, the volume that remains saturated is $V = sh\cos\alpha \approx sh$, since $\cos\alpha \approx 1.0$. Therefore, the rate of volume change is

$$\frac{dV}{dt} = h\frac{ds}{dt} + s\frac{dh}{dt} \tag{11.17}$$

The first term on the right-hand side of Eq. (11.17) denotes the rate at which leachate is drained to the collection pipe, and the second term denotes the rate at which saturated volume decreases due to leakage through the liner. Integrating these two terms separately and normalizing with respect to the initial saturated volume V_0 ($= s_0 h_0$) gives us

$$\frac{V_1}{V_0} = \int \frac{h}{h_0} d\left(\frac{s}{s_0}\right) \tag{11.18}$$

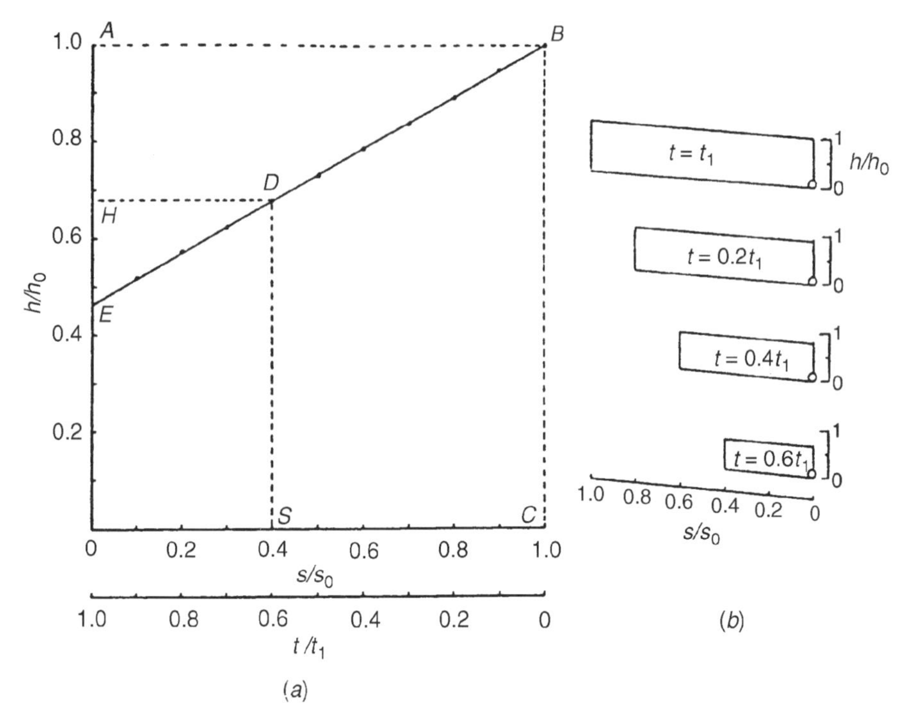

Fig. 11.14. Solution for the coupled barrier leakage and leachate collection problem: (*a*) h/h_0 as a function of t/t_1 and s/s_0; (*b*) schematic showing the physical dimensions of saturated volume on the liner at various times. (Adapted from Wong, 1977.)

and

$$\frac{V_2}{V_0} = \int \frac{s}{s_0} d\left(\frac{h}{h_0}\right) \tag{11.19}$$

where V_1 is the volume drained away by the collection system and V_2 is the volume leaked through the liner. Equations (11.18) and (11.19) imply that the area *SDBC* in Fig. 11.14 represents V_1/V_0, the fraction of the initial saturated volume drained away by the leachate collection system, while the area *ABDH* represents V_2/V_0, the fraction of the initial saturated volume that has leaked through the liner. Thus, Fig. 11.14 serves as a graphical solution for the apportionment of saturated volume at time $t = 0.6t_1$ (point *S*). The area *OHDS* represents the fraction of the initial saturated volume that remains on the liner. V_2/V_0 is often referred as the collection efficiency of the drains. Thus, the closer the point *E* is to point *A*, the greater is the collection efficiency. In reality, *E* will never coincide with *A* because of the leakage of the liner. From a design standpoint, however, the parameters involved in Eq. (11.14) must be chosen such that *E* is as close to *A* as possible. Despite the numerous assumptions made in Wong's model, it is one of the most useful models available at present to apportion leachate. A detailed evaluation of the model and the effects of various design parameters on the apportionment are provided in Wong (1977) and Kmet et al. (1981).

Example 11.2 Assuming that the quantity of leachate on the bottom liner is equal to percolation from the top cover (as estimated in Example 11.1), determine the drain spacing L that would yield a leachate mound less than 1 m high. Assume that the permeability of the drainage layer $k_d = 10^{-3}$ cm/s.

SOLUTION: From Example 11.1, maximum percolation is in the month of January; therefore,

$$q = 59.3 \text{ mm}/31 \text{ days} = 1.9 \text{ mm/day}$$

Using Eq. (11.9), we have

$$L = 2h_{\max}\left(\frac{k_d}{q}\right)^{0.5} = 2 \times 1\,\text{m} \times \left(\frac{10^{-3}\,\text{cm/s}}{1.9\,\text{mm/day}} \times \frac{10\,\text{mm}}{1\,\text{cm}} \times \frac{86400\,\text{s}}{1\,\text{day}}\right)^{0.5} \approx 43\,\text{m}$$

Example 11.3 Redo Example 11.2 with a graded barrier ($\alpha = 10°$).

SOLUTION: Using Eq. (11.10) yields

$$L = \frac{2h_{\max}}{\sqrt{c}} \frac{1}{(\tan^2\alpha/c) + 1 - (\tan\alpha/c)\sqrt{\tan^2\alpha + c}}$$

for

$$c = \frac{q}{k_d} = \frac{1.9\,\text{mm/day}}{10^{-3}\,\text{cm/s}} \times \frac{1\,\text{cm}}{10\,\text{mm}} \times \frac{1\,\text{day}}{86{,}400\,\text{s}} \approx 0.0022$$

$$L = \frac{2 \times 1\,\text{m}}{\sqrt{0.0022}} \frac{1}{(\tan^2 10°/0.0022) + 1 - (\tan 10°/0.0022)\sqrt{\tan^2 10° + 0.0022}} \approx 83.9\,\text{m}$$

11.5 SEEPAGE THROUGH LINERS

As stated in Section 11.4, the barrier leakage and lateral flow in collection system constitute a coupled problem. However, it is customary to separate the leachate collection system design from the barrier design. While designing the barrier, the head of leachate mound is assumed to be constant to simplify the flow problem. For a conservative estimate of leakage through the barrier, the leachate mound may be assumed to be at its maximum elevation since the leakage estimate will be higher under this assumption.

When considering leakage through the barrier, it is important to keep in mind that both quality and quantity of leakage govern the impact of the waste containment system on the surrounding soil and groundwater resources. Thus, a significant leakage of "relatively clean" water might be more acceptable than a small leakage of highly contaminated water. Earlier regulations of waste containment systems were based on limiting the quantity of leakage through the barrier (i.e., the leakage rate should not exceed 1.65 in./yr, the hydraulic conductivity of the barrier should not exceed 1×10^{-7} cm/s, and so on). As such, these quantity-based criteria value the advective transport of contaminants through the barrier with little regard to the effects of dispersion and diffusion mechanisms. More recent criteria for the performance of waste containment systems are based on all mass transport mechanisms.

For a simple configuration of a soil barrier underlain by subsoil (Fig. 11.15), Darcy's law can be used to estimate the flow rate. The pore velocity (or seepage velocity) through the barrier v_s will be maximum under fully saturated conditions; therefore,

$$v_s = \frac{k_s i}{n} \tag{11.20}$$

where k_s is the saturated hydraulic conductivity of the barrier (L/T), i the hydraulic gradient (dimensionless), and n the porosity of the barrier (dimensionless). The transit time through the barrier t may therefore be expressed as

$$t = \frac{d}{v_s} = \frac{dn}{k_s i} \tag{11.21}$$

where d is the thickness of the barrier (L). From a design standpoint, if transit time t is regulated, the required thickness of the barrier is

$$d = \frac{k_s i t}{n} \tag{11.22}$$

Caution should be exercised in identifying the gradient i. While the pressure head at the top of the liner may be well defined, the condition at the bottom of the liner is uncertain. A zero-pressure head is often assumed at the bottom of the liner. Under these conditions, the hydraulic gradient is

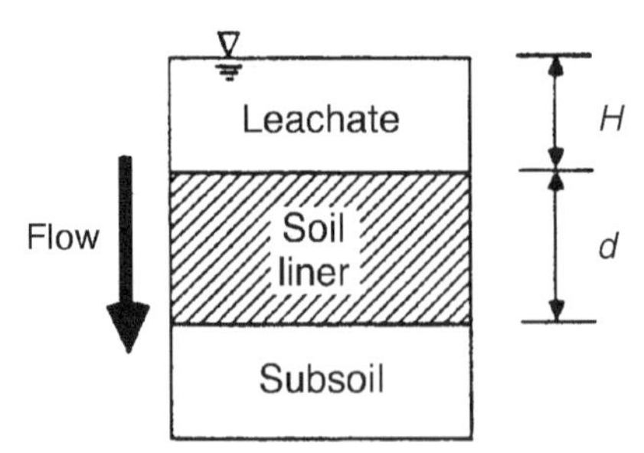

Fig. 11.15. Schematic of a soil liner underlain by subsoil.

$$i = \frac{H + d}{d} \tag{11.23}$$

where H is the height of leachate mound on the barrier. The expanded expressions for t and d then become

$$t = \frac{dn}{k_s} \frac{d}{H + d} \tag{11.24}$$

and

$$d = 0.5 \left\{ \frac{k_s t}{n} + \left[\left(\frac{k_s t}{n} \right)^2 + \frac{4 k_s t H}{n} \right]^{0.5} \right\} \tag{11.25}$$

The equations above are valid for barriers under fully saturated conditions. Although it may be conservative from a design perspective to assume that the barrier is fully saturated, it may sometimes be necessary to estimate the transit time through an initially unsaturated barrier. The Green–Ampt wetting front model is suitable for this purpose. It is a phenomenological model based on the concept of a sharp wetting front, shaped like a square wave, moving down through the unsaturated soil (Fig. 11.16). Above the wetting front, the soil is assumed to be fully saturated, and below it, the moisture content is assumed to be at its initial level. The energy required for the movement of the wetting front is provided by the water suction below the front.

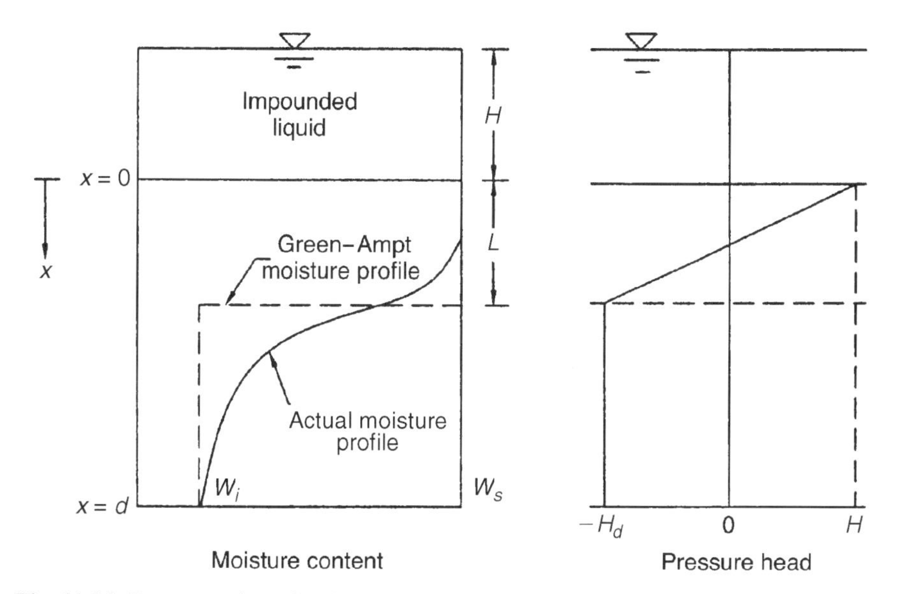

Fig. 11.16. Representation of moisture profile in the barrier using the Green–Ampt wetting front model.

The flow rate per unit area of the barrier q_u may be obtained using Darcy's law and the mass conservation principle. Thus,

$$q_u = k_u \frac{H + L + H_d}{L} \tag{11.26}$$

$$q_u = (w_s - w_i)\frac{dL}{dt} \tag{11.27}$$

where k_u is the unsaturated hydraulic conductivity at the wetting front, L the location of the wetting front, H_d the capillary suction head below the wetting front, w_s the saturated moisture content of the barrier soil, w_i the initial moisture content of the barrier soil, and t the time. Combining Eqs. (11.26) and (11.27) and integrating by parts over $L = 0$ to d, the transit time t may be expressed as

$$t = \frac{w_s - w_i}{k_u}\left[d - (H + H_d)\ln\left(1 + \frac{d}{H + H_d}\right)\right] \tag{11.28}$$

It should be noted that the approach above is a simple idealization of unsaturated flow and as such is only approximate. More accurate estimates of transit time through the barrier are possible only through a rigorous mathematical solution of the unsaturated flow.

The permeability of the bottom liner is a crucial parameter in the design of waste containment systems. As discussed in Chapter 2 (Figs. 2.9 to 2.11), the permeability of compacted clays depends strongly on the compaction variables (i.e., molding water content and energy of compaction). It is generally established that compacting wet of optimum yields low permeability because of the dispersed structure associated with the high molding water contents. Using laboratory results on compaction curves and permeability data corresponding to a number of points on the compaction curves, an acceptable range may be defined for moisture-density requirements (Fig. 11.17*a*). The acceptable zone provides guidance on the compaction variables during liner placement. In cases where shrinkage and cracking potential exists and the shear strength of the liner is also an important consideration to ensure structural integrity of the liner, the acceptable zone may be further restricted, as shown in Fig. 11.17*b*. In addition to meeting the moisture-density requirements on compaction curves, care must be taken during construction phase to destroy clods, eliminate lift interfaces (which may become flow paths), and avoid desiccation cracking of the clay.

When synthetic membranes (geomembranes) are used in combination with a compacted clay layer, many factors must be considered for a successful geomembrane design and installation, including:

1. Selection of proper membrane materials
2. Proper subgrade preparation
3. Membrane transportation, storage, and placement
4. Proper installation conditions (weather, temperature, etc.)

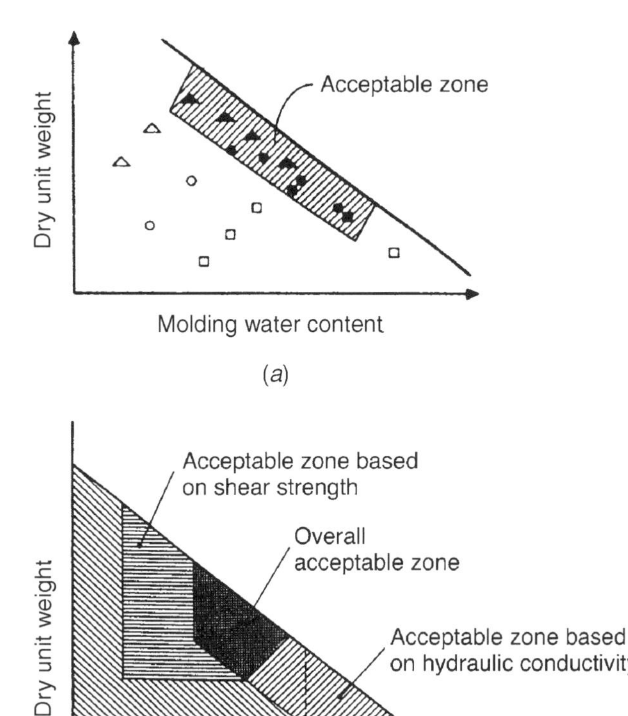

Fig. 11.17. Acceptable zones of molding water content and dry density relationships: (*a*) zone based on low permeability requirement of the liner; (*b*) zone based on low hydraulic conductivity, low desiccation-induced shrinkage, and high unconfined compressive strength. [(*b*) Adapted from Daniel and Wu, 1993; copyright © ASCE; reprinted with permission.]

5. Seaming and testing
6. Application of construction quality assurance

These factors are elaborated in U.S. EPA (1994). In particular, it is important to prevent imperfect seams or pinholes in geomembranes, as they can greatly increase the amount of leakage. The leakage rate through a hole in the geomembrane of a composite liner can be calculated empirically using

$$Q = 3a^{0.75}h^{0.75}k^{0.5} \tag{11.29}$$

where Q is the leachate leakage through the hole in the geomembrane of the liner system (m^3/s), a the area of hole (m^2), h the hydraulic head of liquid applied to

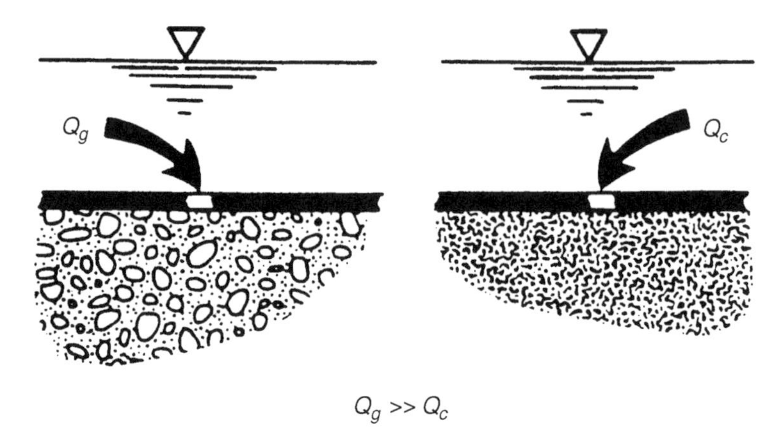

Fig. 11.18. Advantage of a composite liner system with the membrane underlain by a low-permeability soil. (Adapted from U.S. EPA, 1989.)

the membrane (m), and k the hydraulic conductivity of soil layer underlying the geomembrane (m/s). The impact of penetrations in the membrane can be reduced by many orders of magnitude using a low-permeability clay liner underneath the geomembrane (Fig. 11.18). When geomembranes are used alone as liners, a 1-cm^2 hole in the membrane with 1 ft of head can lead to a leakage rate as high as 3300 gpd. The presence of a clay layer with a conductivity of 10^{-7} cm/s underneath the membrane can reduce this rate to 0.2 gpd.

11.6 SEEPAGE THROUGH SLURRY WALLS

The purpose of slurry walls is to cut off seepage for either temporary or permanent control of seepage. Traditionally, they are used to cut off seepage into excavations for foundation construction and other purposes. At waste containment facilities, they are constructed as continuous walls at the periphery of a buried waste or landfill to prevent outward seepage of pollutants (Fig. 11.19). By providing drainage at the interior edge of the isolated zone, the seepage may be directed to be inward. In designing and constructing a slurry wall, the key consideration is to achieve a low permeability. The structural strength of the wall may also be an important consideration.

Slurry walls are constructed by excavating narrow vertical trenches, typically 2 to 5 ft wide, through the pervious materials to impervious strata lying as deep as 100 ft. To keep the trench open, it is filled with bentonite suspension to its top. A minimum viscosity of 40 sec-Marsh is generally recommended to assure trench stability and to provide a filter cake at the interface of slurry wall and native soil (D'Appolonia, 1980). This corresponds to about 5 to 7% bentonite by weight. With the slurry level at the top of the trench and the groundwater level several feet below the top of the trench, trenches more than 100 ft deep and 1000 ft long could be kept open for several weeks (D'Appolonia, 1980).

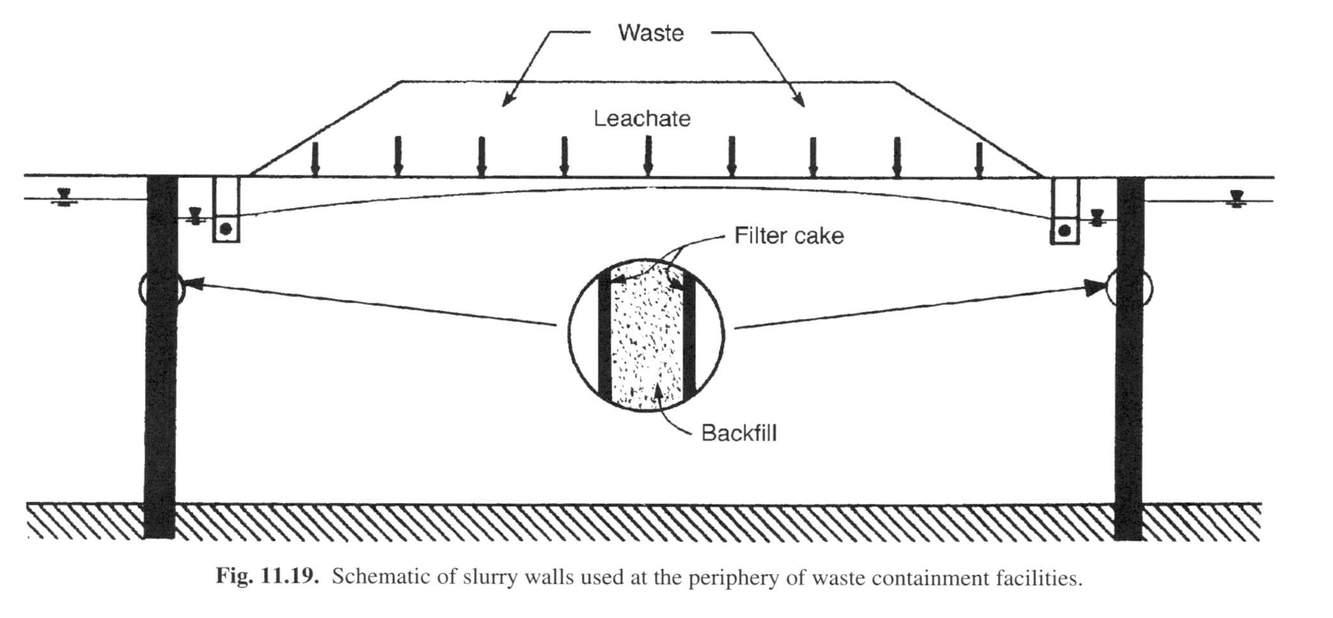

Fig. 11.19. Schematic of slurry walls used at the periphery of waste containment facilities.

After excavation, the slurry in the trench is displaced using a backfill material, which is typically prepared at the site by sluicing the soil material with either fresh slurry or slurry pumped from the trench (in special cases, expensive cement–bentonite and concrete grouts are also used as backfill material). The unit weight of the backfill material is typically between 105 and 120 pcf. For easy displacement, it is recommended that the slurry unit weight is about 15 pcf less than the backfill unit weight. To minimize both permeability and compressibility, granular materials containing 20 to 40% fines are recommended to prepare the backfill with a minimum bentonite content of 1% by dry weight. For detailed construction methods, the reader is referred to D'Appolonia (1980).

The horizontal permeability of the cutoff walls constructed as above is determined as an effective permeability of three layers: the backfill material in the trench and the two filter cakes formed on either side of the backfill material at the interface with the native soil. Using Darcy's law yields

$$q = ki = k\frac{\Delta h}{2t_c + t_b} = k_c\frac{\Delta h_c}{2t_c} = k_b\frac{\Delta h_b}{t_b} \qquad \Delta h = \Delta h_c + \Delta h_b \tag{11.30}$$

where q is the flow rate, k the effective permeability of the cutoff wall, k_c and k_b the permeability of the cake formed and the backfill material, respectively, Δh the total

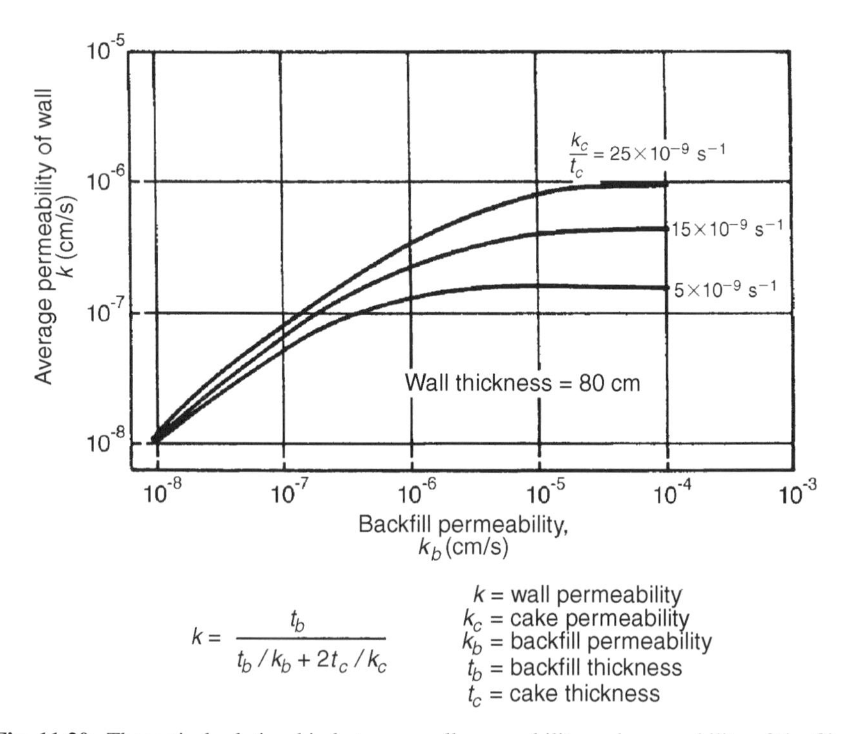

Fig. 11.20. Theoretical relationship between wall permeability and permeability of the filter cake and backfill. (Adapted from D'Appolonia, 1980; copyright © ASCE; reprinted with permission.)

head loss, Δh_c and Δh_b = head loss through cake and backfill material, respectively, and t_c and t_b the thickness of the cake and backfill, respectively. Considering that the thickness of the cake is much less than that of the backfill wall (i.e., $t_c \ll t_b$),

$$k = \frac{t_b}{t_b/k_b + 2t_c/k_c} \tag{11.31}$$

D'Appolonia states that for a wide variety of practical applications, the ratio k_c/t_c varies between the relatively narrow limits of 5×10^{-9} to $25 \times 10^{-9} s^{-1}$. For this range the relationship between the backfill permeability and the effective permeability of the wall is shown in Fig. 11.20. It is seen that an effective permeability of 10^{-6} cm/s could be achieved even for backfill materials with permeability as high as 10^{-4}cm/s. The effective permeability is controlled by the backfill when the backfill permeability is low and by the filter cake when the backfill permeability is high.

Because of the important contribution of the filter cake to the effective permeability of the wall, a filter-press test is often conducted in the laboratory (Fig. 11.21). To simulate field conditions during slurry trench construction, a filter cake is formed by placing slurry over the filter paper and applying a specific head to the slurry. Excess slurry is decanted and replaced with water, and the flow through filter cake is then

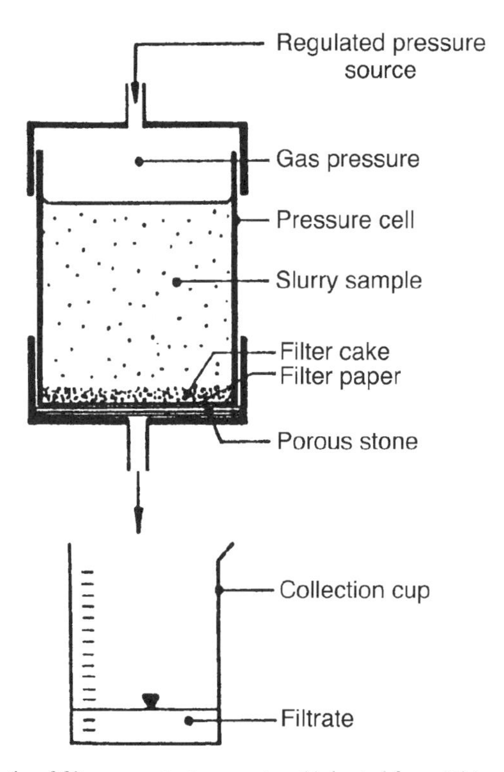

Fig. 11.21. Schematic of filter-press test apparatus. (Adapted from D'Appolonia, 1980; copyright © ASCE; reprinted with permission.)

TABLE 11.11 Filter Cake and Backfill Permeability Increase Due to Leaching with Various Pollutants[a]

Pollutant	Filter Cake	SB Backfill (Silty or Clayey Sand), 30 to 40% fines
Ca^{2+} or Mg^{2+} at 1000 ppm	N	N
Ca^{2+} or Mg^{2+} at 10,000 ppm	M	M
NH_4NO_3 at 10,000 ppm	M	M
HCl (1%)	N	N
H_2SO_4 (1%)	M	N
HCl (5%)	M/H[b]	M/H[b]
NaOH (1%)	M	M
CaOH (1%)	M	M
NaOH (5%)	M	M/H[b]
Seawater	N/M	N/M
Brine (SG = 1.2)	M	M
Acid mine drainage ($FeSO_4$; pH $\sim$ 3)	N	N
Lignin (in Ca^{2+} solution)	N	N
Alcohol	H (failure)	M/H

Source: Adapted from D'Appolonia (1980).

[a]N, no significant effect, permeability increase by about a factor of 2 or less at steady state; M, moderate effect, permeability increase by factor of 2 to 5 at steady state; H, permeability increase by factor of 5 to 10.

[b]Significant dissolution likely.

measured applying a head. The ratio k_c/t_c is determined as the flow rate divided by the head and the sample area.

At waste containment facilities, the presence of contaminants in the pore fluid may alter the permeability of the backfill material and the filter cake. Laboratory tests should be conducted with the pollutants expected at the site, to determine the exact nature of permeability variation. Table 11.11 provides a qualitative guidance on the likely effect of pollutants on the permeability of filter cake and backfill material.

11.7 FILTER DESIGN

In addition to the filter criteria described in Chapters 7 and 8, filters used at waste containment facilities must be designed to ensure that chemical and biological clogging is minimized. Clogging of filters and drainage layers may increase the hydraulic head over the liners to an undesirable level. According to research conducted by Koerner and Koerner (1990), potential clogging of filters (either natural soil or geotextile) may occur due to precipitates in the leachate, biological microorganisms, or to a synergistic combination of both. Reddi et al. (2000) and Hajra et al. (2002) studied the extent of clogging due to the presence of suspended colloidal material and changes in chemical composition of pore fluid (ionic strength and pH). They documented permeability reductions of filters and drainage materials greater than an order of magnitude due to these factors (Figs. 7.19 and 7.20).

Biological growth in filters and drainage layers occurs because of the high biological oxygen demand (BOD) level of the leachate being removed. The potential for biological clogging can be reduced using coarse stones around collection pipes and providing cleanouts for the primary leachate collection pipes. Koerner et al. (1994) proposed a design formulation linking the permeability and clogging criteria together and introduced an overall factor of safety against excessive clogging. The factor of safety (FS) against excessive long-term filter clogging was expressed as

$$\mathrm{FS} = \frac{k_{\mathrm{allow}}}{k_{\mathrm{reqd}} \times \mathrm{DCF}} \tag{11.32}$$

$$\mathrm{FS} = \frac{\psi_{\mathrm{allow}}}{\psi_{\mathrm{reqd}} \times \mathrm{DCF}} \tag{11.33}$$

where k_{allow} is the allowable permeability, k_{reqd} the required permeability, DCF the drain correction factor, ψ_{allow} the allowable permittivity, and ψ_{reqd} the required permittivity. From the definitions of permeability and permittivity (see Chapter 8), the reader will note that Eqs. (11.32) and (11.33) are equivalent. The required permeability or permittivity is usually known from the design of LCRS discussed in Section 11.4. To determine the allowable permeability or permittivity of a given material, the biological clogging test standardized by ASTM D 1987-91 may be performed. This test allows one to measure permeability, or permittivity, of filter materials over an extended period of time using landfill leachates loaded with particulates and microorganisms. Using results from 144 separate tests involving a total of 12 filters and

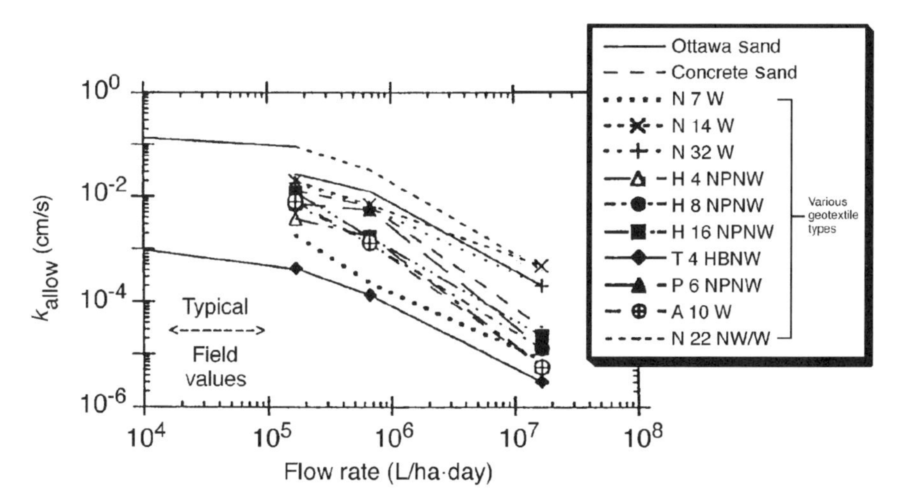

Fig. 11.22. k_{allow} versus flow rate for the 12 filters tested. (Adapted from Koerner et al., 1994; copyright © ASCE; reprinted with permission.)

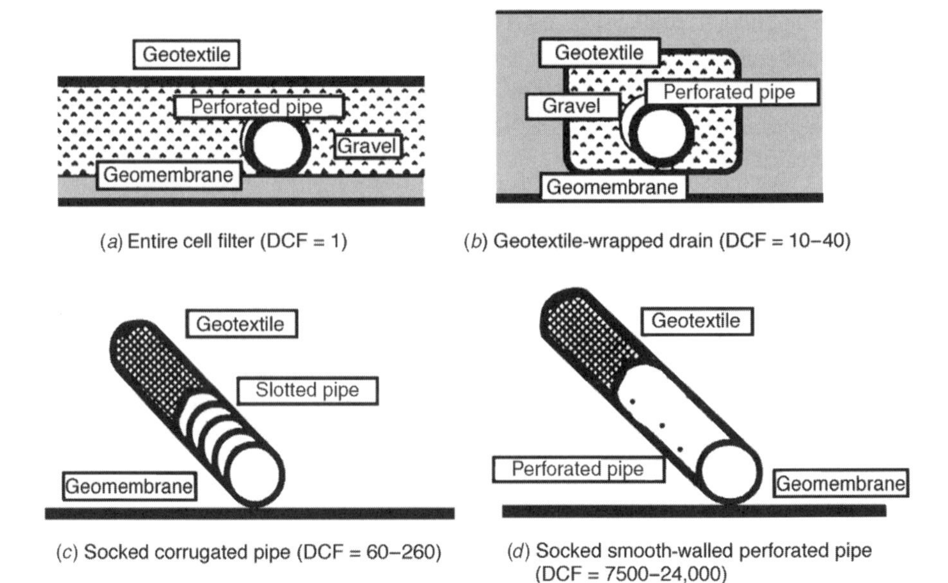

Fig. 11.23. Range of drain correction factors (DCF) for different leachate collection drain configurations. (Adapted from Koerner et al., 1994; copyright © ASCE; reprinted with permission.)

three different municipal solid waste landfill leachates, Koerner et al. (1994) developed a plot showing how k_{allow} varied as a function of flow rate in the filter materials (Fig. 11.22). This figure may be used to determine k_{allow} given the anticipated flow rate in the filter. The drain correction factor (DCF) needed to use Eqs. (11.32) and (11.33) is defined as the total cell area divided by the available filter flow area for the downstream drains of the leachate collection system. Its numeric value depends on the drain spacing, circumference of the drainage pipe, number of holes in the pipe, hole size, and so on. The ranges of DCF for different drainage configurations is shown in Fig. 11.23. As seen in Eqs. (11.32) and (11.33), high values of DCF result in lower FS and defeat the purpose of filters. Using results from four exhumed municipal solid waste leachate collection systems, Koerner et al. (1994) showed that the design formulation above could have prevented failures of leachate collection systems at waste containment facilities.

PROBLEMS

11.1. Derive the analytical solutions expressed in Eqs. (11.8), (11.10), and (11.11) from fundamental principles.

11.2. Derive an equation relating $h_{\max}$ and $h_{\min}$ for flow toward leachate collecting drains in Fig. P11.2.

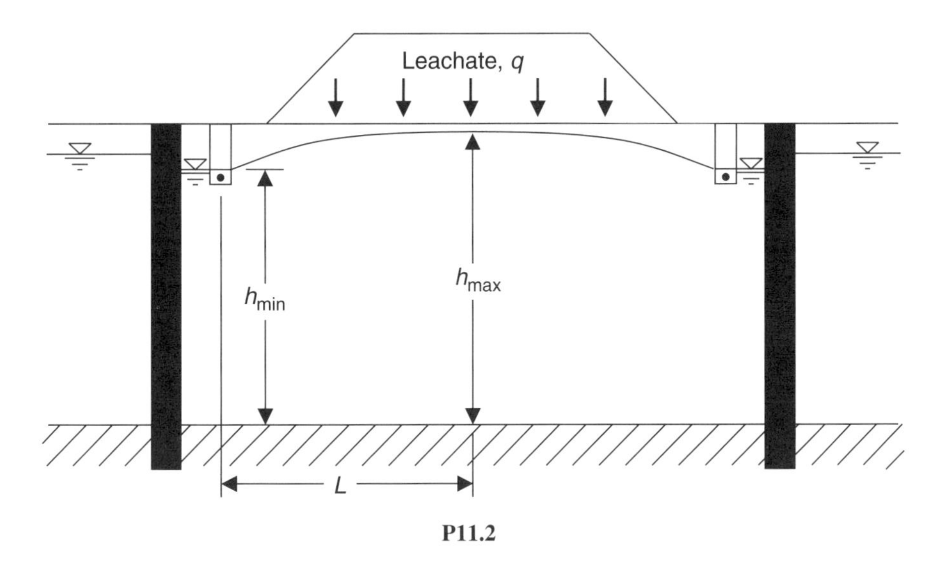

P11.2

11.3. Repeat Example 11.1 for the case of a thinner (1 ft) silt loam cover. Compare the two solutions and make a conclusion on the role of cover thickness.

11.4. Repeat Example 11.1 for the case of a 2-ft sandy loam cover. Compare the two solutions and make a conclusion on the effect of the type of topsoil used in cover systems.

11.5. Revisit Example 11.2. Find combinations of k_d (saturated hydraulic conductivity of the drainage layer) and L (drain spacing) that would yield a leachate mound less than 1 m high. Show the combinations using a graph.

11.6. Repeat Problem 11.5 for the case of a graded barrier. Show the combinations of k_d, L, and α that would yield a leachate mount less than 1 m high using a three-dimensional graph.

11.7. Show the relationship using a three-dimensional graph between the transit time t and thickness d of the compacted clay liner for a range of k_s (saturated permeability of the liner). Assume that $H = 1$ m and $n = 0.3$.

11.8. If seepage through a compacted clay liner ($k_s = 10^{-7}$ cm/s) subjected to a leachate mound of 3 ft should not exceed 1.65 in./yr, what should be the thickness of the liner?

11.9. What should be the thickness of a natural clay layer ($k_s = 10^{-6}$ cm/s), which would yield seepage no more than 1.65 in./yr under a mound height of 3 ft when used as a bottom liner "equivalent" to the one in Problem 11.8?

11.10. Using Wong's (1977) model, determine the collection efficiency of drains for the following parameters: permeability of the drainage layer $k_1 = 10^{-3}$ cm/s, permeability of the clay liner $k_2 = 10^{-7}$ cm/s, $\alpha = 5°$, thickness of the liner $d = 1$ m, and s_0 and $h_0 = 10$ m and 1 m, respectively.

REFERENCES

Al-Hussaini, M. M., and Perry, E. B. (1996). "Corps of Engineers Guide Specification for Use of Geotextiles as Filters," *Proc. Geofilters '96,* École Polytechnique de Montreal, Montreal, Quebec, Canada.

Ardaman & Associates, Inc. (1989). "Top Cover Design Considerations," Short Course presented at the University of Florida, "Design, Construction and Performance of Liner Systems for Environmental Protection," TREEO Center, Gainesville, FL.

Babbitt, H., and Caldwell, D. H. (1948). *The Free Surface Around, and Interference Between Gravity Wells,* University of Illinois Engineering Experimental Station Bulletin 374.

Bakker, K. J. (1987). "Hydraulic Filter Criteria for Banks," presented at the European Conference on Soil Mechanics and Foundation Engineering, Dublin, Ireland.

Bakker, K. J., Breetler, M. K., and Den, A. H. (1990). "New Criteria for Filters and Geotextile Filters under Revetments," presented at the International Conference on Coast Engineering, Delft, The Netherlands.

Bear, J. (1972). *Dynamics of Fluids in Porous Media,* American Elsevier, New York.

Bear, J. (1979). *Hydraulics of Ground Water,* McGraw-Hill, New York.

Bell, J. R., and Hicks, R. G. (1980). *Evaluation of Test Method and Use Criteria for Geotechnical Fabrics in Highway Applications,* Interim Report FHWA/RD-80/021, Federal Highway Administration, Washington, DC.

Benson, C. H., and Daniel, D. E. (1990). "Influence of Clods on Hydraulic Conductivity of Compacted Clay," *ASCE Journal of Geotechnical Engineering*, Vol. 116, No. 8, pp. 1231–1248.

Bhatia, S. K. (1994). "Geosynthetics in Filtration and Drainage Applications," IFAI Professor Training Course for Geosynthetics, course notes.

Bhatia, S. K., and Smith, J. L. (1995). "Application of the Bubble Point Method to the Characterization of the Pore-Size Distribution of Geotextiles," *ASTM Geotechnical Testing Journal,* Vol. 18, No. 1, pp. 94–105.

Bhatia, S. K., Huang, Q., and Smith, J. L. (1993). "Application of Digital Image Processing in Morphological Analysis of Geotextiles," *Proc. Conference on Digital Image Processing: Techniques and Applications in Civil Engineering,* American Society of Civil Engineers, New York, pp. 71–80.

Biswas, A. K. (1970). *History of Hydrology*, American Elsevier, New York.

Boulton, N. S. (1951). "The Flow Pattern near a Gravity Well in a Uniform Water Bearing Medium," *Journal of the Institution of Civil Engineers*, London, Vol. 36, p. 534.

Boynton, S. S., and Daniel, D. E. (1985). "Hydraulic Conductivity Tests on Compacted Clay," *ASCE Journal of Geotechnical Engineering,* Vol. 111, No. 4, pp. 465–477.

Calhoun, C. C. (1972). *Development of Design Criteria and Acceptance of Specifications for Plastic Filter Cloth,* Technical Report 5-72-7, U.S. Army Corps of Engineers Waterways Experiment Station, Vicksburg, MS.

Carroll, R. G. (1983). "Geotextile Filter Criteria," *Transportation Research Record*, Vol. 916, pp. 46–53.

Casagrande, A. (1937). "Seepage through Dams," *Journal of the New England Water Works Association,* Vol. LI, No. 2, pp. 295–336.

Cedergren, H. R. (1960). "Seepage Requirements of Filters and Pervious Bases," *ASCE Journal of the Soil Mechanics and Foundations Division,* Vol. 86, No. SM5, pp. 15–33.

Cedergren, H. R. (1977). *Seepage, Drainage and Flownets*, Wiley, New York.

Cedergren, H R. (1989). *Seepage, Drainage and Flownets*, 2nd ed., Wiley, New York.

Cedergren, H. R. (1994). "America's Pavements: World's Longest Bathtubs," *Civil Engineering*, Vol. 64, No. 9, pp. 56–58.

Cedergren, H. R., O'Brien, K. H., and Arman, J. A. (1972). *Guidelines for Design of Subsurface Drainage Systems for Highway Structural Sections,* Final Report (Summary) FHWA-RD-72-30, Federal Highway Administration, Washington, DC.

Cedergren, H. R., Arman, J. A., and O'Brien, K. H. (1973). *Development Guidelines for the Design of Subsurface Drainage Systems for Highway Pavement Structural Sections*, Report FHWA-RD-73-14, Office of Research and Development, Federal Highway Administration, Washington, DC.

Chapman, T. G. (1956). "Ground Water Flow to Trenches and Well Points," *Journal of the Institution of Engineers, Australia,* Vol. 28, pp. 275–280.

Christopher, B. R., and Fischer, G. R. (1992). "Geotextile Filtration Principles, Practices and Problems," *Journal of Geotextiles and Geomembranes*, No. 11, pp. 337–353.

Christopher, B. R., and Holtz, R. D. (1985). *Geotextile Engineering Manual,* Report FHWA-TS-86/203, Federal Highway Administration, Washington, DC.

Christopher, B. R., and Holtz, R. D. (1989). *Geotextile Construction and Design Guidelines,* Report HI-89-050, prepared for the Federal Highway Administration, Washington, DC.

Christopher, B. R., Gill, S. A., Giroud, J. P., Juran, I., Scholsser, F., Mitchell, J. K., and Dunnicliff, J. (1990). *Reinforced Soil Structures,* Vol. I, *Design and Construction Guidelines,* Report FHWA-RD-89-043, Federal Highway Administration, Washington, DC.

Criddle, W. D. (1958). "Methods of Computing Consumptive Use of Water," Journal of the Irrigation and Drainage Division, ASCE, Vol. 84, IR. 1, pp. 1507-1–1507-27.

Dacosta, J. A., and Bennett, R. R. (1960). "The Pattern of Flow in the Vicinity of a Recharging and Discharging Pair of Wells in an Aquifer Having a Real Parallel Flow," *International Association of Scientific Hydrology*, International Union of Geodesy and Geophysics, General Assembly of Helsinki, USGS Publication 52, pp. 524–536.

Daniel, D. E. (1987). "Earth Liners for Land Disposal Facilities," in R. D. Woods (ed.), *Geotechnical Practice for Waste Disposal*, Geotechnical Special Publication 13, American Society of Civil Engineers, New York, pp. 21–39.

Daniel, D. E. (1994). "State-of-the-Art: Laboratory Hydraulic Conductivity Tests for Saturated Soils," in D. E. Daniel and S. J. Trautwein (eds.), *Hydraulic Conductivity and Waste Contaminant Transport in Soil,* ASTM STP 1142, American Society for Testing and Materials, Philadelphia.

Daniel, D. E., and Benson, C. H. (1990). "Water Content-Density Criteria for Compacted Soil Liners," *ASCE Journal of Geotechnical Engineering*, Vol. 116, No. 12, pp. 1811–1830.

Daniel, D. E., and Wu, Y. K. (1993). "Compacted Clay Liners and Covers for Arid Sites," *ASCE Journal of Geotechnical Engineering*, Vol. 119, No. 2, pp. 223–237.

D'Appolonia, D. J. (1980). "Soil–Bentonite Slurry Trench Cutoffs," *ASCE Journal of the Geotechnical Engineering Division*, No. GT4, pp. 399–417.

Das, B. M. (1983). *Advanced Soil Mechanics*, McGraw-Hill, New York.

Day, S. R., and Daniel, D. E. (1985). "Hydraulic Conductivity of Two Prototype Clay Liners," *ASCE Journal of Geotechnical Engineering*, Vol. 111, No. 8, pp. 957–970.

Dematrocopoulos, A. C., Korfiatis, G. P., Bourodimos, E. L., and Nawy, E. G. (1984). "Modeling for Design of Landfill Bottom Liners," *ASCE Journal of Environmental Engineering*, Vol. 110, No. 6, pp. 1084–1098.

Donnon, W. W. (1959). "Drainage of Agricultural Lands Using Interceptor Line," *ASCE Journal of the Irrigation and Drainage Division*, No. IR1, Mar.

Ehrig, H. J., and Scheelhaase, T. (1993). "Pollution Potential and Long Term Behavior of Sanitary Landfills," *Proc. 4th International Landfill Symposium*, Cagliari, Italy, pp. 1204–1225.

Fenn, D. G., Hanley, K. J., and DeGrace, T. V. (1975). *Use of Water Balance Method for Predicting Leachate Generation from Solid Waste Disposal Sites,* EPA/530/SW-168, U.S. Environmental Protection Agency, Cincinnati, OH.

FHWA (1992). *Drainage Pavement Systems Participant Notebook,* Report FHWA-SA-92-008, Office of Technology Applications and Office of Engineering, Federal Highway Administration, Washington, DC.

Fischer, G. R. (1994). "The Influence of Fabric Pore Structure on the Behavior of Geotextile Filters," Ph.D. dissertation, University of Washington, Seattle, WA.

Fischer, G. R., Christopher, B. R., and Holtz, R. D. (1990). "Filter Criteria Based on Pore Size Distribution," *Proc. 4th International Conference on Geotextiles*, The Hague, The Netherlands, Vol. I, pp. 289–294.

Fischer, G. R., Holtz, R. D., and Christopher, B. R. (1996). "Characteristics of Geotextile Pore Structure," in S. K. Bhatia and L. D. Suits (eds.), *Recent Developments in Geotextile Filters and Prefabricated Drainage Geocomposites,* ASTM STP 1281, American Society for Testing and Materials, Philadelphia.

Forchheimer, P. (1930). *Hydraulik.* 3rd ed., Teuber, Leipzig.

Foreman, D. E., and Daniel, D. E. (1986). "Permeation of Compacted Clay with Organic Chemicals," *ASCE Journal of Geotechnical Engineering*, Vol. 112, No. 7, pp. 669–681.

Foster, M. A., and Fell, R. (1999). *Assessing Embankment Dam Filters Which Do Not Satisfy Design Criteria,* UNICIV Report R-376, University of New South Wales, Sydney, Australia.

Freeze, R. A. (1994). "Henry Darcy and the Fountains of Dijon," *Ground Water*, Vol. 32, No. 1, pp. 23–30.

Freeze, R. A., and Cherry, J. A. (1979). *Ground Water,* Prentice-Hall, Englewood Cliffs, NJ.

French Committee on Geotextiles and Geomembranes (1986). *Recommendations for the Use of Geotextiles in Drainage and Filtration Systems,* Institut Textile de France, Boulogne-Billancourt, France.

Garcia-Bengochea, I., Lovell, C. W., and Altschaeffl, A. G. (1979). "Pore Distribution and Permeability of Silty Clays," *ASCE Journal of Geotechnical Engineering*, Vol. 105, No. GT7, pp. 839–856.

Giroud, J. P. (1982). "Filter Criteria for Geotextiles," *Proc. 2nd International Conference on Geotextiles*, Las Vegas, NV, Vol. I, Industrial Fabric Association International, St. Paul, MN, pp. 103–108.

Hajra, M. G., Reddi, L. N., Marchin, G. L., and Mutyala, J. (2000). "Biological Clogging in Porous Media," in T. F. Zimmie (ed.), *Environmental Geotechnics,* Geotechnical Special Publication 105, American Society of Civil Engineers, New York, pp. 150–165.

Hajra, M. G., Reddi, L. N., Glasgow, L. A, Xiao, M., and Lee, I. M. (2002). "Effect of Ionic Strength on Fine Particle Clogging of Soil Filters," *ASCE Journal of Geotechnical Engineering*, Vol. 128, No. 8, pp. 631–639.

Haliburton, T. A., and Wood, P. D. (1982). "Evaluations of the U.S. Army Corps of Engineers Gradient Ratio Test for Geotextile Performance," *Proc. 2nd International Conference on Geotextiles*, Las Vegas, NV, Vol. I, Industrial Fabric Association International, St. Paul, MN, pp. 97–102.

Haliburton, T. A., Lawmaster, J. D., and McGuffey, V. E. (1982). *Use of Engineering Fabric in Transportation Related Applications,* FHWA Training Manual, FHWA Contract DTFH-80-C-0094, Federal Highway Administration, Washington, DC.

Hansen, V. E. (1949). "Evaluation of Unconfined Flow to Multiple Wells by Membrane Analogy," thesis, Iowa State University, Ames, IA.

Harr, M. E. (1962). *Ground Water and Seepage,* McGraw-Hill, New York.

Hausmann, M. R. (1990). *Engineering Principles of Ground Modification,* McGraw-Hill, New York.

Hazen, A. (1930). "Water Supply"*American Civil Engineers Handbook,* Wiley, New York.

Holtz, R. D., Tobin, W. R., and Burke, W. W. (1982). "Creep Characteristics and Stress–Strain Behavior of a Geotextile-Reinforced Sand," *Proc. 2nd International Conference on Geotextiles*, Las Vegas, NV, Vol. 3, pp. 805–809.

Holtz, R. D., Christopher, B. R., and Berg, R. R. (1998). *Geosynthetic Design and Construction Guidelines Participation Notebook,* Publication FHWA HI-95-038, NHI Course 13213, National Highway Institute, McLean, VA.

Hubbert, M. K. (1940). "The Theory of Groundwater Motion," *Journal of Geology*, Vol. 48, pp. 785–944.

ICOLD (1994). *Embankment Dams, Granular Filters, and Drains,* Commission Internationale des Grands Barrages 151, Paris.

Iterson, F. K. Th. Van. (1919). "Eenige theoretische beschouwingen over kwel," *De Ingenieur.*

Keller, J., and Robinson, A. R. (1959). "Laboratory Research on Interceptor Drains," *ASCE Journal of the Irrigation and Drainage Division*, No. IR3, September.

Kenney, T. C., Lau, D., and Ofoegbu, G. I. (1984). "Permeability of Compacted Granular Materials," *Canadian Geotechnical Journal,* Vol. 21, No. 4, pp. 726–729.

Kenney, T. C., and Lau, D. (1985). "Internal Stability of Granular Filters," *Canadian Geotechnical Journal,* Vol. 22, No. 2, pp.215–225.

Kenney, T. C., and Lau, D. (1986). "Internal Stability of Granular Filters," Reply to Discussions, *Canadian Geotechnical Journal,* Vol. 23, No. 3, pp. 420–423.

Khosla, R. B. A. N., Bose, N. K., and Taylor, E. M. (1954). *Design of Weirs on Permeable Foundations*, Central Board of Irrigation, New Delhi, India.

Kirkham, D. (1958). "Seepage of Steady Rainfall through Soil into Drains," *Transactions, American Geophysical Union*, Vol. 39, No. 5, pp. 892–908.

Kirkham, D. (1960). "Seepage into Ditches from a Plane Water Table Overlying a Gravel Substratum," *Journal of Geophysical Research*, Vol. 65, No. 4, pp. 1267–1272.

Kmet, P., et al. (1981). "Analysis of Design Parameters Affecting the Collection Efficiency of Clay-Lined Landfills," *Proc. 4th Annual Madison Conference of Applied Research and Practice on Municipal and Industrial Waste.*

Koerner, R. M. (1998). *Designing with Geosynthetics,* Prentice Hall, Upper Saddle River, NJ.

Koerner, R. M., and Ko, F. (1982). "Laboratory Studies on Long Term Drainage Capability of Geotextiles," *Proc. 2nd International Conference on Geotextiles*, Las Vegas, NV, Vol. 1, pp. 91–96.

Koerner, G. R., and Koerner, R. M. (1990). *Biological Activity and Remediation Involving Geotextile Landfill Leachate Filters,* ASTM STP 1081, American Society for Testing and Materials, Philadelphia, pp. 313–334.

Koerner, R. M., and Soong, T. Y. (1995). "Use of Geosynthetics in Infrastructure Remediation," *ASCE Journal of Infrastructure Systems,* Vol. 1, No. 1, pp. 66–75.

Koerner, G. R., Koerner, R. M., and Martin, J. P. (1994). "Design of Landfill Leachate-Collection Filters," *ASCE Journal of Geotechnical Engineering*, Vol. 120, No. 10, pp. 1792–1803.

Korfiatis, G. P., and Demetracopoulos, A. C. (1986). "Flow Characteristics of Landfill Leachate Collection Systems and Liners," *ASCE Journal of Environmental Engineering*, Vol. 112, No. 3, pp. 538–550.

Kozeny, J. (1931). Grundwasserbewegung bei freiem Spiegel, Fluss- und Kanalver-sicherung, *Wasserkraft und Wasserwirtschaft,* No. 3.

Lambe, T. W., and Whitman, R. V. (1969). *Soil Mechanics,* Wiley, New York.

Lafleur, J., Mlynarek, J., and Rollin, A. L. (1989). "Filtration of Broadly Graded Cohesionless Soils," *ASCE Journal of Geotechnical Engineering*, Vol. 115, No. 12, pp. 1747–1768.

Lu, J. C. S., Eichenberger, B., and Stearns, R. J. (1985). *Leachate from Municipal Landfills: Production and Management,* Pollution Technology Review 119, Noyes Publications, Park Ridge, NJ.

Luettich, S. M, Giroud, J. P., and Bachus, R. C. (1992). "Geotextile Filter Design Guide," *Journal of Geotextiles and Geomembranes*, Vol. 11, pp. 355–370.

Luthin, J. N. (1966). *Drainage Engineering,* Wiley, New York.

Lutton, R. J., Regan, G. L, and Jones, L. W. (1979). *Design and Construction of Covers for Solid Waste Landfills,* EPA-600/2-79-165, Office of Research and Development, U.S. Environmental Protection Agency, Cincinnati, OH.

Mansur, C. I., and Kaufman, R. I. (1962). "Dewatering," Chapter 3 in G. A. Leonards (ed.), *Foundation Engineering,* McGraw-Hill, New York.

Matrecon, Inc. (1980). *Lining of Waste Impoundment and Disposal Facilities,* EPA 530/SW-870, U.S. Environmental Protection Agency, Cincinnati, OH.

McBean, E. A., Poland, R., Rovers, F. A., and Crutcher, A. J. (1982). "Leachate Collection Design for Containment Landfills," *ASCE Journal of the Environmental Engineering Division*, Vol. 108, No. EE 1, pp. 204–209.

McEnroe, B. M., and Schroeder, P. R. (1988). "Leachate Collection in Landfills: Steady Case," *ASCE Journal of Environmental Engineering*, Vol. 114, No. 5, pp. 1052–1062.

McGown, A. W. (1976). "The Properties and Uses of Permeable Fabric Membranes," *Proc. Residential Workshop on Materials and Methods for Low Cost Roads and Reclamation Works*, Leura, Australia, pp. 663–710.

McGown, A., Andrawes, K. Z., Yeo, K. C., and DuBois, D. D. (1982). "Load Extension Testing of Geotextiles Confined in Soil," *Proc. 2nd International Conference on Geotextiles*, Las Vegas, NV, Vol. 3, pp. 793–798.

Millar, P. J., Ho, K. W., and Turnbull, H. R. (1980). *A Study of Filter Fabrics for Geotechnical Applications in New Zealand,* Central Laboratories Report 2-80/5, Ministry of Works and Development, New Zealand.

Mitchell, J. K., Hooper, D. R., and Campanella, R. G. (1965). "Permeability of Compacted Clay," *ASCE Journal of the Soil Mechanics and Foundations Division,* Vol. 91, No. SM4, Proceedings Paper 4392.

Moore, C. A. (1980). *Landfill and Surface Impoundment Performance Evaluation,* US EPA SW-869, U.S. Environmental Protection Agency, Cincinnati, OH.

Moulton, L. K. (1979). *Groundwater, Seepage and Drainage*, Notes in preparation for a textbook.

Moulton, L. K. (1980). *Highway Subdrainage Design*, Report FHWA-TS-80-224, Offices of Research and Development, Federal Highway Administration, Washington DC.

Muskat, M. (1937). *The Flow of Homogeneous Fluids through Porous Media*, McGraw-Hill, New York.

Ogink, H. J. M. (1975). *Investigations on the Hydraulic Characteristics of Synthetic Fabrics,* Publication 146, Delft Hydraulics Laboratory, Delft, The Netherlands.

Oweis, I. S., and Khera, R. P. (1990). *Geotechnology of Waste Management,* Butterworth, London.

Patton, F. D., and Hendron, A. J., Jr. (1974). "General Report on Mass Movements," *Proc. 2nd International Congress, International Association of Engineering Geology* São Paulo, Brazil, Vol. 2, pp. V-GR.1 to V-GR.57.

Pavlovsky, N. N. (1956). *Collected Works*, Akademia Nauk USSR, Leningrad.

Perry, R. H. (1976). *Engineering Manual*, 3rd ed.,McGraw-Hill, New York.

Philip, J. R. (1995). "Desperately Seeking Darcy in Dijon," *Soil Science Society of American Journal,* Vol. 59, pp. 319–324.

Polubarinova-Kochina, Ya. P. (1952). *Theory of Motion of Ground Water,* Gostekhizdat, Moscow.

Polubarinova-Kochina, Ya. P. (1962). *Theory of Ground Water Movement,* Princeton University Press, Princeton, NJ.

Rankilor, P. R. (1981). *Membranes in Ground Engineering,* Wiley, New York.

Reddi, L. N., and Inyang, H. I. (2000). *Geoenvironmental Engineering: Principles and Applications,* Marcel Dekker, New York.

Reddi, L. N., Xiao, M., Hajra, M. G., and In, M. L. (2000). "Permeability Reduction of Soil Filters Due to Physical Clogging," *ASCE Journal of Geotechnical Engineering*, Vol. 126, No. 3, pp. 236–246.

Ridgeway, H. H. (1976). "Infiltration of Water through the Pavement Surface," *Transportation Research Record,* Vol. 616, pp. 98–100.

Rouse, H., and Ince, S. (1979). *History of Hydraulics.* Iowa Institute of Hydraulic Research, State University of Iowa, Iowa City.

Rowe, R. K. (1991). "Containment Impact Assessment and the Contaminating Lifespan of Landfills," *Canadian Journal of Civil Engineering,* Vol. 18 No. 2, pp. 1204–1213.

Sansone, L. J., and Koerner, R. M. (1992). "Fine Fraction Filtration Test to Assess Geotextile Filter Performance," *Journal of Geotextiles and Geomembranes,* Vol. 11, No. 4–6, pp. 371–393.

Schaffernak, F. (1917). Ueber die Standsicherheit durchlaessiger geschuetteter Daemme, *Allgemeine Bauzeitung.*

Schober, W., and Teindl, H. (1979). "Filter Criteria for Geotextiles," *Proc. 7th European Conference on Soil Mechanics and Foundation Engineering*, Brighton, England, Vol. 2, pp. 123–129.

Schroeder, P. R., Morgan, J. M., Walski, T. M., and Gibson, A. C. (1984a). *The Hydrologic Evaluation of Landfill Performance (HELP) Model,* Vol. I, *User's Guide for Version 1,* Technical Resource Document, EPA/530-SW-84-009, U.S. Environmental Protection Agency, Cincinnati, OH.

Schroeder, P. R., Gibson, A. C., and Smolen, M. D. (1984b). *The Hydrologic Evaluation of Landfill Performance (HELP) model,* Vol. II, *Documentation for Version 1,* Technical Resource Document, EPA/530-SW-84-010, U.S. Environmental Protection Agency, Cincinnati, OH.

Sharma, H. D., and Lewis, S. P. (1994). *Waste Containment Systems, Waste Stabilization, and Landfills,* Wiley, New York.

Sherard, J. L., and Dunnigan, L. P. (1989). "Critical Filters for Impervious Soils," *ASCE Journal of Geotechnical Engineering*, Vol. 115, No. 7, pp. 927–947.

Sherard, J. L., Dunnigan, L. P., and Talbot, J. R. (1984a). "Basic Properties of Sand and Gravel Filters," *ASCE Journal of Geotechnical Engineering*, Vol. 110, No. 6, pp. 684–700.

Sherard, J. L., Dunnigan, L. P., and Talbot, J. R. (1984b). "Filters for Silts and Clays," *ASCE Journal of Geotechnical Engineering*, Vol. 110, No. 6, pp. 701–718.

Spangler, M. G., and Handy, R. L. (1973). *Soil Engineering,* Harper & Row, New York.

Stone, R. (1974). *Disposal of Sewage Sludge into a Sanitary Landfill,* EPA-SW-71d, U.S. Environmental Protection Agency, Cincinnati, OH.

Sweetland, D. B. (1977). "The Performance of Non-woven Fabrics as Drainage Screens in Subdrains," M.Sc. thesis, University of Strathclyde, Glasgow, Scotland.

Taylor, D. W. (1948). *Fundamentals of Soil Mechanics,* Wiley, New York.

Terzaghi, K. (1922). "Der Grundbruch an Stauwerken und seine Verhutung," *Die Wasserkraft*, (the failure of dams by piping and its prevention), Vol. 17, pp. 445–449.

Terzaghi, K. (1929). Discussion on Paper by Julian Hinds on: "Upward Pressures Under Dams," *ASCE Transactions of the American Society of Civil Engineeers,* p. 1563.

Terzaghi, K. (1929). "Effect of Minor Geological Details on the Safety of Dams," *American Institute of Mining and Metallurgical Engineers,* Technical Publications 215, p. 30.

Terzaghi, K. (1939). "Soil Mechanics—A New Chapter in Engineering Science," *Journal of the Institution of Civil Engineers,* Vol. 12, pp. 106–141.

Terzaghi, K. (1942). Quoted by O. E. Meinzer (1942) in: *Hydrology.* Dover Publications, Inc., New York, p. 23.

Thornthwaite, C. W., and Mather, J. R. (1957). "Instruction and Tables for Computing Potential Evapotranspiration and the Water Balance," *Publications in Climatology*, Vol. 10, No. 3, pp. 185–311.

Toksoz, S., and Kirkham, D. (1971). "Steady Drainage of Layered Soils: Nomographs," *ASCE Journal of the Irrigation and Drainage Division*, Vol. 97, No. IR1, pp. 1–18.

U.S. Army Corps of Engineers (1948). "Laboratory Investigation of Filters for Enid and Grenada Dams," Technical Memorandum No. 3-245, Waterways Experiment Station, Vicksburg, Mississippi, pp. 1–50.

U.S. Department of Agriculture (1986). "Soil Mechanics Note No. 1," Guide for Determining the Gradation of Sand and Gravel Filters, Soil Conservation Service.

U.S. Department of the Interior (1955). *The Use of Laboratory Tests to Develop Design Criteria for Protective Filters,* Earth Laboratory Report EM-425, Bureau of Reclamation, U.S. Government Printing Office, Washington, DC.

U.S. Department of the Interior (1977). *Ground Water Manual,* A Water Resources Technical Publication, Bureau of Reclamation, U.S. Government Printing Office, Washington, DC.

U.S. Department of the Interior (1987). "Design Standards No. 13—Embankment Dams," Chapter 5—Protective Filters, Bureau of Reclamation.

U.S. Department of the Interior (1990). *Earth Manual,* Part 2, A Water Resources Technical Publication, Bureau of Reclamation, Materials Engineering Branch Research and Laboratory Services Division, Denver, CO.

U.S. Department of the Navy (1971). *Design Manual: Soil Mechanics, Foundations, and Earth Structures,* NAVFAC DM-7, U.S. Government Printing Office, Washington, DC.

U.S. Department of the Navy, Naval Facilities Engineering Command (1982). *Soil Mechanics,* NAVFAC DM-7.1, U.S. Government Printing Office, Washington, DC.

U.S. EPA (1980). *Lining of Waste Impoundment and Disposal Facilities*, Matrecon Inc. Oakland, CA, Municipal Environmental Research Laboratory, Office of Research and Development, U.S. Environmental Protection Agency, Cincinnati, OH.

U.S. EPA (1986). *Subtitle D Study: Phase I Report,* EPA/530-SW-86-054, U.S. Environmental Protection Agency, Cincinnati, OH.

U.S. EPA (1989). *Requirements for Hazardous Waste Landfill Design, Construction, and Closure,* Seminar Publication, EPA/625/4–89/022, Center for Environmental Research Information, Office of Research and Development, U.S. Environmental Protection Agency, Cincinnati, OH.

U.S. EPA (1991). *Design and Construction of RCRA/CERCLA Final Covers,* EPA/625/4-91/025, U.S. Environmental Protection Agency, Cincinnati, OH.

U.S. EPA (1994). *Design, Operation, and Closure of Municipal Solid Waste Landfills,* EPA/625/R-94/008, Seminar Publication, Office of Research and Development, U.S. Environmental Protection Agency, Washington, DC.

Vaughan, P. R., and Soares, H. F. (1982). "Design of Filters for Clay Cores of Dams," *ASCE Journal of the Geotechnical Engineering Division*, Vol. 108, No. GT1, pp. 17–31.

Visser, W. C. (1954). "Tile Drainage in The Netherlands," *Netherlands Journal of Agricultural Science*, Vol. 2, No. 2, pp. 69–87.

Wates, J. A. (1980). "Filtration, an Application of a Statistical Approach to Filters and Filter Fabrics," *Proc. 7th Regional Conference for Africa on Soil Mechanics and Foundation Engineering*, A.A. Balkema, Rotterdam, The Netherlands, pp. 433–440.

Wilson, S. D., and Marsal, R. J. (1979). *Current Trends in Design and Construction of Emabankment Dams*, American Society of Civil Engineers, New York.

Wong, J. (1977). "The Design of a System for Collecting Leachate from a Lined Landfill Site," *Water Resources Research*, Vol. 13, No. 2, pp. 404–410.

Wooding, R. A., and Chapman, T. G. (1966). "Ground Water Flow over a Sloping Impermeable Layer," *Journal of Geophysical Research*, Vol. 71, No. 12, pp. 2895–2910.

Wu, T. H. (1981). *Soil Mechanics*, T.H. Wu, Worthington, OH.

Xanthakos, P. P., Abramson, L. W., and Bruce, D. A. (1994). *Ground Control and Improvement*, Wiley, New York.

INDEX